PLANT BREEDING: Classical to Modern

P. M. Priyadarshan

PLANT BREEDING: Classical to Modern

Springer

P. M. Priyadarshan
Erstwhile Deputy Director
Rubber Research Institute of India
Kottayam, Kerala, India

ISBN 978-981-13-7094-6 ISBN 978-981-13-7095-3 (eBook)
https://doi.org/10.1007/978-981-13-7095-3

This Springer imprint is published by the registered company Springer Nature Singapore Pte Ltd.
The registered company address is: 152 Beach Road, #21-01/04 Gateway East, Singapore 189721, Singapore

This book is dedicated to Nobel Laureate Dr. Norman E. Borlaug (1914–2009) who, as a plant breeder, strived benevolently to eradicate hunger and poverty.

Foreword

Plant breeding is an art and a science. It is an art for selecting suitable phenotype from variable plant populations. Primitive plant breeders started selecting crop varieties from the variable wild and semiwild populations. The selection was based on the judgement and keen eyes of plant breeders. Diverse crop varieties were selected for 10, 000 years on the basis of empirical observations. The scientific basis of plant breeding started after the rediscovery of Mendel's laws of inheritance during the beginning of the last century. These laws elucidated the mechanism of segregation and recombination. Through hybridization, multiple genotypes were produced, and desired phenotypes were selected. Numerous improved varieties were developed on scientific basis during the last century.

Many plant breeders advanced world agriculture through the development of new crop varieties. Foremost, among them was Dr. Norman Borlaug who received Nobel Peace Prize for developing high-yielding varieties of wheat. Similarly, high-yielding varieties of rice developed at the International Rice Research Institute (IRRI) had a comparable impact on food production and poverty elimination.

The present world population of 7.5 billion is likely to reach 9 billion by 2050. This will require 50% more food. This additional food must be produced under constraints of less land, less water and more importantly under changing climate. Thus, we need environmentally resilient varieties, with higher productivity and better nutrition. Fortunately, breakthroughs in cellular and molecular biology have provided new techniques for crop improvement which will help us meet the challenges of feeding nine billion people.

I am happy Dr. Priyadarshan has taken the initiative to prepare this text, *Plant Breeding: Classical to Modern*. As the title suggests, it discusses the conventional methods of plant breeding as well as the application of advanced techniques. It has 25 chapters arranged into 5 parts. It starts with a general introduction followed by plant development aspects, such as modes of crop reproduction and breeding systems. The next part has an excellent discussion of breeding methods. Specialized breeding methods, such as hybrid breeding, mutation breeding, polyploid breeding and distant hybridization, are in the fourth part. The final part has an excellent discussion of advanced techniques of plant breeding, such as tissue culture, genetic engineering, molecular breeding and application of genomics.

I wish to congratulate Dr. Priyadarshan for his labour of love in assembling voluminous information in this book. It will be useful for teachers and students of plant breeding alike.

Davis, CA, USA Gurdev S. Khush

Preface

Plant breeding is the science that derives new crop varieties to farmers. Based on the principles of genetics, as laid down classically by Gregor Johann Mendel during 1866, which were "rediscovered" in 1900 by Hugo de Vries, Carl Correns and Erich von Tschermak, this science has taken the world forward through firmly addressing hunger, famine and catastrophe. Plant breeding began when agriculture commenced centuries back, but the real science of plant breeding took shape when Mendel's principles of genetics came to light during 1900. The year 2015 commemorated 150 years of Mendelian principles. No nation thrives without agriculture, and plant breeding is the integral part of that science. The researchers of Tel Aviv, Harvard, Bar-llan and Haifa Universities say that agriculture began some 23,000 years ago. If this is true, plant breeding also commenced by then, since farmers must have surely nurtured best cultivars. Centuries of breeding programmes finally culminated in Sonora 64 (wheat) and IR 8 (rice) in the 1960s. While Dr. Norman E. Borlaug of CIMMYT exploited Norin 10 genes to derive semidwarf wheat, in rice, the crosses between Peta (Indonesia) and Dee-geo-woo-gen (DGWG, China) produced IR 8. Peter Jenning, Henry Beachell and Surajit Kumar De Datta of IRRI spearheaded this. This saga continues worldwide in producing thousands of varieties in all edible crops.

The explosive advancements in modern plant breeding enrich traditional breeding practices accomplished through inculcating various "omics", advanced computing and informatics, ending with robotics. The application of systems biology for genetic fine-tuning of crops meant for varied environments is the emerging new science that will soon assist plant breeding. The aim of this book is to narrate both conventional and modern approaches of plant breeding. *Principles of Plant Breeding* by R.W. Allard is a classic. However, referring this requires prior knowledge of the basics of plant breeding. This book is authored with the view to assist BS and MS students.

The TOC is set to address both conventional and modern means of plant breeding like history, objective, centres of origin, plant introduction, reproduction, incompatibility, sterility, biometrics, selection, hybridization, breeding both self- and cross-pollinated crops, heterosis, induced mutations and polyploidy, distant hybridization, resistance breeding, breeding for resistance to stresses, GE interactions, tissue culture, genetic engineering, molecular breeding and genomics. The book extends to 25 chapters dealing the subject in a comprehensive and perspective manner, and care has been taken to include almost all topics as required under the curricula of MS course being taught worldwide.

Striking a balancing chord between narrating fundamentals and inclusion of the latest advancements is an arduous task. I have strived my best to pay justice. Earnest efforts were incurred to correct "typos"/errors and possible misstatements. I owe full responsibility for any remaining errors and pledge to correct them in future editions.

Special thanks to my wife, Mrs. Bindu, and my children, Vineeth and Sandra, for extending their unflinching support and warm counsel.

The long cherished dream of authoring a book on plant breeding for students is fulfilled now. This first edition will further be revised during the years to come. I would appreciate receiving the invaluable comments from the readers, by which I can improve further editions.

Finally, hearty thanks to Springer for publishing this book.

Thiruvananthapuram, Kerala, India P. M. Priyadarshan

Acknowledgements

The guidance and suggestions rendered by my teacher, Prof. P.K Gupta, Professor Emeritus, Chaudhary Charan Singh University, Meerut, India, is gratefully acknowledged. He has been my guide and mentor for all these years.

I place on record a sincere thanks to Prof. M.S. Kang, adjunct professor, Kansas State University, USA, for reviewing the chapter on GE interactions.

Dr. K. Kalyanaraman, adjunct faculty, National Institute of Technology, Tiruchirappalli, India, reviewed the chapter on Basic Statistics. I am extremely indebted to him.

Karen A. Williams, National Germplasm Resources Laboratory, USDA-ARS, Beltsville, and Joseph Foster, Director, Plant Germplasm Quarantine Program, USDA-ARS, Beltsville, gave some details of germplasm conservation and utilization. Their help is duly acknowledged.

Dr. Amelia Henry, Dr. Kshirod Jena and Dr. Arvind Kumar of the International Rice Research Institute, Manila, Philippines, gave me details of drought-tolerant rice varieties. I am extremely thankful to them.

Dr. Ravi Singh, Head of bread wheat improvement, CIMMYT, and Dr. B.P.M. Prasanna, Director, CIMMYT's Global Maize Programme, Nairobi, Kenya, gave me details of drought tolerance in wheat and maize, respectively. My sincere thanks are due to them.

Prof. Lawrence B. Smart, School of Integrative Plant Science, Cornell University, and Prof. Jeff J. Doyle, Professor and chair, Plant Breeding & Genetics, Cornell University, helped me to reconstruct the Table of Contents with the details of the curricula on plant breeding being followed at Cornell University. My sincere thanks to them.

Prof. Dionysia A. Fasoula of the Department of Plant Breeding, Agricultural Research Institute, Nicosia, Cyprus, reviewed the honeycomb design narration. I am extremely thankful to him for this gesture. My Special thanks with indebtedness to Dr. Gurdev S. Khush for providing the foreword to this book.

Contents

About the Author

Dr. P. M. Priyadarshan is a prominent *Hevea* rubber breeder. He began his research career by breeding triticale and wheat. During the 1980s, he focused on the *in vitro* culture of spices. He joined the Rubber Research Institute of India (Rubber Board, Ministry of Commerce, Govt. of India) as a plant breeder in 1990 and specialized in breeding *Hevea* rubber for sub-optimal environments. In 2009, he became the Institute's Deputy Director, and managed its Central Experiment Station until 2016. As a scientist, he has been involved in breeding cereals, spices and *Hevea* rubber for the past 32 years. During that time, he has published several research papers and chapters in journals and books of international repute. He has authored articles for several important journals, e.g. *Advances in Agronomy, Advances in Genetics* and *Plant Breeding Reviews*, and has edited books such as *Breeding Plantation Tree Crops, Breeding Major Food Staples* and the *Genomics of Tree Crops*, as well as a book on the biology of *Hevea* rubber.

Part I

Generalia

Introduction to Plant Breeding

1

Keywords

Scientific basis of plant breeding · World food scenario · Contributions of conventional plant breeding · International Research Centres · Plant domestication · Pre-Mendelian · Post-Mendelian · Norman Borlaug and green revolution · Semi-dwarf varieties of wheat and rice · Facets of plant breeding · Omics · Genetic diversity · Germplasm grouping · Quantitative variation · Mapping traits · Genotype-by-environment interactions · Phenotyping · Phenomics · Future challenges

David Allen Sleper and John Milton Poehlman gave the definition for plant breeding as: "Plant Breeding is the art and science of improving heredity of plants for the benefit of humankind". Above all others, this is the best-suited definition for plant breeding. There are several others as:

Plant breeding is the art and science of changing the genetics of plants in order to produce desired characteristic.

Plant breeding, science of altering the genetic pattern of plants in order to increase their value.

The application of genetic analysis to development of plant lines better suited for human purposes.

By definition, plant breeding is the purposeful manipulation of certain species of plants in order to create desired varieties to achieve specific purposes. The manipulation may be done in several ways.

The application of genetic analysis to development of plant lines better suited for human purposes.

P. M. Priyadarshan, *PLANT BREEDING: Classical to Modern*,
https://doi.org/10.1007/978-981-13-7095-3_1

Man started using selected plant species some 10,000 years ago for his day-to-day needs and knowingly or unknowingly exercised the option of domesticating the plants. This exercise is known as *plant domestication*. Plant domestication is the earliest way of plant breeding. Since then, plant breeding experienced explosive advancements in serving man with newer sources of food, fibre, feed and fuel. All our food crops were derived from domesticated plants (Table 1.1). Among the more than 300,000 plant species under existence now, fewer than 200 are being commercially exploited, and only 3 of them – rice, wheat and maize – contribute to calories and proteins consumed by human.

A plant raised through intentional human activity is called a *cultigen*. Ancestors of cultigen are normally not known. A cultivated crop species evolved from wild populations as a result of selection by farmers is a *landrace*, suited to a particular region or environment. An example is the landraces of rice, *Oryza sativa* subspecies *indica*, which was developed in South Asia, and *Oryza sativa* subspecies *japonica*, which was developed in China. The International Treaty on Plant Genetic Resources for Food and Agriculture (2001) says that a *variety* is a "plant grouping within a single botanical taxon of the lowest rank, defined by the reproducible expression of its distinguishing and other genetic characteristics".

The breeding methods can be streamlined into three categories:

(a) Selection based on observed natural variants
(b) Controlled mating of parents and selection of recombinants
(c) Selection of marker profiles, using molecular tools

The last category is the non-conventional way of breeding plants. It is a fact that relying upon only traditional breeding methods could lead to narrowing of gene pool that ultimately makes the species vulnerable to biotic and abiotic stresses. Non-conventional techniques will lead to more desirable variation. A collection of all such variants (conventional and non-conventional) of a given species is known as *germplasm*.

Scientific Basis of Plant Breeding On the advent of the twentieth century, the principles put forth by Darwin and Mendel established the scientific basis for plant breeding and genetics (see Sections 1.2 and 1.3). Similarly, the twenty-first-century crop improvement is revolutionized by molecular plant breeding that integrates molecular marker applications and genomic research with conventional plant breeding practices. A journey through various milestones of genetics from 9000 BC to till date has taken the humankind to explosive advancements of plant genetics and breeding (Table 1.2). DNA, the seed of life, was first identified and isolated by Friedrich Miescher in 1869 (which Miescher called *nuclein*), and the double helix structure of DNA was first discovered by James Dewey Watson and Francis Harry Compton Crick in 1953. Since then, the science of genetics has taken unstoppable journey aiding the basic principles of plant breeding on which crop improvement is totally based upon.

Table 1.1 Landraces and their domestication

Plant	Where domesticated	Date
Peas	Near East	9000 BC
Barley	Near East	8500 BC
Chickpea	Anatolia	8500 BC
Rice	Asia	8000 BC
Potatoes	Andes Mountains	8000 BC
Beans	South America	8000 BC
Maize	Central America	7000 BC
Bread wheat	Near East	6000 BC
Cassava	South America	6000 BC
Date palm	Southwest Asia	5000 BC
Avocado	Central America	5000 BC
Grapevine	Southwest Asia	5000 BC
Cotton	Southwest Asia	5000 BC
Bananas	Island Southeast Asia	5000 BC
Beans	Central America	5000 BC
Chilli peppers	South America	4000 BC
Amaranth	Central America	4000 BC
Watermelon	Near East	4000 BC
Olives	Near East	4000 BC
Pomegranate	Iran	3500 BC
Garlic	Central Asia	3500 BC
Soybean	East Asia	3000 BC
Cocoa	South America	3000 BC
Squash (*Cucurbita pepo*)	North America	3000 BC
Sunflower	Central America	2600 BC
Rice	India	2500 BC
Sweet potato	Peru	2500 BC
Pearl millet	Africa	2500 BC
Sesame	Indian subcontinent	2500 BC
Sorghum	Africa	2000 BC
Sunflower	North America	2000 BC
Coconut	Southeast Asia	1500 BC
Rice	Africa	1500 BC
Tobacco	South America	1000 BC
Eggplant	Asia	First century BC

In addition to classical breeding, plant breeding in the recent years has achieved commendable strides integrating various tools of biotechnology. Marker-assisted selection or marker-aided selection (MAS) is a process whereby a marker (morphological, biochemical or one based on DNA/RNA variation) is used for indirect selection of a genetic determinant or determinants of a trait of interest (i.e. productivity, disease resistance, abiotic stress tolerance and/or quality). Genetic

Table 1.2 Milestones in genetics and plant breeding

9000 BC: First evidence of plant domestication in the hills above the Tigris river
3000 BC: Domestication of all important food crops in the Old World completed
1000 BC: Domestication of all important food crops in the New World completed
700 BC: Assyrians and Babylonians hand pollinate date palms
1694: Camerarius of Germany first to demonstrate sex in plants and suggested crossing as a method to obtain new plant types
1716: Mather of the USA observed natural crossing in maize
1717: Thomas Fairchild – Developed the first interspecific hybrid between sweet William and carnation species of *Dianthus*
1727: Vilmorin Company of France introduced the pedigree method of breeding
1753: Linnaeus published *Species Plantarum.* Binomial nomenclature born
1761–1766: Kölreuter of Germany demonstrated that hybrid offspring received traits from both parents and were intermediate in most traits and produced the first scientific hybrid using tobacco
1800: Knight, T.A. (English) – First used artificial hybridization in fruit crops
1840: John Le Couteur – Developed the concept of progeny test for individual plant selection in cereals
1847: "Reid's Yellow Dent" maize developed
1866: Mendel published his discoveries in *Experiments on Plant Hybridization*, cumulating in the formulation of laws of inheritance and discovery of unit factors (genes)
1899: Hopkins described the ear-to-row selection method of breeding in maize
1856: de Vilmorin (French biologist) – Further elaborated the concept of progeny test and used the same in sugar beet
1890: Rimpu (Sweden) – First made inheritance cross between bread wheat (*Triticum aestivum*) and rye (*Secale cereale*), which later on gave birth to triticale
1900: de Vries (Holland), Correns (Germany) and von Tschermak (Austria) – Rediscovered Mendel laws of inheritance independently
1900: Nilson, H. (Swedish) – Elaborated individual plant selection method
1903: Chromosome theory of inheritance by Sutton 1903
1903: Johannsen, W.L. – Developed the concept of pure line
1904–1905: Nilsson-Ehle proposed the multiple-factor explanation for inheritance of colour in wheat pericarp
1905: Linkage theory by Bateson and Punnet
1908: Shull, G.H. (USA) and East, E.M. (USA) – Proposed overdominance hypothesis independently working with maize
1908: Davenport, C.B.: First proposed dominance hypothesis of heterosis
1908–1909: Hardy of England and Weinberg of Germany developed the law of equilibrium of populations
1908–1910: East published his work on inbreeding
1909: Shull conducted extensive research to develop inbreds to produce hybrids
1910: Chromosome theory of inheritance by Morgan
1910: Bruce, A.B.; Keable, F.; and Pellew, C. – Elaborated the dominance hypothesis of heterosis proposed by Davenport
1913: First ever linkage map created by Sturtevant
1914: Shull, G.H. – First used the term heterosis for hybrid vigour

(continued)

Table 1.2 (continued)

1917: Donald Forsha Jones invented the double-cross method of hybrid seed production, which helped produce the first commercial hybrid corn in the 1920s. Jones developed first commercial hybrid maize
1919: Hays, H.K. and Garber, R.J. – Gave initial idea about recurrent selection. They first suggested the use of synthetic varieties for commercial cultivation in maize
1920: East, E.M. and Jones, D.F. also gave initial idea about recurrent selection
1925: East, E.M. and Mangelsdorf, A.J. – First discovered the gametophytic system of self-incompatibility in *Nicotiana sanderae*
1926 Pioneer Hi-Bred Corn Company established as the first seed company
1926: Vavilov, N.I. – Identified eight main centres and three sub-centres of crop diversity. He also developed concept of parallel series of variation or law of homologous series of variation
1928: Stadler, L.J. (USA) – First used X-rays for induction of mutations
1934: Dustin discovered colchicines
1935: Vavilov published *The Scientific Basis of Plant Breeding*
1936: East, E.M. – Supported overdominance hypothesis of heterosis proposed by East and Shull in 1908
1939: Goulden, C.H. – First suggested the use of single-seed descent method for advancing segregating generations of self-pollinating crops
1940: Jenkins, M.T. – Described the procedure of recurrent selection
1940: Harlan used the bulk breeding selection method in breeding
1941: One gene encodes on protein by Beadle and Tatum
1944: Avery, MacLeod and McCarty discovered DNA is hereditary material
1945: Hull proposed recurrent selection method of breeding
1945: Hull, F.H. – Coined the terms recurrent selection and overdominance working with maize
1950: Hughes and Babcock – First discovered sporophytic system of self-incompatibility in *Crepis foetida*
1950: McClintock discovered the *Ac-Ds* system of transposable elements
1952: Jensen, N.F. – First suggested the use of multilines in oats
1953: Borlaug, N.E. – First outlined the method of developing multilines in wheat
1953: Watson, Crick and Wilkins proposed a model for DNA structure
1962: Murashige-Skoog developed the MS media in 1962 containing nutrition factors that allowed the in vitro growth of many tissue types
1964: Borlaug, N.E. – Developed high-yielding semi-dwarf varieties of wheat which resulted in Green Revolution
1965: Grafius, J.E. – First applied single-seed descent (SSD) method in oats
1970: Borlaug received Nobel Prize for the Green Revolution
1973: Paul Berg, Stanley Cohen and Herbert Boyer introduced the recombinant DNA technology
1976: Yuan Longping et al. – Developed the world's first rice hybrid (CMS based) for commercial cultivation in China
1983: Beckmann and Soller – RFLPs for genome-wide QTL detection and breeding
1987: Monsanto – Developed world's transgenic cotton plant in the USA
1964: Maheshwari and Guha – Produced haploid plant in vitro from pollen grain
1991: ICRISAT – Developed the world's first pigeon pea hybrid (ICPH 8) for commercial cultivation in India
1994: "FlavrSavr" tomato developed as first genetically modified food produced for the market
1995: *Bt* corn developed

(continued)

Table 1.2 (continued)

1996: Roundup Ready® soybean introduced
1998: Potatoes, genetically engineered by Charles Arntzen and Hugh Mason, are used in the first ever clinical trial of a genetically engineered food to deliver a pharmaceutical. The trial determines the safety and efficacy of an edible vaccine
1999: Andrew Hamilton and David Baulcombe discover a short antisense RNA that can induce gene silencing
2000: *Arabidopsis* genome sequenced by *Arabidopsis* Genome Initiative
2000: Tasios Melis and Liping Zhang of UC Berkeley along with Maria Ghiardi and Marc Forestier of the National Renewable Energy Laboratory discover a metabolic "switch" in algae that allows the plant to produce hydrogen gas. The finding has the potential to create a commercial source of hydrogen gas produced by photosynthesis
2001: Meuwissen et al. – Genomic selection proposed
2001: Ingo Potrykus and Peter Beyer succeed in developing "golden rice", a modified rice plant yellowish in colour that contains beta-carotene, a building block of vitamin A. The crop could help prevent blindness in malnourished children. However, a lack of awareness concerning GMOs curtails production of the crop for over a decade
2002: Production of golden rice (through genetic engineering) that can biosynthesize beta-carotene, a precursor of vitamin A
2002: Rice genome sequenced by the International Rice Genome Sequencing Project
2003: Researchers at Duke, New York University, and the University of Arizona develop an *Arabidopsis* root gene expression map
2004: Roundup Ready® wheat developed
2005: Aaron Liepman and Kenneth Keegstra characterize enzymes responsible for synthesizing fibrous carbohydrates that make up plant cell walls. The work enables development of plants that provide increased nutrition, cheaper food additives and easily digestible animal feed
2005: US Postal Service honours plant genetics pioneer and Nobel Prize winner Barbara McClintock with a postage stamp. The International Rice Genome Sequencing Project publishes DNA blueprint for the crop in *Nature*. The final "map" reveals the location and sequence of more than 37,500 protein-encoding genes among 389 million base pairs of DNA
2005: The International Rice Genome Sequencing Project publishes DNA blueprint for rice. In a consortium led by the University of California, Davis initiates research to advance technology that rapidly identifies genes that may produce higher-quality wheat
2006: Pamela Ronald, Keong Xu, Takeshi Fukao, Abdelbagi Ismail and Julia Bailey-Serres identify a gene in rice that renders the crop tolerant to water submergence
2006: X. Zhang and colleagues describe the first genome-wide high-density methylation map of an entire genome using *Arabidopsis thaliana*
2006: Clone from Wild Wheat Alters Content in the Grain. Researchers clone a gene from wild wheat that increases the protein, zinc and iron content in the grain
2007: Nanotechnology Penetrates Plant Cell Walls. Kan Wang, Victor Lin, Brian Trewyn and Francois Torney demonstrate the first use of nanotechnology to penetrate plant cell walls and simultaneously deliver a gene and a chemical that triggers its expression with controlled precision
2008: iPlant forms, the first national cyber infrastructure centre dedicated to tackling global "grand challenge" questions in plant biology. University of Arizona researchers led by Richard Jorgensen initiate the effort. Supported by NSF, iPlant aims to identify problems in the plant sciences that could benefit from cyber infrastructure and develop methods to coordinate delivery of hardware and software to solve those problems

(continued)

Table 1.2 (continued)

2008: The BioCassava – A Day's Worth of Nutrition in a Single Meal. The BioCassava Plus project genetically modifies the cassava plant to fortify it with enough vitamins, minerals and protein to provide a day's worth of nutrition in a single meal
2008: Next-generation sequencing (NGS) by Schuster
2009: The corn genome published by a consortium led by Richard Wilson. The maize sequence contains more than twice as many genes as the human genome2009
2011: Over 1 million farmers plant Sub1 rice. The new variety could increase food security for 70 million of the world's poorest people
2012: Tomato genome published
2012: Draft genome of pigeon pea (*Cajanus cajan*) published

modification is yet another technique done through adding a specific gene or genes to a plant (interspecific and intergeneric) or by knocking out a gene with RNAi (RNAi is a molecule that inhibits gene expression through destruction of specific mRNA molecules). Genes are normally introduced through *Agrobacterium tumefaciens,* a soil plant pathogenic bacterium. It has the ability to transfer a specific DNA segment (tumour-inducing T-DNA). T-DNA is introduced into the nucleus of infected cells that gets integrated into the host genome. Such genetically modified plants are referred to as *transgenic plants*. Such genetic modification can produce a plant with the desired trait or traits faster than classical breeding. Transgenic plants commercially released are generally resistant to insect/pests and herbicides. Insect resistance is derived from *Bacillus thuringiensis* (*Bt*) that has a gene encoding toxicity to some insects. The cotton bollworm that feeds on *Bt* cotton will imbibe the toxin and die. Herbicides, on the other hand, bind to specific plant enzymes and inhibit their action leading to death of the plant. Such enzymes are known as herbicide target sites. In herbicide-resistant crops, gene that is not inhibited by the herbicide is expressed. So, the spraying of glyphosate selectively kills weeds only. Transgenic plants that can produce pharmaceuticals (and industrial chemicals) are *pharmacrops*. Genetic engineering has achieved new horizons through site-directed changes in gene sequence without a vector. This latest technology is known as CRISPR/Cas9 system. The CRISPR/Cas9 system uses two key molecules to change DNA. Cas9 known as a pair of "molecular scissors" can cut the DNA at a specific location. The second molecule is the guide RNA or gRNA that is 20 base long located in a longer RNA scaffold. The scaffold part helps to find the right part of the DNA so that the Cas9 enzyme cuts at that point. Nucleotide(s) can be added or deleted at this site, changing the amino acid sequence of the protein thus synthesized.

World Food Scenario Meeting the global demands for food, fibre, feed and fuel will depend upon the development of new varieties with unique genes that enhances yield. They must also have the capacity to grow in periods of drought and to withstand stress due to insects and pathogens. This requires concerted efforts by professionals on plant breeding, plant pathology, entomology, agronomy, statistics and biotechnology. Thus, plant breeding is a continuous process year after year to produce new strains to feed the ever-increasing global population. As of 2017, world population is estimated to be 7.38 billion by the United States Census Bureau (USCB) (world population clock). With the continued increase, the global population is expected to reach 9.7 billion by 2050. Some analysts have questioned the sustainability of further world population growth. The world produced 2241 million tons of grain in 2012. This was lesser than 75 million tons as of 2011. In the USA, one farmer produced enough food for 19 people in 1940, rising to 73 people in 1973 and 155 people in 2010. Corn yields averaged 2.44 t/ha in 1950, rising to 9.60 t/ha in 2000. Progress in plant breeding, in particular, has arguably been the engine of growth in productivity supported by improvements in crop management and mechanization. So, overall consumption did exceed world cereal production in 2017 and is projected at 2597 million tons (Fig. 1.1). Corn, wheat and rice account for most of the world's grain harvest. In 2012, the global corn harvest was 852 million tons, wheat was 654 million tons, and rice was 466 million tons. Nearly half of the world's grains are produced by China, the USA and India. Worldwide, carryover grain stocks (the amount left during the previous year) strikes around 423 million tons that is sufficient for 68 days of consumption.

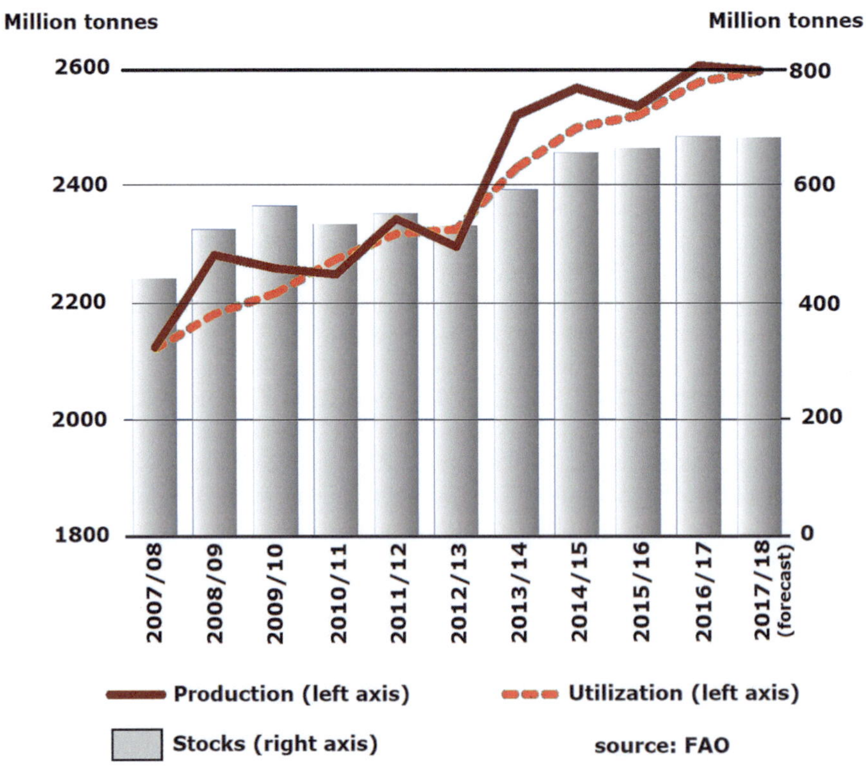

Fig. 1.1 Cereal production, utilization and stocks (source: FAO)

Contributions of Conventional Plant Breeding Conventional plant breeding relies on new genetic combinations derived through sexual hybridization and subsequent selection of phenotypically evaluated genotypes. This could lead to dramatic yield increment that could challenge neo-Malthusian predictions that the food production cannot keep the pace of population growth in the twentieth century. As per FAO statistics, in less than 50 years (1961–2009), the world average of cereal yields has increased from 1.35 to 3.51 t/ha. The new genotypes thus developed could be tested for adaptation to new management practices. This is a clear example of exploitation of genotype x environment (G × E) interactions. The identification of dwarf and semi-dwarf genes in rice (IR-8 in Southeast Asia) and wheat (Sonora 64 in Mexico) made possible the development of non-lodging cultivars with high yield in response to fertilizer application. In the USA, maize yields increased by more than fivefold since 1930 through adopting selection within open-pollinated types, simple F_1 hybrids, development of double and three-way hybrids and GMO F_1 hybrids (GMO=Genetically Modified Organism). This formula was followed in wheat and rice which could be replicated in other crops. Biofortification of grains is the latest trend in plant breeding that can address the nutritional deficiency (see Box 1.1).

Box 1.1: Biofortified Grains

Essential mineral micronutrients are a prerequisite to maintain metabolism in all living organisms, and man obtains these from his diet. But, wheat, rice and maize as staple grains contain suboptimal quantities of micronutrients, especially iron (Fe) and zinc (Zn). However smaller in quantities they are, most of this is removed by milling leading to micronutrient deficiency. Estimates of WHO point that almost 25% of the world population has anaemia. Inadequate Zn intake and Zn deficiency faced by 17.3% of people lead to nearly 433,000 deaths among children aged below 5 years. Also, vitamin A deficiency (VAD) is yet another harmful form of malnutrition causing blindness and weakens the body's immune system causing morbidity and mortality.

Quantity of vitamins and minerals can be increased through biofortification, achieved by means of transgenic techniques. Rice was genetically engineered to produce beta-carotene, a precursor of vitamin A, that finally culminated in the derivation of golden rice (Fig. 1.2). Rice was later biofortified with lysine. Chinese researchers developed a gene-stacking approach capable of delivering many genes at once for rice endosperm to produce high levels of anthocyanin (Fig. 1.3). Purple endosperm holds potential for reducing the risk of certain cancers, cardiovascular disease, diabetes and other chronic disorders. China developed a highly efficient "TransGene Stacking II" that can assemble a large number of genes into a single vector for plant transformation. This system can transform up to eight anthocyanin pathway genes in the endosperm of the *japonica* and *indica* rice varieties. This system could provide a versatile toolkit for transgene stacking. The toolkit possesses a huge potential for synthetic biology (redesigning of existing biological systems).

(continued)

Box 1.1 (continued)

Similarly, wheat is being biofortified with zinc and iron. Maize is with considerable variation in kernel carotenoid composition. Work on biofortification of maize with pro-vitamin A carotenoids (pVAC) is underway.

The Indian Context The implementation of the crop development programmes under various schemes have boosted India's crop production with total food grain production increasing from 217.28 million tons in 2006–2007 to 252.23 million tons in 2015–2016 crop year resulting in almost 18.39% increase in yield of total food grains. Rice increased its yield by 12.29%, wheat by 7.31% and pulses by 14.21%. Horticulture crops increased their production from 191.81 million tons in 2006–2007 to 282.8 million tons in 2015–2016. Also, oil seed production increased from 24.29 million tons in 2006–2007 to 32.9 million tons 2015–2016. Also, production of cotton increased from 521 kg/ha to 568 kg. To improve production and yield of different crops, a number of crop development schemes are being implemented through state governments in the country like the National Food Security Mission (NFSM); Integrated Scheme on Oilseeds, Pulses, Oil Palm and

Fig. 1.2 Golden rice (left) with normal rice (right)

Fig. 1.3 Genetically engineered rice that produce high levels of anthocyanin. The purple endosperm holds potential for decreasing the risk of certain cancers, cardiovascular disease, diabetes and other chronic disorders

Table 1.3 Members of the CGIAR (Consultative Group on International Agricultural Research), a Consortium of International Agricultural Research Centres

Active CGIAR centres	Headquarters location
Africa Rice Centre (West Africa Rice Development Association, WARDA)	Bouaké, Côte d'Ivoire/ Cotonou, Benin
Bioversity International	Maccarese, Rome, Italy
Centre for International Forestry Research (CIFOR)	Bogor, Indonesia
International Centre for Tropical Agriculture (CIAT)	Cali, Colombia
International Centre for Agricultural Research in the Dry Areas (ICARDA)	Beirut, Lebanon
International Crops Research Institute for the Semi-Arid Tropics (ICRISAT)	Hyderabad (Patancheru), India
International Food Policy Research Institute (IFPRI)	Washington, D.C., USA
International Institute of Tropical Agriculture (IITA)	Ibadan, Nigeria
International Livestock Research Institute (ILRI)	Nairobi, Kenya
International Maize and Wheat Improvement Centre (CIMMYT)	El Batán, Mexico State, Mexico
International Potato Centre (CIP)	Lima, Peru
International Rice Research Institute (IRRI)	Los Baños, Laguna, Philippines
International Water Management Institute (IWMI)	Battaramulla, Sri Lanka
World Agroforestry Centre (International Centre for Research in Agroforestry, ICRAF)	Nairobi, Kenya
World Fish Centre (International Centre for Living Aquatic Resources Management, ICLARM)	Penang, Malaysia

Maize (ISOPOM); Technology Mission on Cotton (TMC); etc. All these advancements are made possible through introducing newer and high-yielding varieties raised by various research institutes under the auspices of the Indian Council of Agricultural Research.

International Research Centres Plant breeding scenario on the international front is under the auspices of the Consultative Group on International Agricultural Research (CGIAR). There are 15 future harvest research centres that are actively engaged in agricultural research along with plant breeding (Table 1.3). CGIAR research aims at reducing rural poverty, increasing food security, improving human health and nutrition and ensuring sustainable management of natural resources. The membership of CGIAR includes country governments, such as the USA, Canada, the UK, Germany, Switzerland and Japan, the Ford Foundation, the Food and Agriculture Organization (FAO) of the United Nations, the International Fund for Agriculture Development (IFAD), the United Nations Development Programme (UNDP), the World Bank, the European Commission, the Asian Development Bank, the African Development Bank and the Fund of the Organization of the Petroleum Exporting Countries (OPEC Fund). CGIAR was established on May 19, 1971. In 2014, CGIAR revenue was almost US $1057 million.

The CGIAR originally supported four centres: CIMMYT (*Centro Internacional de Mejoramiento de Maíz y Trigo* – International Maize and Wheat Improvement Center), IRRI (International Rice Research Institute), CIAT (International Center for Tropical Agriculture) and the IITA (International Institute of Tropical Agriculture). The initial focus was on the staple cereals, rice, wheat and maize, and this was further widened including cassava, chickpea, sorghum, potato, millet and other food crops. Again, this was encompassed by livestock, fishes, farming systems, the conservation of genetic resources, plant nutrition, water management, policy research and services to national agricultural research centres in developing countries. There were 13 research centres in 1983, and by the 1990s, the number of centres grew to 18. Mergers between institutions reduced the total to 15.

1.1 Plant Domestication

Domestication is a process by which plants with desirable traits are selected over time by humans (knowingly or unknowingly) for traits that are more advantageous or desirable to him. For instance, by deliberately caring a particular genotype, and through selecting plants for a particular trait, he may choose seed from that plant so that the progeny is likely to inherit that trait. Ancestor of maize, *Teosinte*, is a fine example for domestication. Teosinte had more rows of bigger kernels. Man also selected for desirable traits as non-shattering, exposed kernels and higher yield. Eventually, a new type corn was born. However, this leads to genetic erosion because only certain types were propagated and cultivated. As such, domestication tends to decrease the genetic diversity. However, diversity is available in wild relatives that can be exploited through intentional breeding. The first steps of domestication probably occurred in the Sumerian region between the Tigris and Euphrates Rivers and in Mexico and Central America.

According to National Geographic, agriculture began 12,000 years ago and was firmly established in Asia, India, Mesopotamia, Egypt, Mexico, Central America and South America some 6000 years ago. Some of the crops like corn, rice and wheat were domesticated here before recorded history. These areas also domesticated fibre crops like cotton, flax and hemp. Wheat is believed to have grown wild in the Tigris and Euphrates Valleys and spread from there to the rest of the Old World. Stone Age Europeans grew wheat and China produced wheat as early as 2700 BC. For 35% of the world population, wheat is a staple crop now. The history of corn dates back to 5200 BC and was first cultivated in the high plateau region of central or southern

Mexico. Rice is believed to be originated in Southeast Asia. India cultivated rice as early as 3000 BC, and it got spread throughout Asia and Malaysia. Today, rice feeds almost half of the world population. Cultivation of cotton spread to Egypt and then to Spain and Italy as early as 1500 BC. Other species that were made domestic since antiquity are dates, figs, olives, onions, grapes, bananas, lemons, cucumbers, lentils, garlic, lettuce, mint, radishes and various melons.

Aforesaid is the story generally available in literature that believed farming was invented some 12,000 years ago when civilization took shape in Iraq, Turkey and Iran. Recently, an international collaboration of Universities of Tel Aviv, Harvard, Bar-llan and Haifa offered evidence that trial plant cultivation began some 23,000 years ago. Lineages of *Brassica oleracea* stand as a fine example on how enterprising farmers contributed to the domestication of crops (see Box 1.2).

Box 1.2: Domestication of *Brassica oleracea*

Many crop plants have undergone the domestication process multiple times. Each of these efforts has focused on producing a new variant that could be used as a new vegetable. As such, a spectrum of different vegetables could be derived from the same wild progenitor. *Brassica oleracea stands* as an excellent example for this biological process. Wild progenitor is a weedy herb that grows on limestone in the Mediterranean region. Domestication of several distinct lineages of *B. oleracea* produced several vegetable varieties or cultivar groups or subspecies ("ssp."): kale and collard greens (ssp. *acephala*), Chinese broccoli (ssp. *alboglabra*), red and green cabbages (ssp. *capitata*), savoy cabbage (ssp. *sabauda*), kohlrabi (ssp. *gongylodes*), Brussels sprouts (ssp. *gemmifera*), broccoli (ssp. *italica*) and cauliflower (ssp. *botrytis*). Though these varieties look dramatically different, they are considered the same species since they are all inter-fertile, capable of mating with one another and producing fertile offspring (see Fig. 1.4).

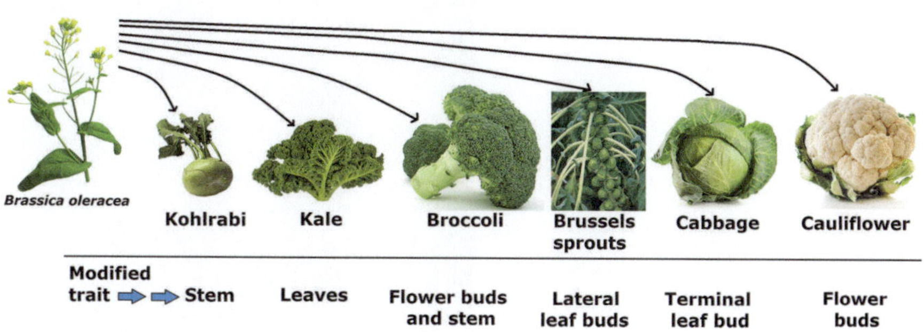

Fig. 1.4 Distinct lineages of *Brassica oleracea*

1.2 Plant Breeding: Pre-Mendelian

With domestication as the most basic method, plant breeding began 10,000 years ago. Domestication can happen at the level of genes also. Movement of nomadic tribes brought about the movement of these selected plant species. Introduction of new plant species/varieties into new areas is an integral part of plant breeding. Transfer of specific genes (say for disease resistance) from wild species to cultivated genotypes through *genetic engineering* can be regarded as domestication.

Man exercised plant breeding for his day-to-day needs. There is evidence to show that Babylonians and Assyrians exercised artificial pollination of date palm as early as 700 BC. Several varieties of "heading lettuce" were developed in France during the seventeenth century that were still in cultivation even during the 1990s. In 1717, Thomas Fairchild (Fig. 1.5) produced the first artificial hybrid, popularly known as "Fairchild" (*Dianthus caryophyllus barbatus)*, a cross between a sweet William and a carnation pink. Louis de Vilmorin established the first plant breeding company in France in 1727. Joseph Gottlieb Kölreuter, a German (Fig. 1.6), made extensive crosses in tobacco between 1760 and 1766. Knight (1759–1835) was the first to develop several new fruit varieties. Le Couteur and Patrick Sheriff developed some useful cereal varieties, and Sheriff published these results in 1873. Sheriff explained that variation of heritable nature responded to selection. This principle was exploited by Vilmorin in 1856 to develop several varieties of sugar beets (*Beta vulgaris*).

Fig. 1.5 Thomas Fairchild (1997–1729)

Fig. 1.6 Joseph Gottlieb Kölreuter (1733–1806)

Nilsson-Ehle and his associates of Svalöf, Sweden, developed individual plant selection methods during 1900. Wilhelm Johannsen proposed the pure-line theory during the early twentieth century that provided the genetic basis for individual plant selection.

Modern genetic mapping techniques seem to indicate that agriculture began in the Shia Crescent in the Middle East, particularly with regard to cereal breeding. However, other scholars, using the same techniques, have concluded that the cultivation of rice originated from various centres in the East (China). Genetic markers show that over the last 10,000 years, cultivated plants have not been modified.

1.3 Plant Breeding: Post-Mendelian

The science of genetics emerged with the rediscovery of the work of Gregor Johann Mendel (July 20, 1822–January 6, 1884) in 1900 (Box 1.3), which was originally published in *Versuche über Pflanzenhybriden* (*Experiments on Plant Hybridization*) and presented at two meetings of the Natural History Society of Brünn in Moravia in 1865. Mendel's laws of inheritance are the foundation for the science of genetics. Mendel's laws explained how traits are passed from one generation to the next. His work was rediscovered in 1900, with confirmation by E. von Tschermak, C. Correns and H. de Vries paving way to the principles of modern genetics. The earliest applications of genetics to plant breeding were made by the Danish botanist, Wilhelm Ludvig Johannsen (February 3, 1857–November 11, 1927) (Fig. 1.7), who while working with garden bean in 1903 developed the *pure-line theory*. His work confirmed that through repeated selfing, selection can produce highly homozygous lines (true breeding). Such lines were hybridized to produce hybrids. These hybrids outperformed either parent with respect to the trait of interest (the concept of *hybrid vigour*). Hybrid vigour (or *heterosis*) is the basis for modern hybrid crop

Fig. 1.7 Wilhelm Ludvig Johannsen (1857–1927)

production. Johannsen demonstrated the constancy of the biological type, which led him to formulate his essential distinction between *genotype* (the genetic makeup of a cell, an organism or an individual) and *phenotype* (expression of a particular trait, e.g. skin colour, height, behaviour, etc.). According to Johannsen, environmental factors that influenced the phenotype could not be transmitted to the genotype and the offspring. It was Theodor Boveri during the 1880s who gave the definitive demonstration that chromosomes are the vectors of heredity. The application of genetics in plant breeding gave explosive advancements. Among them, the derivation of dwarf and environmentally responsive varieties of wheat and rice is extremely notable. Such new varieties transformed world food production dramatically.

Box 1.3: Gregor Johann Mendel

Gregor Johann Mendel was born on July 22, 1822, to Anton and Rosine Mendel at what was then Heinzendorf bei Odrau in Austria, now a part of the Czech Republic. Mendel's parents were small farmers who financially struggled to educate Mendel. After schooling, he joined University of Olomouc in 1840 to learn physics, mathematics and philosophy. Due to financial difficulties, Mendel was compelled to join the Abbey of St. Thomas in Brünn as a monk and became Gregor Johann Mendel

(continued)

Box 1.3 (continued)

(Fig. 1.8). Later, he joined University of Vienna for learning chemistry, biology and physics. He wanted to qualify himself as a high school teacher. He returned to the monastery in 1854 and became a physics teacher at a school at Brünn. He taught there for next 16 years. During this time, Mendel could associate himself with two university professors: Friedrich Franz, a physicist, and Johann Karl Nestler, an agricultural biologist. Nestler was interested in heredity. These professors encouraged Mendel to conduct experiments on garden pea in the 2-ha garden attached with the monastery. Mendel presented the results of his research at sessions of the Natural Research Society of Brünn on Feb. 8 and March 8, 1865.

Mendel's most important conclusions were:

- The inheritance of each trait is determined by something (which we now call genes) passed from parent to offspring unchanged. In other words, genes from parents do not "blend" in the offspring.
- For each trait, an organism inherits one gene from each parent.
- Although a trait may not appear in an individual, the gene that can cause the trait is still there, so the trait can appear again in a future generation.

The rediscovery of Mendelism during 1900 by E. von Tschermak, C. Correns and H. de Vries is only an ensuing story. Totally unaware that a new science of genetics will be born later, Mendel died of a kidney disease, aged 61, on January 6, 1884.

Fig. 1.8 Gregor Johann Mendel (1822–1884)

1.4 Food Scarcity, Norman Borlaug and Green Revolution

"Almost certainly, however, the first essential component of social justice is adequate food for all mankind" – Norman E. Borlaug – the man who saved one billion lives. He also told "Food is the moral right of all who are born into this world".

Since time immemorial, humanity has been facing problems like famines and food scarcity. Foremost among them is the Irish potato famine of the 1840s that led to the death of about one million people. The Gujarat famine of 1899 and the Bengal famine of 1943 which led to the death of about three million are the most devastating famines witnessed in India. According to Thomas Malthus, in 1798, the population shall grow geometrically, while the food production shall increase arithmetically. He could not visualize that technological advancements could make a tremendous difference in the food production to keep pace with the population curve. With the arrival of the Rockefeller Foundation, the Green Revolution took shape.

Henry Wallace, the then US vice president, approached the Rockefeller Foundation to launch a programme of crop breeding in Mexico. Wallace, founder of Pioneer Hi-Bred seed company, was a successful crop breeder who developed first sterile hybrid in corn in the 1920s. The Rockefeller Foundation in 1943 launched Mexican Agricultural Program with the aim of developing high-yielding varieties (HYVs) with higher response to agrochemicals. Initial results of the programme were very encouraging. So, the Rockefeller Foundation established CIMMYT (*Centro Internacional de Mejoramiento de Maíz y Trigo*) in Mexico for international research for wheat and maize. The production of double-cross hybrids in maize significantly improved the yield in the 1960s. Also, concurrently, Green Revolution programmes were introduced in developing countries (India, the Philippines and Indonesia) in the 1960s. Soon after in the same year, the Rockefeller and Ford Foundations together with the Government of the Philippines established the International Rice Research Institute (IRRI) in Manila for the production of high-yielding rice to feed over one billion poor people across the world.

1.4.1 Semi-dwarf Varieties of Wheat and Rice

The derivation and introduction of new semi-dwarf varieties of wheat and rice were the success story of the Green Revolution. According to Borlaug, their wide adaptation, short stature, high responsiveness to inputs and disease resistance are the attributes to their success (see Box 1.4). It all started when Japanese scientists developed the semi-dwarf wheat variety Norin 10 using Daruma as the donor of the semi-dwarfing trait. The recessive genes responsible for dwarfing were named *rht1* and *rht2*. Daruma was a Japanese semi-dwarf variety that was crossed to Fultz, which was a high-yielding US winter wheat. This cross gave Fultz-Daruma. Fultz-Daruma was later crossed with Turkey Red which was also a high-yielding US winter wheat. This cross led to the production of Norin 10 which was a semi-dwarf and high-yielding variety. Norin 10 was later brought to the USA and subjected for crossings with local varieties. These crossed varieties led to the production of

Gaines. This was done by Dr. Orville Vogel in the 1950s. Dr. Borlaug later used the Gaines to develop modern semi-dwarf wheat varieties. Dr. M. S. Swaminathan, the doyen of Indian agriculture, used the *shuttle breeding technology* (coined by Borlaug – wherein alternate generations were grown at two diverse locations) that led to the production of Sonora 64. As these locations differed in terms of soil, temperature, rainfall and photoperiod, this effort resulted in the production of strains possessing wide disease resistance and insensitivity to photoperiod.

Box 1.4: Norman Ernest Borlaug ***(March 25, 1914, to September 12, 2009)***

The credit for the success of the Green Revolution goes to Dr. Norman E. Borlaug who is honoured as "Father of the Green Revolution". Dr. Borlaug spent his entire life striving to alleviate poverty (Fig. 1.9). In 1970, he was awarded with a Nobel Peace Prize for his exemplary work. Born in 1914, in Cresco, Iowa, he earned a PhD in Plant Pathology from the University of Minnesota in 1941. From 1944 to 1960, he worked at the Rockefeller Foundation attached with the Cooperative Mexican Agricultural Program. In 1963, he became the leader of the Wheat Program at CIMMYT. He held this position till his retirement in 1979. He could spread this successful model of shuttle breeding technology to other developing nations like India and Pakistan in the mid-1960s. Between 1964 and 2001, the wheat production in India increased from 12 to 75 million tons, while in Pakistan, it increased from 4.5 to 22 million tons. Thus, the work of Dr. Borlaug revolutionized agriculture in the developing countries and saved millions of people from starvation.

He received the Congressional Gold Medal in 2006, America's highest civilian honour, becoming one of only five individuals to receive the Nobel Prize, the Presidential Medal of Freedom and the Congressional Gold Medal.

The genesis of dwarf rice varieties started with introduction of recessive gene, *sd1* (for short height), from a Chinese variety *Dee-geo-woo-gen* (meaning short-legged). The IRRI team (Peter Jennings, Henry Beachell and S.K. De Datta) developed a semi-dwarf variety IR8 in 1962 by using tall *Peta* as female (from Indonesia) and *Dee-geo-woo-gen* as male. *Dee-geo-woo-gen* has stiff straw augmenting for semi-dwarf nature. IR8 had stiff straw and resistance to lodging and was insensitive to photoperiod. These attributes made IR8 a preferred variety among farmers with good adaptability. Thus, IR8 became the *miracle rice*. While the earlier varieties had a harvest index of 0.3 (ratio of grain to straw as 30:70 with 10–12/ha biomass), with a maximum yield of 4 t/ha, the improved Green Revolution semi-dwarf varieties of wheat and rice had a harvest index of 0.5. The improved varieties owned total biomass potential of 20 t/ha with a yield potential of 10 t/ha with 120 kg of nitrogen per hectare. According to Gurdev Singh Khush, a well-known rice breeder, the

Fig. 1.9 Norman Ernest Borlaug (1914–2009)

improvement of harvest index is responsible for increasing yield potential. From 1950 to 1990, the worldwide irrigated land area increased from 94 million ha to 240 million, while fertilizer usage increased from 14 million tons to 140 million tons. It is the contribution of great plant breeders that made significant strides towards nurturing the humankind over the years. A list of prominent plant breeders and their contributions are available in Table 1.4. Many institutions like Cornell University, Ithaca; University of Georgia, Athens; Texas A&M University; Iowa State University; Washington State University; John Innes Centre (formerly Plant Breeding Institute, Cambridge), Norwich, UK; and University of California, Davis, and USDA research centres, along with international research centres of CGIAR, took active role in these advancements.

1.5 Facets of Plant Breeding

Plant breeding met with consummate success during the twentieth century as it engaged in crossing parents with desired traits to generate genetic variation through recombination. Further, the selection of best combinations based on the phenotypes across locations, over time, gave the substantial impact. Research investments in cell and molecular biology grew significantly during the end of the 1980s, and in the

Table 1.4 Some prominent plant breeders (list neither exclusive nor exhaustive)

André Gallais	French specialist in quantitative genetics and breeding methods theory
Andrew H. Paterson	US geneticist, research leader in plant genomics
Barbara McClintock	American cytogeneticist, Nobel Prize for genetic transposition
Bernard Dutrillaux	French cytogeneticist, chromosome banding, comparative cytogenetics
Berwind P. Kaufmann	US botanist, did research in basic plant and animal cytogenetics
C.C. Li	Eminent Chinese-American population geneticist and human geneticist
C.M. Rick	Botanist who pioneered research on the origins of tomato
Charles Leonard Huskins	English-born Canadian cytogeneticist at McGill University and University of Wisconsin-Madison
Christian Jung	German plant geneticist and molecular biologist
D.S. Falconer	Scottish quantitative geneticist, wrote textbook to the subject
David Catcheside	UK plant geneticist, expert on genetic recombination, active in Australia
Derald Langham	American agricultural geneticist, the "Father of Sesame"
Dronamraju Krishna Rao	Indian-born geneticist, founder of the Foundation of Genetic Research
E.B. Babcock	US plant geneticist, pioneered genetic analysis of genus *Crepis*
E. Baur	German geneticist, botanist, discovered inheritance of plasmids
Edgar Anderson	Eminent US plant geneticist
Edward H. Coe, Jr.	US maize (corn) geneticist
Emmy Stein	German botanist and geneticist
Erich von Tschermak	Austrian agronomist and one of the rediscoverers of Mendel's laws
Ernie Sears	Wheat geneticist who pioneered methods of transferring desirable genes from wild relatives to cultivated wheat in order to increase wheat's resistance to various insects and diseases
Floyd Zaiger	Fruit geneticist and entrepreneur
Frank Stahl	American molecular biologist, the Stahl half of the Meselson-Stahl experiment
G.H. Shull	American geneticist, made key discoveries including heterosis
G. Ledyard Stebbins	American botanist, geneticist and evolutionary biologist
George Beadle	US *Neurospora* geneticist and Nobel Prize winner
Guido Pontecorvo	Italian-born Scottish geneticist and pioneer molecular biologist
Gurdev S. Khush	An agronomist and geneticist who, along with mentor Henry Beachell, received the 1996 World Food Prize for his achievements in enlarging and improving the global supply of rice during a time of exponential population growth
Harriet Creighton	US botanist who with McClintock first saw chromosomal crossover
Hugo de Vries	Dutch botanist and one of the rediscoverers of Mendel's laws in 1900
Ivan Vladimirovich Michurin	Russian plant geneticist, scientific agricultural selection
James Birchler	*Drosophila* and maize geneticist and cytogeneticist
James F. Crow	US population geneticist and renowned teacher of genetics
J.B.S. Haldane	Brilliant British human geneticist and co-founder of population genetics

(continued)

Table 1.4 (continued)

Jean-Baptiste Lamarck	French naturalist, evolutionist, "inheritance of acquired traits"
Jens Clausen	Danish-US botanist, geneticist and ecologist
John C. Sanford	American horticultural geneticist and intelligent design advocate
Karl Sax	American botanist and cytogeneticist, research on the effects of radiation on chromosomes
Keith Downey	Canadian agricultural scientist and, as one of the originators of canola, became known as the "Father of Canola"
L.J. Stadler	Eminent American maize geneticist
Luther Burbank	US botanist, horticulturist, pioneer in agricultural science
M.S. Swaminathan	Indian agricultural scientist, geneticist, leader of Green Revolution in India
Marcus Rhoades	Great maize (corn) geneticist and cytogeneticist
Massimo Pigliucci	Italian-US plant ecological and evolutionary geneticist. Winner of the Dobzhansky prize
Nazareno Strampelli	Italian agronomist and plant breeder. He was the forerunner of the so-called Green Revolution
Niels Ebbesen Hansen	A Danish-American horticulturist
Nikolai Vavilov	Eminent Russian botanist and geneticist
Nina Fedoroff	US plant geneticist, cloning of transposable elements, plant stress response
Norman Ernest Borlaug	American agronomist and humanitarian who led initiatives worldwide that contributed to the extensive increases in agricultural production termed the Green Revolution
Oliver Nelson	US maize geneticist, profound impact on agriculture and basic genetics
Peter Michaelis	German plant geneticist, focused on cytoplasmic inheritance
R.L. Phillips	US plant geneticist; genetics and genomics of cereal crops
R.A. Brink	Canadian-US plant geneticist and breeder, studied paramutation, transposons
R.A. Emerson	American plant geneticist, pioneer of corn genetics
R.A. Fisher	British stellar statistician, evolutionary biologist and geneticist (to be seen)
R.C. Punnett	English geneticist, discovered linkage with William Bateson
Richard Goldschmidt	German-American, integrated genetics, development and evolution
Richard Jefferson	US molecular plant biologist in Australia, reporter gene system GUS
Susan R. Wessler	US plant molecular geneticist, transposable elements regenetic diversity
T.H. Morgan	Head of the "fly room", first geneticist to win the Nobel Prize
Theodosius Dobzhansky	Noted Ukrainian-US geneticist and evolutionary biologist
Thomas Andrew Knight	British horticulturalist and botanist known for his work on geotropism
W. Gottschalk	Worked on mutation breeding
William Bateson	British geneticist who coined the term "genetics"

academic scenario, conventional plant breeders were replaced by cell and molecular biologists. This can reduce the time taken in releasing varieties, developing segregating populations or producing genetic stocks, which were the main tasks of plant breeding. This fact was realized in the last decade. Now, conventional crossbreeding and usage of tools from omics and transgenic research go hand in hand. Thus, plant breeding is multifaceted. A summary of facets of plant breeding is presented here.

Society

Plant breeding derives crops that address human needs. Due to enhancement of genetic potential, after World War II, crop yields increased steadily. Otherwise, prices for all crops should have been 35–66% higher in 2000 against their actual prices. In the absence of high-yielding varieties, there would have been 13.3–14.4% lower per capita calorie intake and an increase of malnourished children between 6.1 and 7.9% in the developing world. Nearly, 18–27 million ha was saved by the Green Revolution from being brought into agriculture. The twenty-first century is expected to make explosive advancements. Annual breeding gains must increase by 2.5 that can double crop yields by 2050.

Omics

DNA "fingerprints" will introduce new genetic variation, and DNA markers will decrease the dependability on field trials. Genetic engineering introduces new traits from other species/genera, thereby supplementing novel diversity for plant breeding. Farmers have been growing transgenic crops since the 1990s. Marker-aided breeding (MAB) was extensively used in the last two and half decades. In recent years, omics research has greatly contributed towards identification and functional analysis of genes. DNA sequencing today unravels the relationships among alleles and traits.

Population

As per Hardy-Weinberg law, the frequency of alleles and genotypes remains constant through generations. Crop domestication had significantly affected allele frequency and genetic segregation of those genes that produce striking morphological changes. Alleles at these loci were fixed during early crop domestication, thereby reducing the genetic diversity for traits. The evolution of cultivated plants is believed to have disrupted Hardy-Weinberg equilibrium through selection, non-random mating, genetic drift, migration through gene flow, mutation and meiotic drive favouring transmission of allele regardless of its phenotypic expression.

Genetic Diversity

Genetic diversity depends on the richness of alleles. Allelic richness refers to the total number of distinct alleles. The coefficient of gene diversity is the probability of how two distinct gametes are randomly chosen from a population. There are several measures like Wright's fixation index F, heterozygosity level, the degree of population divergence F_{ST} or G_{ST} and the degree of linkage disequilibrium to judge genetic diversity level. Total heterozygosity can be estimated by adding the allelic diversity within and among populations. While F measures the deviation of genotypic frequencies from an expected random mating or panmictic population, the F_{ST} measures population differentiation ensuing from population structure using biallelic DNA markers. The G_{ST} is a quantitative index of the degree of genetic differentiation between subgroups or population divergence considering multiple alleles.

Distance Measures

The degree of similarity can be measured by DNA markers. Genetic relationships in plant germplasm and defining heterotic groups among breeding populations can be judged with this exercise. However, DNA markers are yet to prove their ability in predicting heterosis. Measurements for genetic distance can be done with through Euclidean or statistical means. The Euclidean metric between two plants is a straight line measuring the "ordinary distance" as defined by the difference of the frequency of alleles between them. While calculating statistical distances, DNA marker data, especially single-nucleotide polymorphisms (SNP), can be taken into account because they increase the precision of relatedness.

Germplasm Grouping

When several traits are under study in one individual or in a population, multivariate techniques are useful for categorizing germplasm as several groups. While univariate analysis considers the variation on each trait independently, multivariate variate analysis delineates traits and their relationships that determine how the plants vary while considering all traits together. Non-hierarchical principal component analysis (PCA) is yet another tool that determines patterns of variation among groups and subgroups among germplasm accessions. PCAs are functions of eigenvalues and eigenvectors of the variance/covariance matrix. PCAs and DNA markers follow entirely opposite functions. However, PCAs can be determined based on genetic distances calculated from DNA marker data. Cluster analysis is yet another hierarchical procedure to group gene bank accessions. A cluster diagram represents diagrammatic depictions of eigenvalues that are shown as a dendrogram. A dendrogram is a tree like diagram placing individuals with close distance (see Chapter on GE interactions).

Quantitative Variation

Phenotypic variation is governed by genes, the environment and the genotype-by-environment interaction (GE). Phenotypic variation is measured across locations, seasons or years. Sir Ronald A. Fisher in 1918 and Sewall G. Wright in 1921 were the scientists who gave explanations for the analysis of variance components. The mathematical theory of natural and artificial selection of J.B.S. "Jack" Haldane in 1932 further influenced such models. Maize stands as the best model genetic system. Genetic gains are primarily due to selection of favourable alleles with additive genetic effects. The selected individuals are evaluated in replicated trials. Those with superior breeding values are crossed further and selection is exercised again. The best linear unbiased prediction (BLUP) that was originally devised for animal breeding is a useful technique to learn relationships among the offspring. BLUP is also useful for predicting hybrid performance of cross-pollinated crops as also for modelling GE. A genotype may not be a very accurate predictor of a phenotype when the interaction and the GE are significant. Genetic architecture denotes the underlying basis of a phenotype. Genes can show additive, dominance or epistatic effects and interact with the environment. Effect of each gene may vary in its magnitude significantly.

Mapping Traits

QTL (Quantitative Trait Loci) linkage analysis began in the 1980s. This analysis determines the dissimilarity of phenotypes among genetically related individuals. Microsatellites (SSR=Single Sequence Repeats) and single-nucleotide polymorphism (SNP) determine the understanding of the genetic architecture. Plant genomics and DNA sequencing with the support of friendly software facilitates the analysis of genetic and phenotypic data. Complex quantitative variations could be mapped in this way. Linkage disequilibrium or association mapping provides associations between target traits and polymorphic DNA markers on a historical basis. Association mapping or linkage disequilibrium is a technique that can be done without specific mating. Data from nursery, advanced breeding trials and multi-environment testing can be used for this. Linkage disequilibrium is the distance between loci across chromosomes. This is really a new advancement that can dissect complex quantitative traits. Transcriptomics is another promising area. Transcriptomics (study of complete set of RNA transcripts that are produced, under specific circumstances) can throw light on regulatory genetic factors affecting quantitative variation.

Genotype-by-Environment (GE) Interactions

For the appraisal of the phenotypes, multi-environment testing must be practised. The phenotypic effect as a result of interactions between genotypes and the environments is GE. While testing genotypes under different environments, the ranking of genotypes can change. GE is the change in the ranking of genotypes. Either the genotype or the environment can be fixed. In a linear model, the other should be regarded as random. In a mixed model, the genotypes are usually regarded to be random. The testing environments are often fixed; the environment is repeated across years and locations.

Factorial regression is an ordinary linear model wherein traits from crop husbandry, soil or weather data can be incorporated. These variables could, however, show a high collinearity (linear association between two explanatory variables). This situation complicates the interpretation. However, modelling increases accuracy. The additive main effects and multiplicative interaction (AMMI) model is one used for analysing multi-environment trials involving two-way data tables. It uses main effects first and then uses the PCA (principal component analysis) for analysing the interactions (see Chap. 20). Main effects are in the horizontal axis, and the environments are in the vertical axis. The respective scores are multiplied to calculate the GE interactions for a given genotype and environment. When both G and E have the same sign for these scores, it is positive GE. It is negative when G and E have opposite signs. GGE (genotype main effects and genotype-by-environment interaction effects) is yet another model that delineates which genotype performs better in which environment. It also efficiently defines mega-environments. Mega-environments are those that have similar biotic and abiotic stresses, cropping systems, levels of production and consumer preferences. Full- or half-sibs are related individuals and data taken from them are therefore correlated. A QTL lacking GE will have wider adaptation (i.e. across environments), and QTL with a significant GE will have only specific adaptation. In most crops, QTL × environment interaction is prevalent. Genes perform distinctively and hence their GE interactions will be different. But whole genome approaches can monitor polymorphisms of several hundreds of loci.

Phenomics

Phenomics is the study of gene expression of a given species in a specific environment. Data provided by drones/robotics offers precise information on plant development that relates phenotype with the genotype under controlled environments. Forward phenomics uses high-throughput resolution of valuable physiological traits. High-throughput and cost-effective phenomic platforms are in infancy. If refined further, they can assess the response under stressful environments. Please refer to Table 1.5 for a comprehensive list of new plant breeding techniques.

1.6 Future Challenges

According to FAO, due to higher income levels, about 70% of the world's population will be urban in the future (compared to 49% today). While food production needs to reach 70%, cereal production will have to attain 3 billion tons mark (against 2.5 billion today). If the necessary investments, policies and regulations for agricultural production are undertaken, this target may not be difficult. In developing countries, cropping intensity accounts for 80% of the yield increase. Only 20% comes from the expansion of arable land. This calls for use of improved agricultural technologies and biotechnologies. In addition to caloric demands, food supply must ensure intake of vitamins, essential minerals and other nutritional factors. This can be achieved through production of *biofortified* food that can nourish children in poorer countries.

Climate changes and desertification dramatically affect physiological processes and increase soil erosion. Over the years, atmospheric concentration of CO_2 has increased from approximately 315 ppm (parts per million) in 1959 to a current concentration of approximately 385 ppm. The accompanying increase in greenhouse gases (methane, ozone and nitrous oxide) due to intensified burning of fossil oils and other man-made activities has contributed to higher atmospheric concentration of CO_2. The current global warming is due to increase in the greenhouse effect. This will have an adverse effect on average annual mean warming with an increase of 3–5 °C in the next 50–100 years. Increased desertification in many parts of the world

Table 1.5 Description of some of the new plant breeding techniques

Technique	Summary
Accelerated plant breeding (speed breeding)	Induction of early flowering to accelerate cross-breeding. Also, implemented in in vitro nurseries, which could substantially shorten generation time through rapid cycles of meiosis and mitosis
Agro-infiltration	Use of recombinant *Agrobacterium* to achieve transient expression of genes in plant tissues. Here, a suspension of *Agrobacterium tumefaciens* is introduced into a plant leaf by direct injection or by vacuum infiltration or brought into association with plant cells immobilized on a porous support (plant cell packs), whereafter the bacteria transfer the desired gene into the plant cells *via* transfer of T-DNA
Centromere-mediated genome elimination	Centromeres are points where spindle fibres are attached. Centromeres depend on an epigenetic signal, that is, a persistent DNA modification that does not depend on sequence. This largely mysterious epigenetic signal requires a variant histone H3, called CENH3. The experimental alteration of CENH3, by swapping its amino-terminal region and fusing it to green fluorescent protein (GFP) to produce "Tailswap CENH3", can lead to genome elimination. Genome elimination only occurred when a plant strain with the altered CENH3, referred to as the "Tailswap" haploid inducer, was crossed to a wild-type plant, leading to the elimination of all the Tailswap chromosomes. To date, this event has only been reported in *Arabidopsis*, but given the conserved nature of the perturbed mechanism, it is likely to also apply to crop plants
Cisgenesis	Transformation of plants with genes derived from the same or from a sexually compatible species and present in their natural orientation; have their own introns and are flanked by their native promoters and terminators
Grafting on GM rootstocks	Production of chimaeras from GM rootstocks and non-GM scions. Here, only root stocks are genetically modified. Use of *short interfering RNA* (siRNA) is another application which is made in the genetically modified rootstock. They are transported to the graft (scion) where they cause the desired effect. Using this technique, protein production, for example, can be regulated in the upper stem
Induced hypomethylation	Silencing of genes. Loss of the methyl group in the 5-methylcytosine nucleotide, when it is followed by a guanosine (G)
Intragenesis	Transformation of plants with DNA sequences derived from the same or from a sexually compatible species. While cisgenesis involves genetic modification using a complete copy of natural genes with their regulatory elements that belong exclusively to sexually compatible plants, intragenesis refers to the transference of new combinations of genes and regulatory sequences belonging to that particular species
Meganuclease technique	Use of synthetic meganucleases to knock out targeted genes, to correct targeted genes or to insert new genes at a predetermined site in the genome

(continued)

Table 1.5 (continued)

Technique	Summary
Methyltransferase technique	Use of synthetic methyl transferases for targeted methylation of genomic sequences. This will further alter the protein structure and function
Oligonucleotide-directed mutagenesis (ODM)	ODM is a tool for targeted mutagenesis, employing a specific oligonucleotide, typically 20–100 bp in length, to produce a single DNA base change in the plant genome. This oligonucleotide is of a single base pair change. In cultured plant cells, they bind to the corresponding homologous plant DNA sequence. Then, the cell's natural repair machinery recognizes this single-base mismatch and undertakes required repair. Plants carrying the specific mutation are subsequently regenerated by tissue culture and can be used for breeding the desirable trait into elite plant varieties
Reverse breeding	Production of homozygous parental lines from heterozygous plants by suppressing meiotic recombination (see Chap. 13 on recombinant inbred lines)
RNA-directed DNA methylation (RdDM)	Many small interfering RNAs (siRNAs) direct de novo methylation by DNA methyltransferase. DNA methylation typically occurs by RNA-directed DNA methylation (RdDM), which directs transcriptional gene silencing of transposons and endogenous transgenes. RdDM is driven by non-coding RNAs (ncRNAs) produced by DNA-dependent RNA polymerases IV and V (Pol IV and Pol V). The production of siRNAs is initiated by Pol IV, and ncRNAs produced by Pol IV are precursors of 24-nucleotide siRNAs
Seed production technology	Use of transgenic maintainer lines to propagate male sterile female parental lines used in producing hybrid seeds. Hybrid seed production uses cytoplasmic male sterile lines or photoperiod/thermosensitive genic male sterile lines (PTGMS) as female parent. Cytoplasmic male sterile lines are propagated via cross-pollination by corresponding maintainer lines, whereas PTGMS lines are propagated via self-pollination under environmental conditions restoring male fertility. Alternatively, construction of male sterility system using a nuclear gene that encodes a putative glucose-methanol-choline oxidoreductase regulating tapetum degeneration and pollen exine formation. Cross-pollination of the fertile transgenic plants to the non-transgenic male sterile plants produces male sterile seeds of high purity
TALEN technique	Transcription activator-like effector nucleases (TALEN) are restriction enzymes that can be engineered to cut specific sequences of DNA. They are made by fusing a TAL effector DNA-binding domain to a DNA cleavage domain (a nuclease which cuts DNA strands). Transcription activator-like effectors (TALEs) can be engineered to bind to practically any desired DNA sequence, so when combined with a nuclease, DNA can be cut at specific locations. TALEN is a tool in genome editing
Targeted chemical mutagenesis	Use of oligonucleotides coupled to chemical mutagens to trigger mutations at a predetermined site of the genome

(continued)

Table 1.5 (continued)

Technique	Summary
Target mutagenesis with T-DNA	Use of T-DNA to replace an endogenous target gene with a homologous gene with altered DNA sequence
Transformation with wild-type *Agrobacterium*	Use of wild-type *Agrobacterium rhizogenes* for producing transformed plants
Virus-induced gene silencing (VIGS)	This is a technique for using recombinant viruses to achieve transient gene silencing in plants. VIGS is a technology that exploits an RNA-mediated antiviral defence mechanism. It is one of the reverse genetics tools for analysis of gene function that uses viral vectors carrying a target gene fragment to produce dsRNA which trigger RNA-mediated gene silencing. Virus-derived inoculations are performed on host plants using different methods such as agro-infiltration and in vitro transcriptions
Zinc finger nuclease technique	Zinc finger nucleases (ZFNs) are a class of engineered DNA-binding proteins that facilitate targeted editing of the genome by creating double-strand breaks in DNA at user-specified locations. Each zinc finger nuclease (ZFN) consists of two functional domains: (a) A DNA-binding domain comprised of a chain of two-finger modules, each recognizing a unique hexamer (6 bp) sequence of DNA. Two-finger modules are stitched together to form a zinc finger protein, each with specificity of $\geq$24 bp. (b) A DNA-cleaving domain comprised of the nuclease domain of Fok I. When the DNA-binding and DNA-cleaving domains are fused together, a highly specific pair of "genomic scissors" are created (see Chap. 22 on "Genetic Engineering")
CRISPR/Cas9	CRISPR/Cas9 (clustered regularly interspaced short palindromic repeats) was adapted from a naturally occurring genome editing system in bacteria. The bacteria capture snippets of DNA from invading viruses and use them to create DNA segments known as CRISPR arrays. If the viruses attack again, the bacteria produce RNA segments from the CRISPR arrays to target the viruses' DNA. The bacteria then use Cas9 or a similar enzyme to cut the DNA apart, which disables the virus. Small piece of RNA with a short "guide" sequence that attaches to a specific target sequence of DNA in a genome with Cas9 enzyme is made. Cas9 enzyme cuts the DNA at the targeted location. Once the DNA is cut, cell's own DNA repair machinery is used to add or delete pieces of genetic material or to make changes to the DNA by replacing an existing segment with a customized DNA sequence
Single-base editors	Scientists have developed a single-base editing system (base editor) through combining of CRISPR/Cas9 system with cytosine deaminase. Compared with Cas9 system, this base editor can convert cytosine to thymine (C > T) at specific site more efficiently without inducing double-strand breaks to avoid generation of indels (insertion or deletion of bases). However, the base editor can only generate transition of pyrimidine but could not modify purines. Recently, a novel base editing system

(continued)

Table 1.5 (continued)

Technique	Summary
	to convert adenine to guanine (ABEs, adenine base editors) through fusion of Cas9 nickase to a modified deaminase has been evolved through screening of random library based on tRNA adenine deaminase from *E. coli*

is due to the combined effect of climatic changes, global warming, drought and salinity. Around 41% of Earth's surface is dry land and accounts for more than 38% of the total global population. Soil salinization can also be the end result of climate change and desertification. Altogether, net result shall be 30% arable land loss over the next 25 years and up to 50% land loss by 2050.

Challenges to agricultural production and productivity to meet food needs of the rising population and also to raise raw materials for industrial production (e.g. cotton for textiles) are formidable. The added pressure from climate change affecting yield of crops increases this challenge. The mix of increased levels of CO_2, changes in temperature and rainfall are increasingly breaching extremes and changing patterns of crop diseases and pests. This adds uncertainties in crop production that can be addressed only through plant breeding.

Plant breeding in the twenty-first century will focus on producing more yield with less inputs. Farmers have been growing transgenic crops since the 1990s. Marker-aided breeding (MAB) gave way to explosive advancements during the last two and a half decades. Genomics research involve understanding genes and their functions. Today, DNA sequencing helps in unravelling the relationships among alleles controlling traits. All these modern methods are welcome, but they must assist the breeders in deriving varieties that can assist the farmers with higher yield.

Further Reading

Baenziger SP, Al-Otyak SM (2007) Plant breeding in the twenty-first century. Afr Crop Sci Conf Proc 8:1–3

Birchler JA, Han F (2018) Barbara McClintock's unsolved chromosomal mysteries: parallels to common rearrangements and karyotype evolution. Plant Cell 30:771–779

Bouis HE, Saltzman A (2017) Improving nutrition through biofortification: a review of evidence from HarvestPlus, 2003 through 2016. Glob Food Sec 12:49–58

Bradshaw JE (2017) Plant breeding: past, present and future. Euphytica 213:60

Cowling (2013) Sustainable plant breeding. Plant Breed 132:1–9

Ferrante A et al (2017) Plant breeding for improving nutrient uptake and utilization efficiency. Advances in research on Fertilization management of vegetable crops. Part of the Advances in Olericulture book series (ADOL), pp. 221–246

Plant breeding: the art of bringing science to life. Highlights of the 20th EUCARPIA General Congress, Zurich, Switzerland, 29 August–1 September 2016

Schlegel RHJ (2017) History of plant breeding. CRC Press, Boca Raton

Snir A, Nadel D, Groman-Yaroslavski I, Melamed Y, Sternberg M, Bar-Yosef O et al (2015) The origin of cultivation and proto-weeds, long before neolithic farming. PLoS One 10(7): e0131422. https://doi.org/10.1371/journal.pone.0131422
Wesesler J, Zilberman D (2017) Golden rice: no progress to be seen. Do we still need it? Environ Develop Econom 22:107–109

Objectives, Activities and Centres of Origin 2

The main objectives of plant breeding are to improve the qualities of plants in many respects such as:

(a) *To evolve new varieties of crops which have better yielding potential (grains, fodder, fibres, oils, etc.).*

High crop yield: plants that invest a large proportion of their total primary productivity into seeds, roots, leaves or stems must be selected. It must be ensured that all the light that falls on a field is intercepted by leaves so that high primary productivity and efficient final production may be achieved. Greater efficiency in photosynthesis could perhaps be achieved by reducing photorespiration. Native varieties can be sued to derive hybrids that can be evaluated for higher yield.

The classical examples for using native varieties are the utilization of Dee-geo-woo-gen (DGWG) and Taichung Native 1 in rice and Norin 10 in wheat. ADT 27 (*indica* x *japonica* cross-derivative) is the first high-yielding rice variety of Tamil Nadu, India. Dee-geo-woo-gen and wonder rice IR 8 (Peta x DGWG) challenged poverty. Kalyan Sona in India was derived from norin10 wheat genes. The cytoplasmic male sterility (CMS), especially Texas male sterility, resulted in the production of a number of varieties. CMS produces sterile male flowers facilitating the avoidance of removal of male flowers (de-tasselling).

In pearl millet, production increased to manyfold because of breeding with male sterile line Tift 23A at Tifton, Georgia, by Burton. This led to the release of hybrid bajra HB1 to HB4 in India. In jowar (sorghum), the first hybrid CSH 1 (CK 60A x IS 84) was released during the 1970s. Breeding of male sterile line with kafir 60A gene was responsible for this.

(b) *To increase the quality of grains and crop as a whole with respect to size, colour, shape, taste, nutritional content,* etc. *(e.g. aroma and grain colour, milling and cooking quality in rice; gluten content and milling and baking*

P. M. Priyadarshan, *PLANT BREEDING: Classical to Modern*,
https://doi.org/10.1007/978-981-13-7095-3_2

quality in wheat; protein content in pulses; polyunsaturated fatty acids (PUFA) content in oil seeds).

(c) *To produce varieties resistant to fungal and bacterial diseases, insects and pests.*

Crop loss due to diseases is estimated to be between 10% and 30% of the total crop production. Resistant varieties are in advantage for disease and insect management. In the case of rusts in wheat, they offer the only feasible means of control. Resistant varieties offer increased and stabilized yield.

(d) *To produce early- or late-maturing varieties according to our desire. It permits new crop rotation and often extends crop area.*

(e) *To produce varieties accommodative to a particular climate and soil (to produce varieties with a wide range of adaptability).*

An array of attributes come under the umbrella of climate and soil. They are weather fluctuations, pests and pathogens, resistance to weeds and tolerance to heat, cold, drought, wind, soil salinity, acidity or aluminium toxicity.

(f) *To change the growth habit of crops such as dwarfness, few branching and less tillering or tallness with profuse branching so as to increase the straw for fodder.*

(g) *To develop varieties responsive to fertilizers and irrigation.*

To reduce the need for nitrogen fertilizer, cereals can be bred that encourage nitrogen-fixing microorganisms to grow around their roots.

(h) *Development of varieties with tolerance to salt and moisture stress.*

Crop production in India can be improved with the development of varieties for rainfed areas and resistant to saline soils. Nearly 70% of the cropped area in the country is rainfed. A range of 7–20 million ha are saline, of which about 2.8 million ha are alkaline. Much of these categories of soils are in the states of Uttar Pradesh, Haryana and Punjab.

(i) *Some crops have toxic substances like khesari (Lathyrus sativus) that contains a neurotoxin (lathyrogen), β-N-oxalyl-amino-L-alanine, or BOAA, that can cause paralysis. Brassica oil has harmful eruic acid. Nutritional value of these crops can be improved through removal of those toxic substances.*

(j) *Derivation of photo-insensitive varieties.*

Breeding for climate change demands production of varieties that are insensitive to photoperiod and temperature. Such varieties can be cultivated in new areas. Photoperiod insensitivity genes (Ppd1 and Ppd2) are prominent in wheat.

(k) *Biofortifying crops with essential mineral elements like Fe and Zn, vitamins and amino acids that are otherwise lacking in cereals.*

(l) *Plant architecture and adaptability to mechanized farming.*

For mechanical farming and harvesting, plant architecture needs to be modified. Positioning of the leaves, branching pattern, height and positioning of panicle determine/govern mechanical harvesting.

(m) *New cropping systems: contrasting cropping, intercropping and sustainable cropping systems.*

Breeding programme consists of a series of activities like variate, isolate, evaluate, inter-mate, multiply and disseminate. Plant breeders in classical plant breeding generally select the different plants with desirable characters (pure lines) and crossing (hybridization) them to obtain the desired traits in offsprings. The offsprings with desirable traits are then selected, tested, multiplied and then supplied to the farmers or growers.

The following are the various broad steps required for developing new varieties:

(a) *Collection of variability*
(b) *Evaluation and selection of parents*
(c) *Cross-hybridization among the selected parents*
(d) *Selection and testing of superior recombinants*
(e) *Testing, release and commercialization of new cultivars*

The present-day crop plants originated from weed-like wild plants. This was achieved by rigorous plant breeding efforts. This change has been brought about by man through plant breeding. The production of semi-dwarf cereal varieties of wheat and rice has been the spectacular milestone of modern agriculture. The semi-dwarf wheat varieties were developed by N.E. Borlaug and co-scientists of CIMMYT, Mexico. Japanese variety Norin 10 was the source of dwarfing genes. Kalyan Sona and Sonalika produced in India were with Norin 10 genes with lodging resistance, fertilizer responsiveness and higher yield. They are generally resistant to rusts and other major diseases due to the incorporation of resistance genes, thus stabilizing wheat production in the country.

Similarly, the development of semi-dwarf rice varieties from Dee-geo-woo-gen (DGWG), a dwarf, early-maturing variety of *japonica* rice from Taiwan, has revolutionized rice cultivation along with Taichung Native 1 (TN1) and IR8 (Peta from Indonesia x Dee-geo-woo-gen) developed at IRRI (International Rice Research Institute), Philippines. It all began with the Food and Agriculture Organization (FAO) of the United Nations establishing an International Rice Commission to undertake a *japonica-indica* crossing programme at Cuttack in India. Its mission was to undertake crosses involving short *japonica* and taller *indica* to develop short-stature varieties with higher yield. ADT 27 and Mahsuri, selected from such crosses, were widely planted across the Indian subcontinent in the 1960s. Such varieties were later replaced by semi-dwarf varieties like Jaya and Ratna, which are semi-dwarf with lodging resistance, fertilizer responsiveness, high yield and photo-insensitiveness. Photo-insensitivity has a bearing on the introduction of rice to Punjab which is otherwise ideal for cultivation of wheat.

Noblization of sugarcane is yet another achievement. The Indian sugar canes (of *Saccharum barberi* origin) were hardy, but poor in yield and sugar content. The tropical noble canes of *Saccharum officinarum* origin had thicker stem and higher sugar content. Noble canes performed badly in North India primarily due to low winter temperatures. C.A. Barber and T.S. Venkataraman of Sugarcane Breeding Institute, Coimbatore, transferred the thicker stem, higher sugar content and other desirable characters from the noble canes to the Indian canes. This is widely known as *noblization of Indian canes*. They also crossed *Saccharum spontaneum*, a wild species, to transfer disease resistance and other desirable characteristics to the cultivated varieties.

Special mention must be made about the hybrid varieties of maize, jowar or sorghum (*Sorghum bicolor*) and pearl millet or bajra (*Pennisetum glaucum*). Hybrid maize varieties Ganga Safed 2 and Deccan were developed in India with Rockefeller and Ford Foundation funding. A number of corn hybrids were developed by DuPont Pioneer and Syngenta in the USA. Several hybrids of jowar (CSH 1, CSH 2, CSH 3, CSH 4, CSH 5, CSH 6, CSH 9, CSH 10 and CSH 11) and bajra (PHB 1O, PHB 14, BJ 104 and BK 560) are also noteworthy. The Maharashtra Hybrid Seeds Co. Pvt. Ltd. (Mahyco) has been leading in the production of jowar hybrids. DuPont Pioneer has been leading in the production of bajra hybrids. ICRISAT under CGIAR has been the leading international organization in the production of bajra and jowar.

India has achieved the distinction of commercially exploiting heterosis in cotton. The first hybrid variety of cotton which was H4 developed by the Gujarat Agriculture University was released for commercial cultivation in 1970. Several other hybrid varieties, like Godavary, Sugana, H6 and AKH 468 (all within *Gossypium hirsutum*) and Varalaxmi, CBS 156, Savitri and Jayalaxmi (all G. *hirsutum* x G. *barbadense*), have been released for cultivation. The hybrid varieties are high yielding and have good fibre quality.

2.1 Centres of Origin

An understanding of the origin of most major crop species is vital for crop improvement programmes. The brilliant Russian agronomist and geneticist Nikolai I. Vavilov (1887–1943) undertook such a work between the 1920s and 1940s (Fig. 2.1). A large amount of information was collected from the then Union of Soviet Socialist Republics (USSR). According to Vavilov, the centres of origin of most cultivated plants are those where a concentration of genetically related species or wild relatives occurred with maximum genetic diversity. The variation we know today about these species has been accumulated by human populations inhabited in such areas.

Vavilov is believed to be the first scientist to have gathered such a massive collection of plants in order to fully investigate their unique intrinsic characteristics. During his lifetime, he organized and conducted more than 100 expeditions to collect botanical samples from the world's most important agricultural areas. Vavilov travelled to the sites of ancient agricultural civilizations and various mountainous regions.

Vavilov proposed eight centres of origin of cultivated plants: 1. China; 2. India; 2a. Indo-Malayan region; 3. Central Asia, including Pakistan, Punjab, Kashmir, Afghanistan and Turkestan; 4. Near East; 5. Mediterranean; 6. Ethiopia; 7. Southern

Fig. 2.1 Nikolai I. Vavilov

Mexico and Central America; and 8. South America (8a. Ecuador, Peru, Bolivia; 8b. Chile; 8c. Brazil-Paraguay). The eight Vavilovian centres and the crops originated are given in Table 2.1 (see Fig. 2.2).

2.1.1 Vavilov's Original Concepts

According to Vavilov, the centre of origin of a species is that with maximum diversity. This diversity demonstrates subsequent evolution. Vavilov established new concepts like primary and more ancient crops in contrast to secondary ones. He also characterized with good precision the centres where species originated and how such species got dispersed through different pathways.

In 1924, Vavilov wrote: "The history and origin of human civilizations and agriculture are, no doubt, much older than what any ancient documentation in the form of objects and inscriptions reveals to us. A more intimate knowledge of cultivated plants and their differentiation into geographical groups helps us attribute their origin to very remote epochs, where 5000–10,000 years represent but a short moment".

Vavilov, in an attempt to put genetics and plant breeding at the service of the national economy of the USSR, worked out a systematic geographic classification of cultivated plants. He and other Soviet botanists gathered data from 250,000 samples and identified 7 basic geographic centres of origin of cultivated plants.

1. The South Asian tropical centre is the native habitat of about 33% of all cultivated plants, including rice, sugarcane and many tropical and vegetable crops.

Table 2.1 Vavilovian centres and crops originated

1	**Chinese centre:** The largest independent centre which includes the mountainous regions of Central and Western China and adjacent lowlands. A total of 136 endemic plants are listed, among which are a few known to us as important crops	**Cereals and legumes** 1. Broomcorn millet, *Panicum miliaceum* 2. Italian millet, *Panicum italicum* 3. Japanese barnyard millet, *Panicum frumentaceum* 4. Kaoliang, *Andropogon sorghum* 5. Buckwheat, *Fagopyrum esculentum* 6. Hull-less barley, *Hordeum hexastichum* 7. Soybean, *Glycine max* 8. Adzuki bean, *Phaseolus angularis* 9. Velvet bean, *Stizolobium hassjoo* **Roots, tubers and vegetables** 1. Chinese yam, *Dioscorea batatas* 2. Radish, *Raphanus sativus* 3. Chinese cabbage, *Brassica chinensis*, *B. pekinensis* 4. Onion, *Allium chinense*, *A. fistulosum*, *A. pekinense* 5. Cucumber, *Cucumis sativus* **Fruits and nuts** 1. Pear, *Pyrus serotina*, *P. ussuriensis* 2. Chinese apple, *Malus asiatica* 3. Peach, *Prunus persica* 4. Apricot, *Prunus armeniaca* 5. Cherry, *Prunus pseudocerasus* 6. Walnut, *Juglans sinensis* 7. Litchi, *Litchi chinensis* **Sugar, drug and fibre plants** 1. Sugarcane, *Saccharum sinense* 2. Opium poppy, *Papaver somniferum* 3. Ginseng, *Panax ginseng* 4. Camphor, *Cinnamomum camphora* 5. Hemp, *Cannabis sativa*
2	**Indian centre:** This area has two sub-centres. **a. Main centre (Hindustan):** Includes Assam and Burma, but not Northwest India, Punjab nor Northwest Frontier Provinces. In this area, 117 plants were considered to be endemic	**Cereals and legumes** 1. Rice, *Oryza sativa* 2. Chickpea or gram, *Cicer arietinum* 3. Pigeon pea, *Cajanus indicus* 4. Urd bean, *Phaseolus mungo* 5. Mungbean, *Phaseolus aureus* 6. Rice bean, *Phaseolus calcaratus* 7. Cowpea, *Vigna sinensis* **Vegetables and tubers** 1. Eggplant, *Solanum melongena* 2. Cucumber, *Cucumis sativus* 3. Radish, *Raphanus caudatus* (pods eaten) 4. Taro, *Colocasia antiquorum* 5. Yam, *Dioscorea alata* **Fruits** 1. Mango, *Mangifera indica*

(continued)

Table 2.1 (continued)

		2. Orange, *Citrus sinensis* 3. Tangerine, *Citrus nobilis* 4. Citron, *Citrus medica* 5. Tamarind, *Tamarindus indica* 4 Lecture 5 **Sugar, oil and fibre plants** 1. Sugar cane, *Saccharum officinarum* 2. Coconut palm, *Cocos nucifera* 3. Sesame, *Sesamum indicum* 4. Safflower, *Carthamus tinctorius* 5. Tree cotton, *Gossypium arboreum* 6. Oriental cotton, *Gossypium nanking* 7. Jute, *Corchorus capsularis* 8. Crotalaria, *Crotalaria juncea* 9. Kenaf, *Hibiscus cannabinus* **Spices, stimulants, dyes and miscellaneous** 1. Hemp, *Cannabis indica* 2. Black pepper, *Piper nigrum* 3. Gum arabic, *Acacia arabica* 4. Sandalwood, *Santalum album* 5. Indigo, *Indigofera tinctoria* 6. Cinnamon tree, *Cinnamomum zeylanticum* 7. Croton, *Croton tiglium* 8. Bamboo, *Bambusa tulda*
	b. Indo-Malayan centre: Includes Indo-China and the Malay Archipelago	Fifty-five plants were listed, including: **Cereals and legumes** 1. Job's tears, *Coix lacryma* 2. Velvet bean, *Mucuna utilis* **Fruits** 1. Pummelo, *Citrus grandis* 2. Banana, *Musa cavendishii*, *M. paradisiaca*, *H. sapientum* 3. Breadfruit, *Artocarpus communis* 4. Mangosteen, *Garcinia mangostana* **Oil, sugar, spice and fibre plants** 1. Candlenut, *Aleurites moluccana* 2. Coconut palm, *Cocos nucifera* 3. Sugarcane, *Saccharum officinarum* 4. Clove, *Caryophyllus aromaticus* 5. Nutmeg, *Myristica fragrans* 6. Black pepper, *Piper nigrum* 7. Manila hemp or abaca, *Musa textilis*
3	**Central Asiatic centre:** Includes Northwest India (Punjab, Northwest Frontier Provinces and Kashmir), Afghanistan, Tadjikistan, Uzbekistan and western Tian-Shan. Forty-three plants are listed for this centre, including many wheats	**Grains and legumes** 1. Common wheat, *Triticum vulgare* 2. Club wheat, *Triticum compactum* Lecture 5 5 3. Shot wheat, *Triticum sphaerocoecum*

(continued)

Table 2.1 (continued)

		4. Pea, *Pisum sativum* 5. Lentil, *Lens esculenta* 6. Horse bean, *Vicia faba* 7. Chickpea, *Cicer arietinum* 8. Mungbean, *Phaseolus aureus* 9. Mustard, *Brassica juncea* 10. Flax, *Linum usitatissimum* (one of the centres) 11. Sesame, *Sesamum indicum* **Fibre plants** 1. Hemp, *Cannabis indica* 2. Cotton, *Gossypium herbaceum* **Vegetables** 1. Onion, *Allium cepa* 2. Garlic, *Allium sativum* 3. Spinach, *Spinacia oleracea* 4. Carrot, *Daucus carota* **Fruits** 1. Pistacia, *Pistacia vera* 2. Pear, *Pyrus communis* 3. Almond, *Amygdalus communis* 4. Grape, *Vitis vinifera* 5. Apple, *Malus pumila*
4	**Near-Eastern centre:** Includes interior of Asia Minor, all of Transcaucasia, Iran and the highlands of Turkmenistan. Eighty-three species including nine species of wheat were located in this region	**Grains and legumes** 1. Einkorn wheat, *Triticum monococcum* (14 chromosomes) 2. Durum wheat, *Triticum durum* (28 chromosomes) 3. Poulard wheat, *Triticum turgidum* (28 chromosomes) 4. Common wheat, *Triticum vulgare* (42 chromosomes) 5. Oriental wheat, *Triticum orientale* 6. Persian wheat, *Triticum persicum* (28 chromosomes) 7. *Triticum timopheevi* (28 chromosomes) 8. *Triticum macha* (42 chromosomes) 9. *Triticum vavilovianum*, branched (42 chromosomes) 10. Two-row barleys, *Hordeum distichum*, *H. nutans* 11. Rye, *Secale cereale* 12. Mediterranean oats, *Avena byzantina* 13. Common oats, *Avena sativa* 14. Lentil, *Lens esculenta* 15. Lupine, *Lupinus pilosus*, *L. albus* 6 Lecture 5 **Forage plants** 1. Alfalfa, *Medicago sativa*

(continued)

Table 2.1 (continued)

		2. Persian clover, *Trifolium resupinatum* 3. Fenugreek, *Trigonella foenum-graecum* 4. Vetch, *Vicia sativa* 5. Hairy vetch, *Vicia villosa* **Fruits** 1. Fig, *Ficus carica* 2. Pomegranate, *Punica granatum* 3. Apple, *Malus pumilo* (one of the centres) 4. Pear, *Pyrus communis* and others 5. Quince, *Cydonia oblonga* 6. Cherry, *Prunus cerasus* 7. Hawthorn, *Crataegus azarolus*
5	**Mediterranean centre:** Includes the borders of the Mediterranean Sea. Eighty-four plants are listed for this region including olive and many cultivated vegetables and forages	**Cereals and legumes** 1. Durum wheat, *Triticum durum expansum* 2. Emmer, *Triticum dicoccum* (one of the centres) 3. Polish wheat, *Triticum polonicum* 4. Spelt, *Triticum spelta* 5. Mediterranean oats, *Avena byzantina* 6. Sand oats, *Avena brevis* 7. Canary grass, *Phalaris canariensis* 8. Grass pea, *Lathyrus sativus* 9. Pea, *Pisum sativum* (large-seeded varieties) 10. Lupine, *Lupinus albus*, and others **Forage plants** 1. Egyptian clover, *Trifolium alexandrinum* 2. White clover, *Trifolium repens* 3. Crimson clover, *Trifolium incarnatum* 4. Serradella, *Ornithopus sativus* **Oil and fibre plants** 1. Flax, *Linum usitatissimum*, and wild *L. angustifolium* 2. Rape, *Brassica napus* 3. Black mustard, *Brassica nigra* 4. Olive, *Olea europaea* **Vegetables** 1. Garden beet, *Beta vulgaris* 2. Cabbage, *Brassica oleracea* 3. Turnip, *Brassica campestris*, *B. napus* 4. Lettuce, *Lactuca sativa* 5. Asparagus, *Asparagus officinalis* Lecture 5 7

(continued)

Table 2.1 (continued)

		6. Celery, *Apium graveolens* 7. Chicory, *Cichorium intybus* 8. Parsnip, *Pastinaca sativa* 9. Rhubarb, *Rheum officinale* **Ethereal oil and spice plants** 1. Caraway, *Carum carvi* 2. Anise, *Pimpinella anisum* 3. Thyme, *Thymus vulgaris* 4. Peppermint, *Mentha piperita* 5. Sage, *Salvia officinalis* 6. Hop, *Humulus lupulus*
6	**Abyssinian centre:** Includes Abyssinia, Eritrea and part of Somaliland. In this centre were listed 38 species. Rich in wheat and barley	**Grains and legumes** 1. Abyssinian hard wheat, *Triticum durum abyssinicum* 2. Poulard wheat, *Triticum turgidum abyssinicum* 3. Emmer, *Triticum dicoccum abyssinicum* 4. Polish wheat, *Triticum polonicum abyssinicum* 5. Barley, *Hordeum sativum* (great diversity of forms) 6. Grain sorghum, *Andropogon sorghum* 7. Pearl millet, *Pennisetum spicatum* 8. African millet, *Eleusine coracana* 9. Cowpea, *Vigna sinensis* 10. Flax, *Linum usitatissimum* **Miscellaneous** 1. Sesame, *Sesamum indicum* (basic centre) 2. Castor bean, *Ricinus communis* (a centre) 3. Garden cress, *Lepidium sativum* 4. Coffee, *Coffea arabica* 5. Okra, *Hibiscus esculentus* 6. Myrrh, *Commiphora abyssinica* 7. Indigo, *Indigofera argente*
7	***New World*** **South Mexican and Central American centre:** Includes southern sections of Mexico, Guatemala, Honduras and Costa Rica	**Grains and legumes** 1. Maize, *Zea mays* 2. Common bean, *Phaseolus vulgaris* 3. Lima bean, *Phaseolus lunatus* 4. Tepary bean, *Phaseolus acutifolius* 5. Jack bean, *Canavalia ensiformis* 6. Grain amaranth, *Amaranthus paniculatus leucocarpus* 8 Lecture 5 **Melon plants** 1. Malabar gourd, *Cucurbita ficifolia* 2. Winter pumpkin, *Cucurbita moshata*

(continued)

Table 2.1 (continued)

		3. Chayote, *Sechium edule* **Fibre plants** 1. Upland cotton, *Gossypium hirsutum* 2. Bourbon cotton, *Gossypium purpurascens* 3. Chayote, *Sechium edule* **Miscellaneous** 1. Sweet potato, *Ipomea batatas* 2. Arrowroot, *Maranta arundinacea* 3. Pepper, *Capsicum annuum*, *C. frutescens* 4. Papaya, *Carica papaya* 5. Guava, *Psidium guajava* 6. Cashew, *Anacardium occidentale* 7. Wild black cherry, *Prunus serotina* 8. Cochenial, *Nopalea coccinellifera* 9. Cherry tomato, *Lycopersicum cerasiforme* 10. Cacao, *Theobroma cacao* 11. *Nicotiana rustica*
8	**South American centre:** (62 plants listed). Three sub-centres are found. **a. Peruvian, Ecuadorean, Bolivian centre:** Comprised mainly of the high mountainous areas, formerly the centre of the Megalithic or Pre-Inca civilization. Endemic plants of the Puna and Sierra high elevation districts included:	**Root tubers** 1. Andean potato, *Solanum andigenum* (96 chromosomes) 2. Other endemic cultivated potato species. Fourteen or more species with chromosome numbers varying from 24 to 60 3. Edible nasturtium, *Tropaeolum tuberosum*. Coastal regions of Peru and non-irrigated subtropical and tropical regions of Ecuador, Peru and Bolivia included: **Grains and legumes** 1. Starchy maize, *Zea mays amylacea* 2. Lima bean, *Phaseolus lunatus* (secondary centre) 3. Common bean, *Phaseolus vulgaris* (secondary centre) Lecture 5 9 **Root tubers** 1. Edible canna, *Canna edulis* 2. Potato, *Solanum phureja* (24 chromosomes) **Vegetable crops** 1. Pepino, *Solanum muricatum* 2. Tomato, *Lycopersicum esculentum* 3. Ground cherry, *Physalis peruviana* 4. Pumpkin, *Cucurbita maxima* 5. Pepper, *Capsicum frutescens* **Fibre plants**

(continued)

Table 2.1 (continued)

		1. Egyptian cotton, *Gossypium barbadense* **Fruit and miscellaneous** 1. Passion flower, *Passiflora ligularis* 2. Guava, *Psidium guajava* 3. Heilborn, *Carica candamarcensis* 4. Quinine tree, *Cinchona calisaya* 5. Tobacco, *Nicotiana tabacum*
8	**b. Chile centre (island near the coast of Southern Chile)**	1. Common potato, *Solanum tuberosum* (48 chromosomes) 2. Wild strawberry, *Fragaria chiloensis*
8	**c. Brazilian-Paraguayan centre**	1. Manioc, *Manihot utilissima* 2. Peanut, *Arachis hypogaea* 3. Rubber tree, *Hevea brasiliensis* 4. Pineapple, *Ananas comosa* 5. Brazil nut, *Bertholletia excelsa* 6. Cashew, *Anacardium occidentale* 7. Purple granadilla, *Passiflora edulis*

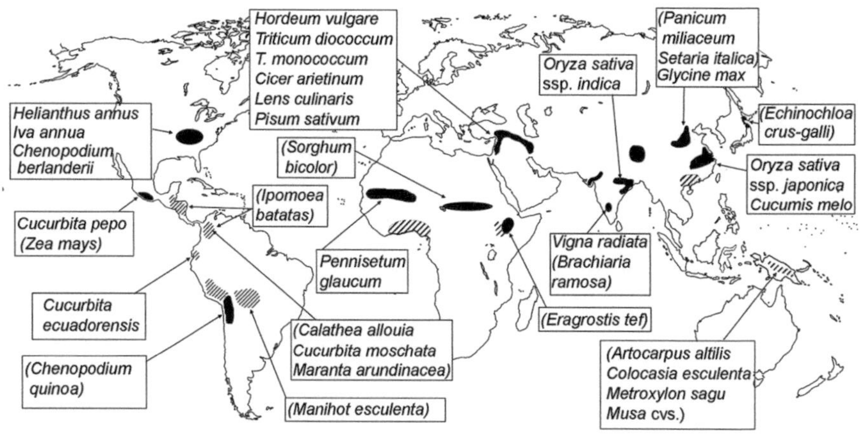

Fig. 2.2 Origin of world's food crops. These were widely redistributed so that today's leading producing countries are not the same as the areas in which these crops were first domesticated

2. The East Asian centre for soybeans and various millet, vegetable and fruit species accounting for 20% of cultivated plants.
3. The Southwest Asian centre for bread grains, legumes, fruit crops and grapes. This centre is home of 4% of all cultivated plants.
4. The Mediterranean centre from where 11% of the species originated. Olive the carob (*Ceratonia siliqua*) is a prominent species of this centre.

5. The Ethiopian centre from where 4% of the cultivated plants originated. This centre is characterized by teff, *Guizotia* (a unique species of banana) and the coffee tree. Endemic species and subspecies of wheat and barley also originated here.
6. The Central American centre where corn, long-fibre cotton species, cacao, beans and squash originated.
7. The Andes centre, home of tuberous species, cinchona and cocoa.

It was formerly believed that the primary centres of the ancient farming cultures were the broad valleys of the Tigris, Euphrates, Ganges, Nile and other large rivers. Vavilov demonstrated that virtually all cultivated plants appeared in the mountain regions of the tropical, subtropical and temperate zones. The main geographic centres of initial cultivation of most of the plants now raised are related the high level of ancient civilizations. The South Asian tropical centre is linked to sophisticated ancient Indian and Indo-Chinese cultures. The Mediterranean centre is tied to the Etruscan, Hellenistic and Egyptian cultures that spanned to more than 6000 years.

Many archaeological investigations in the 1960s and 1970s have confirmed Vavilov's theories concerning the centres of origin of cultivated plants. Numerous scientists, including the Soviet botanists P.M. Zhukovskii, E.N. Sinskaia and A.I. Kuptsov, have continued Vavilov's work and have modified his theories.

Further Reading

Abbo S, Gopher A (2017) Near eastern plant domestication: a history of thought. Trends Plant Sci. https://doi.org/10.1016/j.tplants.2017.03.010

Khoury CK et al Increasing homogeneity in global food supplies and the implications for food security. PNAS. www.pnas.org/lookup/suppl/doi:10

Germplasm Conservation

3

Keywords
Significance of germplasm conservation · In situ conservation · Ex situ conservation · In vitro germplasm preservation · Germplasm regeneration · Characterization · Evaluation · Documentation and distribution · Characterization · Molecular descriptors · Evaluation · Passport data · Characterization · Preliminary evaluation · Documentation · Standards for data preparation · Quarantine information · Passport information · Herbarium information · Field evaluation · Gene bank information · Germplasm collecting missions database · Distribution of germplasm · FAO and plant genetic resources · FAO commission on plant genetic resources · Germplasm – international vs. Indian scenario · Plant introduction · Historical perspective · Plant introduction – the international scenario · Import regulations · Plant germplasm import and export · Plant introduction in India · Conservation of endangered species/crop varieties

Germplasm is a collection of various strains and species that accommodates total of all the genes present in a crop and its related species. Germplasm is the basic indispensable ingredient of all breeding programmes, and hence, collection, evaluation and conservation of germplasm types become an integral part of any breeding programme. Usually, the germplasm accessions are conserved in the form of seeds stored at ambient temperature, low temperature or ultralow temperature.

Significance of Germplasm Conservation

(a) Preservation of genetic diversity of various strains and species is conservation. Such preserved accessions can be used in the future.
(b) The valuable genetic traits present in primitive plants will be lost unless such endangered types are conserved.

P. M. Priyadarshan, *PLANT BREEDING: Classical to Modern*,
https://doi.org/10.1007/978-981-13-7095-3_3

(c) In clonally multiplied species, the seeds are not feasible material to be conserved due to genetic heterogeneity. In this case, their genes are to be conserved.
(d) The preservation of roots and tubers is difficult because they lose viability. Also, they require larger space. Also, GMOs may be unstable. Such accessions are to be conserved carefully following special techniques.

Biodiversity International This is an international apex body under the auspices of CGIAR that leads germplasm conservation. It provides requisite support for collection, conservation and utilization of plant genetic resources. Such germplasm accessions are preserved as both in situ and ex situ.

In Situ Conservation In situ conservation of germplasm is conserving species in their natural environment through establishing biosphere reserves (or national parks/gene sanctuaries). This is accomplished by preserving land plants near natural habitat along with several wild relatives with genetic diversity. The in situ conservation is considered as a high-priority germplasm preservation programme. The limitations are as follows: (a) environmental hazards may endanger the preservations and (b) the cost of maintenance is very high.

Ex Situ Conservation Otherwise known as gene banking, this is a method for the preservation of both cultivated and wild. There are two types of gene banking: in vivo and in vitro. While in vivo gene banks preserve seeds, vegetative propagules, etc., in vitro gene banks preserve cell and tissues. For this, knowledge of sampling, regeneration, maintenance of gene pools, etc. are essential. The limitations are as follows: (a) viability of seeds is reduced or lost with passage of time; (b) seeds are susceptible to insect or pathogen attack, often leading to their destruction; (c) this approach is exclusively confined to seed propagating plants, and therefore, it is of no use for vegetatively propagated plants, e.g. potato, *Ipomoea* and *Dioscorea*; and (d) it is difficult to maintain clones through seed conservation.

3.1 In Vitro Germplasm Preservation

(a) Germplasm can be preserved in vitro through cryopreservation, low-pressure storage and low-oxygen storage. In cryopreservation, the cells are preserved in a frozen state using solid carbon dioxide (at −79 °C), low temperature deep freezers (at −80 °C), vapour phase nitrogen (at −150 °C) and liquid nitrogen (at −196 °C).

Cells stay in completely inactive state. So, they can be conserved for long periods. Tissues like meristems, embryos, endosperms, ovules, seeds, cultured plant cells, protoplasts and callus are usually used for cryopreservation. Cryoprotectants are to be added during cryopreservation. They are DMSO (dimethyl sulfoxide), glycerol, ethylene, propylene, sucrose, mannose, glucose, etc. The damage caused by freezing and thawing will be prevented by cryoprotectants. An outline of the protocol for cryopreservation of shoot tip is depicted in Fig. 3.1.

(b) Germplasm conservation by cold storage is done at low and non-freezing temperature (1–9 °C). Here, only growth of the tissue is slowed down. So, cold storage prevents cryogenic injuries. An example to this method is virus-free strawberry plants that can be preserved at 10 °C for about 6 years. Grape plants can be preserved for 15 years at 9 °C.
(c) In low-pressure and low-oxygen storage, the atmospheric pressure and oxygen concentration are reduced. The lowered partial pressure reduces the in vitro growth of plants. Low oxygen concentration keeps partial pressure of oxygen below 50 mmHg (mmHg is a manometric unit of pressure) which reduces growth. Reduced availability of oxygen leads to reduced photosynthetic activity. This technique can be used in increasing the shelf life of fruits, vegetables and flowers. A comparison of different approaches is available in Table 3.1.
(d) Somatic embryos desiccated by calcium alginate coating (artificial seeds) can be stored at low (4 °C) or ultralow (−20 °C) temperatures. This approach is yet to be evaluated for such an application. This is possible only in species where in vitro somatic embryogenesis is possible.

Fig. 3.1 Protocol for cryopreservation of shoot tip

Table 3.1 Comparison of approaches for in vitro germplasm conservation

Feature	Cryopreservation	Slow growth	DNA clones
Tissue/organ conserved	Shoot tips, zygotic or somatic	Slow-growing shoots	DNA pieces as phage clones
Metabolic activity	Nil	Slow	Nil
Storage temperature	−196 °C	4–9 0r 15°	4 °C in lyophilized state
Storage in	Liquid nitrogen refrigerators	Ordinary refrigerators	Deep freeze
Attention needed during storage	Replenishing liquid nitrogen	Subculture every 6–36 months	Virtually nil

Merits of germplasm storage are as follows: it requires relatively very small space, they are free from diseases, and storage can be over long periods and are ideal for germplasm exchange. The demerits are as follows: requirement of sophisticated facilities for freezing and DNA cloning, requirement of skill and cryopreservation can cause damage.

DNA of plants can also be stored as ex situ germplasm collection (Box 3.1).

Box 3.1: DNA Banks or Gene Banks

Germplasm can also be conserved as DNA segments cloned in a suitable vector like cosmids, plasmids or YACs (yeast artificial chromosomes). This is sophisticated, technically demanding and expensive. Threatened species can thus be conserved. Till date, there are no cases where DNA banks are being used as a replacement to traditional method of conservation. However, due to small sample size, this technique has promising potential for the storage of genetic information.

It has become routine to extract DNA from the nuclei, mitochondria and chloroplasts. Derivatives like as RNA and cDNA are also being extracted. Technologies are available to allow all these to be stored quickly and at low cost in DNA banks as an insurance policy against loss of crop diversity. DNA storage allows genetic material for molecular applications. However, use of DNA in conservation is limited as whole plants cannot be directly reconstituted. The genetic material must be introduced through transgenic means. However, DNA banks have a potential future as new technologies develop day by day.

3.2 Germplasm Regeneration

While regenerating germplasm, there is a risk of genetic integrity loss when regenerating genetically heterogeneous accessions. Germplasm regeneration is also very expensive. Regeneration is done due to two reasons: (a) to increase the quantity of initial seeds or tissues and (b) to recharge or reload seed stocks or tissues. In cross-pollinated species, maintenance of seeds in its originality is a challenge. In the case of tree species, regeneration is time-consuming and the maintenance of genetic integrity is difficult.

Each crop has its own growing environment and agro-management practices. Readers may consult website of crop gene bank for more information.

While regenerating germplasm accessions, the following factors are important:

(a) Best suitable environmental must be selected to avoid natural selection.
(b) It is important to fully understand the breeding system. Cross-pollinated species need proper isolation.

(c) Site must have adequate irrigation facilities and nutritive soil to minimize the loss of plants.
(d) In order to reduce unintentional gene flow, pests and diseases, adequate distance may be maintained.
(e) Adequate number of plants must be grown to maintain genetic integrity.
(f) Due care must be taken to breaking dormancy and induction of flowering.
(g) Optimum spacing has to be followed to ensure good seed set.
(h) To have representative samples, mix equal number of seeds from all plants.

Regenerating germplasm in the ecological region of origin will be advisable to ensure flowering and seed set because day length and vernalization are important factors for seed set. Also, environment is a vital factor that influences the preference of some genotypes getting selected against others. This is essential to maintain genetic integrity. While handling germplasm distributed by gene banks, proper phytosanitary measures must be observed to avoid seed-borne pathogens and pests.

Please see www.genesys-pgr.org for further details on germplasm collections at the world level. Their accession map shows that 482 institutions are involved in maintaining with 3,631,898 plant accessions. CGIAR International Gene Banks, ECPGR EURISCO network (European Cooperative Programme for Plant Genetic Resources-EURISCO is a software development company), USDA-ARS-NPGS, COGENT (coconut germplasm network) and CWR (crop wild relatives) project are the major components of this system.

3.3 Characterization, Evaluation, Documentation and Distribution

3.3.1 Characterization

The description of plant germplasm is germplasm characterization. From morphological or agronomical features to seed proteins and molecular markers, it determines the expression of highly heritable characters. In order to offer information on traits that give maximum utilization, characterization is essential. It also enables the recording and compilation of data on important traits that distinguish accessions within a species. The genetic diversity thus obtained can be used for breeding.

Characterization is being done by growing a representative number of plants following statistically replicated design in a full growing cycle. A minimum three replicates and data from at least ten plants is believed to be acceptable for many crops. Biodiversity International has been coordinating the development and updating of plant descriptors for various crops (see https://www.bioversityinternational.org/). Descriptor lists are available for more than 90 crops. The characterization is done based on the descriptor of the crop in question. A brief sample descriptor for cassava central leaflet is available in Box 3.2. In addition to morphological descriptors, herbarium samples are good records of variation. Digital pictures of samples can be taken to store data of collected germplasm.

Box 3.2: Leaflet Diversity in Cassava

The simple leaves of cassava consist of foliar lamina and the petiole. The foliar lamina is palmate and lobate. Completely developed leaves are in different colours, depending on the cultivar. The basic colours are purple, dark green and light green. The number of leaf lobes ranges from 3 to 9. Central lobes are larger than the lateral ones.

There are primarily ten types of shape of central leaf in cassava. They are ovoid, elliptic-lanceolate, obovate-lanceolate, oblong-lanceolate, lanceolate, straight or linear, pandurate, linear-piramidal, linear-pandurate and linear-hostatilobalate (see Fig. 3.2).

Molecular Descriptors Molecular markers are reliable tools to characterize genetic variation and utilize genetic selection. DNA polymorphism assay is a powerful tool to characterize and investigate germplasm accessions. RFLP and PCR-derived molecular markers are useful for Mendelian gene tagging and QTL mapping (see Chaps. 23 and 24 for details). Molecular characterization of germplasm collections for preservation, identification of phenotypic variants and reduction of genetic erosion are frontier avenues now to breed potential varieties.

Many statistical packages are available to analyse the data collected like analysis of variance for single straight data and multivariate analysis for multiple traits. Cluster analysis and principal component analysis (PCA) can be done to look for natural grouping among the germplasm accessions. Two ways of identifying such clusters are (a) grouping based on hierarchical procedure, separating wild from cultivated types using taxonomic knowledge, and (b) creating groups based on

Fig. 3.2 Leaf shape of cassava

multivariate analysis of genetic markers and principal component analysis see Chap. 20).

3.3.2 Evaluation

Germplasm evaluation deals with a range of activities like (a) receipt of the new samples, (b) growing accessions for seed increment, (c) characterization and preliminary evaluation and (e) documentation. Germplasm are of diverse types:

(a) Those derived from centres of diversity (primitive cultivars, natural hybrids between cultigen and wild relatives, wild relatives) and related species and genera
(b) Those derived from areas of cultivation (commercial types, extinct varieties, primitive varieties)
(c) Those derived from breeding programmes (pure lines, elite varieties/hybrids, breeding lines, mutants, polyploids and intergeneric and interspecific hybrids)

The curator of the germplasm and breeder must work in tandem to ensure the effective utilization of germplasm accessions for breeding new varieties. Germplasm evaluation consists of seed increase, preparation of descriptor list and measurement of data. The components of germplasm evaluation are seed increase, preparation of descriptor list and types of characters and measurement of data.

Seed increase is vital as it involves the risk due to poor germination, lack of adaptation, disease and pest damage and contamination due to admixtures. Seed stocks are to be sufficiently increased in one cycle. Such seeds can be used for evaluation, differentiation and storage. It is wise to keep a portion of seeds as reserve in order to have another planting in case the first planting fails. Quarantine measures can be observed during seed increase.

Preparation of descriptor lists involves four steps, *viz.*, passport data, characterization, preliminary evaluation and further characterization and evaluation. The descriptor lists of IBPGR (International Board of Plant Genetic Resources – a body under Biodiversity International) are very exhaustive and the same are being used by scientists. Descriptors for 62 agri-horticulture crops have already been published by the IBPGR and many more are under preparation.

Passport Data In order to find out duplicates, passport data must include all basic information. The important passport descriptors are the site of collection; type of material; date of collection; collector's number; altitude, latitude and longitude for site of collection; status; growing conditions; and source. This is essential to plan further collections and to set up evolutionary or population genetic research (Box 3.3).

Box 3.3: Sample Passport Data Collection Form

COLLECTION OF xxxxxxx GERMPLASM IN xxxxxxxx

Coll. No. __________ **Latin name** ______________________________

Local name _______________ **Locality data** ______________________________

____

____

Landowner ____

Elev.(m) __________ **Latitude** __________ **Longitude** __________ **Geographic ref.**_____

Make altimeter __________**Make GPS**__________________ **Uncertainty GPS (m)** _____

Site size (m²) _______ **Linear extent (m)** ________ **Herbarium specimen no.**_____

Plant description ____

____

____

Improvement Status: wild weedy landrace other:________

Sample Source: wild pop. field garden market store other:_____

Frequency in area: abundant frequent occasional rare **Pop. Distrib.:** ________________

No. plants found________ **No. plants sampled**________ **Sampling method**______________

Population age/stage class distribution ______________________________

Type Propagule Collected: seed cuttings root plant other:_______ **Propagule maturity**____

Quantity propagules collected ________ **Propagule Source:**

SITE DESCRIPTION

Exposure/aspect ________ **Slope**________

Site physical ____

____

Site vegetative ____

____

OTHER NOTES ____

____

Collectors______________________________________ **Date**____________

source: National Germplasm Resources Laboratory, USDA-ARS, Beltsville.

Characterization Characterization is a process by which all heritable characters are recorded. This must provide a record which together with passport data can provide information that leads to the identification of an accession. Characterization highlights the range of diversity in collections that include taxonomic characters like spike/panicle shape, seed shape and colour, etc.

Preliminary Evaluation Preliminary evaluation consists of recording some additional agronomic physiological characters like vernalization requirement, tillering, time to flowering and maturity. This could help the breeders to narrow down the selection of right genotypes to be used in their breeding programmes. The preliminary evaluation descriptors used are site data, planting data (seed, cutting, grafts), leaf characters (leaf type, petiole type, size, leaflet type), floral characters (position of flowers, type of inflorescence, colour of flower bud, length of pedicel, length of bud, number of stamens, flower aroma, pollination), fruiting characters (number of days from flowering to harvest, main harvest season, yield), fruit characters (number of fruits/cluster, fruit length and width, protein percent, fat percent, shattering habit, seeds/fruit) and seed characters (seed size, hilum size and colour, 100-seed weight).

Further Characterization and Evaluation There are several traits like stress tolerance, disease and pest resistance and quality aspects beyond the ability of a curator of a germplasm collection. Studies on such traits involve subjects like cytogenetics and evolution, physiology, pathology, entomology, biochemistry and agronomy. Many horticultural plants are propagated by means of grafting, and hence, selection and evaluation of root stocks are vital. Further evaluation requires the services of breeders, pathologists, entomologists, agronomists and biochemists as per needs. There are observable and non-observable traits to be scored while evaluating the accessions. Observable characters include morphological, physiological or biochemical characters relating to survival, productivity or quality that can be transferred from an exotic source to an adapted cultivar by repeated backcrossing. On the other hand, non-observable characters are controlled by the environment and are largely polygenic. Qualitative data are easy to score, while quantitative data pose multitude of problems. For this, check lines are raised and the accessions in question are to be evaluated under appropriate field trials. Such check lines are usually locally adapted cultivars familiar to breeders. Check lines are useful to understand comparisons and also are dependable to monitor trial-to-trial variation. A fine example is to score disease resistance in the new accessions against available local check variety.

3.3.3 Documentation

In current days, documentation is information system. Such a system has to be dynamic and must ensure reliability and integrity of the data. Such a system is known as *database management system.*

During the 1970s, TAXIR (Taxonomic Information Retrieval) – a general-purpose and computer-assisted information system, was developed at the Taximetrics Laboratory of the University of Colorado, USA. Later, EXIR (Executive Information Retrieval) system has evolved at the same university to meet data management.

The Nordic Gene Bank at Weibullsholm Plant Breeding Institute in Sweden is the frontrunner in developing software for gene bank documentation. Also, the GRIN (Germplasm Resources Information Network) system developed in the USA (available with USDA, Beltsville) is quite capable of monitoring information on world's largest collection at the National Seed Storage Laboratory (NSSL), Fort Collins (see their web sites for further details).

The presence of voluminous data is a major challenge for managing the data. For instance, the National Plant Germplasm System (NPGS), USA, maintains over 400,000 accessions of germplasm, and 7000 to 15,000 accessions are added every year. The International Rice Research Institute holds nearly 86,000 samples, and data on 75 traits are being stored generating nearly 6.4 million pieces of information.

Two basic types of database management systems can be identified, namely, hierarchical and relational. In the hierarchical system, there is superior-subordinate type of relationship occurring between data and hierarchical structure. In the relational system, data are represented in the form of two-dimensional tables and are simple. Some of the DBMS are dBASE III PLUS, dBASE IV, FOXBASE, FOCUS, ORACLE, UNIFY, INGRESS and SYBASE. While dBASE III PLUS or dBASE IV are appropriate for small databases, Oracle DBMS is a powerful package for handling large databases.

3.3.3.1 Standards for Data Preparation

The data gathered needs to be standardized in terms of terminology and measurement to make the information more meaningful and applicable. There must be an internationally accepted system to record and maintain data. This was duly recognized by IBPGR. For the meticulous handling of data, IBPGR has put forth at least six points that can be exercised: plant introduction reporters and crop inventories, quarantine information, passport information, herbarium information, field evaluation and gene bank information. In India, NBPGR was constituted during 1976. NBPGR initiated a project "Genetic Resources Information Programme (GRIP)" in 1986. NBPGR follows six points included in the IBPGR guidelines.

Plant Introduction and Crop Inventories An exotic introduction to India was made during 1940. After that, NBPGR has registered over 900,000 samples. At the time of its entry, each accession is given EC (Exotic Collection) number, and the other details like botanical name, original identification number/names, source country and address, recipient name and address, number of samples, etc. are entered. The National Register records all accessions. Plant Introduction Reporter (PIR) published as crop inventory includes all such information.

Quarantine Information All plant introductions must undergo quarantine procedure and are given Import Quarantine (IQ) number. A quarantine register is being maintained for this purpose. Normally checklists are prepared to know beforehand risks in importing a plant material.

Passport Information A set of passport descriptors like collection number, scientific name of the crop, common name, provenance data (latitude, longitude, altitude) and habitat are included in these descriptors.

Herbarium Information In India, NBPGR has a National Herbarium of 2200 species covering 950 genera and 180 families. Herbarium information is recorded for a set of descriptors, viz. collector number and name, botanical name, name of identifier, etc.

Field Evaluation NBPGR generated evaluation data in the form of 48 crop catalogues. These catalogues give in detail the complete listing of evaluation data along with the available passport information, details of quantitative and qualitative traits and the estimates of variability. Germplasm Evaluation Information System (GEIS) based on DBASE IIIPLUS handles the data. Eight major groups of crops, viz. grain legumes, cereals and pseudo-cereals, oilseeds, millets and minor millets, vegetables, horticultural crops/plants, medicinal and aromatic plants and miscellaneous crops, have been formed.

Gene Bank Information In India, over 135,000 accessions have been stored in a national repository for long-term conservation at NBPGR. Data is maintained on some of the important descriptors, viz. crop name, genus and species, identification number, germination percentage, moisture content, month and year of storage, etc. Details like gene bank labels and information on cryopreserved samples are also maintained.

Germplasm Collecting Missions Database The Consultative Group on International Agricultural Research (CGIAR) has a Germplasm Collecting Missions Database that extends access to all collections made after 1975. The data include species name (as identified by the collector), the number of samples in each species, time of collection, the country of collection and whether the species was wild or cultivated. The institute's name that received and collected germplasm is coded (please see http://www.ecpgr.cgiar.org/resources/germplasm-databases/).

Some of the international multi-crop databases are Crop Wild Relative Global Portal, SINGER, PGR Forum, GENESYS, Mansfeld's World Database for Agricultural and Horticultural Crops, WIEWS and EU Plant Variety database. In addition to these, there are national multi-crop databases as:

- Australian Plant Genetic Resource Information Service (AusPGRIS)
- Austria – National Inventory of Austria
- Bulgaria – National Seed Gene Bank
- Czech Republic – Information System on Plant Genetic Resources (EVIGEZ)
- France – BRG – collections de ressources génétiques végétales (Collections of Plant Genetic Resources)
- Germany – BIG-Flora, Zentralstelle für Agrardokumentation und – information (ZADI) (Central Office for Agricultural Documentation and Information)
- Germany – Federal Research Centre for Cultivated Plants – Julius Kuhn Institute
- Germany – Leibniz Institute of Plant Genetics and Crop Plant Research (IPK)
- Italy – CRA Consiglio per la Ricerca e Sperimentazione in Agricoltura (Council for Research and Experimentation in Agriculture)
- The Harold and Adele Lieberman Germplasm Bank, Institute for Cereal Crops Improvement (ICCI), Tel Aviv University, The George S. Wise Faculty of Life Sciences

- New Zealand – Arable Crop Gene Bank and Online Database, New Zealand Institute for Crop and Food Research
- Russian Federation – N.I. Vavilov All-Russian Scientific Research Institute of Plant Industry (VIR)
- Spain – INIA – Centro de Recursos Fitogenéticos – Genebank (Center for Plant Genetic Resources – Genebank)
- Sweden – Stored material at the Nordic Genebank
- Switzerland – Conservation of PGRFA – Swiss National Database
- The Netherlands – Centre for Genetic Resources (CGN)
- The USA – National Plant Germplasm System

3.3.4 Distribution of Germplasm

The distribution of germplasm is a vital programme of any genetic resources centre. For this, the following points are important:

(a) Distribution of germplasm is the responsibility of the gene bank centres.
(b) To avoid cumbersome work of book keeping, germplasm samples are generally supplied free of cost.
(c) Seed samples are sent in small quantities.
(d) The receiver is informed of the records maintained on the important traits of accessions.
(e) For acclimatization, germplasm is evaluated for one or two crop seasons.

3.4 FAO and Plant Genetic Resources

Since 1983, FAO has developed a global system on plant genetic resources.

1. With the constitution of **International Undertaking on Plant Genetic Resources**, a flexible legal framework was organized. This is a formal arrangement to ensure that species that holds economic and social importance will be explored, collected, preserved and evaluated. Such collections will be made available for future breeding programmes.
2. **The Commission on Plant Genetic Resources**, an intergovernmental forum, was organized by FAO, where donor countries or users of germplasm can interact on matters of plant genetic resources and monitor implementation.
3. For conservation and promotion of plant genetic resources, FAO constituted an **International Fund for Plant Genetic Resources**. This is to ensure that intergovernmental and non-governmental organizations and private industries and individuals fulfil the conservation of world's plant genetic diversity.

   More than 122 countries cooperate with the aforesaid programmes.

3.4.1 FAO Commission on Plant Genetic Resources

After its constitution during November 1983, the Commission discusses issues like (a) laws relating to Plant Breeders' Rights in developed countries and the restriction of exchange of certain species and (b) streamlining of activities of the Commission and other organizations dealing with plant genetic resources. Plant breeders' rights and farmers' rights were recognized in these meetings. This has a large bearing on recognizing the efforts put forth by both plant breeders and farmers. The Commission formulates modalities on germplasm availability and exchange.

FAO, IBPGR and International Agricultural Research Centres (IARCs) have a collaboration in addressing issues related to germplasm conservation and utilization, and a memorandum of understanding (MOU) between these agencies exists to make the system work. The following are the points in that MOU:

(a) The Commission will strive for the availability of germplasm and for streamlining the guidelines for safer transfer of specific crops.
(b) Organizational network will be formed at the national and regional level to coordinate the activities of MOU.
(c) The IBPGR and the IARCs can provide the scientific inputs in joining FAO and the Commission in mobilizing International Fund for Plant Genetic Resources.
(d) Crop network will be constituted in all member countries.
(e) Avoid duplication in base collections.
(f) In situ crop reserves will be a national responsibility.
(g) The Commission will oversee the strengthening of national capability of germplasm evaluation.

Besides FAO/IBPGR/IARCs collaboration, the following centres are involved in PGR activities:

- The Asian Vegetable Research and Development Centre (AVRDC, Taiwan)
- The International Development Research Centre (IDRC) (for bamboos and rattans, banana, oilseeds, smaller millets)
- International Jute Organisation (IJO) (for jute and kenaf)
- Japanese International Cooperation Agency (JICA)
- German Agency for Technical Cooperation (GTZ)
- United States Agency for International Development (USAID)
- International Network for the Improvement of Banana and Plantain (INIBAP, France)
- Commonwealth Scientific and Industrial Research Organisation (CSIRO, Australia)
- National Plant Germplasm System, USDA
- N.I. Vavilov All-Union Scientific Research Institute of Plant Industry/VIR (USSR)
- For Africa, the Plant Genetic Resources Centre/Ethiopia (PGRC/E)
- For Latin America, CENARGEN, Embrapa (Brazil)

- For East Asia, the Institute of Crop Germplasm Resources under the Chinese Academy of Agricultural Sciences (CAAS), Beijing
- For Southeast Asia, the National Plant Genetic Resources Laboratory, University of the Philippines, at Los Baños, Philippines
- For South Asia, the National Bureau of Plant Genetic Resources (NBPGR), New Delhi, India
- Commonwealth Science Council (CSC), UK (for lesser known plants/traditional useful plants – plants of ethnobotanical interest)

3.5 Germplasm: International vs. Indian Scenario

Globally, CGIAR centres established 11 gene banks in addition to the 1750 individual gene banks available. While 130 gene banks hold more than 10,000 accessions, 8 have more than 100,000 accessions. In order to provide international conservation for PGR, Svalbard Global Seed Vault (SGSV) was established in 2008 in partnership by the Government of Norway, the Nordic Genetic Resources Centre (NordGen) and Global Crop Diversity Trust (GCDT) (Box 3.4; Fig. 3.3). As per FAO records, the four largest gene banks are (a) National Centre for Genetic Resources Preservation (NCGRP) in the USA; (b) Institute of Crop Germplasm Resources, Chinese Academy of Agricultural Sciences (ICGR-CAAS), in China; (c) ICAR-NBPGR in India; and (d) N.I. Vavilov All-Russian Scientific Research Institute of Plant Industry (VIR) in the Russian Federation.

Box 3.4: Svalbard Seed Vault

Though more than 1700 gene banks have collections of food crops around the world, many of them are vulnerable to disasters and catastrophes. A poorly functioning freezer can ruin the entire collection. Any loss of crop variety is irreversible. Norwegian government in 2008 opened a seed vault at Svalbard some 1300 kilometres beyond its border with Arctic Circle. Crates of seeds are sent here for safe and secure long-term storage in cold and dry rock vaults. Svalbard has the capacity of 4.5 million varieties of crops. A maximum of 2.5 billion seeds can be stored. More than 930,000 samples are stored now. The temperature use to be −18 °C which is optimal for storage. The samples are stored in three-ply foil packages. Because of low temperature, low metabolic activity is ensured so as to keep the seeds viable for longer time (see Fig. 3.4).

For more details, visit: https://www.nordgen.org/sgsv/.

Three global international agreements envisage access, exchange, conservation and utilization of PGR: (a) the Convention on Biological Diversity (CBD-1993), (b) the International Treaty on Plant Genetic Resources for Food and Agriculture (ITPGRFA-2004) and (c) the Nagoya Protocol on Access to Genetic Resources and the Fair and Equitable Sharing of Benefits Arising from Their Utilization (NP-2014).

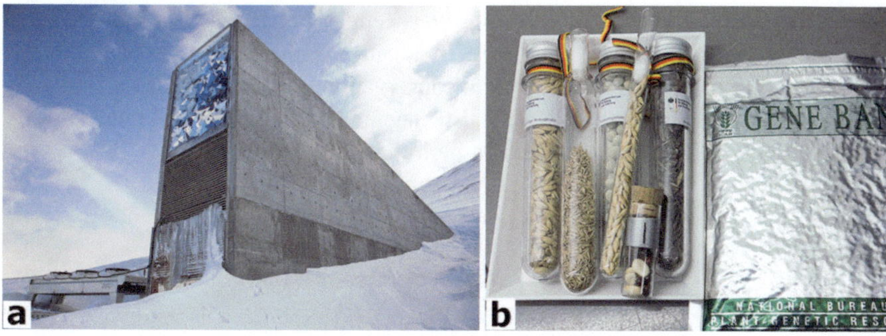

Fig.3.3 (**a**) Svalbard Global Seed Vault; (**b**) samples of preserved seeds

Fig. 3.4 Diversity in seeds of cereals and pulses

The ITPGRFA is the legal instrument for Access to Genetic Resources and Benefit Sharing (ABS) for 64 crops listed in the Treaty. The NP facilitates utilization of all genetic resources. Such policies virtually control germplasm exchange patterns among countries.

India has varied geography and diverse ecosystems that make it genetically rich. With about 46,042 species of flowering and non-flowering plants, India is one of the 12 mega diversity centres of the world. The hot spots are Eastern Himalayas, Western Ghats, Indo-Burma and Nicobar Islands. Besides this, introduced genetic resources have been subjected to natural selection and adaptation leading to heterogeneous gene pools. The introduction and exchange of genetic material were executed by the Division of Plant Introduction at the Indian Agricultural Research Institute (IARI) during the 1960s under the aegis of the Indian Council of

Agricultural Research (ICAR). This division was upgraded to the National Bureau of Plant Genetic Resources (NBPGR) in 1976 housing the National Genebank (NGB), established during 1985–1986 for ex situ conservation. India has ratified all the three treaties (CBD, ITPGRFA and NP) and also enacted its own Biological Diversity Act (BDA-2002). The BDA governs Indian biological resources.

3.6 Plant Introduction

Transport of a species from its native place to a new area is known as plant introduction. According to Frankel (1957), *plant introduction is the transposition of a genetic entity from an environment to which it is attuned to one in which it is untried.* Germplasm is a collection of all genotypes (both indigenous and exotic) of any given species. This is a vital resource for breeding new varieties with increased production since plant breeders need more diversity to be utilized in breeding programmes. Such introduced genotypes are used either as varieties for large-scale cultivation or as sources of useful traits like higher yield and other secondary attributes.

Of the 250,000 higher plant species that are described taxonomically, 115,000 are with PGR (46%) and 35,000 (14%) are cultivated. However, less than a dozen flowering plants provide 80% of calorie intake for man. In the cultivated species alone, the diversity available is enormous (Fig. 3.4).

3.6.1 Historical Perspective

Plant introduction was undertaken by travellers, pilgrims, invaders, explorers or naturalists when agriculture began. Because of geographic contacts, movement of species within the Old World was made possible. Old World was the pioneer at domesticating crops and animals to enhance their well-being, whereas the New World grew their own crops as source of food (Old World is used in the west to refer to Africa, Europe and Asia. They are regarded collectively as part of the world known to Europeans before their contact with the New World: Americas including nearby islands like Caribbean and Bermuda). Only after the discovery of the Americas by Columbus in 1492 and the European colonization soon after, the exchange of plants between the New World and the Old World began. The USA did not have Old World wheat, soybean and rice some 400 years ago and were importing them. Crops like maize, potato, sweet potato, tomato and groundnut (all are New World crops) are source of food for the Old World. During the sixteenth century, Portuguese, British, French and Dutch introduced many plants as a process of colonization.

In India, Mohammedan rulers introduced many species like cherries and grapes from Afghanistan and Iraq. New World crops like maize, groundnut, chilli, potato, sweet potato, guava, custard apple, pineapple, cashew nut and tobacco were introduced by Portuguese during the seventeenth century. Tea, litchi and loquat (all from China) was introduced by British East India Company. Cabbage, cauliflower and other winter vegetables were brought from the Mediterranean region by the British. During the eighteenth century, mangosteen was brought from Malaysia, and annatto (*Bixa orellana* – a source of edible dye) and mahogany came from the West Indies.

In 1926, N.I. Vavilov, a Russian botanist/explorer, identified eight phytogeographical regions where crop diversity was found to be extremely intense for some species. These areas were recognized as "centres of origin" (see Chap. 2). Such areas were further studied by scientists from the USA, the erstwhile USSR, Europe and Australia through explorations. Such species were eventually brought into new areas and further evaluated. This prompted plant breeders all over the world to acquire such materials to be used in further breeding programmes.

3.7 Plant Introduction: The International Scenario

Movement of Plant Genetic Resources envisages an element of risk of spreading of diseases and pests. The International Plant Protection Convention (IPPC) of FAO states that harmful biotic agents like viroids, viruses, bacteria, fungi and pests can pose such threats. Many countries have passed legislations to regulate the movement of plant materials. In the event of plant material passing through international borders, the material needs to be accompanied with phytosanitary certificate stating that the screening standards of the country importing it are met with. This will ensure quality of the plant material.

3.7.1 Import Regulations

There are three categories of import regulations:

(a) Permissible imports (low risk)
(b) Imports that are prohibited
(c) Imports that need to undergo quarantine

Materials that need quarantine are "carriers" of pests that are imported under "Q label". Such materials are monitored through growing them in quarantine station. Institutions that are importing the germplasm are supposed to understand the diseases/pests associated with the material being imported. The importing institution must have the list of diseases and pests associated with the plant species. There are standards adopted under the Intergovernmental Panel on Climate Change (IPPC) with the main objective of spread of pests and diseases. IPCC has formulated technical guidelines on disease indexing to ensure phytosanitary procedures while moving germplasm internationally.

3.7.2 Plant Germplasm Import and Export

Plant germplasm can be moved in the form of as true seed, in vitro cultures or vegetative material. True seed is the best material to be transported, as they pose minimum threat with pests and diseases. In vitro material must undergo quarantine procedures. Such quarantine procedures must be amply documented as germplasm health statement (see Box 3.5 with *Musa* as example). The import of germplasm needs to complete the following formalities:

- Make a formal request to donor organization/country through NPPO (National Plant Protection Organization).
- Generate import conditions through Pest Risk Analysis (PRA).
- NPPO or the organization responsible to screen the plant material at the port of entry shall inform the donor country (through the institute importing the material) the utility of the material being imported.
- The donor country NPPO evaluates conditions of the importing country and confirms compliance of norms.
- If import conditions are met, NPPO of the donor country prepares a phytosanitary certificate.
- The recipient country issues a Plant Import Permit (PIP). While importing a material, PIP and phytosanitary certificate of the donor country must accompany the material.
- Materials with "Q label" are subjected to quarantine formalities.
- There are countries that do not allow transgenic material. If allowed, such materials are subjected for the verification of the National Biosafety Committee.
- Plant breeders' rights are to be protected while importing any material.
- If the material is imported for cultivation directly, then such materials must undergo formalities of variety release system.

Box 3.5: Germplasm Health Statement
Bioversity International Germplasm Health Statement

ITC Accession Number:
Accession Name:
Origin of Accession:

The material designated above was obtained from a shoot-tip cultured in vitro. Shoot tip culturing is used to eliminate the risk of the germplasm carrying fungal bacterial and nematode pathogens and insect pests of *Musa*. However, shoot tip cultures could still carry virus pathogens.

Screening for Virus Pathogens

(continued)

Box 3.5 (continued)

A representative sample of four plants derived from the same shoot tip as the germplasm designated above has been grown under quarantine conditions for at least 6 months, regularly observed for disease symptoms and tested for virus pathogens, as indicated below, following methods recommended in the Bioversity International Technical Guidelines for the Safe Movement of *Musa* Germplasm (2015) for the diagnosis of virus diseases.

PCR-based methods	[ ] BBTV – banana bunchy top virus
	[ ] CMV – cucumber mosaic virus
	[ ] BBrMV – banana bract mosaic virus
	[ ] BSV – banana streak viruses
	[ ] BanMMV – banana mild mosaic virus
Electron microscopy	[ ] isometric virus particles – includes CMV and unknown viruses
	[ ] bacilliform virus particles – includes unknown BSVs
	[ ] filamentous virus particles – includes BBrMV, BanMMV and unknown viruses

[P] = test positive, [N] = test negative, [ ] = test not undertaken

Distribution of Virus Pathogens and Other Information

(Example: BBTV and BBrMV are not known to occur in country of origin)

eBSVs are present in the B genome of *Musa* (banana). Consequently, almost all accessions containing the B genome may develop BSV infection and may express symptoms during any stage of growth.

The information provided in this germplasm statement is based on the results of tests undertaken at Bioversity International's Virus Indexing Centre by competent virologists following protocols current at the time of the test and on present knowledge of virus disease distribution. However, neither Bioversity International nor its Virus Indexing Centre staff assume any legal responsibility in relation to this statement.

Signature Date

This statement provides additional information on the phytosanitary status of the plant germplasm described herein. It should not be considered as a substitute for the official "Phytosanitary Certificate" issued by the plant quarantine authorities of Belgium.

Courtesy: Biodiversity International

The export of germplasm needs to complete the following formalities:

- The donor country provides import conditions of recipient country.
- Some species that are restricted from export are protected plant varieties as per CITES (Convention on International Trade in Endangered Species of Wild Fauna and Flora, Geneva).
- NPPO (National Plant Protection Organization) of the donor country verifies compliance to the import conditions and prepare phytosanitary certificates.
- Under exceptional circumstances, Material Transfer Agreement (MTA) may be required between exporting and importing institutions.

3.8 Plant Introduction in India

In India, NBPGR is the nodal agency for germplasm exchange and research. NBPGR assists the all India crop improvement programmes, ICAR crop-based institutes and state agricultural and horticultural universities. NBPGR also closely collaborates with more than 85 countries besides the Plant Introduction Agencies having headquarters at Beltsville (USA), Canberra (Australia), Leningrad (USSR), Ottawa (Canada), São Paulo (Brazil), Buenos Aires (Argentina), Lisbon (Portugal), Peradeniya (Sri Lanka), Dhaka (Bangladesh), Islamabad (Pakistan), Addis Ababa (Ethiopia), Tápiószele (Hungary), Sofia (Bulgaria), Manila (Philippines), Tsukuba (Japan) and many allied agencies, universities, botanical gardens and private nurseries/organizations. It has cooperating relationship with the International Agricultural Research Centres (IARCs) under the Consultative Group on International Agricultural Research (CGIAR), like IRRI (Philippines), CIMMYT (Mexico), CIAT (Colombia), CIP (Peru), ICRISAT (India), ICARDA (Syria), IITA (Nigeria) as well as other centres like AVRDC (Taiwan) and WARDA (Liberia), besides the Biodiversity International (IBPGR) (see Table 3.2 for details). The first crop imported to India through ICAR-NBPGR (Plant Introduction Unit, IARI) in August, 1940 is Giant Star Grass (*Cynodon plectostachys*) with Exotic Collection number EC 1.

The *Destructive Insects and Pests Act* (*DIP Act*) *of 1914* (Directorate of Plant Protection, Quarantine and Storage, Ministry of Agriculture and Irrigation, 1976) is the legislation for import and export of seeds, plants, plant products and planting material in India. This legislation has undergone revision several times subsequently. Enforcement of the *DIP Act* is the responsibility of the Plant Protection Adviser to the Government of India, Ministry of Agriculture.

The Government of India has approved the following national institutions as nodal agencies for exchange of plant materials:

1. The National Bureau of Plant Genetic Resources (NBPGR), New Delhi (agri-horticultural and agri-silvicultural crops).
2. The Forest Research Institute (FRI), Dehradun (forest plants).
3. The Botanical Survey of India (BSI), Calcutta (for species of botanical interest. See https://cropgenebank.sgrp.cgiar.org/images/file/management/plant%20quarantine.pdf for further details.

Table 3.2 Some promising primary introductions to India

Crop	Variety/(donor country)	Characteristics
Wheat	Ridely (Australia)	Bold amber-coloured grain, resistant to rust, found promising for northern hills of Himachal Pradesh and U.P. hills
	Lerma Rojo-64 (Mexico)	Semi-dwarf, medium late, resistant to all the three rusts
	Sonora-64 (Mexico)	Semi-dwarf wheat with good tillering, resistant to all the three rusts, suitable for sowing under high fertility conditions in Punjab, Delhi, U.P., Bihar, West Bengal, M.P. and Maharashtra
	P.V. 18 (Mexico)	Semi-dwarf, high yielding under high fertility conditions
Barley	L SB 2 (USA)	Hull-less cultivar, selected from USA 95, performed well in northern hills of the Himachal Pradesh
	Dolma (USA)	Hull-less cultivar, selected from USA 115, performed well in Himachal Pradesh
	Clipper (Australia)	Two-rowed hulled variety, which performed well in northern plains
Rice	I.R. 8 (Philippines)	Dwarf, maturing in 135 days, long bold grain, photo-insensitive
	I.R. 50 (Philippines)	Dwarf, very popular in drought-prone areas in Tamil Nadu
Oats	Kent (Australia)	Stiff stemmed, medium early, dual-purpose variety
	Rapida (USA)	Early maturing medium tall, with good protein content (14.2%) suitable for milling industry
Sunflower	Peredovik (USSR)	Early maturing with average oil content (47.9%), released in A.P., Karnataka and Maharashtra
	Aramvirikij (USSR)	Early maturing (95–100 days) with average of 49.1% oil content
Groundnut	Asiriya Mwitunde (Tanganyika)	Useful introduction, performed well in many groundnut growing states of India
	Rehovot 33-1 (Israel)	Selection from Rehovot-33, performed well in southern states of India
	M 13 (USA)	Selection from NC 13, recommended for Punjab State
Soybean	Bragg (USA)	Yellow-seeded cultivar with wider adaptability in southern states of India
	Lee (USA)	High-yielding variety with attractive bright yellow seed colour
	Improved Pelican (USA)	Bold yellow-seeded cultivar
Cowpea	EC 5000 (Rhodesia)	Very high green pod yielder, photo-insensitive, bushy type with attractive light green medium pods
	Pusa Barsati (Philippines)	Selected from an introduction imported from Philippines with light green pods
	EC 1077 155 (PI 194293, USA)	High green pod yielder, performed well in Delhi

(continued)

Table 3.2 (continued)

Crop	Variety/(donor country)	Characteristics
Pea	Harbhajan (EC 33866, Portugal)	Dwarf, early, dual-purpose variety, maturing in 110 days, in northern India
Tomato	Sioux (USA)	Early variety, with large red fruits, suitable for cultivation in both winter and summer
	Labonita (USA)	Dwarf, variety with good fruiting and leaf cover, dual-type variety for use as table as well as paste type, fruits with thick skin, medium in size, stands transportation well, with good keeping quality
	Dwarf Money Maker (EC 108759, Israel)	Dwarf paste type, high yielding, fruits deep red
	Molakai (Australia)	Prolific fruit bearer, good table variety, fruit large in size
	Fire Ball (Canada)	Early-maturing type, found promising in high-altitude areas of India
Cauliflower	Early Snow Ball	Early variety, with white curd
	Snow Ball – 16 (EC 12013, Holland)	Medium duration variety
Cabbage	Golden Acre (Denmark)	Early variety, with compact round white head
	Drum head (Denmark)	Late variety, with flat compact head
	Express (Denmark)	Medium-type variety, very popular in Himachal Pradesh
Water melon	Ashahi Yamato (Japan)	Fruit medium in size/5–8 kg each, flesh deep pink, mid-season type
	Sugar Baby (USA)	Fruits round, fine textured, attractive dull green skin; flesh uniform deep red, very sweet, 10–12% TSS, with average fruit weight 3–5 kg
Banana	Lady Finger (EC 160160, Australia)	Possesses resistance/tolerance to bunchy top virus
	Grand Nain M. S. (EC 27237, France)	High-yielding, disease-tolerant cultivar
	Valery (EC 115363, West Indies)	High yielding, quality variety
Papaya	Sunrise (EC 134371, USA)	Promising high-yielding variety
	Cariflora (EC 300205, USA)	Dioecious, with high degree of tolerance to papaya ring rot virus, fruits yellow with agreeable taste and aroma
	Carite Special (EC 187250, Philippines)	High-yielding variety
Apple	Vered (EC 24349, Israel)	Low chilling cultivar, suited for lower hills and plain areas. It bears small- to medium-sized fruits (45 g), conical flat, of 4.3 cm length and 4.5 cm diameter with 12% TSS, light yellow with green skin splashed with red, sparingly soft flesh, ripens in the middle of June, self-fruitful
	Spur-type Red	Bud sport of red delicious, regular and heavy bearer, with medium large

(continued)

Table 3.2 (continued)

Crop	Variety/(donor country)	Characteristics
	Delicious-II (EC 43974, USA)	Fruits (140 g) with red splashed skin, ripening in the middle of August; semi-dwarf, open, spreading and well suited for high-density planting; performed well in Shimla hills
	Red Baron (EC 115820, USA)	Heavy bearer, fruits medium size, yellow bright red colour, creamish yellow crisp, juicy and very sweet flesh
	Mollies Delicious (USA)	Bears large fruits, red in colour, very sweet, crisp in taste with good keeping quality; matures in the last week of July; has performed well at Solan in Himachal Pradesh
	Skyline Supreme Red Delicious (EC 27801, USA)	Bears medium to large dark red fruits, very sweet, fruits with good keeping quality, mature in the first week of August; has wide adaptability from medium to high altitudes
Pear	Flemish Beauty (EC 27810, USA)	Bears extra large fruit (172 g), conical round in shape, very sweet, 14% TSS, greenish yellow skin with numerous tiny dots, white melting smooth, juicy
	Max Red Bartlet (EC 28386, Italy)	Bears large fruits (135 g), pyriform, very sweet, 14% TSS, dark cranberry red, skin turning to an attractive bright red colour, white flesh, excellent in taste, medium keeping quality; fruits ripen in the first week of August
	Devoe (EC 27811, USA)	Bears pyriform, large light green fruits, flesh white, melting juicy, very sweet
	Manning Elizabeth (EC 27809, USA)	Bears small round yellowish green fruits, with a bright red blush at the blossom end; fruits are very sweet and excellent in taste; fruits ripen in the first week of July
Peach	Stark Early Glo (EC 27791, USA)	Early type, with medium-sized fruit (79 g); round deep yellow skin with bright red splashes; flesh is deep yellow; fine textured, juicy and very sweet; 12% TSS, with free stone; fruits ripen in the second week of June
	Candor (EC 57530, USA)	Promising cultivar for growing in Shimla hills, with medium-sized fruits (83 g), round, TSS 11.9%, bright red blush over rich yellow ground colour, fine textured juicy, semi-free stone; fruits ripen in the second week of June
	Flordasun (USA)	Low chilling cultivar, which gave excellent performance in plains of Uttar Pradesh, Delhi and Rajasthan
Plum	Methley (EC 340450, Kenya)	Promising variety, with medium-sized fruits (18.0 g), very sweet, 20% TSS; fruits ripen in the middle of June
	Kanto-5 (EC 27810, USA)	Promising variety, fruits – medium, large (13.0 g), very sweet, 20% TSS; fruits ripen in the middle of June

(continued)

Table 3.2 (continued)

Crop	Variety/(donor country)	Characteristics
Apricot	Nugget (EC 27791, USA)	Most promising cultivar for hills, with medium to large (52.0 g) round fruits, of bright red colour, quite sweet, 15.3% TSS, free stone, self-fruitful; fruits ripen in the second week of June
	Coninos (EC 28382, Italy)	Promising variety, with medium-sized fruits; fruits ripen in the middle of June
Almond	Nonpareil (EC 28387, USA)	Thin-shelled cultivar, with mean fruit weight of 2.0 g, has been found promising for Shimla hills
Walnut	Lake English (EC 24562, USA)	Medium-shelled, high fruit yielder, nut – medium large with good taste and good filling
	Hansen (EC 26580, USA)	Paper-shelled cultivar, with high percentage of kernel, self-pollinating, winter hardy
	Payne (EC 26890, USA)	Paper-shelled cultivar, with good appearance, kernel – medium sized with excellent taste, mean weight of kernel (4.0 g), fruit shell semi-hard
	Tutle 31 (EC 27484, USA)	Promising cultivar, in both appearance and taste, medium hard shell, with fairly good filling

Source: Biodiversity International

3.9 Conservation of Endangered Species/Crop Varieties

A major threat to the biodiversity is the extinction of species. Five mass extinctions were believed to have occurred during the past 500 million years that has caused over 50% species. We are into the opening phase of a sixth mass extinction, predicted to be human impacted. Plants are extremely important for the conservation of biodiversity from both ecological and human economics viewpoint. However, plant diversity is facing tremendous threat mainly because of unsustainable harvesting for their multifarious utilization and habitat degradation. According to the UN World Conservation and Monitoring Centre (WCMC), Cambridge, UK, it is estimated that more than 8000 tree species are endangered worldwide (www.unep-wcmc.org); however, another estimate predicts this between 22 and 47 percent of the world's plants. The rate of extinction is also approximated to be very fast, and it is estimated that around 1800 populations are being destroyed per hour (16 million annually) in tropical forests alone. The extinction of wild crop varieties is no different from this. The adoption of new high-yielding varieties (HYVs) has only ensured the extinction of traditional/wild crop varieties cultivated by man over the ages.

Further Reading

Reed BM et al (2004) Technical guidelines for the management of field and in vitro germplasm collections. IPGRI handbooks for gene banks no:7

Olson AE, Stepp JR (2016) New perspectives on the health-environment-plant nexus. Springer, Cham

Niklas K (2016) Plant evolution: an introduction to the history of life. University of Chicago Press, Chicago. 560 pp

Murat F et al (2017) Reconstructing the genome of the most recent common ancestor of flowering plants. Nature Genet 49:490–496

Chen C et al (2017) Historical introduction, geographical distribution, and biological characteristics of alien plants in China. Biodivers Conserv 26:353–381

Henry RJ (2007) Genomics strategies for germplasm characterization and the development of climate resilient crops. Front Plant Sci 5:68. https://doi.org/10.3389/fpls.2014.00068

Bioversity International (2007) Guidelines for the development of crop descriptor lists, Biodiversity technical bulletin series. Biodiversity International, Rome

Domaingue et al (2017) Evolution and challenges of varietal improvement strategies. In: Sustainable development and tropical Agri-chains. Springer, Dordrecht, pp 141–152

Flachowsky G, Reuter T (2017) Future challenges feeding transgenic plants. Anim Front 7:15–23

Zargar M, Rai V (2017) Plant omics and crops breeding. In: CRC Press

Thomas JE (2015) MusaNet technical guidelines for the safe movement of musa germplasm, 3rd edn. Bioversity International, Rome

Part II

Developmental Aspects

Modes of Reproduction and Apomixis

4

Keywords
Sexual reproduction · Vegetative (asexual) reproduction · Apomixis · Gametophytic apomixis · Sporophytic apomixes · Genetics of apomixis · Apomixis in agriculture

Flowering plants follow either one of these three fundamentally different modes of reproduction: (a) through cross-pollinated seeds, (b) self-pollinated seeds and (c) asexual (vegetative) means. Mode of reproduction is a decisive factor in moulding population structure and evolutionary potential. All three modes are being used by perennial plants. Apomixis is another way of asexual reproduction. The sexual life cycle of vascular plants follows haploid and diploid generations in an alternate fashion. Haploid spores are produced by diploid sporophytes through meiosis. Haploid egg and sperm are produced by gametophytes through mitosis. Egg and sperm unite to form diploid zygotes from which new sporophytes develop. When offspring are produced through modifications of the sexual life cycle avoiding meiosis and syngamy, the process is asexual reproduction (Fig. 4.1).

4.1 Sexual Reproduction

All flowering plants (angiosperms) practise sexual reproduction. Bisexual flowers have pollen and ovule producing structure together. In monoecious plants, pollen and ovule are seen separately in different flowers. In dioecious species, they are borne on entirely different plants. The angiosperms are the largest taxa in the plant kingdom and dominate most terrestrial environments. They are generally distinguished by key features like presence of flowers with perianth (e.g. petals) around the reproductive organs and ovules that are enclosed in carpels (female sporophylls that after fertilization of the ovule form part of the fruit). During seed formation,

P. M. Priyadarshan, *PLANT BREEDING: Classical to Modern*,
https://doi.org/10.1007/978-981-13-7095-3_4

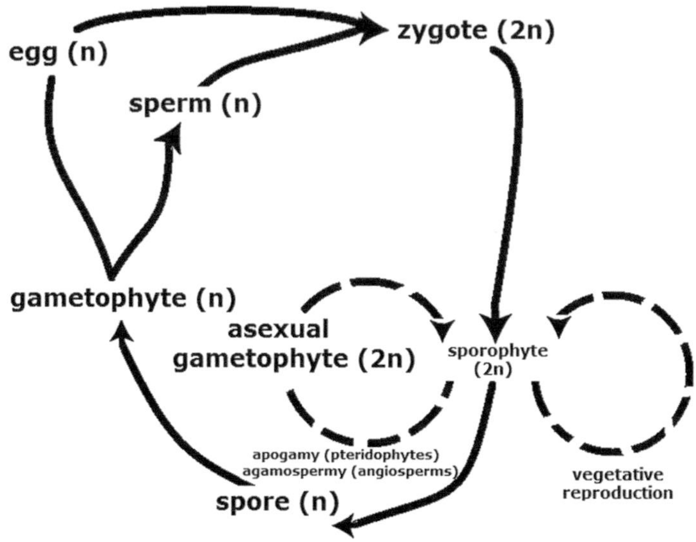

Fig. 4.1 Basic vascular life cycle in plants. Asexual cycles are indicated in dashed lines and sexual cycle is in solid lines

following *double fertilization*, one male gamete unites with the ovum that forms the embryo and the other unites with the secondary nucleus (*triple fusion*) to form triploid endosperm. Triploid endosperm provides additional nutrition to the developing embryo (see Fig. 4.2).

Flowers Flowers are modified shoots meant for sexual reproduction. This part of the shoot is called the **receptacle** that has modified leaves. They can have up to four whorls of "leaves". The first two whorls are the sepals and petals and are modified to attract pollinators. Sepals and petals are otherwise known as **calyx** and **corolla**. The other two whorls are stamens and **carpels** and are fertile. Stamens consist of **filament** and **anther** (**androecium**). While the anthers produce the pollen or male gametophyte (see Chap. 6 for details on microsporogenesis), the carpels are differentiated into **stigma**, to receive pollen, and the **style** that supports the stigma and the ovary (Fig. 4.3). Stigma, style and ovary are together known as **gynoecium**. The **ovules** are inside the ovary. Ovules produce ovum through meiosis, which, after double fertilization, forms the **embryo** and **endosperm**. The ovules attain maturity and form **seeds**. Ovary matures into the **fruit**. Flowers are the organ that spread genes since pollen and seeds can leave the plant. Male and female genes are mixed in a flower through fertilization and contribute to genetic diversity. Fruits help to continue the generations.

The ovary is said to be **inferior** when sepals, petals and stamens are inserted on the top of the ovary and the flower is **epigynous**. If sepals and petals are below, the ovary is **superior** and the flower is **hypogynous**. The flowers are **perigynous** when

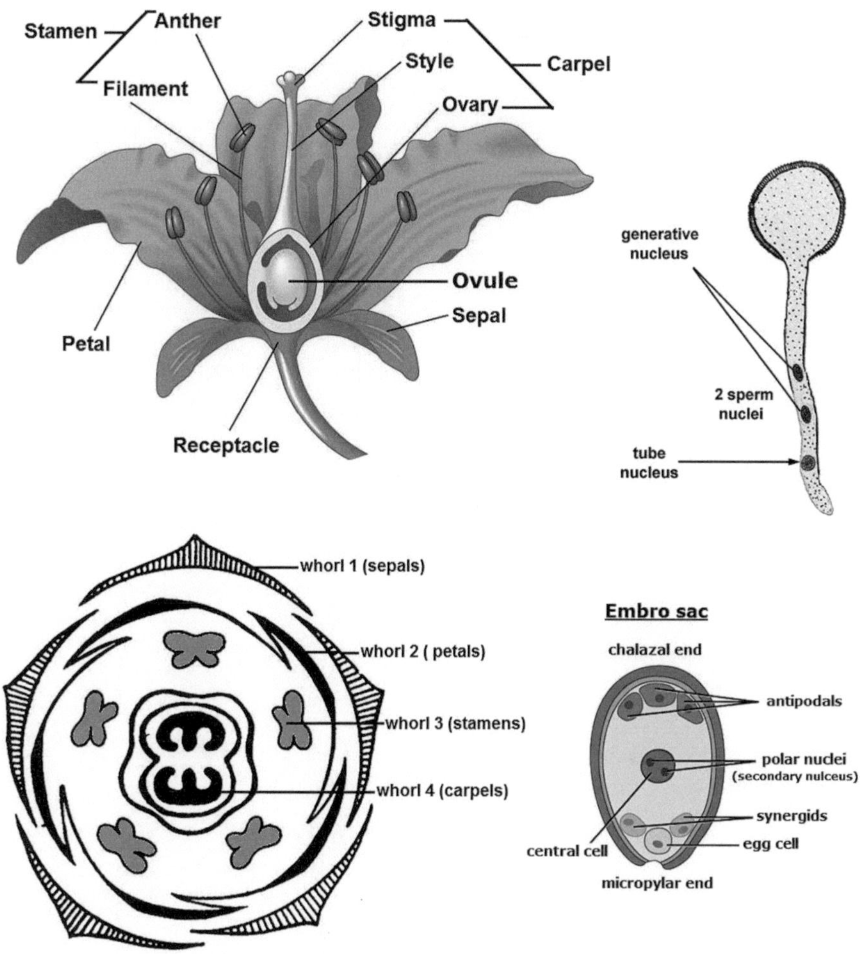

Fig. 4.2 Reproductive organs of angiosperms

the floral parts are fused halfway to the ovary, or fuse to themselves, forming a cup around the ovary. Flower can be radial (**actinomorphic**), with the whorls distributed evenly around the receptacle, or it can be with bilateral symmetry (**zygomorphic**) (Fig. 4.4).

Fruits Ovaries ripen into fruits. After fertilization, ovules develop into seeds and the ovary wall develops into fruit wall. The wall develops from **carpels**. A fruit can develop from either one or many carpels. Depending on the number of carpels, the number of seeds varies. Exceptionally, the fruit may develop in the absence of seeds (as a seedless grape or naval orange), through **parthenocarpy**. The fruit is a **berry** (as in coffee, grape) when the ovary wall is fleshy. If the fruit breaks open upon maturity, it is a **capsule** (as in cotton). When ovary wall is in different layers, with an

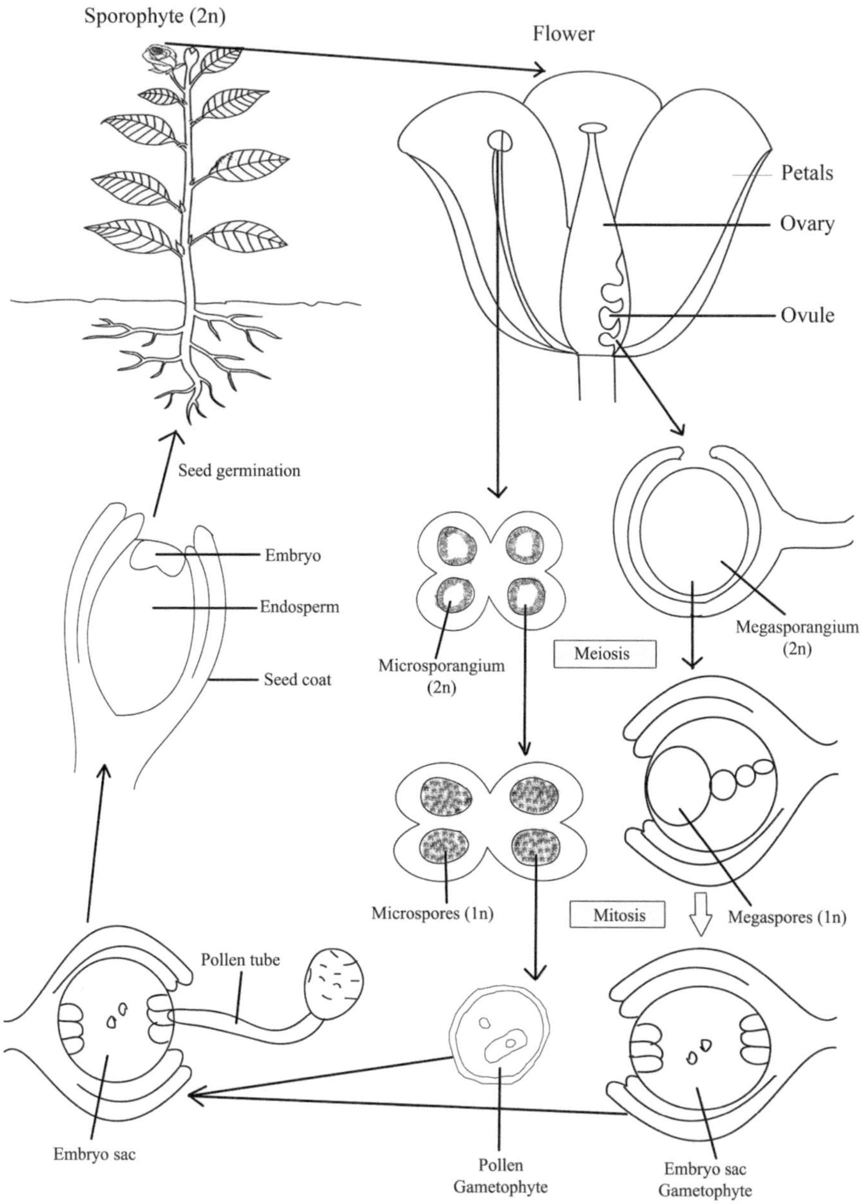

Fig. 4.3 Sexual reproductive cycle of angiosperms

inner most stony layer, it is a **drupe** (coconut, pepper). When additional flower parts form part of the flesh of the fruit, it is an **accessory fruit** (mulberry and strawberry). When the ripening ovaries fuse together, they form **aggregate fruits** (custard). Fruit is **compound or multiple** when ovaries of separate flowers fuse together (pineapple).

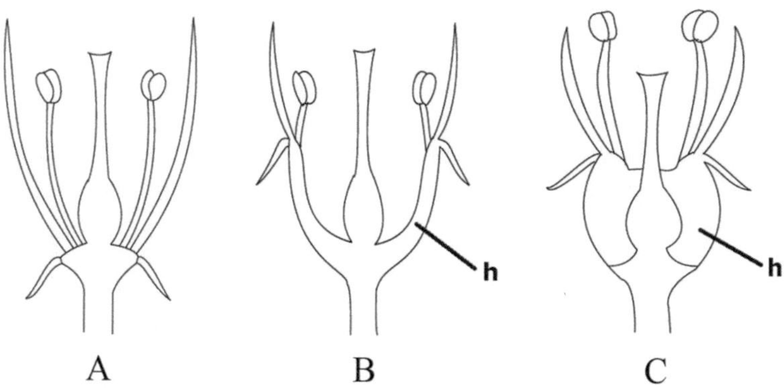

Fig. 4.4 Relative positions of floral appendages. (**a**) Hypogynous flower: superior ovary with ovary above stamens and perianth. (**b**) Perigynous flower: superior ovary, with bases of perianth and stamens united into a hypanthium. (**c**) Epigynous flower: inferior ovary, with stamens and perianth positioned above the ovary on a hypanthium (**h**)

4.2 Vegetative (Asexual) Reproduction

Asexual reproduction (vegetative), or **cloning**, is the propagation through vegetative tissues (i.e. not involving sexual reproduction). It involves only cell divisions by **mitosis** and not by **meiosis**. Vegetation reproduction results in a new plant called a **ramet** that is genetically identical to the original **donor**, also called the **ortet**. Most methods of vegetative propagation, both those occurring in nature and those used by people to clone plants, involve taking part of a plant and re-growing the missing parts, e.g. starting with a shoot and developing **adventitious roots** or starting with a root and producing one or more **adventitious shoots**. Some of the ways of vegetative propagation are summarized here.

Layering When a drooping lower branch comes in contact with the soil, adventitious roots form at the point of soil contact. This method of propagation is layering. Many high-elevation tree species readily reproduce through layering, resulting in expanding tree islands of smaller ortets around a central ramet (e.g. *Picea*, *Abies*). Western redcedar (*Thuja plicata*) and yellow cedar (*Chamaecyparis nootkatensis*) also layer easily.

Sprouting and Suckering When trees are cut down often, new shoots emerge from the stump since the auxin/cytokinin ratio drops. This is popularly known as **coppicing**. Coppicing is for forest regeneration (e.g. coast redwood). Formation of adventitious shoots due to low auxin/cytokinin ratio from roots is **suckering**. As auxin is produced by growing shoot tips and transported down, and cytokinin is produced by roots and transported up, cutting down the stem of a plant results in a low auxin/cytokinin ratio in the stump.

Rooted Cuttings Reproduction through rooting branch cuttings is relatively rare in nature. The branches of black cottonwood (*Populus trichocarpa*) trees along rivers can be broken from the crown by storms, and these branches can float downstream and lodge in the moist riverbank, and the cuttings can then produce adventitious roots. In general, however, the production of adventitious roots from severed stems is much more common as a method used by humans for propagation than a means of natural regeneration.

Rhizomes, Stolons, Bulbs, Corms and Tubers Many of the herbaceous and woody plants propagate through **rhizomes** – horizontal, underground stems. Genetically identical plants emerge from these rhizomes. Small rhizome segments can be planted horizontally. **Corms**, **bulbs** and **tubers** are under the soil vegetative propagules of herbaceous plants. Plants can be regenerated from corms that are vertical underground stems (elephant foot, *Colocasia*). Bulbs are with fleshy scales. Tubers are thickened storage rhizomes. They are with buds that are capable of regenerating plants (onion). **Runners or stolons** are aboveground horizontal shoots as in strawberries (*Fragaria* sp.).

Air Layering Air layering is done by artificially wounding a shoot. The wound is then wrapped with a moist medium (e.g. guava, roses) and covered by a waterproof material (plastic). Adventitious roots arise at the wound site. Such rooted branches can be cut and planted. Air layering is not a popular method but can be practised where other methods fail. Layering is not a practical way to generate inexpensive trees in large numbers.

Grafting is attaching a shoot from one individual to the stem of another plant. The stem on to which the grafting is done is the **root stock**. It produces a genetic mosaic, where most of the stem and crown of a tree or shrub are of one genotype with its root system of a different genotype. Grafting is the only method of propagating older trees. It is vital that xylem, phloem and cambium of stock and scion are in contact and intact. Stock and scion grow together and develop continuous vascular tissue after the initial wound callus formation. Stock and scion are to be genetically compatible. Otherwise, they may not develop properly and eventually die. Grafting is a common method to produce genetically superior trees for horticultural purposes (e.g. *Hevea* rubber tree).

Tissue Culture involves growing an **explant** (piece of leaf, cotyledon or embryo) in a medium that contains hormones, sugars, amino acids and micronutrients. Initially, **callus tissue** and adventitious buds are produced. Adventitious shoots are placed in rooting medium with high auxin concentration to promote root formation and growth. Individual cells from the callus can also be grown in liquid medium to regenerate plants. This is cell culture, a most favoured propagation system following genetic engineering. Though tissue culture has been successful in many species, many forest trees are difficult to be propagated in this way (see Chap. 21).

Somatic embryogenesis is the development of embryos form a callus. These somatic embryos can then be packaged as "artificial seeds" in calcium alginate crystals or **cryopreserved** (stored at very low temperatures) (see Chap. 21).

4.3 Apomixis

Apomixis is the asexual formation of seed from the maternal tissues of the ovule. This is by avoiding meiosis and fertilization that leads to embryo development. The first case of apomixis was in a solitary female plant of *Alchornea ilicifolia* (syn. *Caelebogyne ilicifolia*) from Australia that continued to form seeds when planted at Kew Gardens in England. This was observed by Smith in 1841. Winkler in 1908 introduced the term apomixis to mean "substitution of sexual reproduction by an asexual multiplication process without cell fusion".

Apomixis occurs in around 10% of the 400 families of flowering plants. Apomixis is predominant in Gramineae (the cereal family), Compositae (sunflower family), Rosaceae (which includes many fruit trees) and Asteraceae (the dandelion family). Apomixis can happen in two ways. Apomictic seeds either can arise from sexual cells (which fail undergo meiosis) or can arise from non-sexual (somatic) cells. However, under rare circumstances, both sexual and asexual seeds can develop from the same flower. Pollen of apomictic plants is often viable, presuming that apomixes can also be transmitted through sexual reproduction. Apomixis can ensure production of *clones* through seeds. (See Fig. 4.5 for diagrammatic representation of various kinds of apomixis.)

A systematic classification of apomixis is difficult. However, Maheshwari in 1950 used the following classification:

(a) Non-recurrent apomixis
(b) Recurrent apomixis
(c) Adventive embryony
(d) Vegetative apomixis

In non-recurrent apomixis, a haploid embryo sac (megagametophyte) is formed as per usual procedure. Then the embryo may arise either from the egg (haploid parthenogenesis) or from a cell of the gametophyte (haploid apogamy). Since the process is not repeated from one generation to another, hence it is non-recurrent.

Recurrent apomixis is often called *gametophytic apomixis*, since the megagametophyte will be having the same number of somatic chromosomes because the meiosis is not completed. Recurrent apomixis arises either from archesporial cell or from nucellus.

Adventive embryony is also called *sporophytic apomixis*. Here, the embryos arise from cells of nucellus or the integument. Adventive embryony is important in several species of *Citrus*, *Garcinia* and *Euphorbia dulcis*, *Mangifera indica*.

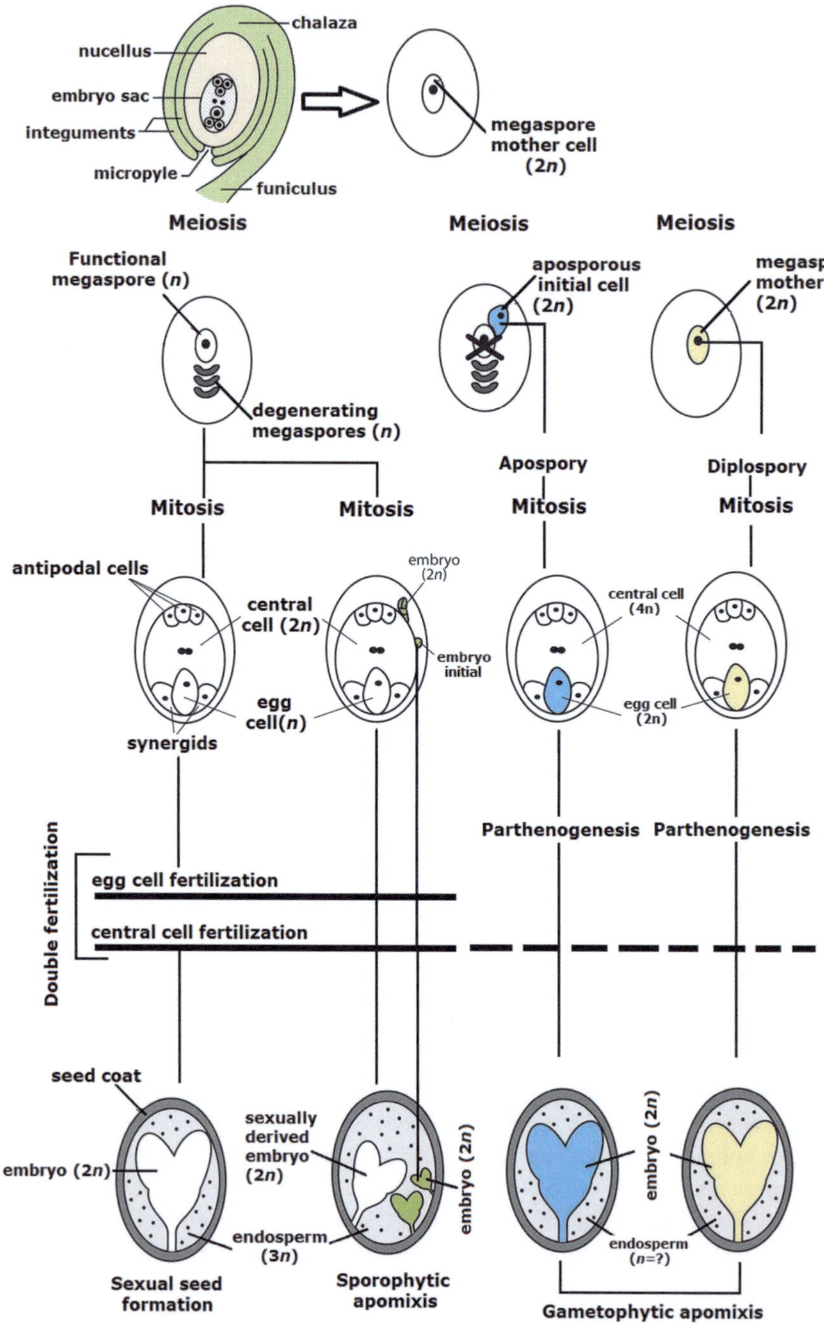

Fig. 4.5 Various kinds of apomixis

In vegetative apomixis, bulbils or other vegetative propagules replace flowers. These bulbils germinate frequently, while they are still on the plant. Vegetative apomixis is seen in *Allium*, *Fragaria*, *Agave* and some grasses.

4.3.1 Gametophytic Apomixis

In gametophytic apomixes, meiosis is bypassed by *apomeiosis*. This unreduced female gametophyte (diploid) leads to gametophytic apomixis. In the absence of fertilization, a cell of the unreduced embryo sac develops into an embryo (parthenogenesis). In gametophytic apomicts, endosperm formation may be independent of fertilization (autonomous endosperm) or may be through fertilization (pseudogamous endosperm). Apomeiosis can occur by two major means, viz. *diplospory* and *apospory*. In diplospory, the megaspore mother cell remains unreduced (mitotic diplospory), or it fails to undergo meiosis (meiotic diplospory) (Fig. 4.6). In apospory, the megaspore mother cell differentiates as usual. However, additional cells, known as aposporous initials (*ai*), differentiate in close proximity to such cells. Such *ai* cells through mitosis lead to unreduced embryo sacs.

4.3.2 Sporophytic Apomixis

In sporophytic apomixes, embryos arise from diploid ovule cells, termed embryo initial (*ei*) cells. This process happens adjacent to a developing female gametophyte. Sporophytic apomixis is common in mango and citrus and otherwise known as *adventitious embryony*. Sometimes, if the embryo sac is not fertilized, multiple embryos arise from *ei* cells. Such polyembryonic seeds are commonly used to generate rootstocks for citrus propagation. Sporophytic apomixis is not studied in detail; however, available research indicate dominant inheritance.

4.3.3 Genetics of Apomixis

Sporophytic and gametophytic apomixis can be categorized into:

(a) Bypassing meiosis to form an unreduced embryo sac having an ovum capable of fertilization
(b) Independent embryogenesis
(c) Production of an endosperm that is either fertilization-dependent or fertilization-independent

The aforementioned categories of apomixis are believed to be controlled by one to five dominant loci. Genetic mapping studies have been conducted in *Pennisetum*, *Paspalum*, *Poa* and *Tripsacum* (all members of the grass family, Poaceae) and in *Hieracium*, *Erigeron* and *Taraxacum*. In *Pennisetum squamulatum*, *Cenchrus*

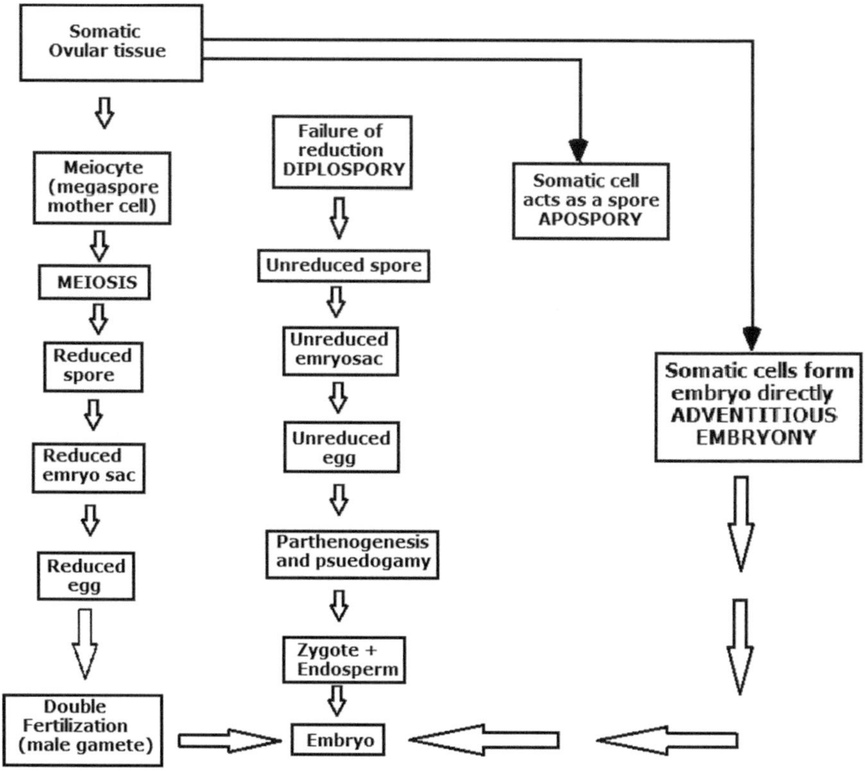

Fig. 4.6 Flow chart showing production of apomixis

ciliaris, *Panicum maximum*, *Tripsacum dactyloides* and *Paspalum* species, simple dominant Mendelian inheritance of apospory or apomixes is predominant.

The dominant locus controlling apospory in *Panicum* spp., *Ranunculus* spp. and *Hieracium* spp. co-segregates with parthenogenesis, indicating thereby that a single locus controls apomixis. The genetic loci controlling apomeiosis, parthenogenesis and functional endosperm formation can be delineated in other apomicts. So, at least three loci are involved in controlling apomixes in these species. However, more than one gene may be involved in controlling each apomictic component (see Box 4.1).

Box 4.1: Molecular Genetics of Apomixis

Molecular markers in *Pennisetum* indicate that there is an apospory-specific genomic region (ASGR) that is physically large and hemizygous (having single copy of a gene instead of two copies) and heterochromatic (tightly coiled, dark attaining). However, evidences suggest that the line between apomixis and sexuality is not clear because both these processes share key

(continued)

Box 4.1 (continued)
regulatory mechanisms. This observation suggests that apomixis might have emerged from deregulation of sexuality, rather than as a novel mode of reproduction.

Comparative gene expression studies using either differential display (a technique to identify changes in gene expression at the mRNA level between two and more cell samples) or subtractive hybridization (this is a powerful technique to study gene expression in specific tissues or cell types or at a specific stage; this is a PCR-based amplification of only cDNA fragments that differ between a control and experimental transcriptome – the mRNA) pointed out differentially expressed genes. In *Poa pratensis*, cDNA-AFLP transcriptional profiling technique could isolate 179 differentially expressed transcripts. Here, two genes, namely, SERK (somatic embryogenesis receptor kinase) and APOSTART, were characterized. APOSTART is potentially associated with apomixis, and its transcripts are detectable specifically in aposporic initials and embryo sacs. These two genes are believed to be involved in cell-to-cell interaction of both the signalling pathway and hormone stimulation. Expression of SERK gene in nucellar cells is the stimulation for embryo sac development. Further, the SERK pathway and the auxin/hormonal pathway controlled by APOSTART may interact with each other. The gene APOSTART has some control over meiosis and programmed cell death.

Apomixis is seen as the result of changes in control of sexual pathway. Here, the omission/changes of key steps and timing of gene expression are the key factors for induction of apomixis. Since most apomicts are polyploid, apomixis could arise from heterochronic expression (changed expression of same gene over different time). The efficiency of apomictic seed set in facultative apomicts (where sexual and apomictic reproduction occur together) is believed to be dependent on how far the dominance and penetrance of apomictic pathway prevail over sexual pathway.

4.3.4 Apomixis in Agriculture

Apomixis ensures genetically uniform populations and carries forward hybrid vigour in successive generations. The following are the advantages of apomixis:

(a) Rapid generation and multiplication of superior genotypes from novel germplasm. This is evident in species multiplied by asexual means. Also in those species which are multiplied through grafting, the apomictic seeds can have true-to-type plants generation after generation.
(b) The reduction time taken for breeding and cost.
(c) The avoidance of complications like cross-incompatibility.

Farmers in the developed world are benefited with new, advanced and high-yielding varieties in mechanized agricultural systems. However, in the developing world, the benefits farmers foresee are the release of high-yielding varieties for specific environments. But, apomixis is poorly understood in crop species. Apomixis is prominent only in tropical and subtropical fruits like mango, mangosteen and citrus and tropical forage grasses such as *Panicum*, *Brachiaria*, *Dichanthium* and *Pennisetum.* The exercise of transferring apomixis into maize from its wild relative *Tripsacum dactyloides* has been actively pursued but not met with success. Once practically utilized, the uses of apomixis in agriculture are immense. Very recently, a process of asexual reproduction has been standardized in rice with the aid of BABY BOOM gene to induce parthenogenesis (see Box 4.2).

Box 4.2: Asexual Reproduction in Rice

The molecular pathways that prevent occurrence embryo without fertilization are not well understood. In rice, a gene called BABY BOOM1 (BBM1), a member of the AP2 family2 of transcription factors that is expressed in sperm cells, is sufficient for parthenogenesis. *BBM1* can bypass the fertilization in the female gamete. Zygotic expression of *BBM1* is initially specific to the male allele but is subsequently biparental, and this is consistent with its observed auto-activation. The knock out (triple knockout) of *BBM1*, *BBM2* and *BBM3* causes embryo arrest and abortion. Upon pollination by male-transmitted *BBM1*, the embryo formation is restored. Scientists at the University of California, Davis, USA, and other institutes at Davis have demonstrated this.

If genome editing to substitute mitosis for meiosis (*MiMe*) is combined with the expression of *BBM1* in the egg cell, clonal progeny can be obtained that retain genome-wide parental heterozygosity. The synthetic asexual propagation trait is heritable through multiple generations of clones. Hybrid crops provide increased yields that cannot be maintained by their progeny owing to genetic segregation. This work establishes the feasibility of asexual reproduction in crops and could enable the maintenance of hybrids clonally through seed propagation.

Further Reading

Holsinger KE (2017) Reproductive systems and evolution in vascular plants. Proc Natl Acad Sci USA 97:7037–7042

Said H, Jan F, David (2016) Male gametophyte development and function in angiosperms: a general concept. Plant Reproduct 29:31–51

Tucker MR, Koltunow AMG (2009) Sexual and asexual (apomictic) seed development in flowering plants: molecular, morphological and evolutionary relationships. Funct Plant Biol 36:490–504

Smet et al (2010) Embryogenesis – the humble beginnings of plant life. Plant J 61:959–970

Koltunow A, Grossniklaus U (2003) APOMIXIS: a developmental perspective. Annu Rev Plant Biol 54:547–574

Hafidh S (2016) Male gametophyte development and function in angiosperms: a general concept. Plant Reprod 29:31–51. https://doi.org/10.1007/s00497-015-0272-4

Khanday I et al (2018) A male-expressed rice embryogenic trigger redirected for asexual propagation through seeds. Nature. https://doi.org/10.1038/s41586-018-0785-8

Self-Incompatibility

5

Keywords
Homomorphic and heteromorphic incompatibility · Gametophytic and sporophytic incompatibility · Mechanism of self-incompatibility · Pollen-stigma-style-ovary interactions · Significance of self-incompatibility · Methods to overcome self-incompatibility

A generalized definition of self-incompatibility by de Nettancourt is "the inability of a fertile hermaphrodite seed plant to produce zygotes after self-pollination".

In a bisexual flower, male and female reproductive organs are in close proximity, and plants have evolved various genetic mechanisms to avoid self-fertilization. Incompatibility is a mechanism that enforces outbreeding in plants. The morphological structure of a flower ensures such outbreeding following two main types: *heteromorphic and homomorphic.* In heteromorphy, flowers may be either *distylic* or *tristylic*. Flowers are *distylic* when two types of flowers, namely, *thrum* with short style and high anthers and *pin* with long style and low anthers, occur. In *tristylic* condition, flowers with long, mid and short styles can occur separately (Fig. 5.1). Distyly is controlled by a single gene with two haplotypes (haplotype is a set of alleles in a single chromosome) S and s. Flowers with short styles (thrums) are generally Ss, whereas flowers with long styles (pins) are ss. Tristyly is generally controlled by two genes, each of which has two haplotypes (S,s and M,m). S is responsible for short style, S and M to medium style and s and m to long style. A 1:1 ratio exists between individuals of each SI type (Table 5.1).

Homomorphic SI can be of two types: gametophytic and sporophytic. In gametophytic self-incompatibility, pistil distinguishes between selfed pollen and non-selfed pollen. Gametophytic SI is of two types: one involving S-RNase system (S-RNase GSI system) and the other without S-RNase. S-RNase system is found in members of the Solanaceae, Rosaceae and Scrophulariaceae. Non-S-RNase is seen in Papaveraceae. Selfed pollen is rejected, and non-selfed pollen is accepted.

P. M. Priyadarshan, *PLANT BREEDING: Classical to Modern*,
https://doi.org/10.1007/978-981-13-7095-3_5

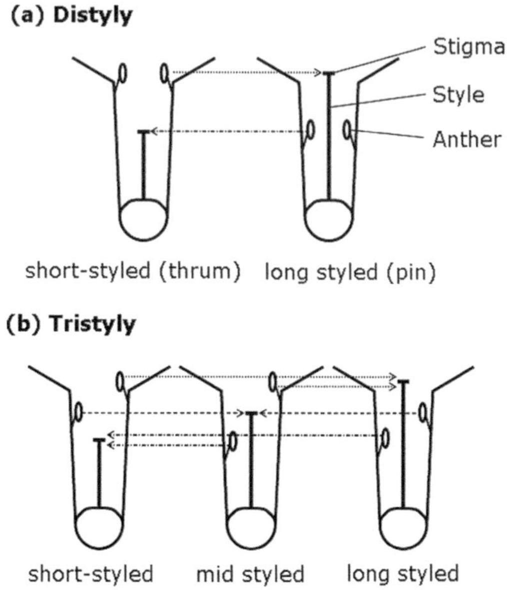

Fig. 5.1 Diagrammatic representation of flowers with pin and thrum type having distyly and tristyly

Table 5.1 Summary of SI mechanisms

Plant family	Type of SI	Genetic locus	Female determinant	Male determinant	Mechanism
Solanaceae, Rosaceae, Scrophulariaceae	GSI	S-locus	*S-RNase*	*SLE/SFB?*	S-RNase-mediated degradation of pollen tube RNA
Papaveraceae	GSI	S-locus	*S-gene*	Unknown	S-protein-mediated signalling cascade in pollen
Brassicaceae	SSI	S-locus	S-locus receptor kinase *SRK*	S-locus cysteine-rich protein *SCR/* S-locus protein-11*SP*11	Receptor kinase-mediated signalling in stigma

Solanaceae family is a model system for molecular and biochemical studies. This is under the control of a single polymorphic locus – the S-locus. S-proteins control the ability of the pistil to reject selfed pollen. The biochemical mechanism of self-rejection is through the action of RNase.

The genetic constitution of gametes controls gametophytic SI. Pollen grains with similar allele of that of stigma will not germinate (Fig. 5.2). Examples are potatoes, wild tomatoes, tobacco, roses, bajra, rye and sugar beet. The diploid genotype of the sporophyte (pollen-producing plant) controls the sporophyte SI. Here, germination or

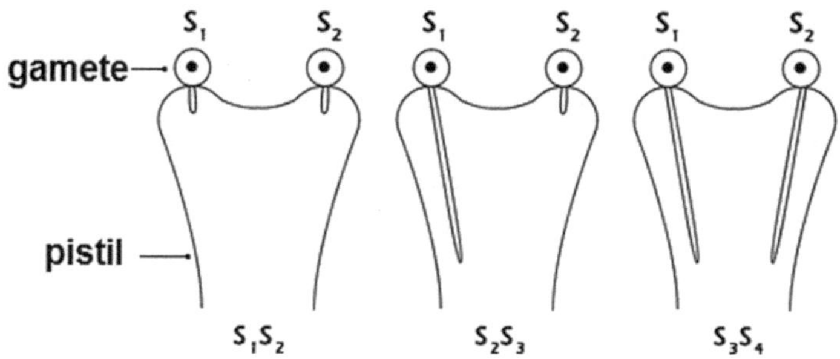

Fig. 5.2 Diagrammatic representation of gametophytic self-incompatibility

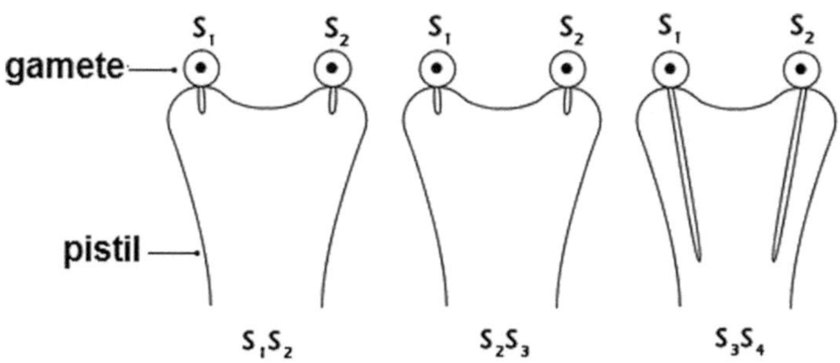

Fig. 5.3 Diagrammatic representation of sporophytic self-incompatibility

pollen tube growth inhibited on the stigma of the same flower. When the pollen contains either of the two alleles that are present in the sporophyte, pollen will not germinate. Pollen grains (S_1 or S_2) produced by S_1S_2 plant will germinate only on S_3S_4 plant not on S_1S_2 or S_1S_3 (Fig. 5.3). Sporophytic SI follows the order of dominance as $S_1 > S_2 > S_3 > S_4$. Examples are Brassicaceae, Caryophyllaceae, Asteraceae, Sterculiaceae and Convolvulaceae. To simplify, S_1S_2 X S_3S_4 is fully compatible; S_1S_2 X S_1S_3 is partially compatible; and S_1S_2 x S_1S_2 is fully incompatible.

5.1 Mechanism of Self-Incompatibility

Of the 383 families of angiosperms, SI has been described in 81 families. Among them, 15 families have been well described as having gametophytic SI, and sporophytic SI has been described in 6 families. 39 families have SI but of an undefined type, and 21 may have SI although it has not been confirmed yet.

Gametophytic SI Inhibition of incompatible pollen is slow and takes hours in S-RNase GSI system. Pollen tubes are arrested at stylar extracellular matrix (ECM). In *Nicotiana alata*, stylar proteins showed an abundant S-glycoprotein (of 30 kDa size). This protein is having genetic linkage with the S-locus. S-locus glycoproteins (SLGs) are ribonucleases (S-RNases), and these are responsible for the rejection of incompatible pollen. S-RNase available in ECM enters the pollen tube cytoplasm, degrading ribonucleic acid (RNA). This will interfere with the growth of incompatible pollen tubes. An F-box gene (SLF, S-Locus F-box, or SFB, S-locus F-Box gene) is responsible for this process. The SLF/SFB gene system led to a new model for the mechanism of S-RNase-based GSI (Fig. 5.4a). S-RNase is taken into the pollen tube cytoplasm and it interacts with SLF/SFB. In a compatible interaction,

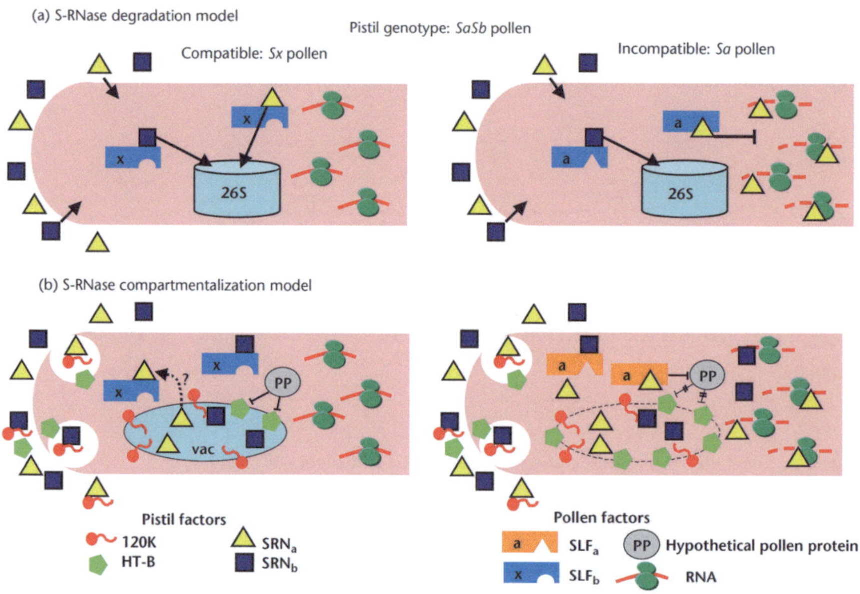

Fig. 5.4 Proposed mechanisms for the self-incompatibility reaction in the S-RNase system. The products of the female S-gene, the S-RNases, which are secreted into style are encountered by pollen. If the pollen carries an S haplotype corresponding to either of the haplotypes present in the style, then inhibition occurs. Two models have been proposed for the inhibition mechanism. Compatible (Sx-, left) and incompatible (Sa-, right) pollinations are shown on an SaSb pistil. Symbols for pistil factors (S-RNase, HT-B (HT-B=high top band proteins) and 120 K) and pollen factor (SLF=*S*-locus F-box proteins) are shown below the figure. (**a**) S-RNase degradation model: S-RNase enters the pollen tube cytoplasm from the extra cellular matrix (ECM) (arrows). A compatible non-self-S-RNase/SLF interaction (left) results in ubiquitylation (post-translational modification process by which ubiquitin is attached via an isopeptide bond to lysine residues on a protein) and degradation of S-RNases by the 26S proteasome, so there is no cytotoxic action and pollen tube growth continues. An incompatible self-S-RNase/SLF interaction (right) does not result in S-RNase degradation; cytotoxicity results in RNA degradation and hence incompatible pollen tube growth is inhibited. (**b**) S-RNase compartmentalization model: S-RNase, 120 K and HT-B are taken up by endocytosis and sorted to a vacuole. In a compatible interaction (left), S-RNase remains compartmentalized, hence, although S-RNase is present, it is not cytotoxic because it is sequestered. Degradation of HT-B in compatible pollen tubes is mediated by a hypothetical pollen protein (PP). How S-RNase gains access to SLF (arrow, question mark) is not known. In an incompatible interaction (right), HT-B is not degraded and the vacuolar compartment containing S-RNases degrades. S-RNase is released into the cytoplasm and RNA is degraded by its cytotoxic action, and pollen tube growth is inhibited. (Courtesy: Springer Science and Business Media)

S-RNase is degraded by the 26S proteasome. Hence, the pollen is "rescued" from cytotoxic S-RNases.

In addition to S-RNases, other pistil components like "HT-B" and "120 K" are also prevalent. These are independent of S-RNase. HT-B is yet another pistil protein taken to pollen tubes. In compatible pollen, massive HT-B degradation occurs that retains an intact vacuole to keep S-RNases compartmentalized and ineffective. This has led to a new model on S-RNase action (Fig. 5.4b).

S-RNase is not always responsible for pollen inhibition in GSI system (e.g. *Papaver rhoeas*). Here, the initial arrest of pollen growth is rapid and it occurs in stigmatic surface. The stigmatic S-proteins are small (~15 kDa). S-protein interacts with pollen S-gene product which is believed to be a plasma membrane receptor. Inhibition is mediated by a Ca^{2+}-dependent signal transduction pathway (see Box 5.1). This pathway is activated by the haplotype-specific interaction of the stigma and pollen S-proteins. Continued pollen tube growth requires pollen-tip-focused Ca^{2+} gradient. This gradient will get reduced by a rapid increase in cytosolic free Ca^{2+}. Such complex events lead to inhibition of the incompatible pollen. Protein phosphorylation transduced by Ca^{2+} signals. A mitogen-activated protein kinase (MAPK) p56 is activated in incompatible pollen during the SI reaction. This p56 is a transducer of SI response. Yet another small cytosolic protein, Pr-p26.1, is also phosphorylated. Both calcium and phosphorylation reduce its activity that becomes a potential mechanism to inhibit pollen tube growth.

Box 5.1: Cell-Cell Signalling and Self-Incompatibility

In plants, the pollen-pistil interactions that precede fertilization give significant insights into the molecular and genetic basis of cell-cell signalling. There are two related polymorphic proteins (SLG and SRK) expressed specifically in the stigmatic papillar cells. SLG, a *S*-locus glycoprotein, is a soluble cell wall-localized protein and SRK is a *S*-locus receptor kinase (plasma membrane-anchored signalling receptor). SRK shares sequence similarity with SLG. The future research on this signalling system will focus on characterizing the molecular interactions between the stigma and pollen determinants of SI. The production of SRK and its interactions with SCR (*S*-locus cysteine-rich protein) will be the new domain of research.

Every aforesaid system follows a mechanism known as *signal transduction*. A series of molecular events enable chemical or physical signal to be transmitted through a cell. This is done by protein phosphorylation catalysed by protein kinases. The stimuli are detected by proteins known as receptors or sensors. Once the receptor senses the signal, it leads to a signalling cascade. This is a chain of biochemical events. There will be changes in transcription and translation that happen at molecular level. These changed molecular events control cell growth, proliferation and metabolism.

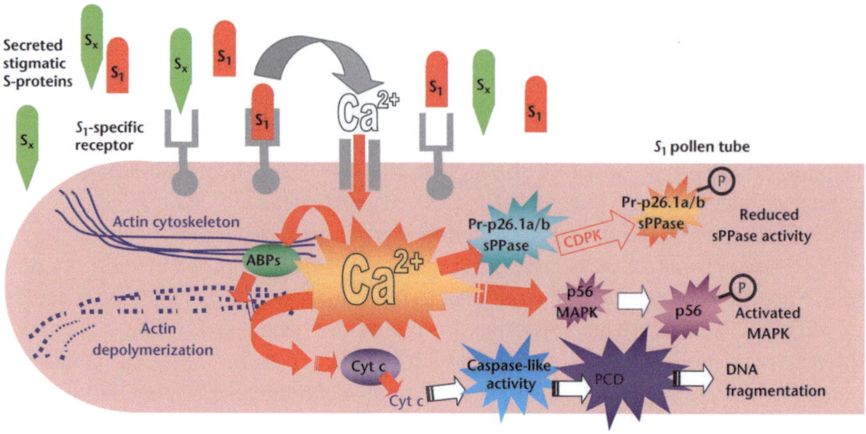

Fig. 5.5 A proposed model for the self-incompatibility mechanism in *Papaver rhoeas*. Incompatible pollen undergoes an S haplotype-specific interaction. Secreted stigmatic S-proteins interact with the pollen S-receptor. An haplotype-specific interaction such as binding S1 protein to S1 pollen results in triggering an intracellular Ca^{2+} signalling cascade(s), involving large-scale Ca^{2+} influx and increases in $[Ca^{2+}]$i. A series of events then occur in the incompatible pollen. Within 1 min, there is a dissipation of the tip-focused calcium gradient that is required for continued pollen growth and the activation of calcium-dependent protein kinase (CDPK). The CDPK phosphorylates Pr-p26.1, a soluble inorganic pyrophosphatase (sPPase). Both calcium and phosphorylation inhibit sPPase activity, resulting in a reduction in the biosynthetic capability of the pollen, thereby inhibiting growth. Dramatic changes to pollen cytoskeleton organization are apparent within 1 min, with extensive depolymerization of the F-actin causing rapid arrest of pollen tube tip growth. p56-Mitogen-Activated Protein Kinase (MAPK) is activated and may signal to programmed cell death (PCD). PCD is triggered, involving key features of PCD including caspase-like activity, cytochrome c leakage and DNA fragmentation. This ensures that incompatible pollen does not start to grow again. ABP=actin binding protein. (Courtesy: Springer Science and Business Media)

In *Papaver* pollen, programmed cell death (PCD) is triggered by SI. A mechanism to kill selfed pollen is through cell death mechanisms like apoptosis/PCD. An increment in Ca^{2+} will mediate PCD that ensure death of incompatible pollen. Hence, in *Papaver* SI, there is complex network of events, leading to PCD (Fig. 5.5).

Sporophytic SI SSI exhibits a dominance relationship unlike GSI. Here, class I haplotypes are strong SI phenotypes that are dominant or co-dominant. The class II haplotypes are recessive and are weaker. Among the pistil proteins is an S-locus glycoprotein (SLG) of 60 kDa. Another homologous to SLG, 120 kDa S-receptor kinase (SRK), is also identified. These proteins are encoded at the S-locus. SRK is serine/threonine kinase and belongs to a large family of plant receptor-like kinases. SCR (S-locus cysteine-rich) (also known as SP11 – S-locus protein 11) is yet another gene involved. Interaction of SCR and SRK triggers a signal transduction cascade that triggers rapid inhibition of pollen tube growth on stigma (Fig. 5.6).

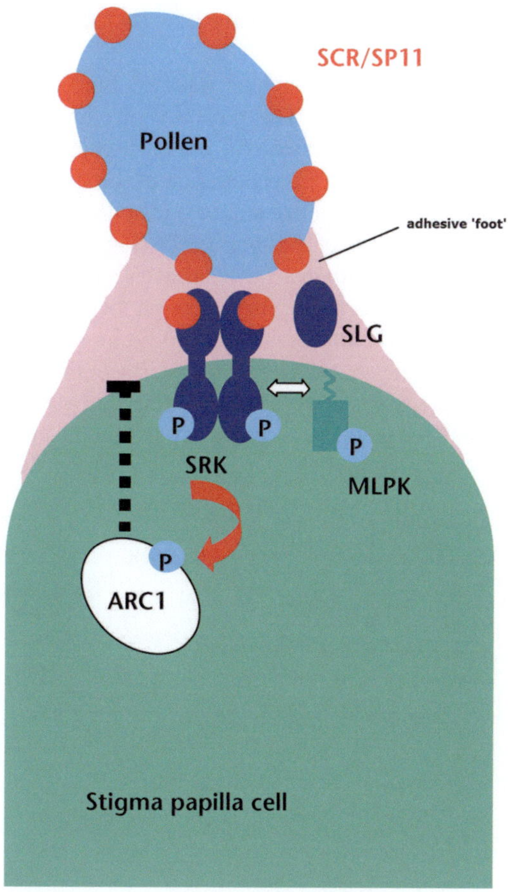

Fig. 5.6 A proposed model for the *Brassica* self-incompatibility reaction. In *Brassica*, the SI response occurs within the stigma. When a pollen grain alights on the papilla surface, the pollen coat flows to form an adhesive "foot", thus making a connection with the surface of the stigmatic papilla. The pollen S-locus cysteine-rich/S-locus (SCR/SP11) protein is carried within this coating, and when this is allelic with the recipient stigma, an incompatible reaction is induced. SCR binds to the extracellular domain of the S-receptor kinase (SRK), which results in the activation of the kinase. The role of the S-locus glycoprotein (SLG) in this recognition event is unclear, as evidence suggests it is not essential for the SI reaction. However, in some S haplotypes, it does appear to enhance the SI response. MLPK (M locus protein kinase), a membrane-localized protein, is a positive effector of SI and may form a complex with SRK. Following activation, SRK interacts with ARC1 in a phosphorylation-dependent manner. This ultimately leads to pollen rejection by an unknown mechanism. ARC1= Armadillo repeat containing 1 protein. ARC1 is a downstream component of SRK, which is located in the cytoplasm, and is phosphorylated by SRK. (Courtesy: John Wiley & Sons)

5.1.1 The Pollen-Stigma-Style-Ovule Interactions

Pollen is the dehydrated male gametophyte released from the anther. It contains 15–35% of water by fresh weight. The pollen-stigma interaction comprises six stages: (a) pollen capture and adhesion, (b) pollen hydration, (c) germination of the pollen to produce a pollen tube, (d) penetration of the stigma by the pollen tube, (e) growth of the pollen tube through the stigma and style and (f) entry of the pollen tube into the ovule and discharge of the sperm cells (Fig. 5.7). Angiosperm stigmas are either wet or dry where wet stigmas have surface secretion. Hydration of pollen appears to be unregulated in all wet stigmas. Though there are variations in pollen-stigma communication, three broad areas seem to be in consensus in most model systems: (a) presence of lipids at pollen-stigma interface; (b) initial directional cue for pollen tube growth is water; and (c) small cysteine-rich proteins, especially lipid transfer proteins (LTPs), are involved. A gradient of water potential is established by

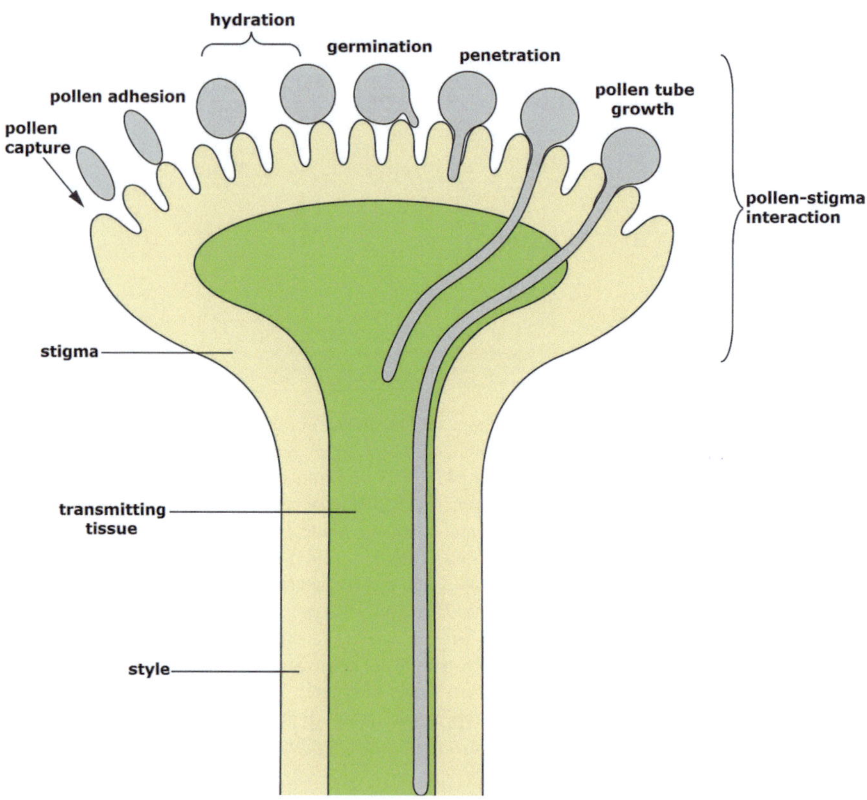

Fig. 5.7 Different stages of the pollen-stigma interaction. The diagram represents a typical stigma of the dry papillate type found in species from the Brassicaceae. Pollen is shown at various stages of development on the stigma and growing into the transmitting tissue of the style

the lipids between pollen and the turgid cells of the stigma, and this makes the pollen tubes to sense and grow.

In both wet and dry stigmas, a range of small cysteine-rich proteins are involved in governing the pollen-stigma interactions. Major players are LeSTIG1 and LAT52 and their receptor kinase partner LePRK2. Stigma/style cysteine-rich adhesion protein (SCA) is also involved in pollen tube adhesion. Lipid transfer proteins (LTPs) and LTP-like cDNAs are identified through transcriptome analysis in pollen coat and stigma. A plantacyanin similar to chemocyanin has been identified in conjunction with SCA which is said to be involved in pollen tube growth.

The pollen tube from the hydrated pollen germinates and grows to penetrating the stigmatic cuticle, inner and outer layers of the cell wall. This is made possible through enzyme modification of these layers. The stigmatic cell wall at the pollen contact point is expanded due the enzymes like polygalacturonases and pectin esterases. The enzymes secreted by the stigmatic papilla and ER and Golgi are responsible for the initial expansion of stigmatic cell wall. Exo70A1, a component of exocyst complex, is also a vital player for pollen tube penetration. The pollen tube grows through the cell wall layers of the stigmatic papillae through producing its own cell wall-modifying enzymes.

Further, the interaction of pollen with ovule is a bit complicated with the involvement of several genes and biochemicals. So, the process is simplified as under:

Pollen tubes grow down to the style and reach the septum (a central tissue that runs to the base of the ovary) and then the funiculus, and finally through micropylar opening, it reaches the ovule to release the sperm cells. One of the first molecules proposed to guide pollen tubes was γ-aminobutyric acid (GABA). In wild-type pistils, GABA is seen in the inner integument of the ovule at a higher concentration that follows a gradient. Pollen tube growth is guided by this gradient.

The female gametophyte with guidance made available from funicular and micropylar systems produces pollen tube guidance cues. The expression of novel *Gamete-Expressed (GEX)3* gene in the egg cell is a vital factor. Reduced *GEX3* expression will hamper locating micropyle by the pollen. ANXUR1 (ANX1) and ANXUR2 (ANX2) are genes expressed at highest levels in the pollen. In a double-recessive (anx1/anx2) mutant, pollen tubes rupture prematurely. ANX1 and ANX2 in conjunction with the FER/SRN receptor kinase signalling in the synergid cells are responsible for coordinating the pollen tube rupture and release of the sperm cells (Fig. 5.8). MYB98 is yet another transcriptional regulator required for pollen tube guidance and the formation of the synergid cell filiform apparatus. Central Cell Guidance (CCG), another transcriptional regulator in the central cell of the ovule, regulates pollen tube growth to the micropyle (Fig. 5.8). The LORELEI (LRE) gene is also expressed in the synergid cells. The recessive lre female gametophyte mutant displays impaired sperm cell release, similar to the fer/srn mutant. RNA processing and metabolism is governed by MAA3 gene. The gradient of pollen-pistil protein (POP-GABA) which starts from the stigma increases its concentration to the inner integument of the ovule guiding pollen tube growth. The pollen tube enters the micropyle and penetrates a synergid cell and then releases the two sperm cells for fertilization. FER/SRN receptor kinase in the synergids controls this process. (FER/SRN=FERONIA/SIRÈNE receptor kinase)

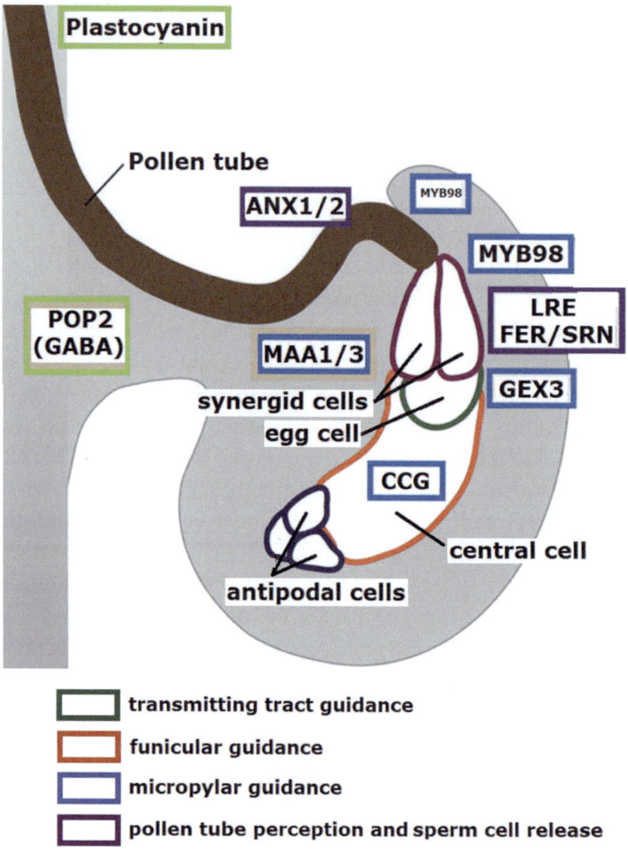

Fig. 5.8 Model of pollen tube guidance to the female gametophyte in *Arabidopsis thaliana*. An illustration of a pollen tube growing to an ovule is shown, with the guidance cues and genes that are proposed to regulate pollen tube guidance and perception overlaid on this diagram. If expression patterns are known, gene names are coloured to match the cells where they are expressed. Coloured boxes indicate steps that are disrupted in mutants (see text for details)

While in GSI, the haploid genome determines the *S* phenotype of the pollen, in SSI the diploid phenotype of the parent determines *S* phenotype. In GSI, incompatible pollen tubes happen within the style. In SSI, inhibition occurs due to pollen-stigma interaction. This happens before pollen tube penetrates the stigma.

5.1.2 Significance of Self-Incompatibility

SI promotes allogamy and prevents autogamy. This is largely used for hybrid seed production in *Brassica* and sunflower. Two self-incompatible lines are planted in alternate rows for hybrid seed production. Also, a self-incompatible line may be planted in inter-row with a self-compatible line. In this scheme, hybrid seeds are harvested from self-incompatible line. In *Brassica*, production of double-cross and triple-cross hybrids has been demonstrated by using self-incompatible lines.

5.1.3 Methods to Overcome Self-Incompatibility

There are 13 different ways by which incompatibility can be overcome. They are (1) bud pollination, (2) mixed pollination, (3) deferred pollination, (4) test tube pollination, (5) stub pollination, (6) intra-ovarian pollination, (7) in vitro pollination, (8) use of mentor pollen, (9) elevated temperature treatment, (10) irradiation, (11) surgical method, (12) application of chemicals and (13) protoplast fusion. These methods are briefly dealt here:

Bud pollination is the most successful method in both gametophytic and sporophytic SI. The best stage to overcome self-incompatibility is 2–7 days before anthesis. In bud stage, the stigma lacks exudates, and if the stigma is self-pollinated at bud stage, when the factor responsible for the exudates has not appeared, the pollen tubes will grow normally and effect fertilization.

In mixed pollination, the stigma is camouflaged with a mixture of chemically treated or irradiated compatible pollen with incompatible pollen. Proteins secreted from the compatible pollen neutralize the inhibition reaction over the stigma.

Deferred pollination is achieved by deferring the pollination for a few days. In *Brassica* and *Lilium*, delayed pollination has been successful in overcoming self-incompatibility.

In test tube pollination, the bare ovules are directly dusted with pollen after removing stigmatic, stylar and ovary wall tissues. Successfully pollinated ovules are cultured in a nutrient medium that supports germination as well as development of fertilized ovules into seeds. This is successfully done in *Papaver somniferum.*

In stub pollination, stigma and part of the style are removed. When stigmatic surface is the primary site of incompatibility, if the stigmatic lobe is removed and the cut surface is pollinated, then the pollen tube grows uninhibited into the ovule (e.g. *Ipomoea trichocarpa*). Similarly, following the removal of a large part of the style from *N. tabacum* and smearing the cut surface with agar-sucrose medium to function as a substrate followed by pollination with the pollen of *N. rustica*, it was observed that in majority of the cases, fertilization was successful.

Intra-ovarian pollination is done by surface sterilizing the ovary followed by injecting the aqueous pollen suspension (with or without specific substance for germination) by a hypodermic syringe followed by sealing the holes with petroleum jelly. The introduced pollen grains germinate and achieve fertilization. The method has also been successful in other members of Papaveraceae, like *Papaver rhoeas* and *P. somniferum.*

In vitro pollination is achieved by removing the stigmatic, stylar and ovary wall tissues and directly dusted with pollen grains and then cultured in a suitable nutrient medium that supported both the germination of pollen and the development of fertilized ovules. A better result is obtained by culturing the ovules within the intact placental tissue, as such the technique is also termed as placental pollination (e.g. *Papaver somniferum*) (see Box 5.2 for in vitro fertilization).

Box 5.2: In Vitro Fertilization in Maize Is Done as Follows

(i) Ears are bagged before emergence of silk to prevent pollination. Ears are collected at a receptive stage when emerged silks reach 12–13 cm in length.

(ii) Egg cells are isolated from ovules dissected from mature ears and are incubated in an enzymatic solution containing 0.5% macerozyme and 0.5% cellulase at pH 5.7. Egg cells are gently picked out from the embryo sac by manual microdissection using an inverted microscope.

(iii) Sperm cells are released from freshly collected pollen grains after an osmotic shock in 12% mannitol.

(iv) The fusion of egg and sperm cell performed in a 3.5-cm-diameter plastic petri dish filled with 2 ml bovine serum albumin (BSA) fusion medium. The fusion process is observed under an inverted microscope. The dish is inserted at the middle of a 3 cm petri dish with 1.5 ml nutrient medium that contains feeder cells obtained from embryogenic suspension cultures of another maize inbred line. The cultures are then incubated under 16 h photoperiod.

(v) Fertilized egg cells are cultured in droplets of the modified basic MS medium. The fertilized egg shows karyogamy within 1 h of fusion and 90% of the fusion products produce mini-colonies. In most cases, a mini-colony grows into an embryo and ultimately into a fertile plant (see Fig. 5.9).

The compatible pollen made ineffective by irradiation or repeated freezing and thawing or treating with chemicals, like ethanol, for fertilization is known as mentor pollen. This has been used successfully to overcome incompatibility by using them along with live incompatible pollen. In *Cosmos*, mentor pollen and their diffusates were effective in overcoming self-incompatibility. It has been successfully used in *Brassica oleracea*, *Petunia*, *Nicotiana*, *Lilium* and pear. The function of mentor pollen is to provide recognition substances to incompatible pollen or to provide pollen growth substance.

High-temperature treatment is done by subjecting style with hot water treatment. Style is kept at 50 °C for 6 min before pollination to overcome self-incompatibility. In species like *Secale cereale*, 30 °C treatment is sufficient. Genetic studies indicate that sensitivity to temperature is due to a dominant gene marked as T-gene. Further, the stress generated by the daily variation in temperature has a positive effect in the strength of self-incompatibility.

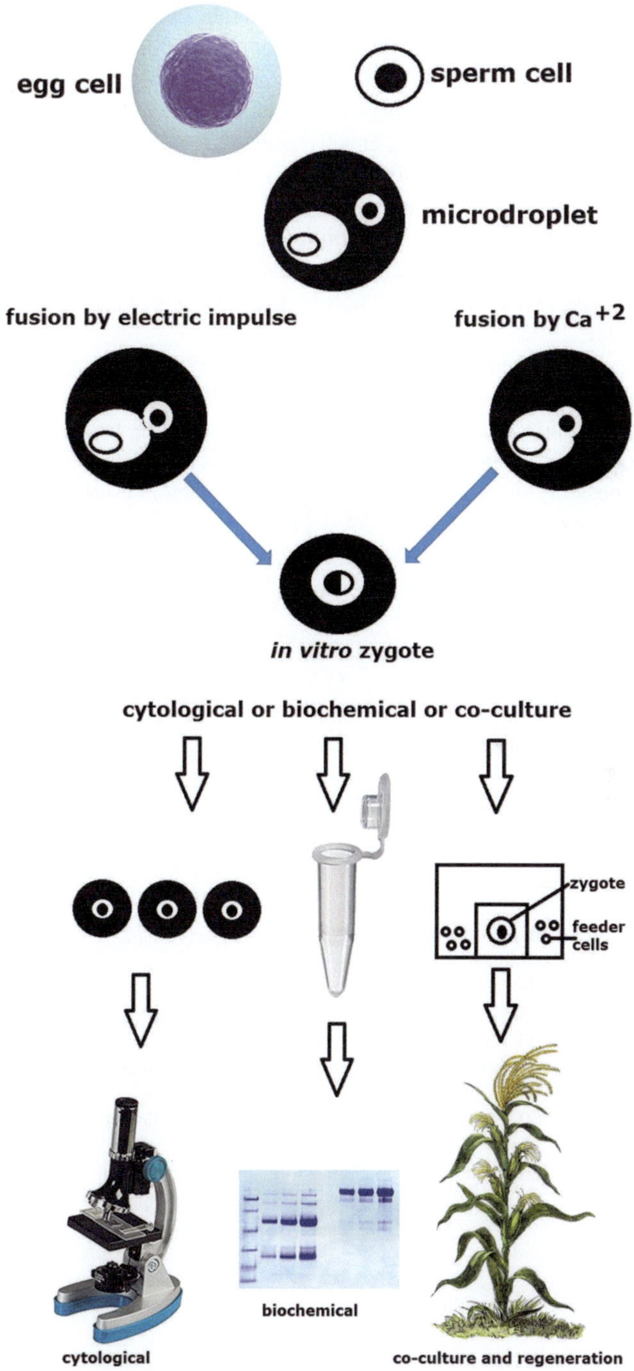

Fig. 5.9 Summary of in vitro fertilization in maize. Isolated egg and sperm cells are placed in microdroplet and covered with thin layer of mineral oil. The gametes are fused electrically (left) or chemically (right). The fusion product is characterized cytologically and biochemically or co-cultured with feeder cells to induce division and plant regeneration

X-ray irradiation of flower buds at pollen mother cell stage helps to overcome self- incompatibility. Irradiation damages the physiological mechanism of self-incompatibility in the style, thus allowing the pollen tube to pass through the style. Studies on S-locus in *Oenothera organensis* and *Prunus avium* have demonstrated that irradiation induces temporary inactivation of the S-allele, thus enabling the pollen tube to pass through the style. The offsprings have incompatibility. Permanent mutation leads to mutated allele (S_A) that can induce growth on all styles, but S_A-style will prevent the growth of a non-mutated S_A allele pollen.

Decapitation of the stigma before pollination or deposition of pollen grains directly into the stylar tissue through a slit has helped in overcoming self-incompatibility.

Chemicals like olivomycin and cycloheximide, the inhibitors of RNA and protein synthesis, could overcome self-incompatibility in *Petunia hybrida*, when injected into the flower buds just 2–3 days before anthesis. The treatment of *Brassica oleracea* stigma before pollination with hexane was found to be effective in fruit set. Hexane possibly inactivates the incompatibility factors on the stigma. Application of p-chloromercuribenzonate, GA_3, indole butyric acid and NAA has been effective in *Petunia*, *Tagetes*, *Trifolium*, *Brassica*, *Lilium* and *Lycopersicon*. Benzylaminopurine is most effective in inducing selfed seed set in the self-incompatible *Lilium*.

Fusion of isolated protoplasts has achieved great success in overcoming incompatibility. Since it involves the fusion of somatic protoplast, the method is described as parasexual hybridization. The technique involves isolation of protoplasts, fusion of the isolated protoplasts and culture of hybrid protoplast to regenerate whole plants.

Further Reading

Ambrosino L (2016) Bioinformatics resources for pollen. Plant Reprod 29:133–147. https://doi.org/10.1007/s00497-016-0284-8

Charlesworth D (2010) Self-incompatibility. Biol Rep 2:68

Erbar C (2003) C pollen tube transmitting tissue: place of competition of male gametophytes. Int J Plant Sci 164(Suppl 5):S265–S277

Lewis D (1949) Incompatibility on flowering plants. Biol Rev. https://doi.org/10.1111/j.1469-185X.1949.tb00584.x

Silva NF, Goring DR (2007) Mechanisms of self-incompatibility in flowering plants. Cell Mol Life Sci 58:1988–2007

Takayama S, Isogai A (2005) Self-incompatibility in plants. Ann Rv Plant Biol 56:467–489

Tovar-Mendez A, McClure B (2016) Plant reproduction: self-incompatibility to go. Curr Biol 26: R102–R124

6 Male Sterility

Keywords

Male sterility · Genetic male sterility · Cytoplasmic male sterility · Genes for CMS and restoration of fertility (cytoplasmic-genetic male sterility) · Mechanisms of restoration · Engineering male sterility · Dominant nuclear male sterility (pollen abortion or barnase/barstar system) · Male sterility through hormonal engineering · Pollen self-destructive engineered male sterility · Male sterility using pathogenesis-related protein genes · RNAi and male sterility · Mitochondrial rearrangements for CMS · mtDNA recombination and cyto-nuclear interaction · Regulation of CMS transcripts via RNA editing · Accumulation of toxic protein products · Chloroplast genome engineering for CMS · Male sterility in plant breeding · Male sterility and hybrid seed production

Flowers are organized into four concentric whorls of organs, namely, sepals, petals, stamens and carpels. Stamens are the sporophytic organ system with male sporogenous (diploid) cells which undergo meiosis and produce haploid male spores or microspores or pollen grains. Stamen consists of anther and the filament (Fig. 6.1), and the filament is a vascular tissue that supplies water and nutrients to the anther. The production of pollen grains involves an array of extraordinary events that are independent of a conventional meristem, with a transition from sporophytic to gametophytic generation (Fig. 6.2). In addition, production of coenocytic tissues (the tapetum and the microsporocyte mass) is part of pollen development. Subsequently, pollen grains that are self-contained units for genome dispersal are made.

There are two phases of anther development. In phase 1, establishment of anther morphology takes place, differentiation of cell and tissue occur, and pollen mother cells undergo meiosis. At the end of this phase, tetrads are available within the pollen sacs. In phase 2, pollen grains get differentiated, and the anther and pollen grain will get released. The cellular mechanisms that regulate anther cell differentiation that

P. M. Priyadarshan, *PLANT BREEDING: Classical to Modern*,
https://doi.org/10.1007/978-981-13-7095-3_6

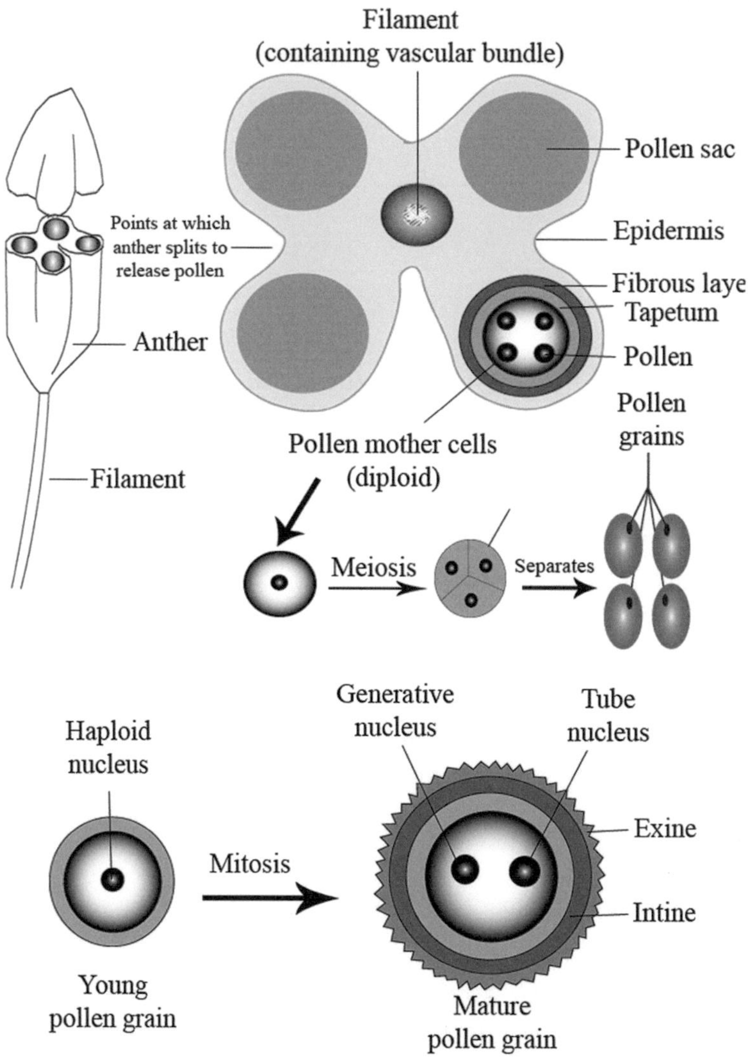

Fig. 6.1 Pollen formation: development of a pollen within pollen sac of anther. Each pollen sac is filled with cells containing large nuclei. These cells go through two meiotic divisions forming a tetrad. These are called microspores. Each microspore becomes pollen grain. Each pollen sac is enclosed by a protective epidermis and fibrous layer. Inside the fibrous layer is the tapetum. The tapetum stores food that provides energy for future cell divisions

makes the anther to switch from phase 1 to dehiscence programme of the anther (phase 2) are not well known (Fig. 6.3).

Sterility is a complex hereditary phenomenon that prevents self-pollination either through lack of pollen grain production or through production of sterile pollen grains. Anther is composed of several tissues, *viz.*, tapetum, endothecium,

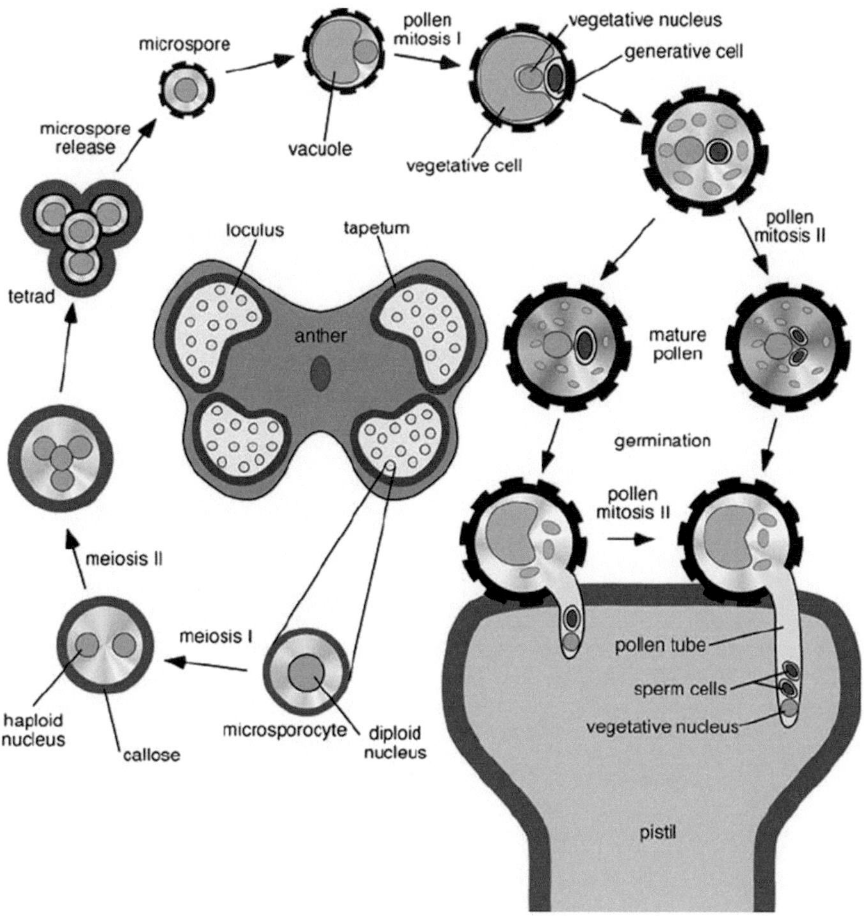

Fig. 6.2 Morphological stages of microsporogenesis and microgametogenesis. During microsporogenesis, microsporocytes undergo two nuclear divisions at meiosis followed by cytokinesis to produce a tetrad of four haploid microspores. During microgametogenesis, microspores undergo two stereotypical mitotic divisions, pollen mitosis I and pollen mitosis II, to produce bicellular (70% of species) or tricellular pollen grains (e.g. *Arabidopsis*). In species with bicellular pollen grains, pollen mitosis II occurs in the growing pollen tube within the pistil

connective tissues, vascular tissues and cell types. Tapetum is a specialized anther tissue that plays a vital role in pollen production. Tapetum gets degenerated towards maturity of anther. Tapetum is responsible for the production of proteins that aid in pollen development. Many male sterility mutations occur in tapetum. Hence, tapetal tissue is essential for the production of functional pollen grains (Fig. 6.4) A diagrammatic representation of the ultrastructure of pollen is available in Fig. 6.5.

Pollen tube contains several zones. The tip-most zone is clear zone since the organelles present there have quite low refractivity. Amyloplasts with starch shall be missing from this clear zone. This clear zone comprises two distinct regions, apical

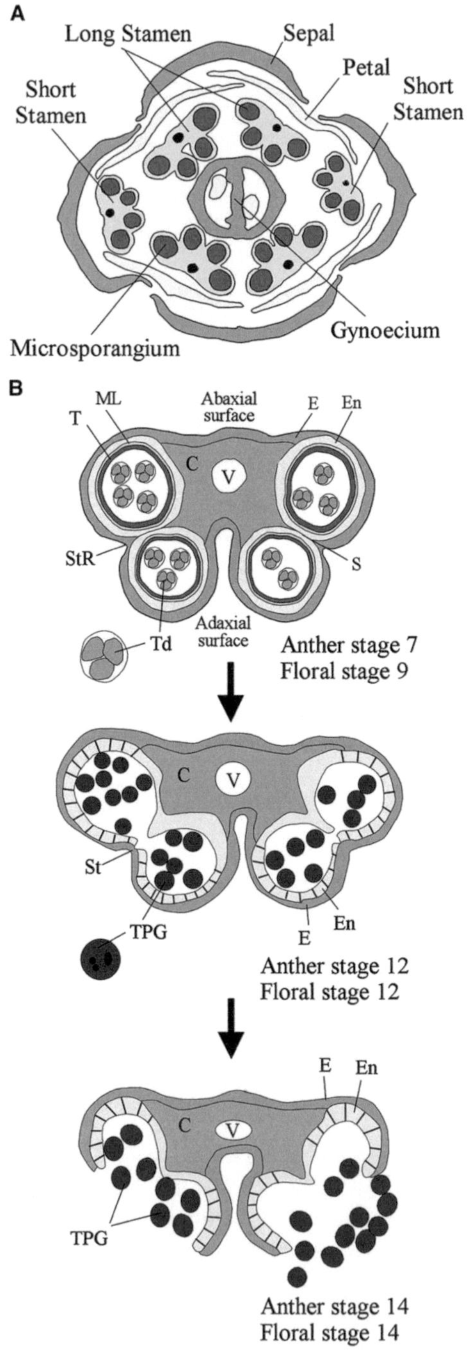

Fig. 6.3 Stamen structure and function. (**a**) Scheme of a transverse section through an *Arabidopsis* floral bud showing the number, position and orientation of the floral organs. (**b**) Schemes of transverse sections through *Arabidopsis* anthers at different stages. *C* connective, *E* epidermis, *En* endothecium, *ML* middle layer, *S* septum, *St* stomium, *StR* stomium region, *T* tapetum, *Td* tetrads, *TPG* tricellular pollen grains, *V* vascular bundle. (Courtesy: American Society for Plant Biologists-Plant Cell)

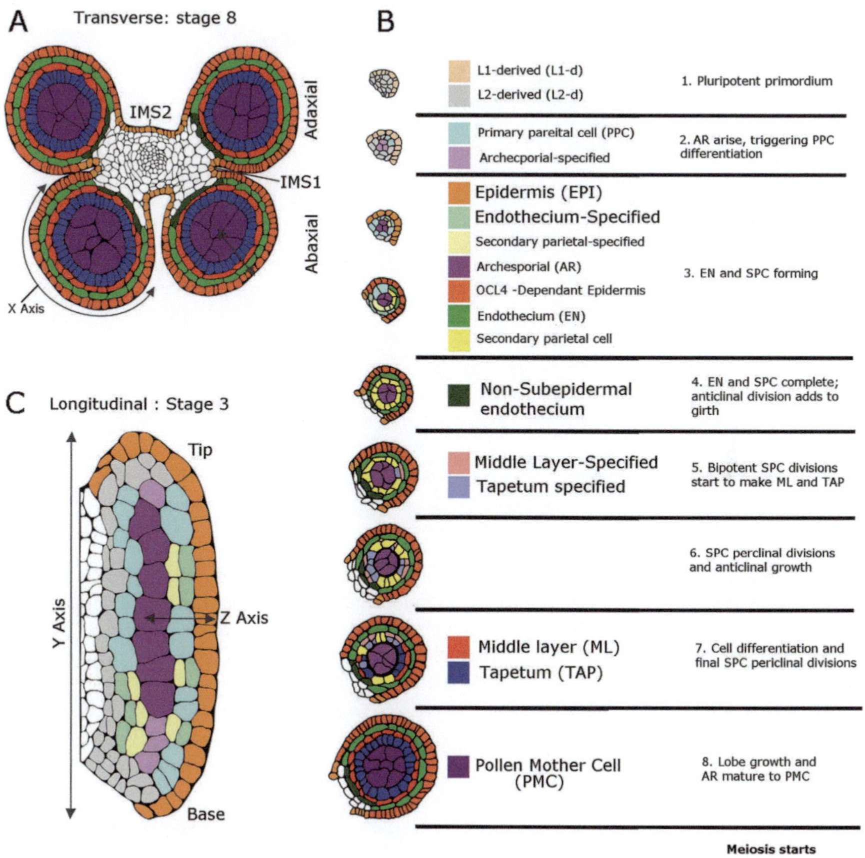

Fig. 6.4 Pre-meiotic anther development: (**a**) The four-lobed anther typical of flowering plants with a central column of vasculature that extends into the stamen filaments surrounded by connective tissue. (**b**) Anther lobe patterning. (**c**) Longitudinal view of an anther lobe. (Courtesy: Prof. Virginia Walbot, Stanford University and Frontiers in Plant Science). (See Box 6.5 for details)

and sub-apical (Fig. 6.6). This region is inverted cone-shaped where endoplasmic reticulum and vesicles are available. Sub-apical region contains Golgi apparatus and mitochondria. Amyloplasts and vacuoles are seen behind the clear zone. This region has a different refractivity which is higher than clear zone.

6.1 Male Sterility

Male sterility is defined as non-function of pollen grain. It can also be defined as the incapability of plants to produce or release functional pollen grains. Male sterility can be successfully used in hybrid seed production since it avoids the cumbersome process of emasculation. Male sterility is of five types:

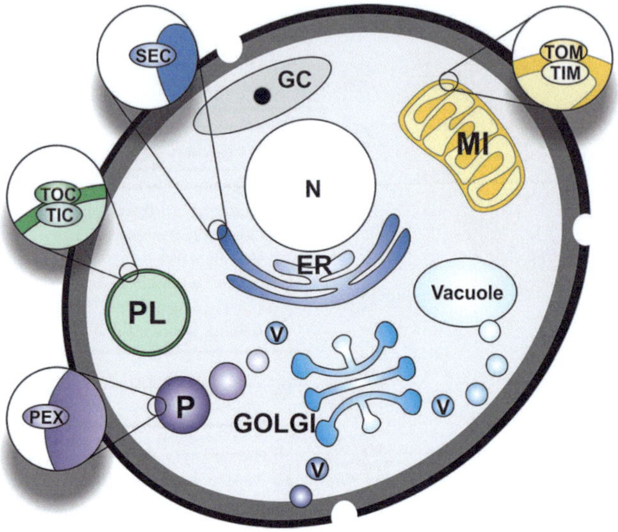

Fig. 6.5 Schematic structure of pollen. Highlighted are the membranes in which protein translocation complexes are hosted. The complexes in mitochondrial membranes (MI) are annotated as translocon of the outer/inner mitochondrial membrane (TOM/TIM) in the membranes of plastids (PL) as translocon of the outer/inner chloroplast envelope (TOC/TIC), in the membrane of endoplasmic reticulum (ER) as SEC translocase and in the membrane of peroxisomes (PEX). Others are nucleus (N), the Golgi system, the vesicles (V) and generative cell (GC). (Courtesy: Springer Publishing International)

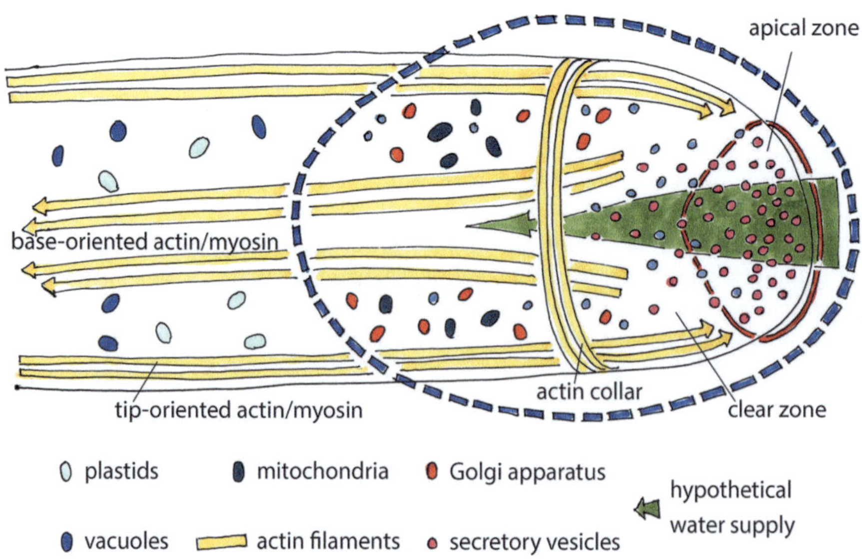

Fig. 6.6 Pollen tube apical region. Lily pollen tube tip showing action cytoskeleton dynamics and pollen tube zonation. (Courtesy: Springer Publishing International)

1. Genetic male sterility
2. Cytoplasmic male sterility
3. Cytoplasmic-genetic male sterility
4. Chemical-induced male sterility
5. Transgenic male sterility

The phenotypic manifestations of male sterility are very diverse like (a) complete absence of male organs, (b) the failure to develop normal sporogenous tissues (no meiosis), (c) the abortion of pollen, (d) the non-dehiscence of stamens and (e) the inability of mature pollen to germinate on stigma. Nuclear (genetic) male sterility is recessive mutation. Nuclear (genetic) male sterility in maize is controlled by several hundred loci. A number of functions like metabolism of plant hormones, biosynthesis of lipid molecules or synthesis of secondary metabolites are known. Cytoplasmic male sterility (CMS) is the maternally controlled inability to produce viable pollen. Mitochondria owes major role in this sterility. Therefore, CMS is resulted from a mitochondrial gene that blocks the production of viable pollen without affecting the other plant functions. The existence of male sterility may lead to gynodioecy (dimorphic reproductive system in which both male sterile and hermaphrodite plants/flowers coexist).

6.1.1 Genetic Male Sterility

Genetic male sterility is usually governed by a single recessive gene (ms or s) or a dominant gene. Male sterility allele either can rise spontaneously or can be artificially induced. It is found in natural conditions in pigeon pea, castor, tomato, lima bean, barley, cotton, etc. In this type, F_1 individuals would be fertile. In the F_2 generation, the fertile/sterile segregation will be in 3:1 ratio (Fig. 6.7). These mutations can regulate proteins involved in male meiosis, plant hormones and biosynthesis of lipid molecules.

6.1.2 Cytoplasmic Male Sterility

CMS is a valuable tool for hybrid seed in self-pollinated crops like maize, rice, cotton and a few vegetable crops. This will assist the production of new hybrid varieties to increase the world's supply. The use of hybrid rice in China reduced rice areas from 36.5 million ha in 1975 to 30.5 million ha in 2000. The total production increased from 128 to 189 million tons, with a yield increase of 3.5 to 6.2 tons/ha.

Progeny of male sterile plants would always be male sterile since cytoplasm of zygote comes primarily from the egg cell (Fig. 6.8). Through using male sterile strain

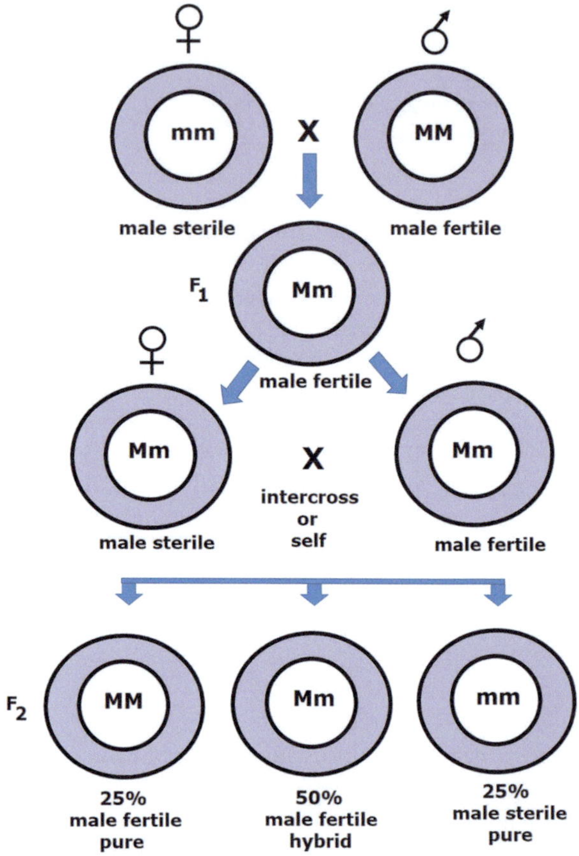

Fig. 6.7 Genetic male sterility

as a pollinator (recurrent parent), CMS may be transferred easily to successive generations of backcross programme. The nuclear genotype of male sterile line would be identical like recurrent pollinator strain after 6–7 backcrosses. The male sterile line is maintained by crossing it with pollinator strain used as a recurrent parent in backcross, since the nuclear genotype of the pollinator is identical with that of the new male sterile line. Such a male fertile line is known as maintainer line or "B" line and male sterile line is also known as "A" line. The control of CMS resides in mitochondria and not governed by any environmental factor. The premature degeneration of the tapetum layer of the anther is the first sign of CMS. In T-cytoplasm (Texas cytoplasm) of maize, mitochondria of the tapetum begin to degenerate soon after meiosis (see Box 6.1).

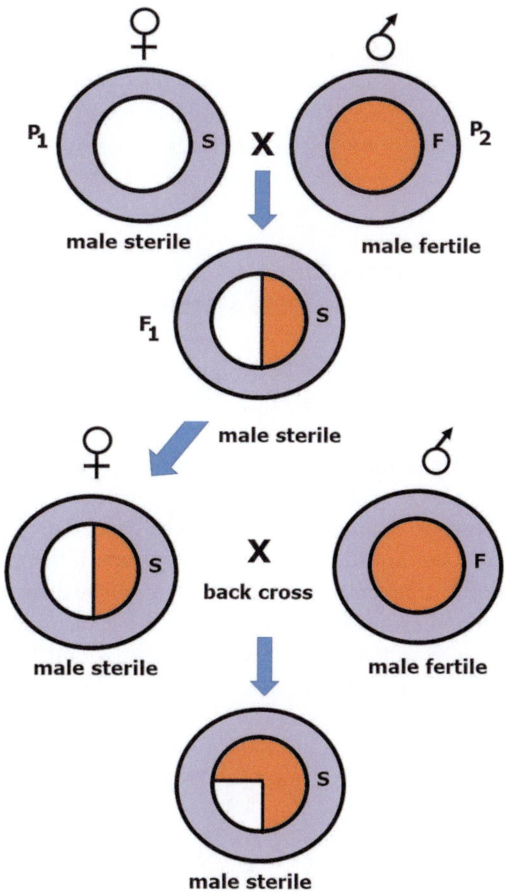

Fig. 6.8 Cytoplasmic male sterility

Box 6.1: Male Sterility in Maize

CMS occurs due to the interaction of nuclear and mitochondrial genomes that suppresses pollen production. In maize, three types of CMS systems, namely, CMS-T (Texas), CMS-S (USDA) and CMS-C (Charrua), have been identified. These types are categorized because of the reaction to restorers, mitochondrial DNA restriction digest patterns and compliments of low molecular weight plasmids. CMS-T is restored fully by *Rf-1* and *Rf-2*, CMS-S by *Rf-3* and CMS-C by *Rf-4*. All restorer genes except *Rf-2* restore fertility through governing the transcript profile of CMS-associated locus. The disorganization of the tapetum and surrounding cell layers causes sterility. In addition to the dysfunction of genes in mitochondria, the chloroplasts have emerged as ideal organelles for engineering male sterility. Recently, polyhydroxybutyrate was

(continued)

Box 6.1 (continued)
identified as a potential candidate gene for engineering male sterility. Moreover, a broad group of proteins called PPR (pentatricopeptide repeat) proteins have also been shown to hold great promise for engineering male sterility.

6.1.3 Genes for CMS and Restoration of Fertility (Cytoplasmic-Genetic Male Sterility)

This is a special type of cytoplasmic male sterility, where nuclear genes could restore fertility in male sterile line. This is achieved by a fertility restorer dominant gene "R" found in certain strains.

CGMS includes A, B and R lines. A is male sterile, B is similar to "A" but it is male fertile and R is restorer line. R restores fertility in the F_1 hybrid (Fig. 6.9). B line is used to maintain the fertility and hence known as the maintainer line. It would be male sterile with male sterile cytoplasm. If the nuclear genotype is rr, it will be male sterile. If the nucleus is Rr or RR, it will be male fertile. New male sterile lines can be derived as in CGMS system, but the nuclear genotype of the pollinator strain used must be with a fertility restorer system. For the development of new restorer strain, a restorer strain (R) is crossed with male sterile line. Then, the F_1 male fertile plants are

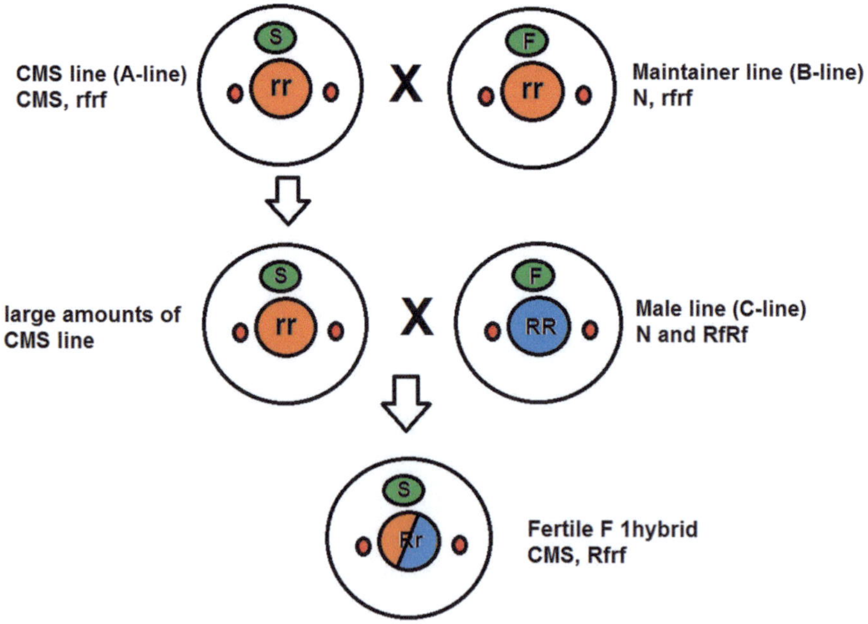

Fig. 6.9 Cytoplasmic-genetic male sterility with restorer genes

used as the female parent to repeatedly backcross with the strain (C) used as the recurrent parent to which transfer of restorer gene is required. Only male fertile plants are used as female for backcrosses, and male sterile plants are discarded in each generation. At the end, a restorer line isogenic to the strain "C" is recovered.

Although male sterility is wholly controlled by cytoplasm, a restorer gene if present in the nucleus will restore fertility. If female parent is male sterile, then genotype (nucleus) of male parent will determine the phenotype of F_1 progeny. The male sterile female parent will have the recessive genotype (*rr*) with respect to restorer gene. If male parent is *RR*, F_1 progeny would be fertile (*Rr*). On the other hand, if male parent is *rr*, the progeny would be male sterile. If F_1 individual (*Rr*) is testcrossed, 50% fertile and 50% male sterile progeny would be obtained.

CGMS is believed to be the result of lesions in the mitochondrial genome (Fig. 6.10). Sequences responsible for CMS are difficult to identify since mitochondrial genomes are large enough (200–2400 kb). Mitochondria are responsible for tricarboxylic acid cycle and ATP synthesis. They have only around 60 genes for the electron transfer chain, ribosomal proteins, transfer RNAs and ribosomal RNAs. Several plant mitochondrial genomes have been sequenced. Genomic studies on CGMS/Rf systems (Rf – fertility restorer) can address difference between mitochondrial and nuclear genomes.

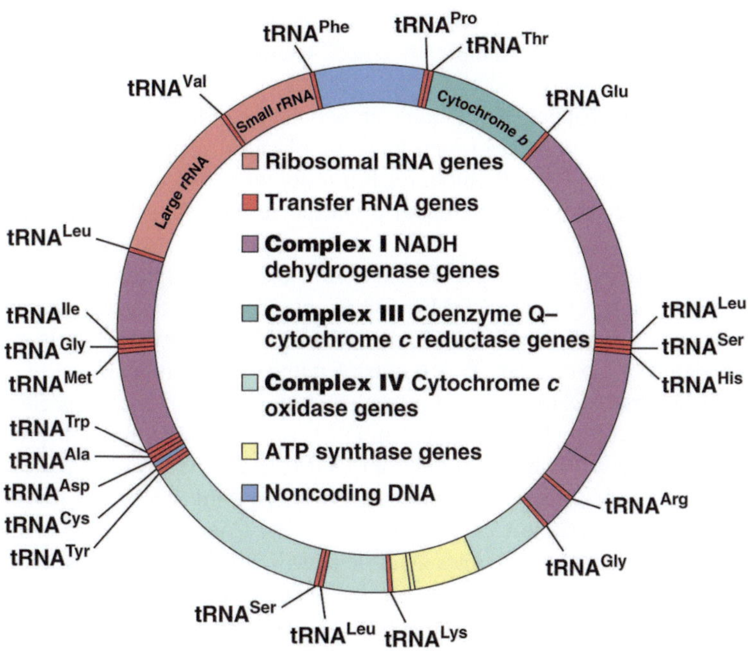

Fig. 6.10 Mitochondrial genome (representative)

CGMS is often associated with unusual open reading frames (ORFs). The differences in mitochondrial gene expression patterns among normal fertile, male sterile, restored fertile and fertile revertant plants have thrown more light into the functions. The key test is the functional assay of a candidate sequence. In sunflower, RFLP analysis of PET1 cytoplasm demonstrated that a 17-kb region of the mitochondrial genome includes 12-kb inversion and 5-kb insertion flanked by 261-bp inverted repeats.

CGMS arises spontaneously because of wide crosses or the interspecific exchange of nuclear and cytoplasmic genomes. For example, CGMS-WA (wild abortive) rice was derived from a male sterile plant among the wild rice *Oryza rufipogon* Griff. A cross between Chinsurah Boro II (*O. sativa* subsp. *indica*) and Taichung 65 (subspecies *japonica*) resulted in CGMS-BoroII. Texas male sterile cytoplasm in maize arose spontaneously in a breeding line. An interspecific cross between *Helianthus petiolaris* and *H. annuus* resulted in CGMS-PET1 cytoplasm of sunflower.

Restoration systems are either sporophytic or gametophytic. Sporophytic restorers act in sporophytic tissues and it occurs prior to meiosis. Gametophytic restorers act after meiosis. A heterozygous diploid plant that carries a male sterile cytoplasm with restorer will produce two classes of pollen grains: those that carry the restorer and those that are not. In sporophytic restorer, both genotypic classes of gametes will be functional. By contrast, in the case of a plant heterozygous for a gametophytic restorer, only those gametes that carry the restorer will be functional. S-cytoplasm maize is an example of a well-characterized CMS system that is restored gametophytically.

Restoration can happen due to one or two major restorer loci or due to the concerted action of a number of loci. In T-cytoplasm of maize, PET cytoplasm of sunflower and T-cytoplasm of onion, for full restoration, two unlinked restorers are required. Some of the systems contain duplicate restorer loci. In maize, *Rf8* can substitute for *Rf1*.

Comparison of cytoplasmic genomes in fertile and CGMS lines is one strategy to identify DNA that encodes CGMS. When we compare two cytoplasms, the differences could be due evolutionary divergence. Yet another strategy is to study the segregation of a particular DNA sequence with the phenotype. Both chloroplast and mitochondrial DNAs are uniparentally inherited in most species. The coinheritance of chloroplast DNA and mtDNA can be broken through protoplast fusion. Cybrids (somatic hybrids) between CGMS and fertile parents indicate that fertility is not associated with chloroplast DNA. A third strategy is to compare proteins of mitochondria in CGMS and fertile lines. Comparison of mitochondrial genes, transcript profiles or genomes in fertile and CGMS lines is the most acceptable way to find recombinant genes. However, this method is also not dependable, since restorer loci that may affect transcript profiles may affect both CGMS-associated genes and normal genes.

6.1.4 Mechanisms of Restoration

The physical loss of a CGMS-associated gene from the mitochondrial genome results in the restoration of fertility. The mitochondrial sequence responsible for CGMS (*pvs*) is lost in *Phaseolus* in the presence of nuclear gene *Fr*. But the actual mechanism governing this process is not understood. Transcriptional studies show that in T-cytoplasm maize, the presence of the *Rf1* restorer greatly enhances the accumulation of 1.6-kb and 0.6-kb T-*urf13* transcripts. On the other hand, accumulation of 13-kDa *urf13* protein is reduced. In many instances, post-transcriptional editing leads to fertility restoration. The CGMS-associated ORFs can have a new start (AUG) and/or stop (i.e. UAA, UAG or UGA) codons. The most prudent editing in plant mitochondrial sequences is C-to-U. Sequence analysis of restorer genes will show more information on their functions.

6.2 Engineering Male Sterility

Hybrids yield 10–30% more than pure inbred line. In many instances, CGMS systems are used to produce F_1 hybrids. A full advantage of this system can be used if a nuclear restorer gene suppresses the male sterility in the hybrid. As an example, in maize, Rf 2 gene encodes an aldehyde dehydrogenase. Rf4 is a fertility restorer gene in rice. A wild abortive type of CGMS (WA-CMS) and its *Rf* genes (a mitochondrial gene *orf352* is responsible for WA-CGMS) have been used in producing 99% of the F_1 hybrid cultivars in rice. In male sterile radish (*Raphanus sativus* L.), heterozygous alleles (*RsRf3–1/RsRf3–2*) encoding pentatricopeptide repeat proteins are governing fertility restoration. However, the increased use of such restoration systems can be vulnerable to insects and pathogens. This has happened in maize. Natural male sterility is available only in limited number of species. *Agrobacterium tumefaciens*-mediated gene transfer is seen as a unique system to tide over this issue.

There are several means by which one can genetically manipulate male sterility and bring male sterility into a specific crop species. They are:

(a) Dominant nuclear male sterility (pollen abortion) or *barnase/barstar* system
(b) Male sterility through hormonal engineering
(c) Pollen self-destructive engineered male sterility
(d) Male sterility using pathogenesis-related protein genes
(e) Silencing gene expression for pollen development with RNAi
(f) Mitochondrial rearrangements for CGMS
(g) Chloroplast genome engineering for CGMS

6.2.1 Dominant Nuclear Male Sterility (Pollen Abortion or *Barnase/Barstar* System)

Barnase (bacterial ribonuclease) is a bacterial protein that consists of 110 amino acids and has ribonuclease activity, secreted by the bacterium *Bacillus amyloliquefaciens.* Without its inhibitor *barstar*, *barnase* is lethal to the cell. *Barstar* binds to and obstructs the ribonuclease active site. This prevents *barnase* from damaging the cell's RNA. The *barnase/barstar* complex is extraordinarily tight protein-protein binding (Fig. 6.11).

A tapetum-specific promoter, a cytotoxic gene and a transcription terminator can be constructed to be a chimaeric gene and is used to transform plants (Fig. 6.12). Cytotoxin can selectively destroy the tissues leading to pollen development. RNase digests RNAs. Two genes encoding RNase-barnase and RNase T1 have been cloned. The gene for RNase and a specific promoter can be linked and transferred into plants to derive male sterility. The tapetum-specific promoter TA29 isolated

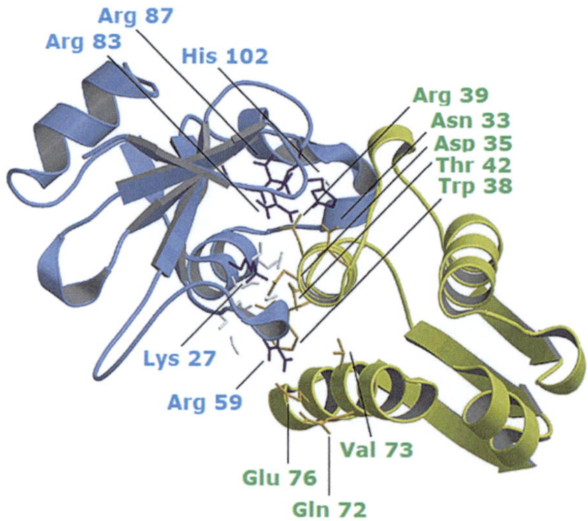

Fig. 6.11 Barnase-barstar complex. The complex between barnase (blue) and barstar (yellow) with 12 interfacial water molecules (grey). Side chains important in binding are indicated

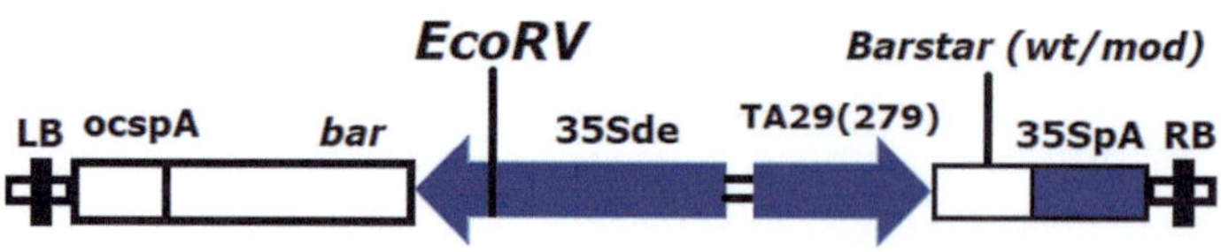

Fig. 6.12 Map of T-DNA region of gene constructs used for the generation of barstar lines. ocspA, polyA signal of octopine synthase gene; 35Sde, CaMV35S promoter with duplicated enhancer; TA29 (279), bp fragment of tapetum-specific TA29 promoter; barstar (wt/mod), wild-type or modified sequence of barstar gene

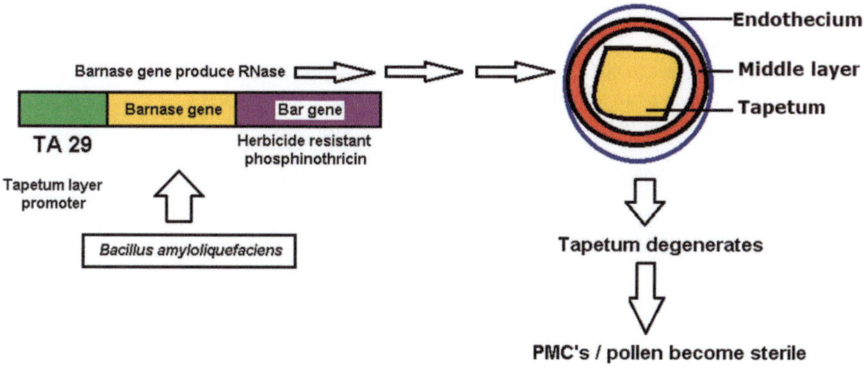

Fig. 6.13 Principle of barnase-barstar system

from tobacco anthers along with *barnase* gene plus RNase T1 gene was introduced through genetic transformation into tobacco and oilseed rape. This selectively destroyed the tapetal cell layer leading to male sterility (Figs. 6.13). The genetic transformation of cauliflower, tomato, cabbage, watermelon and eggplant was achieved in this way. In cabbage, hybrid seeds could be produced when transformed plants were pollinated with normal pollen. Self-pollination never resulted in any seeds. A general scheme being followed for the production of hybrid seeds using barnase/barstar is available in Fig. 6.14.

Tapetal degeneration is a programmed cell death (PCD). This is characterized by cell shrinkage, degradation of mitochondria and cytoskeleton, nuclear condensation, oligonucleosomal cleavage of DNA, vacuole rupture and endoplasmic reticular swelling. Any disruption of the timing of PCD can cause pollen abortion or male sterility. The anther-specific genes involved in these developments include Osc4, Osc6, YY1 and YY2 genes of rice; TA29, TA32 and NTM 19 genes of tobacco; SF2 and SF18 genes of the sunflower; 108 genes of tomato; and BA42, BA112 and A9 genes of *Brassica napus*. Some of these genes are found exclusively in sporophytic tissues of the anthers; others are pollen-specific or are present in both sporophytic and gametophytic tissues of the anthers.

6.2.2 Male Sterility Through Hormonal Engineering

In tomato and tobacco, changes in endogenous level of auxins govern male sterility. In tobacco, "*rol c*" gene of *Agrobacterium rhizogenes* and 35S CaMV promoter flanked with a marker gene were introduced to change hormone system to induce male sterility. Due to an increase in the levels of indole acetic acid and decreased levels of gibberellin, "*rol b*" from *Agrobacterium rhizogenes* affected flower development of transgenic tobacco.

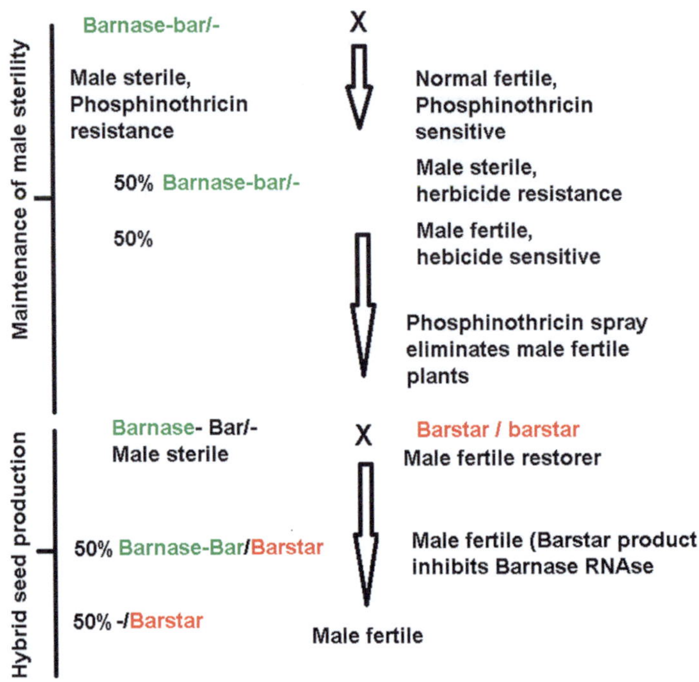

Fig. 6.14 Scheme for the production of hybrid seeds using barnase/barstar system

6.2.3 Pollen Self-Destructive Engineered Male Sterility

It is theoretically feasible to transform plants through genetic engineering to alter levels of endogenous auxin (say indole acetic acid). Such alterations will ensure pollen exhibiting self-destructive mechanisms. A chimaeric gene consisting of pollen-specific promoter (LAT59) and a gene (fins2) that converts indole acetamide (IAM) into IAA can be used for transforming plants. If this is achieved, plants carrying the LAT59-fins2 gene can be sprayed with IAM which can selectively convert IAM into IAA. IAA at very high concentrations can kill the pollen. Yet another route is transformation of plants with chimaeric gene with TA-29 promoter and coding region of β-glucuronidase (GUS). The resultant transformants if prayed with protoxins like sulfonyl urea or maleic hydrazide can cause male sterility. This is achieved through breaking down the tapetum by β-glucuronidase enzyme. If the plants are not sprayed with protoxins, they remain fertile. In this case, a fertility restoration system like TA29-barstar is not required.

6.2.4 Male Sterility Using Pathogenesis-Related Protein Genes

The cell wall is made of callose, a β-1,3-linked glucan. This is seen between cellulose cell wall and plasma membrane. Pathogenesis-related (PR) protein β-1,3-glucanase (callase) is capable of dissolving glucan. Callase can also dissolve tetrads synthesized by microsporocyte. Tapetum secrets callase which can break down callose wall that helps to release free microspores into locular space. Genetic alteration of this process can cause male sterility. This is demonstrated by electron microscopic studies wherein microspore of the tetrad is surrounded by callase in fertile anthers, whereas it was clearly absent in sterile microspores.

6.2.5 RNAi and Male Sterility

Post-transcriptional gene silencing (PTGS) is one upcoming area that can assist in inducing male sterility. Antisense RNA and RNA interference (RNAi) can reduce or silence the expression of target genes (see Box 6.2). In Chinese cabbage and broccoli, through transgenic means, an anti-gene CYP86MF encoding cytochrome P450 (associated with the nuclear male sterility) was transferred, and the resultant plants were male sterile. These male sterile plants set seeds when pollinated with normal pollen. Other genes involved in pollen development are actin gene and DAD1 gene encoding phospholipase A1. Antisense DAD1 gene was introduced into Chinese cabbage that showed male sterility.

Box 6.2: Antisense RNA and RNA Interference

Antisense RNA is single-stranded that is complementary to a protein coding mRNA. This RNA hybridizes with the mRNA and blocks its translation into protein. It is also referred as antisense transcript, natural antisense transcript (NAT) or antisense oligonucleotide. They are long non-coding RNAs (lncRNA), larger than 200 nucleotides. As such, they are having their primary role in gene knock down (see Fig. 6.15).

Gene silencing can be done with the help of microRNA (miRNA). miRNAs are gene regulatory RNAs that are loaded onto the RNA-induced silencing complex (RISC) and interact with partially complementary targets on mRNA to suppress protein expression. The miRNA is originally double-stranded and composed of about 21 nucleotides. Upon loading onto RISC, one strand is degraded, and the other, the "guide" strand, is held on the surface of RISC where it can interact with mRNA. The targets recognized by the guide strand are most commonly on the 3′-untranslated region (UTR) of an RNA. Binding can suppress assembly of an initiation complex on the 5′ cap of an mRNA because the mRNA is bound into a circular shape at the initiation of translation, bringing the 3'-UTR and 5'-UTR close together.

If the RISC loads an RNA and then finds a perfectly complementary target, RISC cleaves the target RNA using the activity of one of the protein components of RISC called Argonaute (Ago2). This property is exploited experimentally by manufacturing small interfering RNAs (siRNA)

(continued)

Box 6.2 (continued)

intentionally targeted to particular target sequences. Once loaded into RISC, these siRNAs might recognize and cleave their perfectly complementary target sequence within an mRNA. The siRNA will also have miRNA-like effects on some partially complementary targets on various mRNAs, leading to the observation that a single siRNA sequence can modulate the expression of hundreds of off-target genes.

RNA interference (**RNAi**) is a biological process in which RNA molecules inhibit gene expression or translation, by neutralizing targeted mRNA molecules. Historically, RNA interference was known by other names, including *co-suppression*, *post-transcriptional gene silencing* (PTGS) and *quelling*. Though these are different techniques, they were all being undertaken by RNAi. Andrew Fire and Craig C. Mello shared the 2006 Nobel Prize in Physiology and Medicine for their work on RNAi. RNAi is now a better technology than antisense RNA technology. RNAi defends cells against parasitic nucleotide sequences like viruses and transposons.

6.2.6 Mitochondrial Rearrangements for CMS

Mitochondria are semi-autonomous and primarily maternally inherited genetic organellar system responsible for producing cellular ATP by oxidative phosphorylation. Plant mitochondrial genomes are known as mitogenomes. Both mitochondrial and the nuclear genomes are responsible for coding mitochondrial proteins. Here, the contribution of nuclear genes is nearly 10%. Mitochondria participate in sending signals to the nucleus to generate various proteins. CMS is associated with rearrangements of mitochondrial genome derived through non-homologous recombination.

Plant mitochondrial genomes may vary enormously in size even within single plant families. For example, in Cucurbitaceae, mitochondrial genomes vary over sevenfold in size, from 379 kb in *Citrullus lanatus* to 2740 kb in *Cucumis melo.* While mitogenomes typically are depicted as single circular rings, many other configurations for plant mitochondrial chromosomes have been reported including diverse linear and circular forms, highly branched and sigma-like morphologies as well as multi-chromosomal structures that are capable of sub-stoichiometric co-occurrence. The mitochondrial genomes of some CMS lines in maize and rice have linear configurations. Repeated sequences are common in plant mitochondrial genomes, with estimates of up to 38% of the mitochondrial genome occupied by repeats of variable size and copy number. The presence of CMS may be associated with the presence of such large repeats. At the molecular level, the development of CMS can be broadly grouped into the following three main categories:

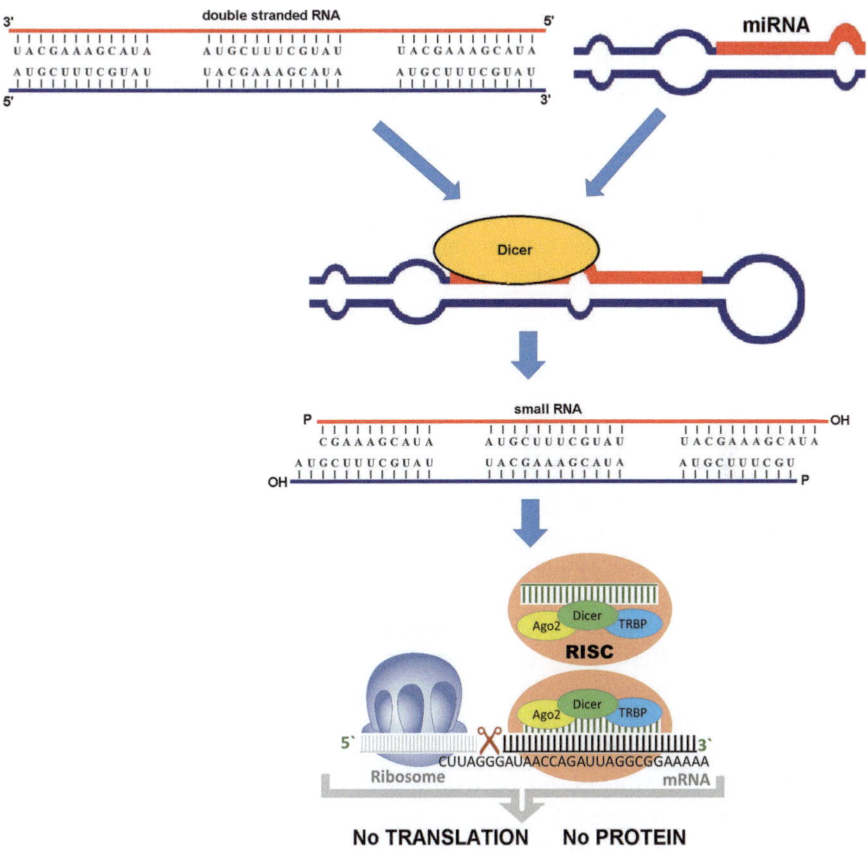

Fig. 6.15 Antisense RNA system. *RSIC* is RNA-induced silencing complex. DICER is a multidomain ribonuclease that processes double-stranded RNAs (dsRNAs) to 21-nucleotide small interfering RNAs (siRNAs) during RNA interference and excises micro RNAs (miRNAs) from precursor hairpins. Ago2 (Argonaute 2) protein is an essential effector protein in miRNA-mediated mechanisms that regulate gene expression. TRBP is a double strand RNA binding protein (dsRBP) that is required for the recruitment of Ago2 to the small interfering RNA (siRNA) bound by DICER

(a) mtDNA recombination and interactions between mitochondrial and nuclear genomes (cyto-nuclear interaction)
(b) Aberrant RNA editing
(c) Accumulation of toxic protein products

mtDNA Recombination and Cyto-nuclear Interaction Mitogenome recombination generates novel chimaeric sequences, and such sequences exhibit co-transcription with upstream or downstream functional genes, such as Turf13 in CMS-T maize and orf352 CMS rice. The modes of action for CMS-related mitochondrial genes appear equally as diverse. In *Brassica napus*, CMS-related orf224/atp6 was found to down-regulate pollen development by causing an energy deficiency. CMS in Chinese cabbage has been associated with retrograde signalling (i.e. signals from the plastid or mitochondrion that control nuclear gene expression) from the mitochondrion that interferes with nuclear gene expression through auxin response and ATP synthesis.

Regulation of CMS Transcripts via RNA Editing Post-transcriptional RNA editing of mitochondrial genes converts specific cytosine residues to uracil (C-to-U). Defects in RNA editing transcripts result ultimately in plant or cell death. The number of RNA editing sites can vary among species, for example, in *Arabidopsis thaliana*, an average 43 different editable sites are there among mitochondrial protein coding regions.

Accumulation of Toxic Protein Products the protein products of CMS genes are the likely agents of CMS. Most CMS-associated proteins possess transmembrane configurations capable of disrupting the mitochondrial membrane structure and/or altering the permeability and potential of mitochondrial membrane. These proteins can directly interfere with energy production, induce the release of cytochrome C *via* accumulation of unusually large numbers of reactive oxygen species (ROS) and stimulate premature programmed cell death in male reproductive tissues. Several CMS proteins have demonstrated toxicity, such as *URF13* in CMS-T maize, *ORFH79* in HL-CMS rice, *Orf507* in CMS chilly and ROS homeostasis-associated protein in cotton. Restoration of fertility can occur at the translational or post-translational level. In many CMS systems, RF genes do not affect accumulation of the CMS transcript, but on the other hand, restored lines are characterized by a marked decrease in toxic CMS protein accumulation. These observations suggest that restoration of fertility occurs *via* reduction in the production of toxic proteins.

Stability of the mitochondrial genome is controlled by nuclear loci. In plants, nuclear genes suppress mitochondrial DNA rearrangements during development. One nuclear gene involved in this process is *Msh1*. *Msh1* appears to be involved in the suppression of illegitimate recombination in plant mitochondria. In tobacco and tomato, experiments show that mitochondrial DNA rearrangements lead to a condition of male (pollen) sterility. The male sterility was heritable and apparently maternal in its inheritance.

6.2.7 Chloroplast Genome Engineering for CMS

A high level of accumulation of polyhydroxybutyrate (PHB) or β-ketothiolase in chloroplasts resulted in male sterility and growth retardation. In transgenic lines with phaA (polyhydroxyalkanoate synthase) gene coding, β-ketothiolase pollen was sterile. Scanning Electron Microscopy (SEM) revealed a collapsed morphology of the pollen grains. Transgenic lines resulted in aberrant tissue patterns. Pollen grains were of irregular shape or of collapsed phenotype. This is due to abnormal thickening of the outer wall and enlarged endothecium. However, more research is needed in genome engineering of chloroplasts for hybrid development.

6.3 Male Sterility in Plant Breeding

Male sterility ensures hybrid seed production. Interspecific crosses in *Nicotiana*, *Dianthus*, *Verbascum*, *Mirabilis* and *Datura* during the eighteenth century by J.G. Kölreuter enthused the concept of hybrid vigour. This was later confirmed by Darwin in vegetables and W.J. Beal in maize. The first male sterility system was developed in onion in 1943. The cases of sugar beet, maize, sorghum, sunflower, rice, rapeseed and carrot followed. The successful breeding efforts in the twentieth century are that of maize (from the 1930s in the USA) and of rice (since 1976 in China). A sixfold increase in yield was observed in corn between 1930 and 1990 in the USA after hybrid seed production. This was a phenomenal change after 60 years of low productivity. In China, hybrid rice varieties produced 8–15% higher yield than that of the check varieties. Such hybrids produced 12 tons per ha in on-farm demonstration fields. Between 1998 and 2005, China released 34 "super" hybrid rice varieties for 13.5 million ha. This produced an additional 6.7 million tons of rice. In case of CMS-T system maize, the system was unused after 1970 due to the susceptibility of CMS-T corn to "southern leaf blight" (caused by *Bipolaris maydis*). Corn hybrids are now produced by manual or mechanical emasculation. Other species like sugar beet, sunflower, rapeseed and sorghum used CMS. Since these systems are different and cannot be transposed from one to other species, efforts are on at several laboratories to generate new hybrids through transgenic means.

CMS eliminates the need for hand emasculation and ensures the production of male fertile, F_1 progeny. In corn, prior to the epidemic of southern corn leaf blight in 1970, approximately 85% of hybrid seed were produced through male sterile T (Texas)-cytoplasm in the USA. By developing female lines that carry CMS cytoplasm, breeders produced hybrid seeds. F_1 hybrid seed carried the CMS cytoplasm that was produced by the female lines. In the near future, CMS will be manipulated further involving genes for pollination (see Box 6.3).

Box 6.3: Identification of Gene to Eliminate Self-Pollination

A naturally occurring wheat gene when turned off can eliminate self-pollination but still can allow cross-pollination. The University of Adelaide along with a US-based plant genetics company DuPont Pioneer has notified this achievement. Wheat delivers around 20% of total food calories and protein to the world's population. Hybrid wheat results from crosses of pure wheat lines. The production of hybrid wheat seed requires large-scale cross-pollination as wheat is a self-pollinator. A gene *Ms1* has been identified in the production of large-scale, low-cost production of male sterile (*ms*) lines.

The use of recessive male sterility was first proposed in the 1950s through a cytogenetic 4E-*ms* system. This system utilizes mutant allele *ms1 g* and a fertility-restoring chromosome from *Agropyron elongatum* ssp. *ruthenicum*

(continued)

Box 6.3 (continued)
Beldie (4E). However, the residual pollen transmissibility of chromosome 4E gave rise to selfed seeds. This has reduced the purity of the hybrid seeds.

The isolation of recessive alleles of *Ms1* gene was utilized to develop a male sterile female-inbred seed (*ms1/ms1*). This was done in the line of seed production technology (SPT) in maize by DuPont Pioneer in the USA. This can overcome the seed purity issues inherent to the 4E-*ms* system. When attached with a functional α-amylase gene for wheat pollen disruption, the system induces male sterility. The identification of *TaMs1* gene sequence can completely restore viable pollen production in *ms1* plants. If this system is made possible, SPT for wheat could become a reality.

CMS-based hybrid seed technology uses a three-line system, which requires three different breeding lines: the CMS line, the maintainer line and the restorer line (Fig. 6.16a). The CMS line has male sterile cytoplasm with a CMS-causing gene (hereafter termed a CMS gene) and lacks a functional nuclear *restorer of fertility* (*Rf* or restorer) gene or genes and is used as the female parent. The maintainer line is with normal fertile cytoplasm but has the nuclear genome as that of CMS line. The restorer line has *Rf* gene (s) and is used as male parent in crosses with the CMS line to produce F_1s. *Rf* gene restores male fertility in F_1s. The combination of nuclear genomes and restorers produces hybrid vigour. Male sterility traits of most GMS mutants cannot be efficiently maintained. However, the advent of EGMS mutants has to be used for hybrid crop breeding. The pollen fertility changes in response to environmental cues (day length and temperature) in EGMS lines. The first photoperiod-sensitive GMS (PGMS) mutant in rice, Nongken 58S (NK58S), was discovered in *japonica* rice (*Oryza sativa* ssp. *japonica*) in 1973. NK58S is completely male sterile when grown under long-day conditions but male fertile when grown under short-day conditions. A temperature-sensitive GMS (TGMS) mutant, Annong S-1, was found in *indica* rice (*O. sativa* ssp. *indica*) in 1988. Annong S-1 is completely male sterile when grown at high temperatures but male fertile at low temperatures. The PGMS and TGMS are featured in Fig. 6.16b. The two-line system thus eliminates the requirement of crossing to propagate the male sterility line. All normal varieties have wild-type fertility alleles which can restore male fertility. So, they can be used as the male parents. Hence, a two-line system reduces costs. In China, production of two-line hybrid rice based on PGMS or TGMS occupies 20% of the total hybrid rice planting area.

Of late, it is revealed that non-coding RNAs are expected to have a decisive role in governing male sterility. The participation of non-coding RNAs is slowly unfurling, and in due course of time, more details will be made available (see Box 6.4).

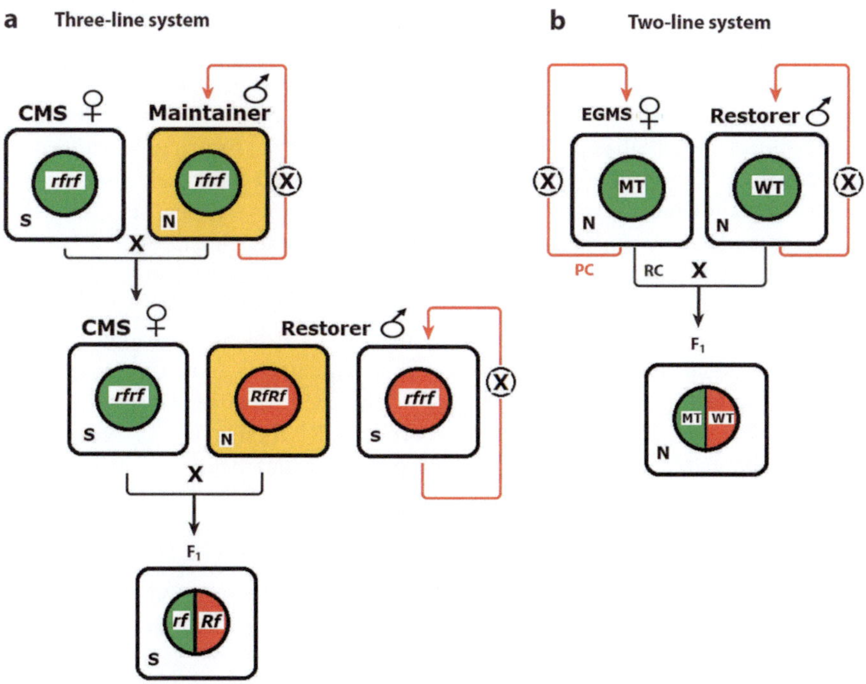

Fig. 6.16 Application of cytoplasmic male sterility (CMS) and environment-sensitive genic male sterility (EGMS) for hybrid seed production in a three-line system and a two-line system. (**a**) The three-line system requires a CMS line, containing sterile cytoplasm (S) and a non-functional (recessive) restorer (*rf*) gene or genes; a maintainer line, containing normal cytoplasm (N) and a nuclear genome identical to that of the CMS line; and a restorer line, with normal (N) or sterile (S) cytoplasm and a functional (dominant) restorer (*Rf*) gene or genes. The CMS line is propagated by crossing with the maintainer line; the maintainer and restorer lines can produce seeds by self-pollination. The CMS line is crossed with the restorer line to produce male fertile hybrids. (**b**) In the two-line system, an EGMS [photoperiod-sensitive GMS (PGMS), reverse PGMS or temperature-sensitive GMS (TGMS)] mutant (MT) line is propagated by self-pollination when grown under permissive conditions (PC) (short-day conditions for PGMS, long-day conditions for reverse PGMS or low-temperature conditions for TGMS). The EGMS line is male sterile under restrictive conditions (RC) (long-day conditions for PGMS, short-day conditions for reverse PGMS or high-temperature conditions for TGMS) and thus serves as the female parent for crossing with a wild-type (WT) line to produce hybrid seeds

Box 6.4: Non-coding RNAs and Male Sterility

Pollen development is a complex process. The release of fertile pollen is vital for breeding. Pollen development is regulated by multigenes and mutations might induce male sterility. Non-coding RNAs (ncRNAs) constitute a large proportion of genetic information. During evolution, several organellar genes were transferred to the nuclear genome. So, biogenesis of plant organelles is

(continued)

Box 6.4 (continued)
governed by both nuclear and organelle genes. Non-coding RNAs are significant among the moieties that regulate plant organ biogenesis. Non-coding RNAs are differentiated based on the length of transcript and functional specificity. Two primary types are small ncRNAs and the long non-coding RNAs (lncRNAs). The family of small ncRNAs in plants is further categorized as microRNAs (miRNAs); heterochromatic small interfering RNAs (hc-siRNAs); phased, secondary, small interfering RNAs (phasiRNAs); and natural antisense transcript small interfering RNAs (NAT-siRNAs), based on their origin and biogenesis. ncRNAs function at transcriptional and post-transcriptional levels. They can also exert influence over a long distance including post-transcriptional silencing or epigenetic changes because of its mobility. Dicer-like (DCL) proteins cleave long RNAs into small fragments. Such fragments get incorporated into the Argonaute family proteins for targeting the complementary nucleotide sequences.

Double-stranded RNAs (dsRNAs) are processed by DCL proteins to derive 21–24 nucleotide small ncRNAs. Such RNAs govern RNAi pathway. These ncRNAs are coupled with Argonaute proteins (AGOs) to form complexes that trigger sequence-dependent RNA silencing through RNA cleavage or DNA methylation. RNA silencing is classified into transcriptional gene silencing (TGS) and post-transcriptional gene silencing (PTGS). In TGS, suppression of transposable elements (TEs) occurs and blocks their way to the next generation. PTGS, on the other hand, inhibits the gene expression *via* target RNA cleavage and/or translational repression.

Only presence of small RNAs does not ensure their involvement in the induction of sterility. Pollen of *A. thaliana* and rice make sure miRNAs on the target mRNAs by cleaving the gene targets. In some cases, translational inhibition happens because of phasiRNAs that influence inflorescence and anther development. A few miRNAs target transcription factors (TFs) instead of mRNAs.

Box 6.5: Pre-meiotic Anther Development (Detailed Legend for Fig. 6.4)
(A) The four-lobed anther typical of flowering plants with a central column of vasculature that extends into the stamen filament surrounded by connective tissue. **(B)** Progression of cell fate specification and anther lobe patterning. At stage 1, the lobe consists of pluripotent Layer 1- and Layer 2-derived cells, coloured in beige and light grey, respectively. For all cell types, just-specified cells are coloured in a pale shade, which gradually darkens as the cells acquire stereotyped differentiated shapes, volumes and staining properties. The first

(continued)

Box 6.5 (continued)
specification event results in visible archesporial (AR) cells centrally within each lobe. In maize, the glutaredoxin encoded by *Msca1* responds to growth-generated hypoxia to initiate AR differentiation, marked by secretion of the MAC1 protein, which is required for cell specification of the subepidermal L2-d cells to primary parietal cells (PPC) [stage 2]. PPC divide periclinally generating the subepidermal endothecium (EN) and the bipotent secondary parietal cells (SPC). In the same time frame, epidermal (EPI) cells differentiate; signals controlled by expression of the OCL4 epidermal-specific transcription factor suppress excess periclinal divisions in the EN [stage 3]. Following these early patterning events that result in a three-layered wall surrounding the AR, there is a period of anticlinal division that expands anther cell number and organ size [stage 4]. Subsequently, each SPC divides once periclinally to generate the ML and TAP, and the final four somatic walled architecture of the pre-meiotic anther lobe is achieved [stages 5–7]. Prior to meiosis, anticlinal divisions occur to increase anther size, and the individual cell types acquire differentiated properties [stages 6–8], including dramatic enlargement of AR as they mature into pollen mother cells (PMC) capable of meiosis [stage 8]. IMS1 and IMS 2 are intermicrosporangial stripes.

Further Reading

Birchler JA, Han F (2018) Barbara McClintock's unsolved chromosomal mysteries: parallels to common rearrangements and karyotype evolution. Plant Cell 30:771–779

Budar F, Pelletier G (2001) Male sterility in plants: occurrence, determinism, significance and use. CR Acad Sci Paris Sciences de la vie / Life Sciences 324:543–550

Chen L, Liu YG (2014) Male sterility and fertility restoration in crops. Annu Rev. Plant Biol 65:579–606

Eckardt NA (2006) Cytoplasmic male sterility and fertility restoration. Plant Cell 18:515–517

Havey MJ (2004) The use of cytoplasmic male sterility for hybrid seed production. In: Daniell H, Chase CD (eds) Molecular biology and biotechnology of plant organelles. Springer, Dordrecht, pp 623–634

Schnable PS, Wise RP (1998) The molecular basis of cytoplasmic male sterility and fertility restoration. Trends Plant Sci 3:175–180

Touzet P, Meyer EH (2014) Cytoplasmic male sterility and mitochondrial metabolism in plants. Mitochondrion 19:166–171

Basic Statistics

7

Keywords

Genetic variation · Measures of variation · Coefficient of variation · Probability · Normal distribution · Statistical hypothesis · Standard error of the mean · Correlation coefficient (*r*) · Regression analysis · Heritability · Principles of experimental design · Completely Randomized Design (CRD) · Randomized Complete Block Design (RCBD) · Latin square design · Tests of significance · Chi-Square Test (for Goodness of Fit) · *t*-Test · Analysis of variance · Multivariate statistics · Cluster analysis · Principal Component Analysis (PCA) and Principal Coordinate Analysis (PCoA) · Multidimensional scaling · Path analysis · Hardy–Weinberg equilibrium

An outline of application of biometrics in plant breeding is dealt here, as envisaged in syllabi of several universities. However, for an in-depth knowledge of the subject, one may consult advanced books.

As per Mendelian principles, the early geneticists investigated the pattern of transmission of hereditary factors at family level. The criterion adopted was the similarity or dissimilarity of phenotypes between the progeny and their parents. Since the population of individuals is deciding the future of genes, the behaviour of genes in the population is very vital. For example, reproductive ability of individuals carrying a given gene may depend upon fitness of this gene, frequency of this gene in the population, size of the population and genotypes of other individuals in the population. Thus, the fate of individuals and consequently the fate of genes contained in them are strongly tied to the factors influencing the population as a whole. Studies of such populations need a strong background of the subject *statistics*.

Statistics is to collect, organize, analyse and interpret numerical information from data. There are two categories: *descriptive statistics* and *inferential statistics*. In descriptive statistics, numerical facts are collected, organized and analysed. The

P. M. Priyadarshan, *PLANT BREEDING: Classical to Modern*,
https://doi.org/10.1007/978-981-13-7095-3_7

primary objective is to describe information gathered. Inferential statistics collects data from relatively small groups of a population. It uses inductive reasoning to make generalizations and inferences. Some of the basic terms commonly used in statistics are defined below.

7.1 Common Biometrical Terms

Population is a complete set of items/members under study. The set may refer to people, objects or measurements that have a common characteristic. Examples of a population are hybrids of an F_1 generation borne out of a cross between two parents, offsprings of a backcross between F_1 and a parent and so on.

Sample is a small group of individuals selected from a population. If every member of the population has an equal chance of being selected for the sample, it is called a *random sample.*

Data are numbers or measurements that are collected. Data may include yield of plants, height of plants, total seeds per fruit, total fruits per plants, temperatures in an area during a given period of time, etc.

Variables are characteristics/attributes/traits that are distinguished between each other. Different individuals will have different values. Some of the variables are height, weight, age and price. Variables are opposite to *constants* which never change.

Phenotype and genotype: Phenotype is the physical manifestation of an organism. It is determined by its genetic constitution, the environment where grown and the interaction of genotype with environment. Genotype is the set of inheritable genes. The information written as genetic code is copied during cell division or reproduction and is passed over future generations. They control everything from the formation of protein macromolecules to the regulation of metabolism and synthesis. The physical result of the genotype is the phenotype. The challenge plant breeders face is to identify and select those plants that have genotypes conferring desirable phenotypes, rather than plants with favourable phenotypes due to environmental effects. As a rule, traits with greater heritability can be modified more easily by selection and breeding than traits with lower heritability.

7.1.1 Genetic Variation

Genetic variability refers to the variation of a given genotype within a population. As the genetic variability of a population increases, its resistance to environmental influences increases. So, the genetic variability is directly related to biodiversity and evolution. In terms of evolutionary biology, if a population lacks sufficient genetic variability, it also lacks the potential to evolve and adapt. In terms of genetics, variability among population genotypes can explain why different plants can have different responses to various treatments and environmental influences. Increased variability increases fitness. The evolutionary adaptations actually observed in nature are described in terms of *variation* rather than *variability*. The differences between these two terms are very subtle. Variability denotes how much a genotype tends to vary between individuals (the ability to vary) and in response to environmental and genetic factors, whereas variation is used to indicate the variation between and within species. Simply put, variability studies genotypes at the level of individuals and populations, and variation studies genotypes in and between species.

In asexual organisms, sources of variability are limited because the genetic code is the same for the parent and offspring. Similar limitation occurs when inbreeding is practised, because the genetic material from the parents is less variable. The lack of variability within a population can lead to genetic problems such as mutation and drift. If a new individual joins the population, then the potential for variation increases.

7.1.2 Measures of Variation

Range The range for a set of data items is the difference between the largest and smallest values. Although the range is the easiest of the numerical measures of variability to compute, it is not widely used because it is based on only two of the items in the data set and thus is influenced too much by extreme data values. The range is simply the highest score minus the lowest score. Let's take a few examples. For instance, if we see the range of the following group of numbers, 10, 2, 5, 6, 7, 3 and 4, the range is $10 - 2 = 8$. Obviously, there are limitations in using range as a measure of variability. Variance and standard deviation are being considered as authentic measures of variability.

Variance The variance and the closely related standard deviation are measures of how spread out a distribution is. They are measures of variability. Variance is computed as the average squared deviation of each number from its mean. For example, for the numbers 1, 2 and 3, the mean is 2 and the variance is:

$$\sigma^2 = \frac{(1-2)^2 + (2-2)^2 + (3-2)^2}{3} = 0.667$$

The formula (in summation notation) for the variance in a population is:

$$\sigma^2 = \frac{\sum (X - \mu)^2}{N}$$

where μ is the mean and N is the number of scores.

Standard Deviation The standard deviation formula is very simple: it is the square root of the variance. It is the most commonly used measure of spread.

7.1.3 Coefficient of Variation

The coefficient of variation is a statistic that is the ratio of the standard deviation to the mean expressed in percentage and is denoted CV. The coefficient of variation essentially is a relative comparison of a standard deviation to its mean. Suppose 5 weeks of average yield of a tree is 57, 68, 64, 71 and 62. To compute a coefficient of variation for these prices, first determine the mean and standard deviation $\mu = 64.40$ and $\sigma = 4.84$. The coefficient of variation is:

$$\mathrm{CV}_A = \frac{\sigma_A}{\mu_A}(100) = \frac{4.84}{64.40}(100) = 0.075 = 7.5\%$$

The standard deviation is 75% of the mean.

7.1.4 Probability

Statistical probability is a procedure for predicting the outcome of events wherein it may range from 0 (an event is certain not to occur) to 1.0 (an event is certain to occur). Genetic ratios may be expressed as probabilities. Consider a heterozygous plant (Rr). The probability that a gamete will carry the R allele is ½. In a cross, Rr × Rr (selfing), the probability of a homozygous recessive (rr offspring) is ½ × ½ = ¼. Using the cross Rr × Rr, the F2 will produce RR:Rr:rr in the ratio ¼ : ½ : ¼. In using probabilities for prediction, it is important to note that a large population size is needed for accurate prediction. For example, in a dihybrid cross, the F2 progeny will have 9:3:3:1 phenotypic ratio, indicating 9/16 will have the dominant phenotype. However, in a sample of exactly 16 plants, it is unlikely that exactly 9 plants will have the dominant phenotype. For accurate prediction, a larger sample is needed.

7.1.5 Normal Distribution

A continuous random variable has an infinite number of possible values that can be represented by an interval. Its probability distribution is called a continuous probability distribution. The continuous probability distribution in statistics is the normal distribution. Normal distributions can be used to model many sets of measurements like height of the plants in a heterogeneous population, length of the leaves in a plant, petal length of flowers and so on. Such variables are normally distributed random variables (Fig. 7.1).

A normal distribution is a continuous probability distribution for a random variable *x*. The graph of a normal distribution is called the normal curve. A normal distribution has the following properties:

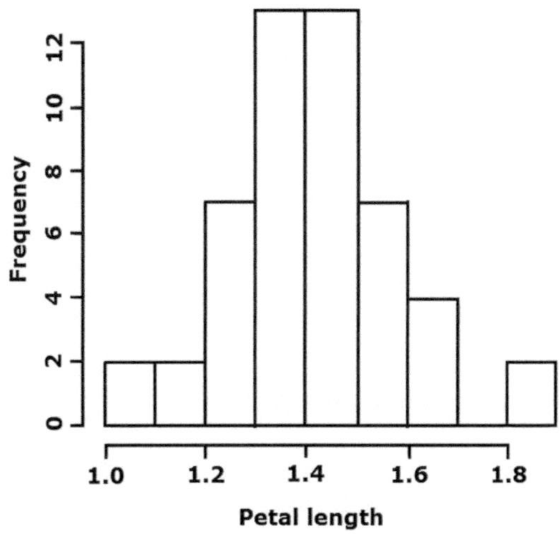

Fig. 7.1 Continuous probability distribution. Normal distributions can be used to model many sets of measurements like height of the plants in a heterogeneous population, length of the leaves, petal length of flowers and so on. Such variables are normally randomly distributed

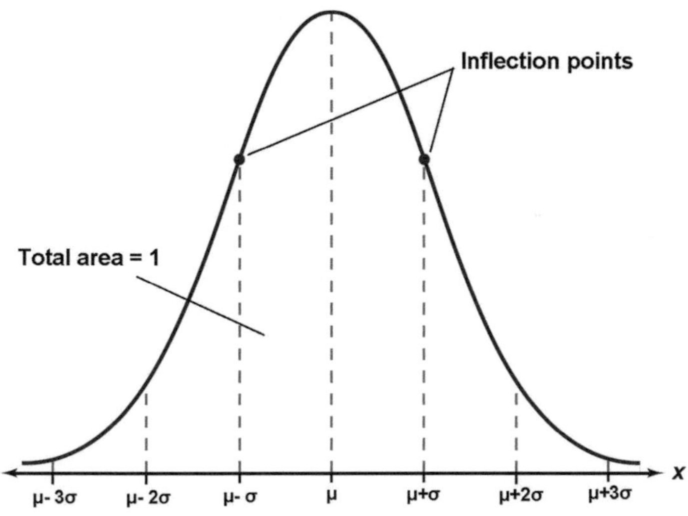

Fig. 7.2 A normal distribution with a continuous probability distribution for a random variable *X*

(a) The mean, median and mode are equal.
(b) The normal curve is bell shaped and is symmetric about the mean.
(c) The total area under the normal curve is equal to 1.
(d) The normal curve approaches, but never touches, the *x*-axis as it extends farther and farther away from the mean.
(e) Between $\mu - \sigma$ and $\mu + \sigma$ (in the centre of the curve), the graph curves downwards. The graph curves upwards to the left of $\mu - \sigma$ and to the right of $\mu + \sigma$. The points at which the curve changes from curving upwards to curving downwards are called *inflection points* (see Fig. 7.2).

If there is a continuous random variable having a normal distribution with mean μ and standard deviation σ, you can graph a normal curve using the equation:

$$y = \frac{1}{\sigma\sqrt{2\pi}} e^{-(x-\mu)^2/2\,\sigma^2}$$

$e \approx 2.718$ and $\pi \approx 3.14$

7.1.6 Statistical Hypothesis

Hypothesis testing is a kind of statistical inference that involves asking a question, collecting data and then examining what the data tells us. There are always two hypotheses. The hypothesis to be tested is called the null hypothesis and given the symbol H_0. The null hypothesis states that there is no difference between a hypothesized population mean and a sample mean. It is the status quo hypothesis. For example, to test a hypothesis that an awn of wheat contains 20 spikelets, the null hypothesis is $H_0 : \mu = 20$. The alternate hypothesis (H_a) is just the opposite of the null hypothesis and can be expressed as $H_a : \mu \neq 20$.

The alternative hypothesis can be supported only by rejecting the null hypothesis. To reject the null hypothesis means to find a large enough difference between your sample mean and the hypothesized (null) mean. It raises real doubt that the true population mean is 20. If the difference between the hypothesized mean and the sample mean is very large, we reject the null hypothesis. If the difference is very small, we do not reject the null hypothesis. In each hypothesis test, we have to decide how much difference must be allowed to reject the null hypothesis (Fig. 7.3). Note that if we fail to find a large enough difference to reject, we fail to reject the null hypothesis.

One must first choose a level of significance or alpha (α) level for their hypothesis test. The most frequently used levels of significance are 0.05 and 0.01. An alpha level of 0.05 means that we will consider our sample mean to be significantly different from the hypothesized mean if the chances of observing that sample mean are less than 5%. Similarly, an alpha level of 0.01 means that we will consider

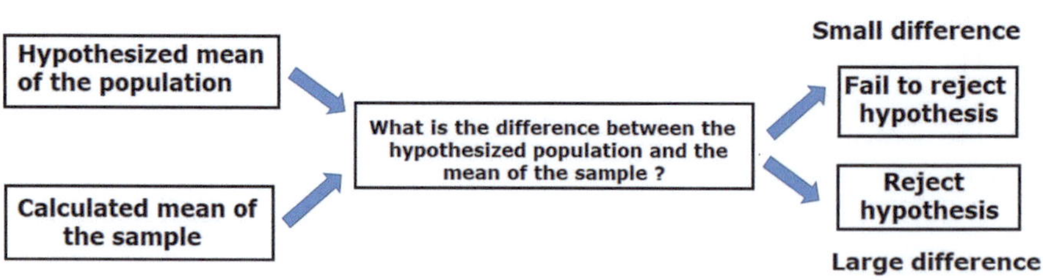

Fig. 7.3 Acceptance and rejection of hypothesis

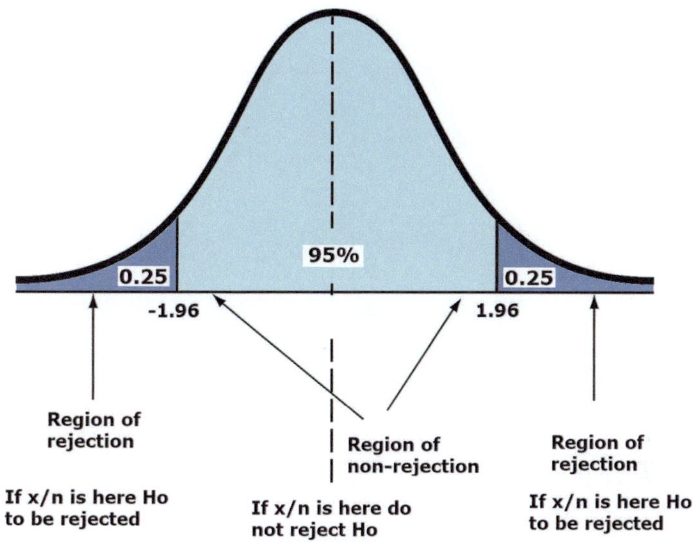

Fig. 7.4 Hypothesis testing. If the difference between the hypothesized mean and the sample mean is very large, we reject the null hypothesis. If the difference is very small, we do not reject the null hypothesis

our sample mean to be significantly different from the hypothesized mean if the chances of observing that sample mean are less than 1%.

A hypothesis test can be one-tailed or two-tailed. In a two-tailed test, the null hypothesis will be rejected if the sample mean falls in either tail of the distribution. For this reason, the alpha level (let's assume 0.05) is split across the two tails. The curve in Fig. 7.4 shows the critical regions for a two-tailed test. These are the regions under the normal curve with a probability of 0.05. Each tail has a probability of 0.025. The z-scores that designate the start of the critical region are called the critical values. If the sample mean taken from the population falls within these critical regions, or "rejection regions", it can be concluded that difference is too much and the null hypothesis will be rejected. If the mean from the sample falls in the middle of the distribution (in between the critical regions), the null hypothesis will not be rejected. When the direction of the results is anticipated or we are only interested in one direction of the results, one can use a single-tail hypothesis. In single-tail hypothesis test, the alternative hypothesis looks a bit different. Symbols of greater than or less than are used here. When a wheat awn contains more than 20 spikelets, it will be considered as *greater* than 20. Then the null hypothesis is $H_0 : \mu \leq 20$. The alternate hypothesis (H_a) is just the opposite of the null hypothesis and can be expressed as $H_a : \mu > 20$. In single-tail hypothesis, there is only one critical region because we put the entire critical region into just one side of the distribution. When the alternative hypothesis is that the sample mean is greater, the critical region is on the right side of the distribution. When the alternative hypothesis is that the sample is smaller, the critical region is on the left side of the distribution (Fig. 7.5).

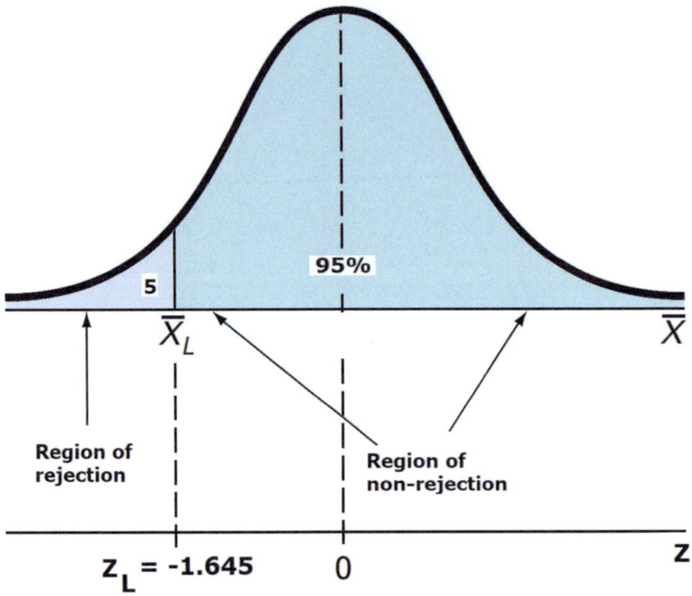

Fig. 7.5 Determining the lower critical value for a one-tail *Z* test for a population mean at the 0.05 level of significance

Table 7.1 Four possible outcomes of hypothesis testing

Decision made	Null hypothesis is true	Null hypothesis is false
Reject null hypothesis	Type I error	Correct decision
Do not reject null hypothesis	Correct decision	Type II error

While rejecting the null hypothesis, we have four possible scenarios: (a) a true hypothesis is rejected; (b) a true hypothesis is not rejected; (c) a false hypothesis is not rejected; and (d) a false hypothesis is rejected. We exercise correctness when options b and d are accepted. But when we accept options a and c, we make an error. Two types of errors can occur in hypothesis testing: type I and type II (Table 7.1).

7.1.7 Standard Error of the Mean

This is a statistic which represents an estimate of the standard deviation that would be present within a sampling distribution of means if it was constructed based on information drawn from a single sample. This estimate of the standard deviation is known as the standard error of the mean. The formula for the standard error of the mean is as follows:

$$s\bar{x} = \frac{s}{\sqrt{n-1}}$$

$s\bar{x}$ = standard error of the mean
s = standard deviation of the sample
$\sqrt{n-1}$ = square root of the number of observations in the sample minus 1

7.2 Correlation Coefficient (*r*)

In statistics, the word correlation refers to the relationship between two variables. One variable might be the number of seeds per panicle and the other could be length of panicle. Perhaps as the number of seeds increases, the length of panicle increases. This is an example of a *positive correlation*. When one variable increases and other decreases, it is *negative correlation*. The correlation coefficient is a measure of how well the predicted values from a forecast model "fit" with the real-life data. The correlation coefficient is a number between 0 and 1. If there is no relationship between the predicted values and the actual values, the correlation coefficient is 0 or very low (the predicted values are no better than random numbers). As the strength of the relationship between the predicted values and actual values increases, so does the correlation coefficient. A perfect fit gives a coefficient of 1.0. Thus, the higher the correlation coefficient, the better will be the relationship between two variables.

The correlation coefficient is calculated as:

$$r = \frac{\sum xy}{\sqrt{\left(\sum x^2\right)\left(\sum y^2\right)}}.$$

For calculating *r*, let us take the following example of total anthocyanin and total pigments per leaf in the leaves of a plant (Table 7.2).

Compute means, corrected sums of squares and corrected sum of cross products as follows:

$$\bar{x} = \frac{\sum_x}{n}$$

$$\bar{y} = \frac{\sum_y}{n}$$

$$\sum_{x^2} = \sum_{i=1}^{n} (x_i - \bar{x})^2$$

$$\sum_{y^2} = \sum_{i=1}^{n} (y_i - \bar{y})^2$$

$$\sum_{xy} = \sum_{i=1}^{n} (x_i - \bar{x})(y_i - \bar{y})$$

where (x^1, y^1) represents the *i*th pair of the *x* and *y* values.

Table 7.2 Computation of correlation coefficient between anthocyanin and total pigments in leaves

Sample number	Total anthocyanin (mg/leaf)	Total pigments (mg/leaf)	Deviation from mean		Square of deviation		Product of deviations
	x	y	X	Y	X^2	Y^2	$(X^2)\ (Y^2)$
1	0.60	0.44	−0.37	−0.38	0.1369	0.1444	0.1406
2	1.12	0.96	0.15	0.14	0.0225	0.0196	0.0210
3	2.10	1.90	1.13	1.08	1.2769	1.664	1.2204
4	1.16	1.51	0.19	0.69	0.0361	0.4761	0.1311
5	0.70	0.46	−0.27	−0.36	0.0729	0.1296	0.0972
6	0.80	0.44	−0.17	−0.38	0.0289	0.1444	0.0646
7	0.32	0.04	−0.65	−0.78	0.4225	0.6084	0.5070
Total	6.80	5.75	0.01	0.01	1.9967	2.6889	2.1819
Mean	0.97	0.82					

Correlation coefficient r is computed as:

$$r = \frac{2.1819}{\sqrt{(1.9967)\ (2.6889)}} = 0.942$$

After calculation of r, compare the r value to the tabular r values from the correlation table with $(n = 2) = 5$ degrees of freedom, which are 0.754 at the 5% level of significance and 0.874 at the 1% level. Since the r value exceeds both the tabular r values, we can conclude that the correlation coefficient is significant at 1% level. This indicates that total anthocyanin and total pigment in the leaves are highly associated. Leaves with high anthocyanin contain high pigments and vice versa.

7.2.1 Regression Analysis

Regression analysis is a statistical procedure that allows a researcher to estimate the linear, or straight line, relationship that relates two or more variables. This linear relationship summarizes the amount of change in one variable that is associated with change in another variable or variables. Such a relationship can also be tested for statistical significance, to test whether the observed linear relationship could have emerged by mere chance. Linear regression explores relationships that can be readily described by straight lines or their generalization to many dimensions. A large number of problems can be solved by linear regression. Also, more analysis can be done by means of transformation of the original variables that result in linear relationships among the transformed variables.

When there is a single continuous dependent variable and a single independent variable, the analysis is called a *simple linear regression analysis*. *Multiple*

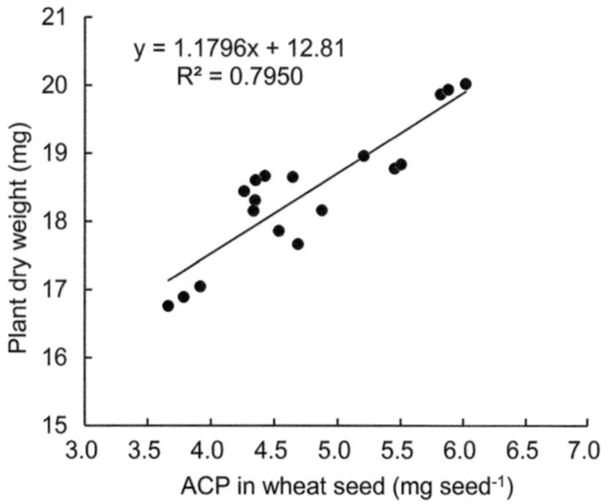

Fig. 7.6 Regression analysis of absolute content of protein (ACP) in wheat seed and plant dry weight at seedling stage

regression is the relationship between several independent or predictor variables and a dependent or criterion variable. *Independent variables* are characteristics that can be measured directly, and *dependent variable* is a characteristic whose value depends on the values of independent variables.

Simple linear regression allows to study relationships between two continuous (quantitative) variables (Fig. 7.6). In a cause and effect relationship, the independent variable is the cause, and the dependent variable is the effect. Least squares linear regression is a method for predicting the value of a dependent variable y, based on the value of an independent variable x. One variable, denoted (x), is regarded as the predictor, explanatory or independent variable. The other variable, denoted (y), is regarded as the response, outcome or dependent variable. Mathematically, the regression model is represented by the following equation:

$$y = \beta_0 \pm \beta_1 x_1 \pm \varepsilon_1$$

where x is independent variable; y is dependent variable; n is number of cases or individuals; Σ_{xy} is sum of the product of dependent and independent variables; β_1 is the slope of regression line; β_0 is the intercept point of the regression line and the y-axis; Σ_x is sum of independent variable; Σ_y is sum of dependent variable; and Σx^2 is sum of square independent variable.

$$\beta_1 = \frac{n\sum_{xy} - \sum_x \sum_y}{n\sum x^2 - \left(\sum_x\right)^2}$$

$$\beta_0 = \overline{y} - \beta_1 \overline{x}$$

Table 7.3 Calculation of linear regression of awn length (x) and grain weight (y) (hypothetical)

Observation	Awn length x	Grain weight y	xy	x^2	Required calculation
1	35	112	3920	1225	$\Sigma x = 491$
2	40	128	5120	1600	
3	38	130	4940	1444	$\Sigma y = 1410$
4	44	138	6072	1936	
5	67	158	10,586	4489	$\Sigma xy = 71{,}566$
6	64	162	10,368	4096	
7	59	140	8260	3481	
8	69	175	12,075	4761	$\Sigma x^2 = 26{,}157$
9	25	125	3125	625	
10	50	142	7100	2500	
Total	491	1410	71,566	26,157	

Calculation of regression from a hypothetical data is available in Table 7.3.

$$\text{average of } x = 49.1$$
$$\text{average of } y = 141$$
$$\beta_1 = \frac{715660 - 692310}{261570 - 241081}$$
$$\beta_1 = \frac{23350}{20489} = 1.140$$
$$\beta_0 = 141 - 1.140 \times 49.1$$
$$\beta_0 = 141 - 55.974$$
$$\beta_0 = 85.026$$

Substitute the regression coefficient into the regression model

$$\text{Estimated grain weight } (\hat{y}) = 85.026 + 1.140\,x$$

7.3 Heritability

Heritability is the variation which is transferred from parents to their offspring. Heritability is a concept that summarizes how much of the variation in a trait is due to variation in genetic factors. The remaining variation is usually attributed to environmental factors. Often, this term is used in reference to the resemblance between parents and their offspring. In this context, high heritability implies a strong resemblance between parents and offspring with regard to a specific trait, while low heritability implies a low level of resemblance.

Phenotypes that vary between the individuals in a population do so because of both environmental factors and the genes that influence traits and various interactions between genes and environmental factors. Unless they are genetically identical (e.g. monozygotic twins in humans, inbred lines in experimental populations or clones), the individuals in a population tend to vary in the genotypes they have at the loci affecting particular traits. The combined effect of all loci, including possible allelic interactions within loci (dominance) and between loci (epistasis), is the genotypic value. This value creates genetic variation in a population when it varies between individuals. In fact, heritability is formally defined as the proportion of phenotypic variation (V_P) that is due to variation in genetic values (V_G).

Broad-sense heritability, defined as $H^2 = V_G/V_P$, captures the proportion of phenotypic variation due to genetic values that may include effects due to dominance and epistasis. On the other hand, narrow-sense heritability, $h^2 = V_A/V_P$, captures only that proportion of genetic variation that is due to additive genetic values (V_A). Often, no distinction is made between broad- and narrow-sense heritability; however, narrow-sense h^2 is most important in animal and plant selection programmes, because response to artificial (and natural) selection depends on additive genetic variance. Moreover, resemblance between relatives is mostly driven by additive genetic variance. Given its definition as a ratio of variance components, the value of heritability always lies between 0 and 1.

7.3.1 Heritability and the Partitioning of Total Variance

Population parameters: Observed phenotypes (P) of a trait of interest can be partitioned, according to biologically plausible nature-nurture models, into a statistical model representing the contribution of the unobserved genotype (G) and unobserved environmental factors (E):

$$\text{Phenotype } (P) = \text{Genotype } (G) + \text{Environment } (E)$$

The variance of the observable phenotypes ($\sigma^2 P$) can be expressed as a sum of unobserved underlying variances:

$$\sigma^2 P = \sigma^2 G + \sigma^2 E$$

Heritability is defined as a ratio of variances, by expressing the proportion of the phenotypic variance that can be attributed to variance of genotypic values:

$$\text{Heritability (broad sense)} = H^2 = \frac{\sigma^2 G}{\sigma^2 P}$$

The genetic variance can be partitioned into the variance of additive genetic effects (breeding values; σ^2 A), of dominance (interactions between alleles at the same locus), of genetic effects ($\sigma^2 D$) and of epistatic (interactions between alleles at different loci) genetic effects ($\sigma^2 I$):

$$\sigma^2 G = \sigma^2 A + \sigma^2 D + \sigma^2 I$$

and heritability (narrow or strict sense) $= h^2 = \frac{\sigma^2 A}{\sigma^2 P}$

In general, $\sigma^2 E$ can be broken down into any number of identifiable, but random, contributing factors that can be specific to the phenotype. Examples include the environmental variance that is common to specified groups, for example, siblings and litters (σ^2CE), and the non-genetic variance that is common to repeated measures of individuals (σ^2PE). We define the remainder of the environmental variance, which cannot be attributed to other factors, as the environmental residual variance, which includes individual stochastic error variance and measurement error (σ^2RE):

$$\sigma^2 E = \sigma^2 \mathrm{CE} + \sigma^2 \mathrm{PE} + \sigma^2 \mathrm{RE}$$

7.4 Principles of Experimental Design

For successful execution of a trial on plant breeding, randomization, replication and local control are vital principles. For instance, when we lay a trial to find out the best variety for a particular location, and the analysis is to identity the best variety from a set of varieties, the experiment needs to be done in a large area, and the aforesaid principles are vital for meaningful data collection and interpretation.

7.4.1 Randomization

The first principle of an experimental design is randomization. This is a random process of assigning treatments to the experimental units. It means that every possible allotment of treatments has the same probability. An experimental unit is the smallest division of the experimental material. A treatment means an experimental condition whose effect is to be measured and compared. The purpose of randomization is to remove bias and other sources of extraneous variation which are not controllable. For example, when we conduct experiment in a large area, randomization can nullify the effect due to soil heterogeneity. Randomization forms the basis of any valid statistical test. Hence, the treatments must be assigned at random to the experimental units. Randomization is usually done by drawing numbered cards from a well-shuffled pack of cards, by drawing numbered balls from a well-shaken container or by using tables of random numbers.

7.4.2 Replication

The second principle is replication, which is a repetition of the basic experiment. In all experiments, experimental units such as individuals or plots of land in breeding experiments cannot be physically identical. This type of variation can be removed by using a number of experimental units. So, the experiment needs to be performed more than once, i.e. we repeat the basic experiment. An individual repetition is called a replicate. The number, the shape and the size of replicates depend upon the nature of the experimental material.

Thus, a replication is:

(a) To secure a more accurate estimate of the experimental error
(b) To decrease the experimental error and thereby increase precision

7.4.3 Local Control

We need to choose a design in such a manner that all extraneous sources of variation are brought under control. For this purpose, we make use of local control, a term referring to the amount of balancing. Balancing means that the treatments should be assigned to the experimental units in such a way that the result is a balanced arrangement of the treatments. The main purpose of the principle of local control is to increase the efficiency of an experimental design by decreasing the experimental error. For example, in an analysis of several varieties to find out the best variety for a particular location, a high-yielding local variety is introduced in the experiment so that when we select the best high-yielding variety, that variety must have significantly better yield than local control.

Experiments are many like single-factor experiment, two-factor experiments and three- or more factor experiments. Such experimental layouts will be briefly explained here.

In single-factor experiments, the treatments consist solely of the different levels of the single-variable factor. All other factors are applied uniformly to all plots at a single prescribed level. There are two groups of experimental designs that are applicable to a single-factor experiment, viz. complete block designs and incomplete block designs. Complete block design is a group of designs which is suited for experiments with small number of treatments and is characterized by blocks, each of which contains at least one complete set of treatments. Incomplete block designs are suited for experiments with a large number of treatments and are characterized by blocks, each of which contains only a fraction of the treatments to be tested. Incomplete block designs are out of scope of this book, and hence, only complete block designs will be covered here. Complete block designs are (a) completely randomized design (CRD), (b) randomized complete block design (RCBD) and (c) Latin square design (LS).

7.4.4 Completely Randomized Design (CRD)

This is done when there is no significant variation in the area or environment. Generally, CRD is applicable for laboratory or greenhouse experiments only. The advantage of CRD is that it can be used for experiments with equal or unequal number of treatments or vice versa and can be used for treatments with uncqual number of replications. The main disadvantage is the restriction of providing uniform condition in the whole experimental area (Fig. 7.7). For data analysis, the data has to be arranged in a simplified manner that will allow easy reading of values of each treatment. A two-way table is constructed putting together in one row all the observations for a particular treatment (Table 7.4a). After arranging all the values, the total of each treatment, total of each replication and mean of each treatment are computed as shown in Table 7.4b.

Degrees of Freedom (df):

$$\text{Treatment} = t - 1 = 6 - 1 = 5$$
$$\text{Error} = t\,(r - 1) = 6\,(6 - 1) = 30$$
$$\text{Total} = tr - 1 = 6 \times 6 - 1 = 35$$

The formula for the sum of squares of each source of variation can be computed.

Correction factor:

$$\text{C.F.} = \frac{GT^2}{tr} = \frac{(551)^2}{6 \times 6} = 8433.3611$$

Total sum of square (ToSS):

$$\begin{aligned} &= \Sigma\Sigma\,(TR)^2 - \text{C.F.} = (T_1R_1)^2 + (T_1R_2)^2 + \cdots + +(T_6R_6)^2 - \text{C.F.} \\ &= \left(17^2 + 20^2 + \cdots + 16^2\right) - 8{,}433.3611 \\ &= 8.745.0000 - 8{,}433.3611 = 311.6389 \end{aligned}$$

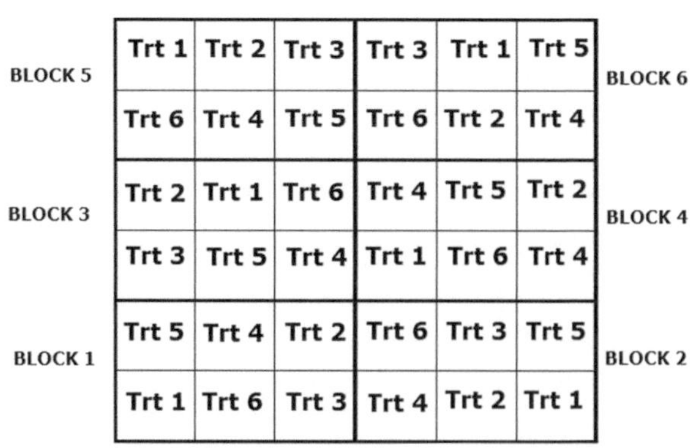

Fig. 7.7 Completely randomized design

Table 7.4a Two-way table constructed by putting together in one row all the observations for a particular treatment

Treatment	Rep1	Rep 2	Rep3	Rep4	Rep5	Rep6
Treatment 1	17	20	17	18	16	17
Treatment 2	18	14	19	11	15	17
Treatment 3	18	22	18	14	11	18
Treatment 4	16	22	14	12	13	14
Treatment 5	15	12	12	11	11	13
Treatment 6	13	15	13	14	15	16

Table 7.4b Total of each replication and mean of each treatment

Treatment	Rep1	Rep 2	Rep3	Rep4	Rep5	Rep6	Total	Mean
Treatment 1	17	20	17	18	16	17	105	17.5
Treatment 2	18	14	19	11	15	17	94	15.7
Treatment 3	18	22	18	14	11	18	101	16.8
Treatment 4	16	22	14	12	13	14	91	15.2
Treatment 5	15	12	12	11	11	13	74	12.3
Treatment 6	13	15	13	14	15	16	86	14.3
Total	97	105	93	80	81	95	551	15.3

Treatment sum of squares (TrSS):

$$\frac{\sum T^2}{r} - \text{C.F.} = \frac{T1^2 + T2^2 + \cdots T6^2}{r} - \text{C.F.}$$
$$= 105^2 + 94^2 + \cdots 86^2 - 8{,}433.3611$$
$$= 8535.8333 - 8{,}433.3611 = 102.4722$$

Error sum of squares (ESS):

$$\sum\sum TR^2 - \frac{\sum T^2}{r} = (T_1R_1)^2 + (T_1R_2)^2 + (T_6R_6)^2 - \frac{T1^2 + T2^2 + \cdots T6^2}{r}$$
$$= \left(17^2 + 19^2 + \cdots 16^2\right) - \frac{\left(105^2 + 94^2 + 86^2\right)}{6}$$
$$= 8{,}745.0000 - 8535.8333 = 209.1667$$

Total sum of squares (ToSS):

$$\sum\sum(TR)^2 - \text{C.F.} = (T_1R_1)^2 + (T_1R_2)^2 + (T_6R_6)^2 - \text{C.F.}$$
$$= \left(17^2 + 19^2 + \cdots 15^2\right) - \frac{105^2 + 94^2 + 86^2}{6} - 8433.3611$$
$$= 8{,}745.0000 - 8433.3611 = 311.6389$$

Block (replication) sum of squares (RSS):

$$\frac{\sum R^2}{t} - \text{C.F.} = \frac{R_1^2 + R_2^2 + \cdots R_6^2}{t} - \text{C.F.}$$
$$= \frac{97^2 + 105^2 + \cdots 95^2}{6} - 8{,}433.3611$$
$$= 8511.5000 - 8433.3611 = 78.1389$$

Treatment sum of squares (TrSS):

$$\frac{\sum T^2}{r} - \text{C.F.} = \frac{T_1^2 + T_2^2 + \cdots T_6^2}{r} - \text{C.F.}$$
$$\frac{105^2 + 94^2 + \cdots 86^2}{6} - 8433.3611$$
$$= 8{,}535.8333 - 8{,}433.3611 = 102.4722$$

Error sum of squares (ESS):

$$\sum\sum TR^2 - \frac{\sum T^2}{r} - \frac{\sum R^2}{t} + \text{C.F.}$$
$$= \left(17^2 + 20^2 + \cdots 15^2\right) - \frac{\left(105^2 + 94^2 + \cdots 85^2\right)}{6}$$
$$- \frac{\left(97^2 + 105^2 + \ldots\ldots 94^2\right)}{6} + 8{,}433.3611$$
$$= 8{,}745.0000 - 8{,}535.8333 - 8{,}511.5000 \quad + 8{,}433.3611 = 131.0278$$

Mean squares:
Treatment mean square (TrMS):

$$\frac{\text{TrSS}}{\text{Trdf}} = \frac{102.4722}{5} = 20.4944$$

Block mean square (RSS):

$$\frac{\text{RSS}}{\text{R df}} = \frac{78.1389}{5} = 15.6278$$

Error mean square (ESS):

$$\frac{\text{ESS}}{\text{E df}} = \frac{131.0278}{25} = 5.2411$$

F computed:
Block F computed (RFc):

$$\frac{\text{RMS}}{\text{EMS}} = \frac{15.6278}{5.2411} = 2.98$$

Treatment F computed (TrFc):

$$\frac{\text{TrMS}}{\text{EMS}} = \frac{204944}{5.2411} = 3.91$$

To double-check the correctness of computation of sum of squares the treatment SS and error SS and to compare them with the total SS in the example, the calculation would be:

TrSS + ESS = 102.4722 + 209.1667 = **311.6389** so computation is correct.

$$\text{Treatment Mean Square (TrMS)} = \frac{\text{TrSS}}{\text{Tr df}} = \frac{102.4722}{5}$$

= 20.4944

$$\text{Error Mean Square (ESS)} = \frac{\text{ESS}}{\text{E df}} = \frac{209.1667}{30}$$

= 6.9722

$$\text{Treatment F Computed (TrFc)} = \frac{\text{TrMS}}{\text{EMS}} = \frac{20.4944}{6.9722} = 2.94$$

Analysis of variance (ANOVA) table can be constructed as given in Table 7.5. The significance of *F* value can be judged through verifying with the *F* table.

7.4.5 Randomized Complete Block Design (RCBD)

Experiments in the open field are conducted using randomized complete block design (RCBD) since condition is not under control. Variation may be due to the soil fertility and type, slope or gradient, wind direction, water direction, etc. Through RCBD, blocking is introduced which will help to reduce such factors. RCBD is considered to be powerful because it is able to partition the total variance into the effect of the treatment, the effect of the block and the unexplained error. Blocking is a method of improving accuracy by arranging the experimental materials into groups so that the units in each group are as homogeneous (uniform) as possible, thereby eliminating the variability between groups. If the fertility of the area is not known, the blocks and plots may be arranged as given in Fig. 7.8. Let us take the data of Tables 7.4a and 7.4b for ANOVA.

Table 7.5 Analysis of variance (ANOVA) **CRD** table can be constructed as given in Tables 7.4a and 7.4b

Source	df	SS	MS	Fc	Ft 1%	Ft5%
Treatment	5	102.472	20.4944	2.94*	3.70	2.53
Error	30	209.1667	6.9722			
Total	35	311.6389				

The significance of F value can be judged through verifying with the F table
*significant at 5% level; **significant at 1% level

C.V. $= \frac{\sqrt{\text{EMS}}}{\text{mean}} + \frac{\sqrt{6.9722}}{15.4} \times 100$

For F-computed values, it is enough to maintain two decimal places because the values in the F table (F_t) are up to two decimal places only

Fig. 7.8 Randomized complete block design

T5 R1	T2 R6	T4 R3	T2 R5	T3 R5	T1 R1
T2 R4	T4 R4	T5 R2	T6 R5	T5 R3	T3 R6
T6 R1	T1 R4	T1 R3	T4 R2	T6 R3	T1 R2
T3 R2	T4 R5	T6 R2	T6 R4	T6 R6	T5 R6
T1 R5	T3 R3	T5 R6	T2 R3	T5 R4	T4 R1
T3 R1	T2 R1	T4 R6	T1 R6	T3 R4	T2 R2

Degrees of Freedom (df):

$$\begin{aligned} \text{Treatment} &= t - 1 = 6 - 1 = 5 \\ \text{Block} &= (r - 1) = 6 - 1 = 5 \\ \text{Error} &= (t - 1)\ (r - 1) = (6 - 1)\ (6 - 1) = 25 \\ \text{Total} &= tr - 1 = 6 \times 6 - 1 = 35 \end{aligned}$$

The formula for the sum of squares of each source of variation can be computed.

Sum of Squares

Correction factor:

$$\text{C.F.} = \frac{GT^2}{tr} = \frac{(551)^2}{6 \times 6} = 8433.3611$$

Total sum of square (ToSS):

$$\begin{aligned} &= \Sigma\Sigma\ (TR)^2 - \text{C.F.} = (T_1R_1)^2 + (T_1R_2)^2 + \cdots + +(T_6R_6)^2 - \text{C.F.} \\ &= \left(17^2 + 20^2 + \cdots + 16^2\right) - 8{,}433.3611 \\ &= 8.745.0000 - 8{,}433.3611 = 311.6389 \end{aligned}$$

Treatment sum of squares (TrSS):

$$\frac{\sum T^2}{r} - \text{C.F.} = \frac{T1^2 + T2^2 + \cdots T6^2}{r} - \text{C.F.}$$
$$= 105^2 + 94^2 + \cdots 86^2 - 8{,}433.3611$$
$$= 8535.8333 - 8{,}433.3611 = 102.4722$$

Error sum of squares (ESS):

$$\sum\sum TR^2 - \frac{\sum T^2}{r} = (T_1 R_1)^2 + (T_1 R_2)^2$$
$$+ (T_6 R_6)^2 - \frac{T1^2 + T2^2 + \cdots T6^2}{r}$$
$$= \left(17^2 + 19^2 + \cdots 16^2\right) - \frac{\left(105^2 + 94^2 + 86^2\right)}{6}$$

$$= 8{,}745.0000 - 8535.8333 = 209.1667$$

Total sum of squares (ToSS):

$$\sum\sum (TR)^2 - \text{C.F.} = (T_1 R_1)^2 + (T_1 R_2)^2 + (T_6 R_6)^2 - \text{C.F.}$$
$$= \left(17^2 + 19^2 + \cdots 15^2\right) - \frac{\left(105^2 + 94^2 + 86^2\right)}{6} \qquad - 8433.3611$$
$$= 8{,}745.0000 - 8433.3611 = 311.6389$$

Block (replication) sum of squares (RSS):

$$\frac{\sum R^2}{t} - \text{C.F.} = \frac{R_1^2 + R_2^2 + \cdots R_6^2}{t} - \text{C.F.}$$
$$= \frac{97^2 + 105^2 + \cdots 95^2}{6} - 8{,}433.3611$$
$$= 8511.5000 - 8433.3611 = 78.1389$$

Treatment sum of squares (TrSS):

$$\frac{\sum T^2}{r} - \text{C.F.} = \frac{T_1^2 + T_2^2 + \cdots T_6^2}{r} - \text{C.F.}$$
$$\frac{105^2 + 94^2 + \cdots 86^2}{6} - 8433.3611$$
$$= 8{,}535.8333 - 8{,}433.3611 = 102.4722$$

Error sum of squares (ESS):

$$\sum\sum TR^2 - \frac{\sum T^2}{r} - \frac{\sum R^2}{t} + \text{C.F.}$$
$$= \left(17^2 + 20^2 + \cdots 15^2\right) - \frac{\left(105^2 + 94^2 + \cdots 85^2\right)}{6}$$
$$- \frac{\left(97^2 + 105^2 + \cdots 94^2\right)}{6} + 8{,}433.3611$$
$$= 8{,}745.0000 - 8{,}535.8333 - 8{,}511.5000 \quad + 8{,}433.3611 = 131.0278$$

Mean squares:
Treatment mean square (TrMS):

$$\frac{\text{TrSS}}{\text{Trdf}} = \frac{102.4722}{5} = 20.4944$$

Block mean square (RSS):

$$\frac{\text{RSS}}{\text{Rdf}} = \frac{78.1389}{5} = 15.6278$$

Error mean square (ESS):

$$\frac{\text{ESS}}{\text{Edf}} = \frac{131.0278}{25} = 5.2411$$

F computed:
Block F computed (RFc):

$$\frac{\text{RMS}}{\text{EMS}} = \frac{15.6278}{5.2411} = 2.98$$

Treatment F computed (TrFc):

$$\frac{\text{TrMS}}{\text{EMS}} = \frac{204944}{5.2411} = 3.91$$

See Table 7.6 for ANOVA.

Table 7.6 ANOVA for RCBD

Source	df	SS	MS	Fc	Ft 1%	Ft5%
Block	5	78.1389	15.6278	2.98*	3.86	2.60
Treatment	5	102.4722	20.4944	3.91**	3.86	2.60
Error	25	131.0278	5.2411			
Total	35	311.6389				

$\text{C.V.} = \frac{\sqrt{\text{EMS}}}{\text{mean}} + \frac{\sqrt{5.2411}}{15.3} \times 100$

7.4.6 Latin Square Design

LSD is useful when the direction of soil fertility/heterogeneity is bidirectional. RCBD will take care of only one gradient, while the other gradient will be confounded (or added) to the treatment effect. Latin square is the more appropriate design because the two-directional blocking, commonly referred to as row blocking and column blocking, is accomplished by ensuring that every treatment occurs only once in each row block and once in each column block. LSD also detects differences due to rows and columns and not due to blocks alone. An LSD layout is available in Fig. 7.9. The same set of hypothetical data used in CRD and RCBD involving six treatments (designated by letters in parenthesis) will be used with the assigned columns and rows included in Tables 7.4a and 7.4b.

Degrees of Freedom (df)

$$\begin{aligned} \text{Treatment} &= t - 1 = 6 - 1 = 5 \\ \text{Column} &= c - 1 = 6 - 1 = 5 \\ \text{Row} &= r - 1 = 6 - 1 = 5 \\ \text{Error} &= (t - 1)(t - 2) = (6 - 1)(6 - 2) = 20 \\ \text{Total} &= tr - 1 = 6 \times 6 - 1 = 35 \end{aligned}$$

In Latin square, the number of treatments (t) equals the number of columns (c) equals the number of rows (r), only t will be used as divisor in the formula to find the sums of squares.

Sums of squares:

$$\text{C.F.} = \frac{(GT^2)}{t^2} \frac{(551^2)}{6^2} = 8{,}433.3611$$

Total sum of squares (ToSS):

Fig. 7.9 Latin square design

	COL 1	COL 2	COL 3	COL 4	COL 5	COL 6
ROW 1	Trt 3	Trt 2	Trt 4	Trt 1	Trt 6	Trt 5
ROW 2	Trt 2	Trt 1	Trt 3	Trt 6	Trt 5	Trt 4
ROW 3	Trt 5	Trt 4	Trt 6	Trt 3	Trt 2	Trt 1
ROW 4	Trt 4	Trt 3	Trt 5	Trt 2	Trt 1	Trt 6
ROW 5	Trt 6	Trt 5	Trt 1	Trt 4	Trt 3	Trt 2
ROW 6	Trt 1	Trt 6	Trt 2	Trt 5	Trt 4	Trt 3

ToSS can be computed using the sequence of treatment × column, treatment × row, column × row or row × column. Here, row × column is used.

$$\begin{aligned}\sum\sum(\text{RoCo})^2 - \text{C.F.} &= (\text{Ro}_1\,\text{Co}_1)^2 \\ &+ (\text{Ro}_1\,\text{Co}_2)^2 + \ldots\ldots(\text{Ro}_6\,\text{Co}_6)^2 - \text{C.F.} \\ &= \left(17^2 + 15^2 + \ldots.. + 17^2\right) - 8433.3622 \\ &= 8{,}745.0000 - 8433.3611 = 311.6389\end{aligned}$$

Treatment sum of squares (TrSS):

$$\begin{aligned}\frac{\sum}{t} - \text{C.F.} &= \frac{T_1^2 + T_2^2 + \cdots T_6^2}{t} - \text{C.F.} \\ &= \frac{105^2 + 94^2 + \cdots 86^2}{6} - 8{,}433.3611 \\ &= 8{,}535.833 - 8{,}433.3611 = 102.4722\end{aligned}$$

Column sum of squares (CoSS):

$$\begin{aligned}\frac{\sum_{\text{Co}^2}}{t} - \text{C.F.} &= \frac{\text{Co}\,1^2 + \text{Co}\,2^2 + \cdots \text{Co}6^2}{t} - \text{C.F.} \\ &= \frac{97^2 + 105^2 + \cdots 94^2}{6} - 8{,}433.3611 \\ &= 8511.5000 - 8433.3611 = 78.1389\end{aligned}$$

Row sum of squares (RSS):

$$\begin{aligned}\frac{\sum_{\text{Ro}^2}}{t} - \text{C.F.} &= \frac{\text{Ro}1^2 + \text{Ro}2^2 + \cdots \text{Ro}6^2}{t} - \text{C.F.} \\ &= \frac{90^2 + 96^2 + \cdots 78^2}{6} - 8{,}433.3611 \\ &= 8{,}499.1667 - 8{,}433.3611 = 65.8056\end{aligned}$$

Error sum of squares (ESS):

The error df of Latin square, $(t-1)(t-2)$, when expanded is $t^2 - 3t + 2$. The term t^2 is the same as *tr* in CRD or RCBD. The term $3t$ refers to squares of treatments, squares of columns and squares of rows. Therefore, the formula to compute error SS for Latin square is:

$$\sum\sum_{(\text{TR})^2} - \frac{\sum_{\text{T}^2}}{t} - \frac{\sum_{\text{Co}^2}}{t} - \frac{\sum_{\text{Ro}^2}}{t} + 2\text{C.F.}$$

Since all these values have been computed as shown above, the final values are:

$$\begin{aligned}&= 8{,}745.0000 - 8{,}535.8333 - 8{,}511.5000 - 8{,}499.1667 \\ &\quad + 2 \times 8{,}433.3611 = 65.2222\end{aligned}$$

Mean squares:
Row mean squares (RoMS):

$$\frac{\text{RoSS}}{\text{Ro df}} = \frac{65.8056}{5} = 13.1611$$

Column mean squares (CoMS):

$$\frac{\text{CoSS}}{\text{Co df}} = \frac{78.1389}{5} = 15.6278$$

Treatment mean squares (TrMS):

$$\frac{\text{TrSS}}{\text{Tr df}} = \frac{102.4722}{5} = 20.4944$$

Error mean squares (EMS):

$$\frac{\text{ESS}}{\text{E df}} = \frac{65.2222}{20} = 3.2611$$

F computed:
Row F computed (RoFc):

$$\frac{\text{RoMS}}{\text{EMS}} = \frac{13.1611}{3.2611} = 4.04$$

Column F computed (CoFc):

$$\frac{\text{CoMS}}{\text{EMS}} = \frac{15.6278}{3.2611} = 4.79$$

Data and analysis of variance are presented in Tables 7.7a and 7.7b.

Table 7.7a Hypothetical data used in CRD and RCBD involving six treatments (designated by letters in parenthesis) used with the assigned columns and rows (as included in Table 7.6)

	Column							
Row	1	2	3	4	5	6	Row total	Trt total
1	(A)17	(F)15	(C)18	(D)12	(E)11	(B)17	90	(A)105
2	(B)18	(C)22	(E)12	(F)14	(A)16	(D)14	96	(B)94
3	(C)18	(D)22	(B)19	(A)18	(F)15	(E)13	105	(C)101
4	(D)16	(A)20	(F)13	(E)11	(B)15	(C)18	93	(D)91
5	(E)15	(B)14	(A)17	(C)14	(D)13	(F)16	89	(E)75
6	(F)13	(E)12	(D)14	(B)11	(C)11	(A)17	78	(F)86
Column total	97	105	93	80	81	94		551

Table 7.7b Analysis of variance for Latin square

Source	df	SS	MS	Fc	Ft 1%	Ft5%
Row	5	65.8056	13.1611	4.04*	4.10	2.71
Column	5	78.1389	15.6278	4.79**	4.10	2.71
Treatment	5	102.4722	20.4944	6.28**	4.10	2.71
Error	20	65.222	3.2611			
Total	35	311.6389				

$\text{C.V.} = \frac{\sqrt{\text{EMS}}}{\text{mean}} + \frac{\sqrt{3.2611}}{15.3} \times 100$

7.5 Tests of Significance

7.5.1 Chi-Square Test (for Goodness of Fit)

Chi-square test is used to determine whether the association between two qualitative variables is statistically significant. The following are the steps:

(a) Formulate hypotheses

Null hypothesis:
H_0: There is no significant association between total grains in an awn of wheat and awn length.
Alternative hypothesis:
H_a: There is a significant association between total grains in an awn of wheat and awn length.

(b) Specify the expected values for each cell of the table (when the null hypothesis is true). The formula for computing the expected values requires the sample size, the row totals and the column totals.

$$\text{Expected value} = \text{row total} - \text{column total}/\text{table total}$$

(c) If the data give convincing evidence against the null hypothesis, compare the observed counts from the sample with the expected counts, assuming H_0 is true.
(d) Compute the test statistic:

The chi-square statistic compares the observed values to the expected values. This test statistic is used to determine whether the difference between the observed and expected values is statistically significant. The chi-square statistic is a measure of how far the observed values are different from the expected ones. The formula is:

$$\chi^2 = \sum \frac{(\text{observed} - \text{expected})^2}{\text{expected}}$$

7.5.2 *t*-Test

The *t*-test is a type of inferential statistics. It is used to determine whether there is a significant difference between the means of two populations. A *t*-test can be used if we wish to compare the yield of side-dressed tomatoes and non-side-dressed tomatoes. With a *t*-test, we have one independent variable and one dependent variable. Here, the independent variable is the variety and the dependent variable is the awn length. If the independent had more than two levels, then we would use a one-way analysis of variance (ANOVA).

With a *t*-test, we wish to state with some degree of confidence that the obtained difference between the means of populations is too great to be a chance event and that some difference also exists in the population from which the sample was drawn. In other words, the difference that we might find between the yields of two populations in our sample might have occurred by chance, or it might exist in the population. If our *t*-test produces a *t*-value that results in a probability of 0.01, we say that the likelihood of getting the difference we found by chance would be 1 in a 100 times. We could say that it is unlikely that our results occurred by chance and the difference we found in the sample probably exists in the populations from which it was drawn.

Calculation of the test statistic requires three components:

The average of both samples (observed averages). Statistically, we represent these as:

$$\bar{x}_1 \text{ and } \bar{x}_2$$

The number of observations in both populations, represented as:

$$\text{SD}_1 \text{ and } \text{SD}_2$$

The number of observations in both populations, represented as:

$$n_1 \text{ and } n_2$$

Let's say an analysis of data comparing side-dressed tomatoes and non-side-dressed tomatoes showed the following:

	Side-dressed tomatoes	Non-side-dressed tomatoes
Average weight	3100 g	2750 g
SD	420	425
N	75	75

$$t = \frac{\bar{x}_1 \text{ and } \bar{x}_2}{\sqrt{\frac{\text{SD}_1 \text{ and } \text{SD}_2}{n_1 \quad + \quad n_2}}}$$

$$t = \frac{3100 - 2750}{\sqrt{\frac{420^2}{75} + \frac{425^2}{75}}}$$

$$t = \frac{350}{\sqrt{2352 + 2408.3}}$$

$$t = 5.07$$

7.6 Analysis of Variance

Analysis of variance (ANOVA) is a hypothesis-testing technique used to test the equality of two or more population (or treatment) means by examining the variances of samples. ANOVA allows one to determine whether the differences between the samples are simply due to random error (sampling errors) or whether there are systematic treatment effects that cause the mean in one group to differ from the mean of the other.

ANOVA is based on comparing the variance (or variation) between the data samples. If the between variation is much larger than the within variation, the means of different samples will not be equal. If the between and within variations are approximately the same size, then there will be no significant difference between sample means.

Assumptions of ANOVA:

(a) All populations involved follow a normal distribution.
(b) All populations have the same variance (or standard deviation).
(c) The samples are randomly selected and independent of each other.

For instance, if we wish to test the response of urea on three wheat varieties, *viz.*, PBW 373, PBW 435 and UP 2425 (control), a hypothetical data to be used is available in Table 7.8

Table 7.8 Mean yield (g) of two wheat varieties

	PBW373	PBW 435	UP 2425 (control)
	643	469	484
	655	427	456
	702	525	402
Mean	666.67	473.67	447.33
S	31.18	49.17	41.68

Null and alternative hypotheses:

The null hypothesis for an ANOVA always assumes the population means are equal. Hence, we may write the null hypothesis as

$H_0: \mu_1 = \mu_2 = \mu_3$. The mean yield/plot is statistically equal across the three varieties.

Since the null hypothesis assumes all the means are equal, we could reject the null hypothesis if only mean is not equal. Thus, the alternative hypothesis is:

H_a: At least one mean pressure is not statistically equal.

Calculate the appropriate test statistic:

The test statistic in ANOVA is the ratio of the between and within variation in the data. It follows an F distribution.

Total sum of squares – The total variation in the data. It is the sum of the between and within variation.

Total sum of squares (SST):

$$\sum_{i=1}^{r} \sum_{j=1}^{C} \left(X_{ij} - \bar{\bar{X}}\right)^2$$

where r is the number of rows in the table, c is the number of columns, Σ is the grand mean and $X\ ij$ is the ith observation in the jth column.

Using the data in Table 7.8, we may find the grand mean:

$$\bar{\bar{X}} = \frac{\sum X_{ij}}{N} = \frac{(643 + 655 + 702 + 469 + 427 + 525 + 484 + 456 + 402)}{9}$$
$$= 529.22$$

$$\begin{aligned} \text{SST} \\ &= (643 - 529.22)^2 + (655 - 529.22)^2 + 702 - 529.22)^2 \quad + (469 - 529.22)^2 \\ &\quad + \cdots (402 - 529.22)^2 \\ &= 96303.55 \end{aligned}$$

Between sum of squares (or treatment sum of squares) – Variation in the data between the different samples (or treatments).

Treatment sum of squares (SSTR) $= \sum r_j \left(\bar{X}_j - \bar{\bar{X}}\right)^2$, where r_j is the number of rows in the jth treatment and $\bar{X}_j$ is the mean of the jth treatment.

Using data of Table 7.8,

$$\text{SSTR} = \left[3 * (666.67 - 529.22)^2\right] = \left[3 * (473.67 - 529.22)^2\right]$$
$$= \left[3 * (447.33 - 529.22)^2\right] = 86049.55$$

Within variation (or error sum of squares) – Variation in the data from each individual treatment.

$$\text{Error Sum of Squares (SSE)} = \sum\sum (X_{ij} - \bar{X})^2$$

From Table 7.8,

$$\text{SSE} = \left[(643 - 666.67)^2 + (655 - 666.67)^2 + (702 - 666.67)^2\right]$$
$$+ \left[(469 - 473.67)^2 + (427 - 473.67)^2 + (525 - 473.67)^2\right]$$
$$+ \left[(484 - 447.33)^2 + (456 - 447.33)^2 + (402 - 447.33)^2\right] = 10254.$$

Note that SST = SSTR + SSE (96303.55 = 86049.55 = 102554)

Hence, you need only computing any two of the three sources of variation to conduct an ANOVA.

The next step in an ANOVA is to compute the "average" sources of variation in the data using SST, SSTR and SSE.

Note that SST = SSTR + SSE (96303.55 = 86049.55 = 102554)

$$\text{MST} = 96303.55/(9 - 1) = 12037.94$$
$$\text{MSTR} = 86049.55/(3 - 1) = 43024.78$$
$$\text{MSE} = 10254/(9 - 3) = 1709$$
$$\text{F} = \text{MSTR}/\text{MSE} = 43024.78/1709 = 25.17$$

In this example, df1 = 3 − 1 = 2 and df2 = 9 − 3 = 6. $F^{cv}{}_{2,6}$ is 5.14.

Reject the null hypothesis since F (observed value) > F_{cv} (critical value). In this example, 25.17 > 5.14, so we reject the null hypothesis.

7.7 Multivariate Statistics

When breeding materials and germplasm accessions are used in breeding programmes, their classification of genetic variability becomes vital. So, methods to classify and order genetic variability are assuming considerable significance. Use of established multivariate statistical algorithms is one strategy to classify germplasm. Some of these algorithms, such as cluster analysis, principal component analysis (PCA), principal coordinate analysis (PCoA) and multidimensional scaling (MDS), are being used now.

7.7.1 Cluster Analysis

This is an analysis by which individuals with same characteristics are grouped mathematically under one cluster. The resulting clusters of individuals should then exhibit high internal (within cluster) homogeneity and high external (between clusters) heterogeneity. There are broadly two types of clustering methods: (a) distance-based methods, in which a pairwise distance matrix is used as an input for analysis by a specific clustering algorithm leading to a graphical representation in which clusters may be visually identified (see also Chap. 9), and (b) model-based methods, in which observations from each cluster are assumed to be random and entry of each individual is performed jointly using standard statistical methods such as maximum likelihood or Bayesian methods.

Distance-based clustering can be either hierarchical or non-hierarchical. In hierarchical method, there could be as many groups as possible. The most similar individuals are first grouped and these initial groups are merged according to their similarities. UPGMA (unweighted paired group method using arithmetic averages) is the most popularly used algorithm in hierarchical method that involves construction of a dendrogram (Fig. 7.10). Options for performing non-hierarchical clustering are available in statistical packages such as SAS [FASTCLUS] and SPSS [QUICK CLUSTER]. Non-hierarchical clustering methods are rarely used for analysis of intraspecific genetic diversity in crop plants due to a number of clusters that are required for accurate lack of prior information about the optimal assignment of individuals.

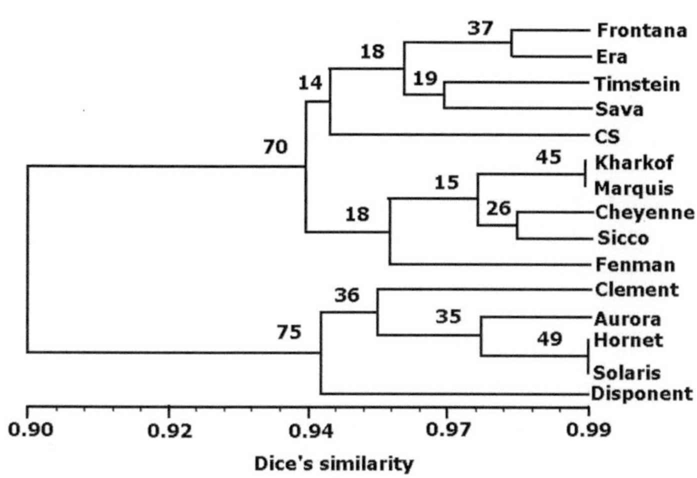

Fig. 7.10 Dendrogram based on similarity values obtained with the UPGMA method. Cultivars were divided into three groups: (**a**) spring wheat (N,S), (**b**) winter wheat (N,W) and (**c**) winter wheat with translocation 1BL/1RS (R,W). Values appearing above the branches are percentage of 1000 bootstrap analysis replicates in which the branches were found

Use of statistical techniques such as bootstrap, MANOVA (multivariate analysis of variance) or discriminant analysis can facilitate determination of optimal number of clusters. In MANOVA, clusters obtained in each cutting point are considered as treatments and individuals falling within that group are considered as replications for that treatment. The analysis is performed individually for each cut point with all characters or variables selected for cluster analysis. The optimal number of clusters or groups will be at that specific point which reveals the highest F value. This is based on the principle that at a proper cut point, within-group variance (error variance) shall be less than between-group variance (between-treatment variance), leading to a higher F value. Similarly, discriminant analysis can be effectively utilized to determine the best possible grouping on the basis of discrimination among groups achieved by different cut points.

7.7.2 Principal Component Analysis (PCA) and Principal Coordinate Analysis (PCoA)

PCA and PCoA are used to derive a two- or three-dimensional scatter plot so that the geometric distances reflect the genetic distances. Wiley in 1981 defined PCA as "method of data reduction to clarify the relationships between two or more characters and to divide the total variance of the original characters into a limited number of uncorrelated new variables". Such an exercise will allow visualization differences among individuals and identify groups. The linear transformation of original variables into uncorrelated variables is known as principal components (PCs). The first step is to calculate eigenvalues that define the total variation that is reflected in principal component axes. While the first PC summarizes most of the variability present in original data, the second PC is not summarized by the first PC. Since PCs are orthogonal and independent of each other, each PC reveals properties of the original data. In this fashion, the total variation in the original data may be separated into components that are cumulative (Fig. 7.11). The proportion of variation accounted for by each PC is expressed as eigenvalue divided by the sum of eigenvalues. The negative eigenvalues can be eliminated through transforming similarity index with the following formula:

$$S'_{ij} = S_{ij} - S_{i.} - S_{.j} + S_{..}$$

where S_{ij} is the coefficient of similarity between individuals i and j, $S_{i\cdot}$ is the mean of the values for the ith row in the similarity matrix, $S_{.j}$ is the mean of the values for the jth column and $S_{..}$ is the overall mean of similarity coefficients.

PCoA aims at producing a low-dimensional graphical plot, where distances between the points are close to original dissimilarities. It gives a matrix of similarities and dissimilarities. On the other hand, PCA uses initial data matrix. An example to this is the presence or absence of alleles in molecular marker data. When the first two or three PCs explain most of the variation, PCA and PCoA become useful techniques for grouping individuals by a scatter plot presentation (Fig. 7.12).

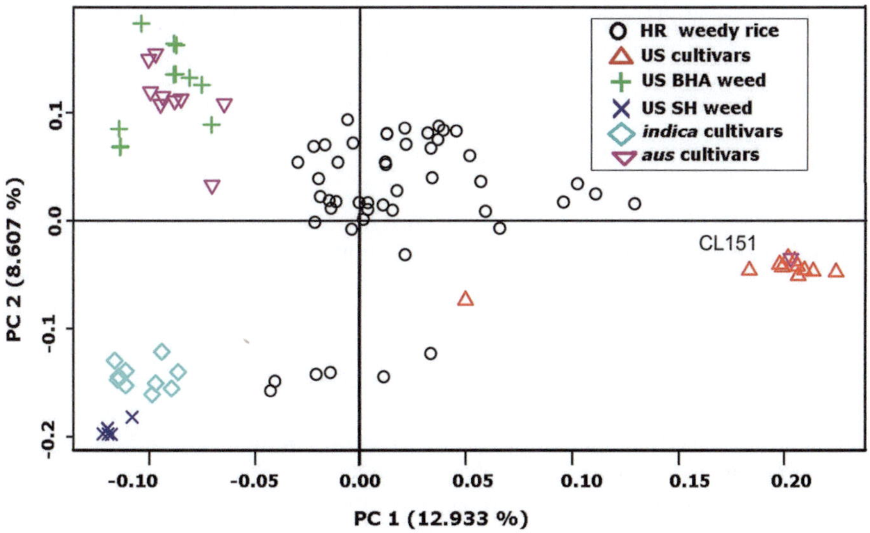

Fig. 7.11 Principal component analysis of HR weedy rice, US cultivated rice, historical SH and BHA weedy rice and Asian *aus* and *indica* cultivars. Principal component 1 (PC1) explains 12.93% of the variance, and PC2 explains 8.61%. The inbred reference Clearfield cultivar, CL151, is labelled

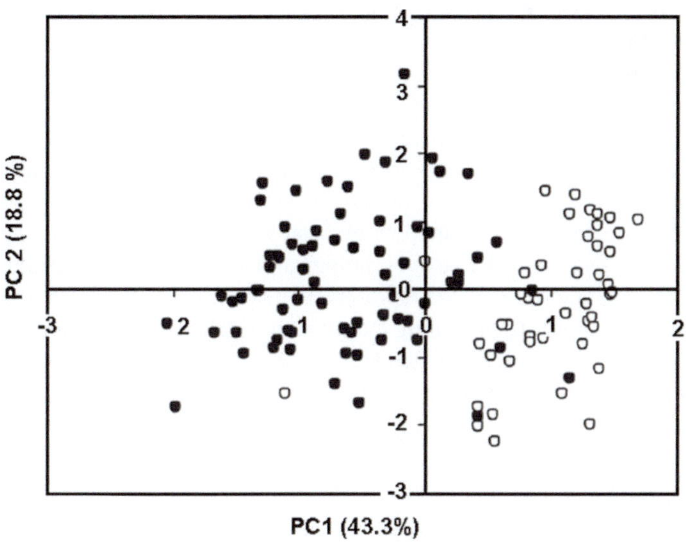

Fig. 7.12 Scatter diagram of the first two principal components (PC) for 45 old (o) and 72 modern (●) winter wheat cultivars evaluated at the experimental field of CRI-Quilamapu (Chile) in 2003. PC1 and PC2 explained 43.3% and 18.8% of the variance, respectively

The eigenvalue of PCs can be used as a criterion to determine how many PCs should be utilized. PCs with eigenvalue >1.0 are considered as inherently more informative.

7.7.3 Multidimensional Scaling

MDS represents a set of genotypes (n) in a few dimensions (m) using a similarity or distance matrix between them in such a way that the inter-individual proximities in the map nearly match the original similarities/distances. It is possible to arrange the n individuals in a low-dimensional coordinate system on the basis of only the rank order of n (n – 1)/2 original similarities-distances and not their magnitude. There are two types of MDS depending on the data input. Qualitative data uses non-metric MDS and quantitative data uses metric MDS. The closeness between original similarities-distances and inter-individual proximities in the map can be tested by different methods. The most commonly used test is a numerical measure of closeness called "stress". Stress indicates the proportion of the variance of the disparities not accounted for by the MDS model. Stress can be measured as:

$$\frac{\left[\left(d_{ij}-\hat{d}_{ij}\right)^{2}\right.}{\left.\left.d_{ij}-\bar{d}\right)^{2}\right]^{1/2}}$$

where $\bar{d}$ is the average distance $\Sigma\, d_{ij}/n$ on the map. Stress value becomes smaller as the estimated map distance approaches the original distance. The interpretation of stress in terms of goodness of fit is as follows: a stress level of 0.05 provides excellent fit, with 0.1 a good fit, 0.2 a fair fit and 0.4 a poor fit. When running MDS analysis with statistical software such as SPSS or Statistical Analysis Software (SAS), the number of dimensions to be extracted from the spatial map must be pre-specified.

In MDS, one can effectively employ the distance matrix obtained among a set of genotypes with data sets, such as morphological, biochemical or molecular marker data as input, to generate a spatial representation of these genotypes in a geometric configuration as output. The resulting multidimensional distance matrices, reflecting the relationships among a set of genotypes, can be presented as a two- or three-dimensional representation that can be more easily interpreted (Fig. 7.13).

7.7.4 Path Analysis

Yield is a complex trait that is known to be associated with a number of interrelated component characters that are highly affected by environmental variations. Such inter-dependence of the contributing characters affects their direct relationship with yield, thereby making correlation coefficients unreliable as selection indices. Thus, specification of causes and measuring the relative importance of each of the yield

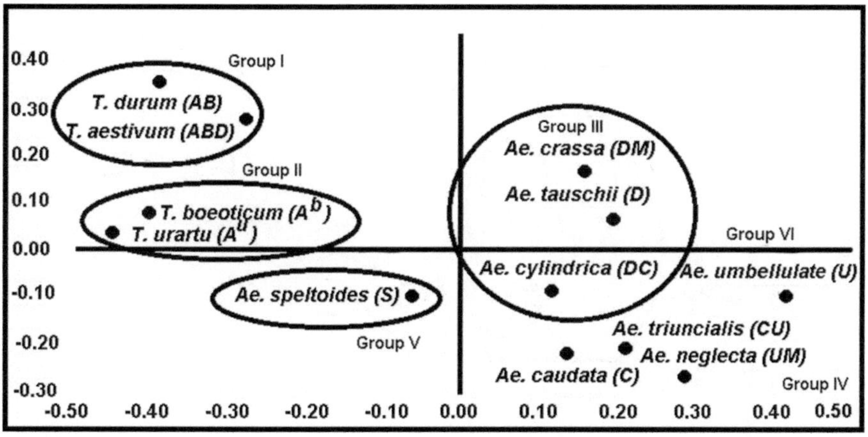

Fig. 7.13 The multidimensional scaling plot of species form of Iranian *Aegilops-Triticum* core collection using Euclidean distance coefficient

components can be achieved by using the method of path analysis, as a mean of separating the direct effects from the indirect ones through other characters. Path analysis was developed by Sewall Wright in 1920.

Breeding and selection programmes often encompass several characters simultaneously. When considering several traits, it is desirable to choose individuals with the best combination of these traits. The basis for such a selection is selection index, which takes into account a combination of traits according to their relative weight. Thus, each individual trait has an index value (score) and selection is based on the sum of the scores (values) of the different traits. Gain from selection for any given trait is expected to decrease as additional traits are included in the index, so the choice for traits to be included must be done objectively.

Path analysis is a multiple regression method that allows to estimate the strength of directional relationships of one trait with multiple dependent variables. A path diagram (Fig. 7.14) is a scheme of causal relationships. Let us consider a plant that grows, flowers, sets seeds and dies. Five traits are measured: cotyledon size ($z1$), time of inflorescence initiation (bolting time; $z2$), number of rosette leaves at flowering initiation ($z3$), inflorescence height ($z4$) and number of fruits ($z5$). In our causal scheme, cotyledon size affects both time of inflorescence initiation and number of leaves, and both of them affect inflorescence height. Inflorescence height in turn influences fruit production. In this scheme, only first-order effects are included.

A path diagram, besides showing the nature and direction of causal relationships, also includes estimates of the strength of those relationships, the path coefficient (p). A path coefficient is the standardized slope of the regression of the dependent variable on the independent variable in the context of the other independent variables. For example, inflorescence height ($z4$) is regressed on bolting time ($z2$).

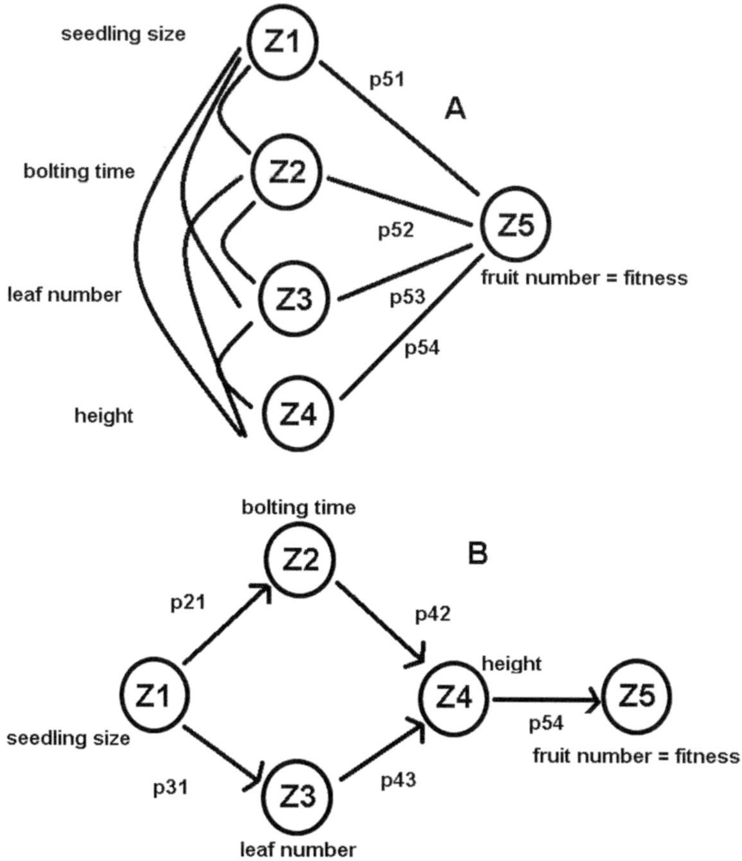

Fig. 7.14 Two different models of trait effects on fitness. (**a**) Multiple regression model showing each trait operating simultaneously on fitness. (**b**) Path analysis model showing five traits at four time periods. Path analysis restandardized regression coefficients. Variation due to error (U) is not included for simplicity

The slope (*b*42) is then standardized (*p*42) by multiplying it by the ratio of the standard deviations of the independent and dependent variables, respectively. If there is only a single independent variable, this standardized coefficient is a Pearson product-moment correlation. If there are additional independent variables, it is a standardized partial regression coefficient. The standardization acts to remove differences in scale among variables. In the model given in Fig. 7.14a, there is no hierarchy of relationships among traits, and all four of the observed traits influence fitness directly and are correlated with each other. This model therefore only allows direct and non-causal effects on fitness, since there is no contrast, in model given in Fig. 7.14b, only one trait (height) has a path leading directly to fitness with no intermediate steps, but all other traits may have indirect (mediated) or non-causal

Table 7.9 Decomposition of the correlation between different traits and fitness under multiple regression and path analysis (see Fig. 7.14a)

Trait	Total selection	Multiple regression		Path analysis	
		Direct selection	Indirect selection	Direct selection	Indirect selection
Seedling size	S1	P51	r21 p52 + r32 p53 + r41p54	P21p42 p54 + p31 p43 p54	
Bolting time	S2	P52	R21 p51 + r32 p53 + r42 p54	P42 p54	P21 p31 p43 p54
Leaf number	S3	P53	R31 p51 + r32 p52 + r43 p54	P43 p54	P31 p21 p42 p54
Height	S4	P54	R41 p51 + r42 p52 + r42 p54	P54	

Direct selection includes both direct and indirect effects, and indirect selection includes non-causal (spurious and correlational) effects. The sum of direct and indirect selection is the total selection accounted for by the model

effects on fitness (Table 7.9). Several computer programs calculate path coefficients automatically [e.g. Procedure CALIS (SAS Institute), LISREL, EQS, RAMONA (SYSTAT for Windows, SPSS, Inc.)].

7.8 Hardy-Weinberg Equilibrium

Hardy and Weinberg in 1908 independently demonstrated that in a large random mating population, both gene frequencies and genotypic frequencies remain constant from generation to generation in the absence of mutation, migration and selection. Such a population is said to be in Hardy-Weinberg equilibrium and remains so unless any disturbing force changes its gene or genotypic frequency.

If we consider single locus, any population will attain its equilibrium after one generation of random mating. Consider one locus with two alleles (A_1 and A_2) in a diploid in a population. In such a population, genotypic frequencies available are given in Table 7.10. The total number of genes relative to locus A in this population is $2N$, i.e. two genes in each diploid individual. Thus, the numbers of A_1 and A_2 alleles are $2_{n1}+_{n2}$ and $2_{n3}+n_2$, respectively, and their frequencies are:

$$p(A_1) = \frac{2n_1 + n_2}{2N} = \frac{n_1 + \left(\frac{1}{2}\right)n_2}{N} = P + \frac{1}{2}Q$$

$$p(A_2) = \frac{2n_3 + n_2}{2N} = \frac{n_3 + \left(\frac{1}{2}\right)n_2}{N} = R + \frac{1}{2}Q$$

Table 7.10 Genotypic frequencies in population with one locus and two alleles

Genotypes	A_1A_1	A_1A_2	A_2A_2	
Number of individuals	n_1	n_2	n_3	$n_1 + n_2 + n_3 = N$
Frequency	$P = n_1/N$	$Q = n_2/N$	$R = n_3/N$	$P + Q + R = 1$

Table 7.11 Genotypic array and its frequencies in the second generation after random mating

	Male gametes			Male gametes	
Genotypes	A_1	A^2	Frequencies	p	q
Female gametes			Female gametes		
A1	A1A1	A1A2	p	p^2	pq
A2	A1A2	A2A2	q	pq	q^2

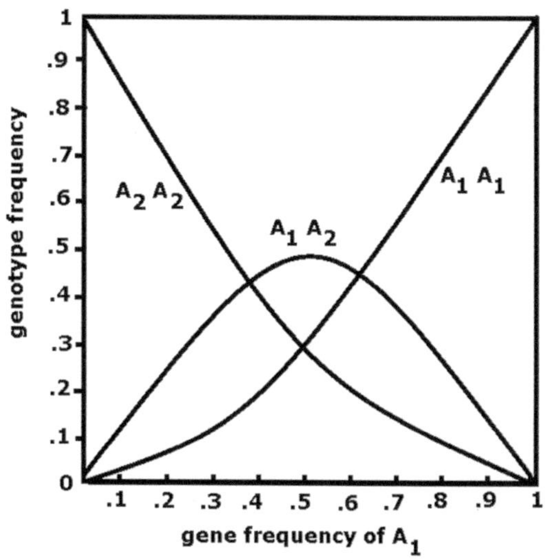

Fig. 7.15 Distribution of genotypic frequencies for gene frequencies ranging from 0 to 1.0 for one locus with two alleles in a population in Hardy-Weinberg equilibrium

Under random mating, since the gametes unite at random, the genotypic array and its frequency in the next generation are given in Table 7.11. Hence, the genotypic frequencies are p^2 (A_1A_1):$2pq$ (A_1A^2):q^2 (A_2A_2), and this population is said to be in Hardy-Weinberg equilibrium because genotypic frequencies are expected to be unchanged in the next generation. The variation of genotypic frequencies for gene frequencies is in the range of 0 to 1 (Fig. 7.15). The Hardy-Weinberg law can also be extended to multiple alleles. In general, if p_i is the frequency of the ith allele at a given locus, the genotypic frequency array can be:

$$\sum_i p^2 i \quad \text{for homozygotes } (A_iA_i)$$

$$\sum_{i<i} p_i p_i' \quad \text{for heterozygotes } \left(A_iA_i'\right)$$

When $p = 0.5$, with two alleles per locus, the gene frequency which gives maximum frequency is heterozygotes ($Q = 2_{pq}$). This is the reason why we find maximum frequency heterozygotes in F_2 populations derived from elite × elite pure-line crosses.

Further Reading

Beurton PJ, Falk R et al (eds) (2000) The concept of the gene in development and evolution. Cambridge University Press, Cambridge

Charmantier A, Garant D (2005) Environmental quality and evolutionary potential: lessons from wild populations. Proc R Soc Biol Sci 272:1415–1425

Falconer DS, Mackay TFC (1996) Introduction to quantitative genetics. Longman, Harlow

Feldman MW (1992) Heritability: some theoretical ambiguities. In: Lloyd EA, Fox Keller E (eds) Keywords in evolutionary biology. Harvard University Press, Cambridge, pp 151–157

Gomez KA, Gomez RA (1984) Statistical procedures for agricultural research. Wiley Inter science, New York

Hill WG et al (2008) Data and theory point to mainly additive genetic variance for complex traits. PLoS Genet 4:e1000008

Lynch M, Walsh B (1998) Genetics and analysis of quantitative traits. Sinauer, Sunderland

Macgregor S et al (2006) Bias, precision and heritability of self-reported and clinically measured height in Australian twins. Hum Genet 120:571–580

Visscher PM et al (2006) Assumption-free estimation of heritability from genome-wide identity-by-descent sharing between full siblings. Public Libr Sci Genet 2:e41

Visscher PM, Hill WG, Wray NR (2008) Heritability in the genomics era – concepts and misconceptions. Nat Rev Genet 9:255–266

Part III

Methods of Breeding

8 Selection

Keywords
History of selection · Genetic effects of selection · Systems of selection and gene action · Selection of superior strains

Selection is a process by which gene frequencies are changed so as to make the genotype suitable for a particular purpose. This is a process by which certain genotypes are preferred over others for further future generations. Selection can be either natural or artificial (by man).

Survival of the fittest is the main force responsible for selection in nature. Under natural selection, tendency is to select against the weaker ones, and only the stronger survived to reproduce. Plant breeding practises artificial selection. Artificial selection is the effort to increase the frequency of desirable genes or combinations of genes that have the ability to produce superior performing offspring. While artificial selection is underway, natural selection also happens silently. When two lines are bred, the seeds obtained are the result of natural selection. Many genetic combinations being tried may not be successful due to selection against them. The gene combination that is not suitable for a particular environment will be selected against so that the embryos thus formed will never get developed or get germinated.

8.1 History of Selection

Several of the modern crop species were domesticated hundreds of thousands of years ago. Wheat was domesticated nearly 10,000 years ago from its wild relatives in the so-called Mesopotamian region of the Near East. Man continued to look for adapted varieties of wheat since farming commenced. A number of varieties of wheat were cultivated by the Swiss of the Neolithic Period. One of Plato's pupils in 300 BC described selection of productivity-oriented wheat in ancient Greece.

P. M. Priyadarshan, *PLANT BREEDING: Classical to Modern*,
https://doi.org/10.1007/978-981-13-7095-3_8

Plant breeding techniques have enabled man to exploit the evolutionary variability in wheat. Genotypes with desirable traits were intercrossed, and their superior progenies were selected prior to 1900. With the advent of Mendel's principles of genetics in 1865, modern plant breeding was born at the beginning of the twentieth century. The basic principles of wheat breeding now include (1) derivation of varieties lodging desired traits like rust resistance and (2) methods of transfer of these traits into adapted cultivars.

8.2 Genetic Effects of Selection

Though selection is not responsible for the creation of new genes, selection increases the frequency of desirable genes. Undesirable gene frequency is reduced. This can be illustrated by the following example. *A* is the desirable gene and *a* the undesirable gene:

P_1 *AA x aa*
F_1 all *Aa* (frequency of allele *A* is 0.5)
F_2 *Aa* x *Aa*
Progeny: 1*AA*: 2*Aa*: 1*aa* (frequency of allele A is still 0.5)

When we cull all *aa* individuals in F_2, the remaining genes shall be four *A* and two *a*. Here, the frequency of the *A* gene is increased to 0.67 and that of *a* gene is decreased to 0.33. The proportion of *AA* individuals in the population will be increased due to increment of *A* gene while culling out *aa* individuals. If the frequency of *A* gene were 0.50 (as per Hardy-Weinberg law), the proportion of *A* individuals would be 0.50 multiplied by 0.50 or 0.25. However, if the frequency of the *A* gene were increased to 0.67, the proportion of AA individual would be 0.67 multiplied by 0.67 or 0.449. The genetic effect of selection is to increase the frequency of the gene selected for and to decrease the frequency of the gene selected against. When the frequency of the desirable gene is increased, the proportion of individuals homozygous for the desirable gene also is increased.

8.3 Systems of Selection and Gene Action

The economic traits of plants are governed by different kinds of gene actions. In traits like plant height and seed colour, only one pair of genes or relatively few genes exert major effect influence. Single pair of genes can also exert major phenotypic effect on quantitative traits. An example for this is semi-dwarfism in rice where *sd-1* gene produces semi-dwarf. This is done through masking the phenotypic expression of many additive genes. In quantitative traits determined by many pairs of genes, they may be expressed in an additive manner or in a non-additive way. Since gene actions for qualitative and quantitative differ, it is essential to differentiate methods used in selecting for or against them.

8.3.1 Selection in Favour of and Against Allele

By all practical means, selection goes in favour of a dominant allele, since traits governed by dominant alleles are desirable under usual circumstances. However, the real issue is to differentiate between homozygous and heterozygous individuals. The heterozygous individuals must be identified by a breeding test or a knowledge of the parental phenotype. Selection for a dominant allele involves the same principle as selection against a recessive allele.

Since the penetrance of the dominant allele is 100%, selection against a dominant allele is relatively easy. Eliminating the dominant allele means that all plants showing the trait should be discarded. When penetrance is low and the alleles are variable in expression, selection against a dominant allele would be much less effective. Attention to the phenotype of the ancestors, progeny and collateral relatives are necessary in order to make the selection more effective.

If penetrance is complete and if the allele does not vary too much in their expression, selection for a recessive allele is relatively simple. Just keeping those individuals which show the recessive trait will make a selection in favour of recessive allele. A fine example would be when one want to have white flowers, one has to make crosses of purple flowers.

8.3.2 Selection for Genes with Epistatic Effects

Epistasis is the interaction between genes. Epistasis can be either complementary or inhibitory. Selection for superior phenotypes among families, lines or breeds would be the desired way of selecting for epistatic gene action. Initially unrelated lines are formed by inbreeding to make them homozygous. Superior F_1 hybrids are made from them. Once two or more lines are found that cross well, such lines can be retained as pure inbred lines and crossed again and again for the production of seeds for commercial purposes. Such procedure is followed in hybrid seed corn.

8.3.3 Selection for a Single Quantitative Trait

Quantitative traits are governed by several pairs of genes having individual phenotypic effects. A phenotype shall be affected by additive or non-additive gene actions or both. Environment also has a pivotal role in the expression of such traits. Heritability (h^2) governs the amount of genetic progress (ΔG) made in one generation of selection of the trait. Heritability multiplied by the selection differential (Sd)

gives the real genetic progress for that trait. Hence, the genetic progress expected in one generation shall be:

$$\Delta G = h^2 \times Sd$$

Selection differential (Sd) is the superiority or inferiority of those selected individuals. This is made in comparison to the average of the population (P) from which the parents were selected. The selection differential is:

$$Sd = \left(P_S - \bar{P}\right)$$

As an example, if the productivity of rice is to be increased by 0.5 tons, the selection differential would be 0.5 tons. The selection differential would be zero if all plants were kept for breeding. In this case, the expected genetic progress would be zero. If the frequency distribution curve for the trait in question is a normal bell-shaped curve, then the selection differential may also be expressed in terms of standard deviation units:

$$Sd = i\sigma_p,$$

where i is the intensity of selection in standard deviation units and σ_p is the phenotypic standard deviation of the trait in the population. If the proportion of plants kept for breeding is known, the selection intensity i may be calculated from the formula $i = z/w$, where z represents the height of the curve and w represents the fraction of the population selected for breeding. The value of z may be obtained from tables showing the ordinates and area of the normal frequency distribution curve.

8.3.4 Selection on the Basis of Individuality

When a genotype is selected or rejected for breeding purposes based on its own phenotype for a particular trait, it is selection based on individuality. This exercise is dependent on the closeness of the genotype with the phenotype. The phenotype is result of the effect of environmental effects or genotype x environment interactions. This phenotypic performance varies throughout its life. The genotype never varies and it is fixed at the time of fertilization. The phenotype of the individual (individuality) is often used to estimate its breeding value. Qualitative traits such as colour and height based on individual's phenotype are more effective only in some instances. Determination of the effect of dominant allele cannot be made from its phenotype since one cannot distinguish the homozygous dominant and the heterozygous dominant individuals. Hence, selection based on individuality for qualitative traits may be useful but not adequate enough to be accurate.

Information on the phenotypes of the close relatives as well as that of the individual makes these estimates of the genotype more accurate. This is true in the case of quantitative traits. Interactions of quantitative traits that are controlled by many genes with various elements of the environment make the individual of the group almost uniform within a group. Quantitative traits are governed by additive gene action or mostly by non-additive gene action or both. The phenotype and genotype of the individual for that trait would be identical if such a trait were 100% heritable. Since environment always affects the phenotype, no quantitative trait is 100% heritable. The phenotypic merit of the individual is determined by comparing the individual's phenotype with that of group average. To make the selection more effective, the comparison with other varieties must be undertaken under controlled environmental conditions.

8.3.5 Selection on the Basis of Pedigrees

A record of a genotype's ancestors is the pedigree. The phenotypic merit of ancestors is to be included in pedigrees. Such pedigrees are *performance pedigrees.* Pedigree may be of importance in detecting carriers of a recessive allele.

One main disadvantage with pedigree selection for a dominant/recessive allele is that there could be unintentional and unknown mistakes. Such mistakes can result in the rejection of an entire family. But for practical reasons, such mistakes must not have occurred. On the other hand, due to incompleteness of the record, the frequency of dominant/recessive allele in a family may be low. Since such records are incomplete, the genotype appears to have a "clean" pedigree. Also, when it is found that the allele is present, which once thought to be free of the dominant/recessive allele, it will be called a "dirty" family. A definite disadvantage of pedigree selection as used in tallness is that all genotypes with the same or similar pedigree are condemned. Nevertheless, the individual still has a questionable pedigree and will be discriminated against by many breeders, either because they are not familiar with the mode of inheritance affecting such a trait or because they are afraid to trust progeny test information. The records on the performance of ancestors can increase the accuracy of determining the probable breeding value of an individual. This would increase the accuracy of predictions. The records of the ancestors would show their merit as compared to that of their varieties. Since the heritability of a trait is not higher, meticulously maintained records on ancestors can make predictions on individual's breeding value more accurate. The attention paid to the records of an ancestor depends on:

(a) The extent of relationship between the ancestor and the individual
(b) The heritability of the trait
(c) Environmental correlations among genotypes used in the prediction
(d) The extent of completeness on the merit of ancestors

On statistical terms, the *accuracy of selection* is an estimate of how accurately the trait of an individual can be predicted from the phenotypic average of its ancestors. There will be more alleles in common in relatives than non-relatives. Superior relatives shall possess superior alleles and such alleles get transmitted to the progeny. Pedigrees may be used to select for traits not expressed early in life. They may also be used to select for traits expressed upon maturity.

8.3.6 Selection on the Basis of Progeny Tests

In this type of selection, the breeder makes a decision to keep or cull a parent based on the average merit of their offspring. Here, selection for both qualitative and quantitative traits is based on progeny tests. Probably the most effective use of progeny tests in selection for qualitative traits is to determine if a dominant phenotype is homozygous or heterozygous. All homozygous recessives and heterozygous genotypes are discarded to produce a pure-breeding line with dominant trait. Though the recessive genotypes can be identified by their phenotypes, heterozygous and homozygous genotypes have similar phenotypes. The genotypes of these two dominant phenotypes must be determined through progeny tests unless it is known that one parent is recessive. One can never be absolutely certain that a genotype is homozygous dominant after it is progeny tested. However, when certain test matings are made, if only homozygous dominant offsprings are produced, then one can be certain that the selected parent is homozygous dominant.

8.3.7 Selection for Specific Combining Ability

Selection for specific combining ability becomes relevant for hybrid vigour when non-additive gene action is vital. Selection based on individuality may not be the efficient method for selecting traits governed by non-additive gene action. Exploitation of hybrid vigour through crossbreeding gives increased merit for such selections. Selection based on individuality will be effective if dominance is considered. Selection is less effective if epistasis and overdominance are important.

In quantitative inheritance, it may not be possible to judge which genotypes are homozygous, where many genes affect the same trait. Formation of several different inbred lines through inbreeding is the first step, where inbreeding increases the homozygosity of all pairs of genes. All individuals within a line must be homozygous, regardless of the phenotypic expression, for all the gene pairs if inbreeding were 100%. However, the breeder may not be sure of which genes are homozygous within an inbred line, which is not necessary. The next step is to test them in crosses to determine which lines combine to produce the best line. In general, the two inbred lines producing the most superior progeny when crossed are the ones giving greater heterozygosity in the progeny. Such inbred lines are kept pure for further crosses in later years to produce commercial hybrids.

8.4 Selection of Superior Strains

Most breeding programmes aim at producing genotypes through large populations with the desired gene combinations. Selection of parents is the initial step in a breeding programme (Fig. 8.1). This step is the most important since it marks the limit on the genetic variability that can be seen in the progeny to exercise selection in subsequent generations. Under normal circumstance, one parent will be an adapted cultivar (say Egyptian wheat variety Gemmiza 7), while the other (say Giza 168) will be rust resistant. Other secondary attributes like drought tolerance, disease resistance, insect tolerance, straw strength, plant height, resistance to shattering, harvestability, seed size, seed shape, seed colour, test weight and grain quality are some of the traits that wheat breeders consider while selecting parents.

A cross between two parents produces hybrid (F_1) seed (Fig. 8.1). Each cross produces fewer than 50 F_1 seeds. Male sterility and fertility restoration systems offer production of large number of F_1 seeds. If the yield potential of F_1 itself is high enough with desirable qualities, then F_1 seed may be sold as hybrid seed. Hybrid winter wheat cultivars are developed for commercial production in the USA for Great Plains in this fashion.

In F_2 generation, the genetic differences between the parents are expressed in a multitude of combinations (Fig. 8.1). Plant breeder can resume selection process at this stage. Plant breeder will reselect within each generation since the progeny from selected individuals does not breed true in early generations (F_2 to F_5). After each succeeding generation, progeny of selected individuals becomes genetically more uniform. The selection for more complex traits, such as yield, normally begins at F_6. The availability of performance data from small plot trials leads to culling of selected strains as they pass from the F_6 to the F_8 generation. Superior selections in F_8 or later are taken to pre-registration trials. In pre-registrations trials, selections are subjected to final evaluation for a minimum of 3 years at 10–20 locations. These multi-environment trials shall be under vivid climatic conditions of the country. Every country follows its own procedure for releasing a crop variety. A procedure being followed in India and Canada is presented for comparative study (Figs. 8.2 and 8.3).

After identification of superior selections, the plant breeder will begin a strain purification process to ensure that acceptable breeder seed stocks are available for distribution to seed growers if the selection/strain is to be successfully registered as a cultivar. This process starts after the first year of evaluation in cooperative trials. Once a cultivar registration is approved, breeder's seed is distributed to seed growers. After this, breeders need to increase the seed stock so that the cultivar is available in sufficient quantities for commercial purpose (Box 8.1) (please see Chap. 25 for details).

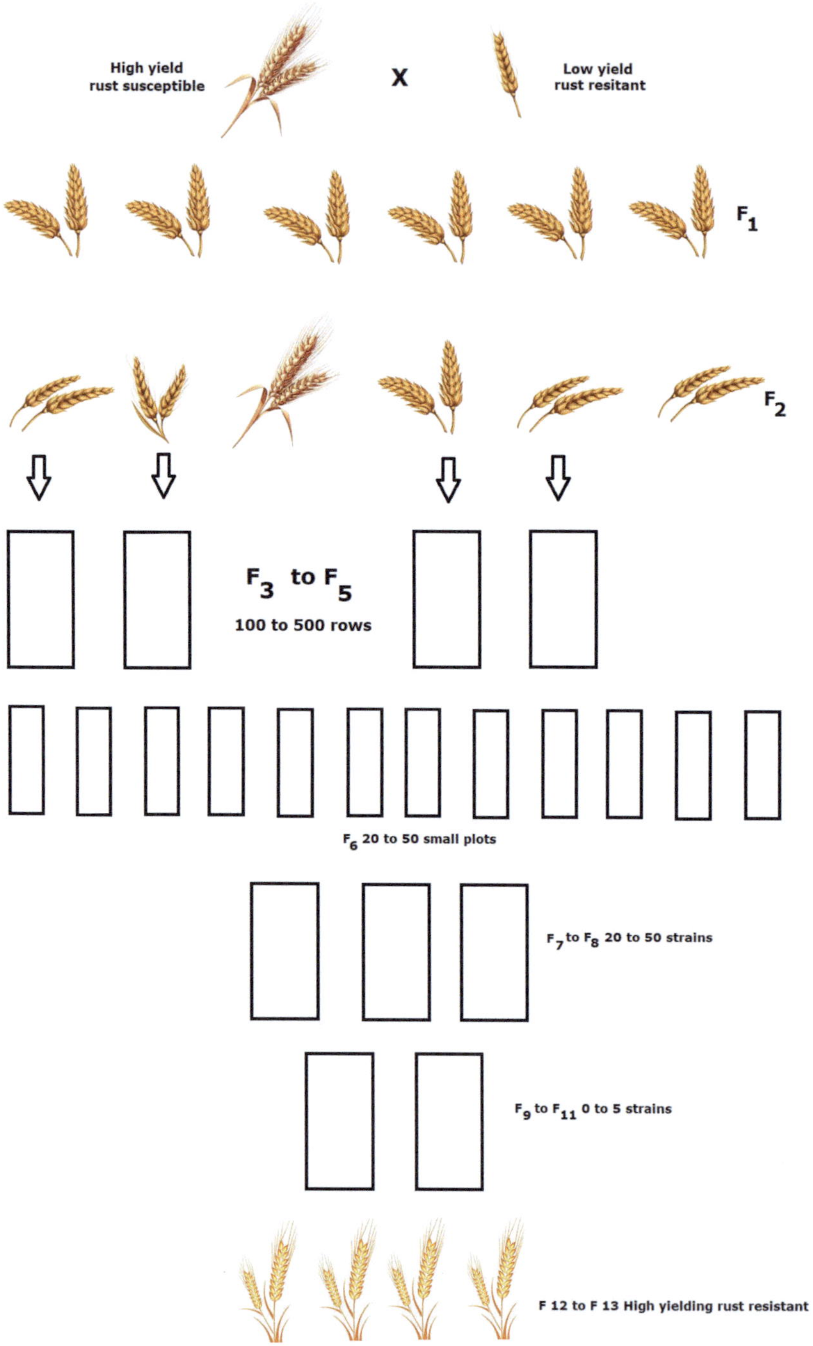

Fig. 8.1 Selection stages in a breeding programme

Fig. 8.2 Procedure of varietal release in India

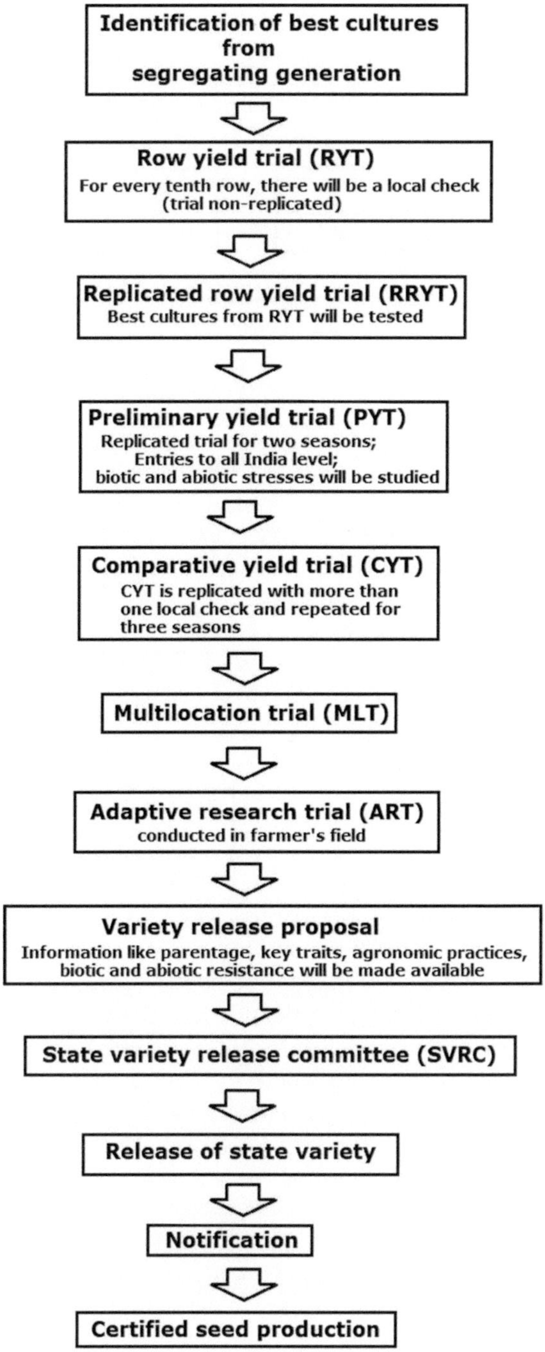

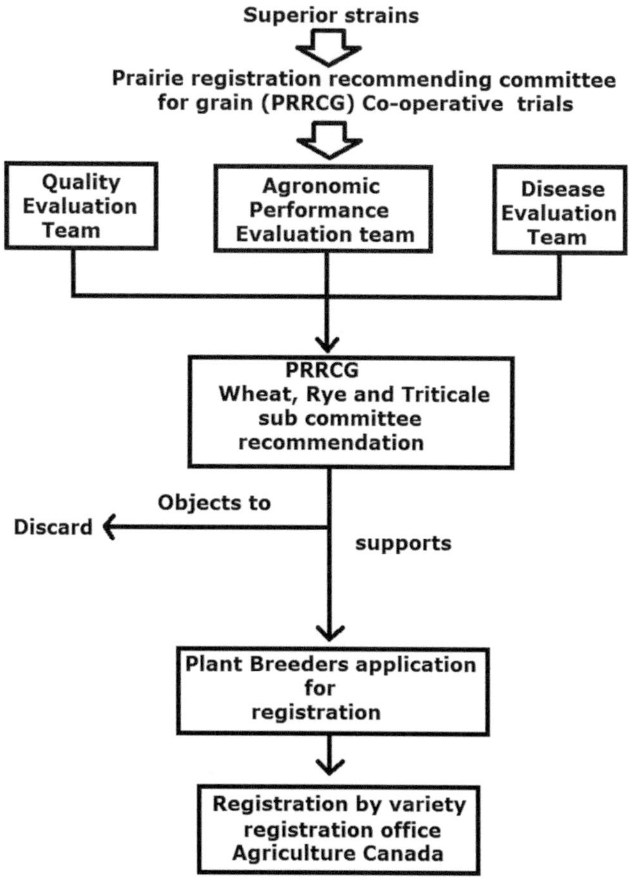

Fig. 8.3 Procedure of varietal release in Canada

Box 8.1: Farmer's Selection and Derivation of Maize

Ancient farmers in Mexico took first steps in domesticating maize. They undertook selections on kernels (seeds) to plant some 8700 years ago. *Balsas teosinte* (*Z. mays* ssp. *parviglumis*), a large wild grass that grows in the Central Balsas River Valley of Mexico, is the closest relative to maize. The farmers saved best kernels for the next season's harvest. This process is known as *selective breeding or artificial selection.* But the abrupt appearance of maize in the archaeological record confused the scientists. The process of genetic archaeology helped geneticists to understand the rearrangements at the DNA level so as to analyse the differences between teosinte and maize. George Beadle was the first scientist to fully appreciate the close relationship between teosinte and maize. He calculated that only about five genes were responsible for the most-notable differences between teosinte and a primitive strain of maize. This contention got acceptance from studies at the molecular level (see Fig. 8.4).

Fig. 8.4 Evolution of maize from teosinte because of farmers' selection

Further Reading

Bos I, Caligari P (2008) Selection methods in plant breeding, 2nd edn. Springer, Dordrecht
Crossa J et al (2017) Genomic selection in plant breeding: methods, models and perspectives. Trends Plant Sci 22:P961–P975

Hybridization

9

Keywords
History · Objectives · Procedure of hybridization · Distant hybridization · Choice and evaluation of parents · Consequences of hybridization

Hybridization involves crossing of two different genotypes that results in a third individual with a different set of traits. Crossing the same species is easy that produces fertile progeny. Because of chromosome-pairing problems during meiosis, wide crosses are difficult and produce sterile progeny. Hybridization is through either insects (oil palm) or wind (maize) under natural conditions. Such plants are referred to as cross-pollinated species. In plants with perfect flowers (*autogamous*, having flowers with both stamens and pistils), cross-pollination rarely occurs in plants (like wheat and rice) since they are normally self-pollinated (Fig. 9.1). Plants that have separate pistillate and staminate flowers on the same plant (such as maize) are *monoecious* (Fig. 9.2). Plants that have male and female flowers on separate plants (such as asparagus) are *dioecious* (Fig. 9.3). Through artificial means, hybrids of both cross-pollinated and self-pollinated plants can be accomplished. The breeder must know the time of development of reproductive structures of the species, treatments to promote and synchronize flowering and pollinating techniques. The concept of hybrid vigour, or heterosis, has resulted from hybridization (see also Chap. 15).

9.1 History

Joseph Gottlieb Kölreuter was the first to report hybrid vigour in interspecific crosses of *Nicotiana* in 1761. He concluded that cross-fertilization was generally beneficial than self-fertilization. In 1799, T.A. Knight concluded that cross-pollination must be the norm as it is widespread in nature. Charles Darwin in 1862 reported his

P. M. Priyadarshan, *PLANT BREEDING: Classical to Modern*,
https://doi.org/10.1007/978-981-13-7095-3_9

Fig. 9.1 Monoecious flower structure (**a**); flowers of wheat (**b**) and rice (**c**)

Fig. 9.2 Male (**a**) and female (**b**) flowers of maize

Fig. 9.3 Male (**a**) and female (**b**) flowers of dioecious *Asparagus*

experiments with maize. He indicated that of the 24 crosses he undertook, an increase in plant height can be attributed to hybridization and that decrease in plant height can be attributed to self-pollination. He also noted that the deleterious effects of selfing or inbreeding could be reversed through crossing. In 1862, Darwin wrote, "Nature tells us, in the most emphatic manner, that she abhors perpetual self-fertilization". William J. Beal evaluated hybrids between maize varieties in the late 1800s. Some of his hybrids yielded 50% more than the mean of their parents. S.W. Johnson provided an explanation for hybrid vigour in 1891. G.W. McClure in 1892 said that hybrids between maize varieties were superior to the mean of the two parents.

The phenomenon of heterosis has been exploited in maize, sorghum, sunflower, onion and tomato. Maize (corn) was the first crop in the USA to have hybrids from inbred lines. George Shull, following the rediscovery of Mendel's laws in 1900, conducted the first experiments on inbreeding and crossing, or hybridizing, of inbred lines. He suggested that inbreeding in maize can result in pure (homozygous) lines.

Crossing of pure lines resulted in hybrid vigour since heterozygosity could be created at many allelic sites. US maize production increased dramatically since hybrid maize was introduced in the late 1920s and early 1930s (see Chap. 15 for details).

Triticale (X *Triticosecale* Wittmack) is the only human-made cereal crop, which is a cross between *Triticum* (wheat) and *Secale* (rye). While the first sterile triticale was reported in 1876 by Scottish botanist Alexander Wilson, the first fertile triticale was made by German botanist Rimpau in 1891. Some of the interspecific and intergeneric barriers should be overcome *via* the newer techniques of gene transfer. It is expected that genes from wild relatives of cultivated plants will continue to be sought to correct defects in otherwise high-yielding varieties. The objectives of hybridization are:

Objectives

- To create genetic variability
- To bring together desired characters found in different plants or plant lines into one plant or plant line, having all the desirable characters, *viz*., high yielding; high resistance to disease, drought or waterlogging; higher food value; better taste; etc.
- To produce useful variations by introducing recombination of characters
- To produce and utilize hybrid vigour, i.e. "the superiority of the hybrid over its parents".

Depending on the nature of plants involved, the cross may be of the following types:

- Inter-varietal: Cross between two varieties of a crop
- Intra-varietal: Cross between different genotypes of the same variety
- Intra-generic: Cross between two species of a genus
- Inter-generic: Cross between two different genera

9.2 Procedure of Hybridization

The aim of hybridization is to bring together desirable genes from two or more different varieties and to produce pure-breeding progeny superior to the parental types. A genotype is a collection of genes. The plant breeder's task is to manage the enormous number of genotypes raised during the generations following hybridization. A cross of 2 wheat varieties differing by only 21 genes can produce more than 10,000,000,000 different genotypes in the second generation. Almost 50,000,000 acres are needed to grow this population. Statistically, 2,097,152 different pure-breeding (homozygous) genotypes can occur; all are new pure-line types. The best option is to follow pedigree, where superior types are selected in successive generations based on a parent-progeny record.

The elimination of genotypes lodging undesirable major genes is done in F_2. In succeeding generations, natural self-pollination leads to pure lines. Normally, one or two superior genotypes are selected within each superior family in these generations.

By F_5, the pure-breeding condition (homozygosity) will be achieved. The pedigree record is useful in making eliminations. To evaluate families for quantitative characters, each selected family is usually harvested in bulk. This is to obtain larger amounts of seeds. Usually by F_7 or F_8, precise evaluation for performance and quality begins. The final evaluation involves:

(a) To detect weaknesses that may not have appeared previously
(b) Precise yield testing
(c) Quality testing

Before releasing for commercial production, derived genotypes are tested for 5 years at five representative locations. The F_2 generation is sown at normal commercial planting rates in a large plot. The crop is harvested in mass, and the seeds are used to establish the next generation in a similar plot. No record of ancestry is kept. By conducting multi-environment trials, the cultivator is subject to natural selection that tends to eliminate poor survivors. Two types of artificial selection are applied: (a) culling out of genotypes with undesirable major genes and (b) make selections for early-maturing plants. Further, single plant selections are made as in the pedigree method.

9.2.1 Techniques

The plant breeder must have first-hand knowledge of the crop botany of the species he is using, i.e. the time of flowering, the stage of flower development at which the anthers burst, stigmatic receptivity and pollen viability. In annual species, stigmas remain receptive for a short period, usually for several hours and very often for not more than a day. In many plants, the stigma becomes receptive at a particular time of the day, as in rice, it becomes receptive in the morning, at around 8 a.m. Stigma receptivity is of utmost importance because if pollination is not done within this period, fertilization normally does not occur. Similarly, if pollination is done with immature pollens or with pollens which have lost their viability, fertilization normally does not take place. In order to prevent any unwanted pollination, the flowers are kept covered by bags long before they open. Necessity of isolation increases with increase in the percentage of natural cross-pollination. The parents are grown in adjacent plots after their due selection based on the trait(s) the plant breeder wants to transfer and a decision will be taken on the usage of male and female parents. In case of rice and wheat, just 10–12 flowers are left on the inflorescence and the rest clipped off in order to facilitate the hybridization better.

In the next step, anthers must be removed before anther dehiscence from the flowers of the female parent to prevent self-pollination through a process known as *emasculation*. In the case of wheat, the middle row of florets that are immature compared to the side rows will be removed with the help of forceps (Fig. 9.4). The florets will be cut in the middle with a pair of scissors. The anthers in the remaining flowers will be removed with fine forceps. Such emasculated panicles are then covered with butter paper bags to prevent any cross-pollination. The next day, the

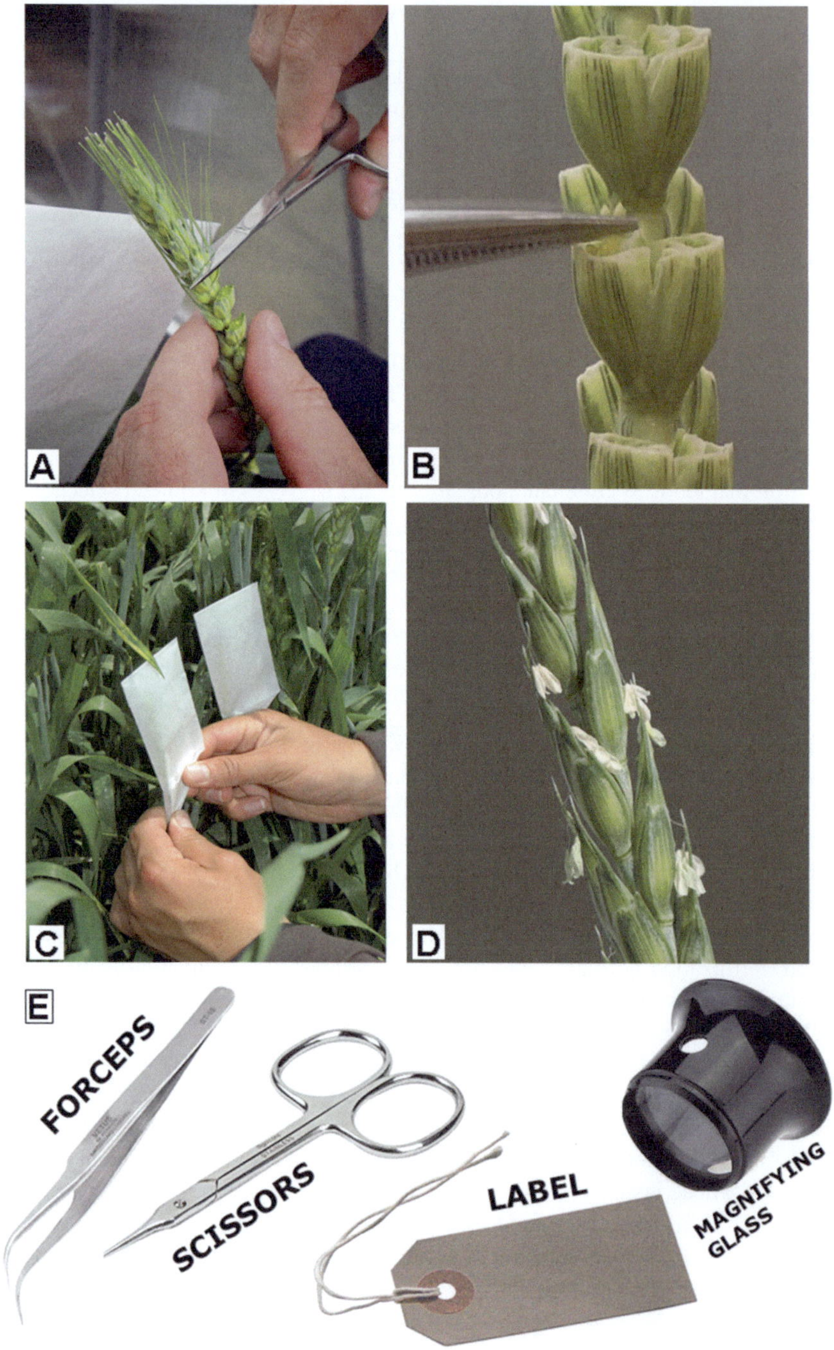

Fig. 9.4 Emasculation in wheat: (**a**) cutting of florets; (**b**) removal of anthers; (**c**) covering of emasculated spike; (**d**) wheat spike with anthesis; (**e**) tools for emasculation

paper bags will be cut on the top with scissors, and the stigmas will be dusted with pollen from male flowers. The whole male flower will be cut and used for dusting pollen over the female stigmas. The pollinated female flowers will be again covered with paper bags, in order to avoid any cross-pollination. The crossed flowers should always be kept properly tagged or labelled showing details of the cross (parentage, date of pollination, etc.). All necessary particulars about the cross should be recorded in the field notebook. Some flowers, which are too small, need a magnifying glass to examine the male and female reproductive organs. Depending on the reproductive biology of the species, the breeder has to modify his pollination procedure, for which he needs to have a first-hand knowledge of the botany of the crop (see Boxes 9.1 and 9.2).

Box 9.1: Pollination

The study of pollination is multidisciplinary that includes botany, horticulture, entomology and ecology. The interaction between flower and vector was first addressed by Christian Konrad Sprengel (German naturalist) in the eighteenth century. Sexual reproduction results in genetically diverse offspring. Pollinators like ants, bats, bees, beetles, birds, butterflies, flies, moths, wasps as well as other unusual animals assist over 80% of the world's flowering plants in their reproduction. Wind (anemophily), gravity and water (hydrophily) are abiotic pollination means. Among these, anemophily is the most common form. About 80% of all plant pollination is biotic. Animal pollinators (most are insects) are around 200,000 in the wild. Pollination by insects (entomophily) is by bees, wasps and occasionally ants (Hymenoptera), beetles (Coleoptera), moths and butterflies (Lepidoptera) and flies (Diptera). Pollination conducted by vertebrates such as birds and bats is popularly known as "zoophily", done by hummingbirds, sunbirds, spider hunters, honeyeaters and fruit bats.

When self-pollination occurs before the flower opens, it is cleistogamy. It is a type of sexual breeding. In contrast to asexual systems such as apomixes, cleistogamy is a mode of sexual reproduction. Some cleistogamous flowers never open. They are in contrast to chasmogamous flowers which open and are pollinated. Cleistogamous flowers are self-compatible or self-fertile. Many plants like apple are self-incompatible.

Plants and their pollinators are mutually evolved systems. The first fossil record for abiotic pollination is from fern-like plants in the late Carboniferous period. The mutual evolution of hymenopterans and angiosperms is indicated by the development of nectary in late Cretaceous flowers. The largest managed pollination event in the world is in California almond orchards. Nearly half (about one million hives) of the USA honey bees are transported to the almond orchards each spring. New York's apple crop requires about 30,000 hives and blueberry crop of Maine State requires about 50,000 hives each year. In

(continued)

Box 9.1 (continued)
commercial plantings of cucumbers, squash, melons, strawberries and many other crops, bees are the pollinators. Apart from honey bees, other bees work as pollinators (e.g. African weevil *Elaeidobius kamerunicus* in oil palm). The alfalfa leafcutter bee is an important pollinator for alfalfa in Western USA and Canada. In greenhouse tomatoes, bumblebees are used as pollinators.

Box 9.2: Emasculation
Emasculation is the process of removing the androecium to avoid self-pollination. Some of the methods followed for this are as follows:

Hand emasculation (forceps and scissor method): In large flowers anthers can be removed with forceps. This is done before anther dehiscence. Anther removal is generally done the previous day (between 4 and 6 p.m.) of anther dehiscence. To avoid self-pollination, it is desirable to remove other young flowers close to the emasculated flower. The corolla of the selected flower is opened with the help of forceps, and the anthers are carefully removed with the help of forceps. In some species like gingelly (*Sesamum indicum*), corolla can be totally removed along with *epipetalous stamens*. In cereals, one-third of the empty glumes will be clipped off with scissors to expose anthers. In any case, gynoecium should not be injured. The breeder must standardize an efficient emasculation technique that prevents self-pollination to facilitate cross-pollination. This method can be used in the case of large flowers, e.g. tomato, cotton and brinjal.

Suction pressure method: This is useful in small flowers. A thin rubber or a glass tube attached to a suction hose is used to suck the anthers from the flowers. The force of suction must be standardized so that it sucks only anthers but not gynoecium. However, self-pollination (up to 10%) is expected to occur. To reduce self-pollination, the stigma can be washed with a jet of water. However, 100% cross-pollination cannot be ensured in this method.

Hot/cold water treatment: This is useful in small flowers where manual removal of anthers is tedious. Pollen grains are more sensitive than female reproductive organs to both genetic and environmental factors. Temperature of water and duration of treatment must be standardized since the sensitivity to temperature varies from crop to crop. For sorghum, 42–48 °C for 10 min is found to be suitable. In the case of rice, 10 min with 40 °C is adequate. Treatment is prior to the opening of the flower. Whole inflorescence is immersed in hot water carried in thermos flask. Cold water or alcohol is also used in sorghum and pearl millet. Cold water treatment kills the pollen grains without damaging gynoecium. In rice, cold water at 0.60 C kills the pollen

(continued)

Box 9.2 (continued)
grains without affecting the gynoecium. However, it is less effective than hot water treatment.

Alcohol treatment: This is done by immersing the inflorescence in alcohol of suitable concentration for a brief period followed by rinsing with water. In Lucerne (*Medicago sativa*), immersion of inflorescence in 57% alcohol for 10 s was highly effective. Compared to suction method, this method is more effective.

Genetic emasculation: Genetic/cytoplasmic male sterility may be used to eliminate the process of emasculation. This is useful in the commercial production of hybrids in maize, sorghum, pearl millet, onion, cotton and rice. In many species with self-incompatibility, emasculation is not necessary. Protogyny will also facilitate crossing without emasculation in pearl millet.

Gametocides: Gametocides are also known as chemical hybridizing agents (CHA). They selectively kill the androecium without affecting the gynoecium, e.g. ethrel, sodium methyl arsenate, zinc methyl arsenate in rice and maleic hydrazide for cotton and wheat.

9.2.2 Distant Hybridization

Distant hybridization is crossing of individuals between species and genera that combine divergent genomes. Wide hybridization breaks species barriers for gene transfer resulting in changes in genotypes and phenotypes of the progenies. The chromosome behaviour of wide hybrids and chromosome constitutions in their progenies give wider opportunities for chromosome manipulations. They can be classified as:

(a) Incorporation of alien chromosome or chromosome fragment of a wild species to enhance crop genetic diversity. This exercise can transfer beneficial characteristics from wild and weedy plants to the cultivated crop species in the form of alien chromosome substitution, addition or translocation.
(b) Production of amphidiploid through incorporation of all alien chromosomes by chromosome doubling. The man-made cereal crop Triticale (X *Triticosecale* Wittmack) is an amphidiploid between wheat (*Triticum turgidum* L. or *Triticum aestivum* L.) and rye (*Secale cereale* L.). Amphidiploids are useful to derive alien gene introgression or alien chromosome substitution, addition and translocation lines.
(c) Induction of crop haploid through elimination of all alien chromosomes. Haploids are used for doubled haploid breeding. As true-breeding crops, wheat and rice can quickly fix genetic recombination through doubled haploids. This enhances breeding efficiency through reducing time taken for a breeding cycle (see Fig. 9.5).

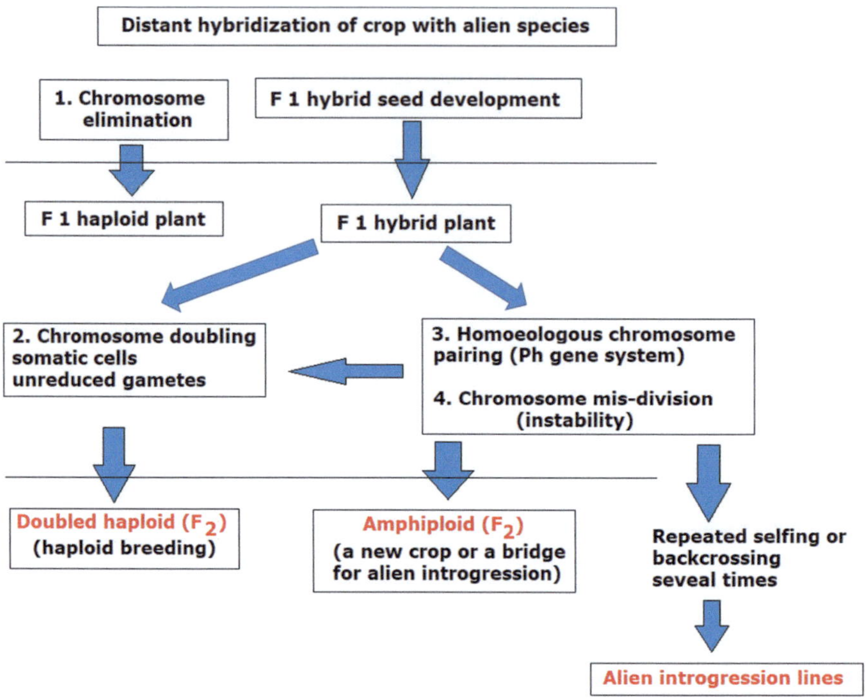

Fig. 9.5 Genetic analysis of recombination. Type 1 is the manipulation for single chromosome, while type 2 and 3 are the genome manipulation by the loss and the addition of alien genome respectively. Chromosome manipulation based on chromosome behaviour in F_1 hybrids. Alien chromosome elimination during the development of F_1 hybrid embryos to produce haploid; chromosome doubling in F_1 hybrid plants to produce amphidiploid; homoeologous chromosome pairing or chromosome mis-dividing in hybrid plants to produce translocation line

Type 1 is the manipulation for single chromosome, while types 2 and 3 are the genome manipulation by the loss and the addition of alien genome, respectively. F_1 hybrid is the first step that arises from the crossing of a crop and an alien species (Fig. 9.5). Crossability is vital to achieve this step. Some genes or QTL for crossability have been found in tetraploid wheat (*T. turgidum* L.) and common wheat (*Triticum aestivum*). By implementing techniques like embryo rescue and hormone treatment, production of F_1 hybrids can be ensured (see Chap. 17 for further details).

9.2.3 Choice and Evaluation of Parents

Breeding self-pollinated plants are performed with single crosses between two parents, followed by production of segregating progeny populations. This method generally results in a reasonable amount of genetic variability needed for selection and attainment of complete homozygosis. However, in cross-pollinated plants, where heterosis leads to superior hybrid genotypes, parental combination is sought to obtain the maximum expression of desirable agronomical traits. Selecting the best hybrid combinations is the initial breeding step that determines the degree of success achieved by the programme because it is fundamental that genetic variability be present in the initial population/progeny to obtain superior genotypes. However, for both self-pollinating and cross-pollinated plants, breeders find it difficult to identify the best parents that when crossed with each other, give rise to hybrid populations of superior performance. It is here that the choice of parents becomes vital for any breeding programme. Mainly, individual's high performance, wider adaptability and yield stability have been the major features taken into account for choosing parental genotypes. There are several strategies by which parents can be selected like individual genotype performance, adaptability and stability, diallel crosses, topcrosses, pedigree data, DNA markers, combined morphological and molecular data analysis and genetic distance measures. These aspects will be briefly dealt here.

Individual Genotype Performance It is still common for the breeder to select parents based on their phenotypic performance regarding specific characteristics. This kind of decision depends on how he could select those genotypes with the best means for targeted characters, such as yield components, grain quality, vegetative and reproductive cycle and pest and disease resistance. However, it is not possible to capture the combining ability among parents based solely on their individual performance. The breeder must obtain crosses and evaluate the progenies or use techniques that allow the prediction of a specific genotype combination before the cross is performed (see Chap. 14 for details).

Adaptability and Stability Parental selection for crosses can take into account high adaptability traits (genotype ability to positively react to environmental stimuli) and yield stability (genotype ability to respond vis-à-vis the environment's yield potential). Considering these points, the selection of parents is also highly important for breeding programmes aiming for a broader area of coverage, mainly for locations that show distinct soil and climate conditions. Many statistical models were developed to make genotype x environment interactions more precise and to facilitate the understanding of adaptability and stability of evaluated genotypes (see Chap. 20).

Diallel Crosses Both general (GCA) and specific (SCA) combining abilities between putative parents can be determined by diallel crosses. Here, one has to cross all the selected genotypes in all possible combinations (complete diallel) and evaluate their progenies, or one can perform part of the crosses (incomplete diallel). Requirement of large number of crosses is the major barrier for their use. Despite these limitations, this type of analysis provides detailed information regarding the genotypes involved, estimates for parameters useful for the selection of the best parental combinations and an understanding of the genetic effects involved in the targeted characters. The most commonly used techniques are as follows: (a) the effects for the general and specific combining ability between parents are estimated; (b) the variety and heterosis are evaluated; and (c) it provides information regarding the character's basic mechanism of inheritance on the genetic values of the parents used and the selection limit. Furthermore, software such as DIALLELSAS05 is available for helping breeders to better design their diallel matings.

Topcrosses This procedure rapidly and precisely tests a large number of high-performance genotypes (elite lines, such as pure lines, open-pollinated or synthetic populations) with a common genotype of wide or narrow genetic base, designated as a tester line. Therefore, it is possible to evaluate the general (GCA) or specific (SCA) combining ability of each genotype against a tester and to estimate the probable outcome of pairwise combinations of the best genotypes by means of progeny tests. Two important aspects of the topcross scheme are relevant for estimating parental performance in pairwise combinations: (a) the contribution of each parent is directly transferred to the progeny mean (**x** parents X $\overline{\mathbf{x}}$ progenies), i.e. through additive gene action, and (b) the reliability of the results being obtained is independent of the quantitative or qualitative nature of the data. This is an efficient technique regardless of the number of genotypes to be tested and its reliability based on the narrow-sense heritability measurements:

$$h_r^2 = \frac{\delta_A^2}{\delta_P^2}$$

where:

h_r^2 = narrow-sense heritability
δ_A^2 = additive variance
δ_P^2 = phenotypic variance

Superior pure lines selected by their combining ability with the tester do not always give satisfactory results when crossed with each other, especially when the tester is proper for evaluating GCA. Therefore, the correlation coefficient (r) between specific crosses involving one parental line and its performance in the testcross is intermediate ($r \leq 0.5$), especially when the tester has a broad genetic base. Thus, the use of a tester with a narrow genetic base can be a favourable alternative to elevate correlation coefficients ($r \leq 0.7$).

Pedigree Data Pedigree data can be studied by Malecot's co-ancestry coefficient and is defined as the probability that two given alleles would be identical by descent in a genotype product of a given cross. This method is described as an easy and affordable alternative to be used for the selection of parental genotypes, and it has been largely employed in genetic distance estimates. On the other hand, pedigree information is not publicly available, and a major barrier for using such a technique is the lack of information at adequate levels for a number of species.

DNA Markers The use of DNA markers in the estimation of genetic distances within and between plant species has grown rapidly. The main types of markers are AFLP (amplified fragment length polymorphism); RFLP (restriction fragment length polymorphism); microsatellites, also known as SSRs (simple sequence repeats); and STS-PCR (sequence-tagged sites-polymerase chain reaction). RAPD (random amplified polymorphic DNA) have been shown to have low reliability and its use has diminished. However, to make more precise inferences about the available gene pool, it is necessary to consider the properties of each marker and the genomic regions they assess. Such kinds of markers are widely being applied in maize and wheat. Hybrid grain yield in maize was correlated with genetic distance based on RAPD markers. The use of quantitative trait loci (QTL) is one of the major goals for breeding programmes during the twenty-first century. Currently, there are studies on the genetic mapping of QTL for many traits related to disease resistance, grain yield as well as main components of grain yield and other traits of agronomic importance. QTL-associated markers, when used in genetic distance studies within species, should increase the chances of finding distant genotypes carrying complementary genes for important agronomic traits.

Combined Morphological and Molecular Data Analysis This is to combine morphological and molecular data into one analysis. This will generate a similarity estimate (index) that ranges from 0 to 1. But this technique has been used by many because the number of data points originating from phenotypic observations is much lower than the ones obtained from molecular markers, resulting in some bias towards the outcome of the molecular analysis. The statistical software developed also does not provide equivalence between the quantitative (phenotypic) and molecular (binary) data when included in different numbers on the combined estimate. In a study on maize, comparisons showed that the total variation was obtained with only 15 polymorphic markers, whereas the initial number used was 131. It has been observed that small distances estimated by molecular markers are consistently associated to small phenotypic distances, while large molecular distances can be associated with either large or small phenotypic distances.

Genetic Distance Measures Multivariate analysis is the tool being used for estimating genetic distances. This analysis has the possibility of gathering many variables into one analysis. In addition to genetic distance studies, it is also necessary that the genotypes selected for crosses possess high individual performance, adaptability and stability for yield. When these requirements are fulfilled, there is a high probability of selecting transgressive genotypes due to the occurrence of heterosis and the action of complementary dominant genes. Genetic distance studies comprise six steps:

(a) Selection of genotypes to be analysed
(b) Data production and formatting
(c) Selection of the distance definition or measurement to be used for the estimations
(d) Selection of the clustering or plotting procedure to be used
(e) Analysis of the degree of distortion caused by the clustering/plotting procedure used
(f) Interpreting the data

The overall distance of Mahalanobis (D^2) and the Euclidean distance are the most used statistical procedures to estimate genetic distances. Since Mahalanobis distance takes into account the environmental effects and allows for obtaining correlations between characters, it has an advantage over Euclidean distance. Once the distance estimates between each genotype pair is obtained, the data display and analysis can be facilitated by the use of a clustering/plotting procedure. An example with 19 wheat genotypes is shown in Table 9.1.

Clustering methods have the goal of separating a pool of observations based on grouping and subgrouping. The hierarchical and optimization methods are employed by plant breeders. In hierarchical methods, genotypes are grouped by a process that repeats itself at many levels, forming a dendrogram (see Fig. 9.6) without concern for the number of groups formed. In this case, three distinct forms of clustering may be used on the basis of genotype pair distances:

Table 9.1 Clustering of 19 wheat genotypes using Tocher's method and the overall distance of Mahalanobis

Groups	Genotypes
I	BRS 119, BRS 120, BRS 177, BRS 192, BRS 194, BRS 208, BR 23, BR 35, BRS 49, CEP 24, ICA 1, PF 950354 and RUBI
II	CEP 29 and ICA 2
III	BR 18 and TB 951
IV	Sonora
V	BH 1146

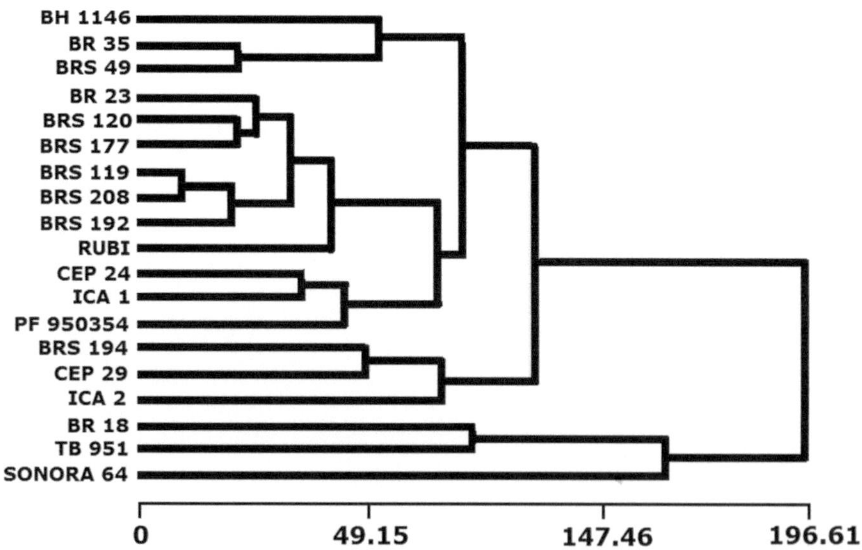

Fig. 9.6 Dendrogram of 19 wheat genotypes obtained by UPGMA using the overall distance of Mahalanobis. The cophenetic correlation coefficient (r) is 0.80. Cophenetic correlation is a measure of how faithfully a dendrogram preserves the pairwise distances between the original unmodelled data points

(a) Using the average of distances between all genotype pairs for the formation of each group, named average linkage analysis or UPGMA (unweighted pair group method with arithmetic mean)
(b) Using the smallest distance between a pair of genotypes known as single linkage or nearest-neighbour analysis
(c) Using the longer distance between a genotype pair, known as complete linkage or farthest neighbour

However, it is at the discretion of the researcher to adopt the procedure that is most suitable for their data set.

For the optimization methods, groups are established according to a fixed clustering criterion, differing from hierarchical methods due to the fact that clusters are mutually exclusive. For the optimization method proposed by Tocher, a criterion of always keeping the average distance within groups smaller than any distance between groups is used. Another way of displaying distances is through a multidimensional scale, which also requires the use of a distance measure. However, the display is obtained by means of dispersion graphics where the dots represent the genotypes evaluated.

The display of distances on a bidimensional plot in the multidimensional scale (MDS) (Fig. 9.7) shows that the longer distance between two genotypes was found between Sonora64 and BH1146, and the results are in agreement with the results from UPGMA and Tocher's analyses (Table 9.1; Figs. 9.6 and 9.7). The bidimensional scale ($r = 0.94$) showed better adjustment between the graphical display and its original matrix, when compared with the UPGMA ($r = 0.80$) analysis. MDS analysis differs from other clustering procedures as it searches for the best adjustment between the original matrix and the graphical display by means of a regression analysis. The best adjustment is then compared with the original distance by a stress function. Thus, although the MDS has shown a cophenetic coefficient higher than UPGMA, the stress value slightly above the accepted level suggests that both techniques are equally efficient in preserving the real distances between the genotype pairs evaluated. A cophenetic correlation for a cluster tree is defined as the linear correlation coefficient between the cophenetic distances obtained from the tree and the original distances (or dissimilarities) used to construct the tree. Thus, it is a measure of how faithfully the tree represents the dissimilarities among observations.

The selection of parents is vital for any breeding programme. Selection of a particular plant ideotype that fulfils market demands is the choice of the breeder. Even though recombination may have its role in amplifying the genetic variability of segregating populations, it is the combining ability between two parents and the high performance in agronomic traits that determine the success of offspring. Phenotypic and DNA marker characterizations, as well as multivariate statistical analyses, are the key components. Biotechnology and bioinformatic tools including DNA marker and software analyses are also important.

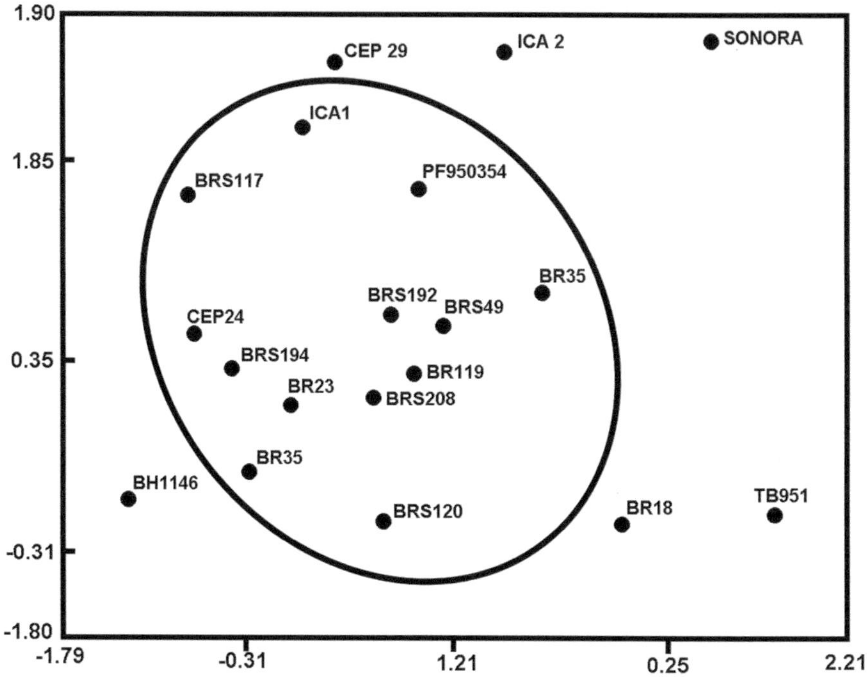

Fig. 9.7 Bidimensional display of 19 wheat genotypes using overall distance of Mahalanobis as a measure of genetic distance (based pon 17 traits). Cophenetic correlation (*r*) is 0.94

9.3 Consequences of Hybridization

Joseph Gottlieb Kölreuter in 1766 observed hybrid vigour and further stated that interspecific hybrids are frequently sterile and difficult to produce. Genetic exchange between species is not possible since the hybrids are sterile. Here, we discuss the phenomena in F_1 hybrids (heterosis), population-level processes like transgressive segregation and adaptive introgression, hybrid speciation and reinforcement.

Heterosis Crossing two genotypes can derive a superior type with hybrid vigour or heterosis. Both Kölreuter and Darwin described heterosis could not offer explanations to the underlying mechanism. Early hypotheses put forth by Jones in 1917 and East in 1936 are dominance and overdominance, respectively. Dominance model explains that recessive deleterious alleles are accumulated at different loci in both parents. In F_1, each of these deleterious alleles is masked by beneficial alleles from the other parent. The overdominance hypothesis postulates that the heterozygous genotype is superior to both homozygous genotypes. Recent advances in genomics have implicated epistatic interactions among alleles at multiple loci, epigenetic modifications to the genome and the activity of small RNAs. It has become clear that multiple causal mechanisms contribute to heterosis.

Quantitative trait locus (QTL) mapping experiments were used to characterize genetic action in heterotic phenotypes in rice, maize, cotton and *Arabidopsis*, which indicated heterosis as the cumulative result of dominant, overdominant and epistatic effects. Recent genomic studies revealed that interactions between divergent epigenetic systems (a system that contribute changes in organisms caused by modification of gene expression rather than alteration of the genetic code itself) lead to heterosis in F_1 hybrids. In *Arabidopsis* and rice, small RNAs, including microRNAs and small interfering RNAs, may be involved in heterosis, as F_1 hybrids often show small RNA expression levels outside of the parental range. If *Arabidopsis*, if F_1 hybrids are treated with a DNA demethylating agent, heterosis can be eliminated. Gene regulation by small RNAs can also be altered by introducing mutations.

Transgressive Segregation Transgressive segregation produces novel phenotypes in F_2 generation and later and may persist indefinitely once established. The best known genetic mechanisms leading to transgressive segregation are complementary gene action and epistasis. In complementary gene action, both parents harbour additive alleles of opposing sign at different loci affecting a multilocus trait (some + and some −), which then sort in favour of one direction in the segregating hybrids. In epistasis model, non-additive interactions between loci from different parents can cause extreme trait values. The small interfering RNAs can also govern such interactions.

Adaptive Introgression Introgression of new genes through hybridization may serve as an evolutionarily creative force by introducing new, possibly adaptive, genetic variation into a population. Introgression can introduce large blocks of novel variation into a population. This was suggested by Anderson and Stebbins in1954. Alleles contributing to an adaptive phenotype are introgressed as demonstrated by genomic analysis. Compared to traits controlled by many loci, adaptive introgressed traits are easier to detect.

Hybrid Speciation As suggested by Linnaeus in1760, new species may arise by hybridization. A new hybrid lineage may be formed through allopolyploidy or through homoploid (an infertile hybrid when becomes fertile after doubling of chromosome number) hybrid speciation. Fusion of unreduced gametes or genome doubling following hybridization can give rise to allopolyploid lineages. Homoploid hybrid speciation describes the formation of a new, reproductively isolated hybrid lineage without change in ploidy. Almost 11% of species across 47 plant genera were likely of allopolyploid origin.

Reinforcement The process of increased reproductive isolation because of selection to decrease hybridization (due to sterility) is called reinforcement. Reinforcement begins with mating between closely related taxa. Here, the hybridization is costly due to low hybrid fertility. Costly hybridization leads to selection favouring new traits that increase assortative mating (mating of similar phenotypes). These novel trait values are selected in sympatric populations (populations in the same geographic area) because they decrease hybridization, but they are not necessarily favoured in allopatry (populations that are unable to interbreed due to geographic separation), thus generating a pattern of character displacement. Thus, hybridization is both the source of reinforcing selection and a major hindrance to the success of reinforcement (Box 9.3).

Box 9.3: Hybrid Rice

Rice is the staple food for more than half of the world's population. The increased demand for rice is expected to exceed production in many countries in Asia, Africa and Latin America. While land, water and labour are all decreasing, world rice production needs to increase. Since 1961, at varying rates, rice production has increased. This is due to improvement in productivity. As per FAO estimates, the annual growth rate of yields declined from 3.5% in the 1960s to about 1.1% in the 1990s. There is stagnation and deceleration of rice yields in many Asian countries.

Prof. Yuan Longping, known as the "Father of Hybrid Rice", started working on hybrid rice in 1964. In 1974, Chinese scientists produced cytoplasmic-genetic male sterile rice. This was done by transferring a gene for male sterility from wild rice. The first generation of hybrid rice varieties are three-line hybrids. They produce 15–20% greater yield compared to varieties with same growth duration. Hybrid rice technology produced two-line hybrids with 5–10% more yield than three-line hybrids. In China, the area under hybrid rice is around 30 million ha that produces 210 million tons of rice. 50% of this area is under hybrid rice. Over the last decade, FAO, the International Rice Research Institute (IRRI), the United Nations Development Programme (UNDP) and the Asian Development Bank (ADB) have provided strong and consistent support to improve national capacity in hybrid rice breeding. Increasing attention has been given to the development of transgenic rice. As on date, hybrid rice technology has a yield advantage of 15–20% (or more than 1 ton of paddy per hectare) over the best varieties. China has diversified agricultural production through hybrid rice production. Chinese rice areas steadily decreased from 36.5 million ha in 1975 to 30 million ha now. But China could feed more than 1 billion people, through hybrid rice programme. The national productivity was increased from 3.5 to 6.7 tons/ha.

Further Reading

Fridman E (2015) Consequences of hybridization and heterozygosity on plant vigour and phenotypic stability. Plant Sci 232:35–40

Hoskin CJ, Higgie M (2013) Hybridization: its varied forms and consequences. J Evol Biol 26:276–278

Liu et al (2014) Distant hybridization: a tool for interspecific manipulation of chromosomes in: alien gene transfer in crop plants, Innovations, methods and risk assessment, vol 1. Springer, New York, pp 25–42

López-Caamal A, Tovar-Sánchez E (2014) Genetic, morphological, and chemical patterns of plant hybridization. Rev Chil Hist Nat 87:16

Backcross Breeding

10

Keywords

Genetic consequences of backcrossing · Procedure of backcross · Recovery rate of RP genes · Molecular marker-assisted backcrossing · Recurrent selection in backcross · Transfer of quantitative characters · AB-QTL in cross-pollinated crops · Merits and demerits of backcross breeding

A cross between F_1 hybrid and one of its parents is known as a backcross. Harlan and Pope in 1922 first proposed backcrossing as an appropriate breeding method for cereal crops. Since then, backcrossing became a widely accepted breeding strategy in diverse crops. This is used to transfer one or a few traits into an adapted/elite variety. Mostly, the elite variety used for backcrossing (called the "recurrent parent" or "recipient parent") used to have a large number of desirable attributes but may be deficient in a few traits. The other parent, called the "donor parent" (or "non-recurrent parent"), lodges one or more traits that is lacking in the elite variety, but with poor agronomic traits.

The following requirements are to be fulfilled for backcrossing:

(a) Availability of a recurrent parent that lacks one or two traits.
(b) Availability of a donor parent having traits to be transferred.
(c) The traits to be transferred must be with high heritability.
(d) Backcrosses must be up to F_7 or F_8 in order to recover recurrent parent with the traits of donor parent.

The following are the utilities of backcross breeding that can be applied for both self- and cross-pollinated crops:

(a) Traits with simple inheritance like disease resistance, seed colour, plant height, etc. can be practised.

P. M. Priyadarshan, *PLANT BREEDING: Classical to Modern*,
https://doi.org/10.1007/978-981-13-7095-3_10

(b) Quantitative traits like earliness, seed size and seed shape can be transferred.
(c) To transfer simply inherited traits like disease resistance from allied species (e.g. transfer of leaf and stem rust resistance from *Triticum monococum* to *Triticum aestivum*).
(d) Transfer of cytoplasm from one variety or species to another (cytoplasmic male sterility).
(e) Utilization of transgressive segregation (it is derivation of extreme phenotypes among segregants compared to parents). They can be either positive or negative.
(f) Production of isogenic lines (individuals with same genotype irrespective of their homo- or heterozygous nature). Vegetatively propagated clones are isogenic. Isogenic lines are achieved through repeated self-fertilization.
(g) When backcross is practised in cross-pollinated crops, a larger number of plants (200–3000) are used to be crossed with recurrent parent.

The following are the genetic consequences of backcrossing:

(a) Increases homozygosity.
(b) Progeny shall be similar to recurrent parent.
(c) Gene under transfer will be maintained by selection in backcross generations. In each backcross generation, there are chances that crossing over can occur between the gene being transferred and tightly linked genes.

10.1 Procedure of Backcross

Recurrent parent and donor parent are crossed to produce an F_1 hybrid. This F_1 is crossed with the recurrent parent to produce the first backcross generation (BC_1F_1). After phenotypic screening for target trait, the selected BC_1 plants are crossed with the recurrent parent to produce the BC_2. Subsequent crosses of BC plants are made with the recurrent parent. Selection must be exercised in each round of backcrossing. Though there is no absolute number of backcrosses needed, 6–8 backcross generations are required to get the trait transferred. After final backcross, selected genotypes are self-pollinated to achieve homozygous lines for the target trait (Fig. 10.1).

In the end, breeder wishes to keep only the individuals homozygous for the resistance gene. To obtain them, self Rr plants from BC_4. The resulting offspring will be 1RR:2Rr:1rr. Progeny testing would be needed to identify RR from Rr plants. Progeny testing is where the genotype of a parent plant is determined by genotypes of the line's progeny. In the case of an RR plant, the progeny will all be RR (no segregation for the gene/trait). However, in the case of an Rr plant, the progeny will segregate 1/4 RR:1/2 Rr:1/4 rr. Therefore, the progeny of RR plants will be uniformly resistant to leaf rust, while the progeny of Rr plants will segregate for resistance and susceptibility (Fig. 10.2).

In contrast, if the genes for rust resistance had been recessive (i.e. rr = resistant) rather than dominant, then the introduced resistant gene is only carried in the heterozygote and would not be detected throughout the backcross programme.

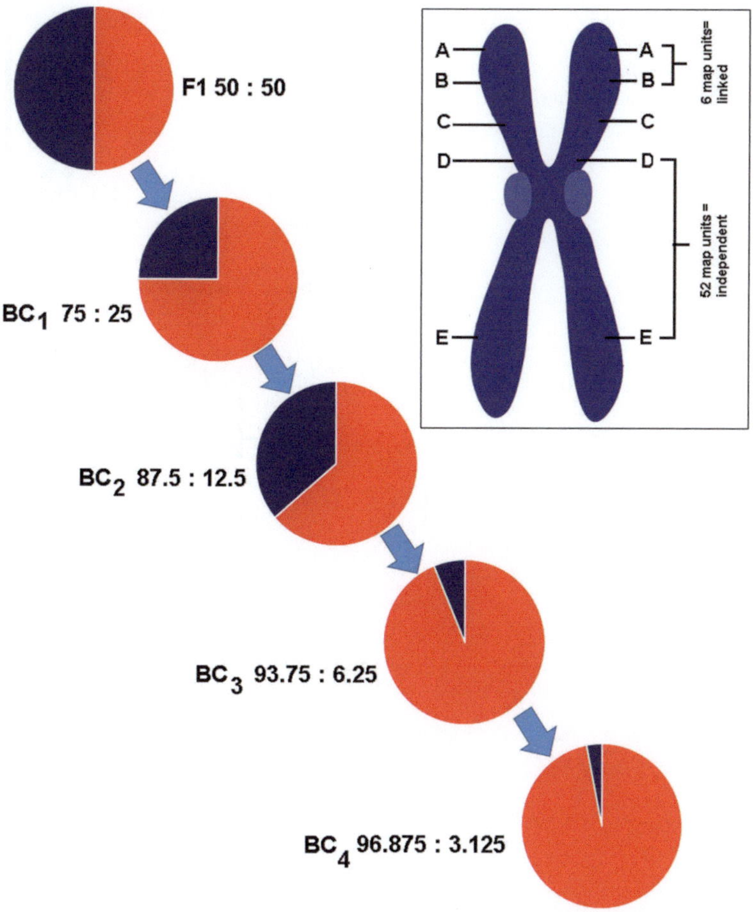

Fig. 10.1 The contribution of the donor parent genome is reduced by half with each generation of backcrossing. Percentages of recurrent parent (red) are expressed as a ratio to percentages of donor parent (blue). (Courtesy: David M. Francis, Ohio State University)

After each backcross, heterozygote (Rr) shall be self-pollinated to produce resistant plants (rr) in the progeny. Resistant plants (rr) are backcrossed to the recurrent parent (RR) (Fig. 10.3). While working with recessive traits, Allard in 1960 proposed advancing the first backcross to the F_2 generation followed by selection for the desirable character from the donor parent (rr) and the general features of the recurrent parent. The second and third backcrosses are then made in succession after which the inbreeding with selection phase for rr is repeated. This is followed by the fourth and fifth backcrosses in succession. The BC_5 F_2 that are resistant (rr) are crossed to recurrent parent (RR) for the BC_6F_1 which is Rr. The BC_7F_1 is selfed to get in the BC_6F_2:1/2 RR (susceptible):1/2 Rr (susceptible):1/2 rr (resistant) backcross with intense selection for both the desired character (ss) and the recurrent parent plant phenotype. You have successfully transferred the gene.

Fig. 10.2 Backcross procedure to transfer leaf rust resistance (RR,Rr) from resistant variety to susceptible (rr)

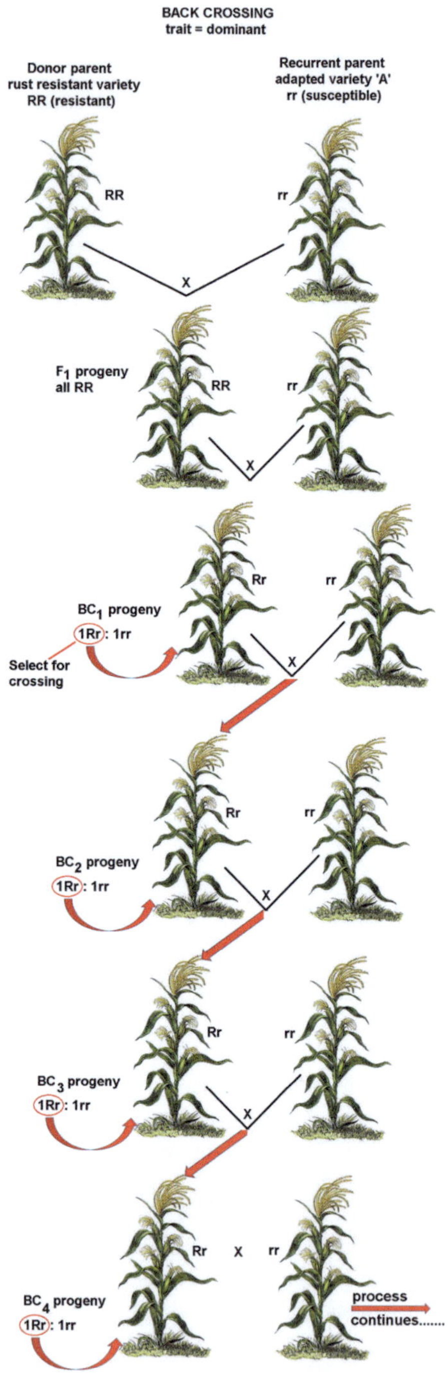

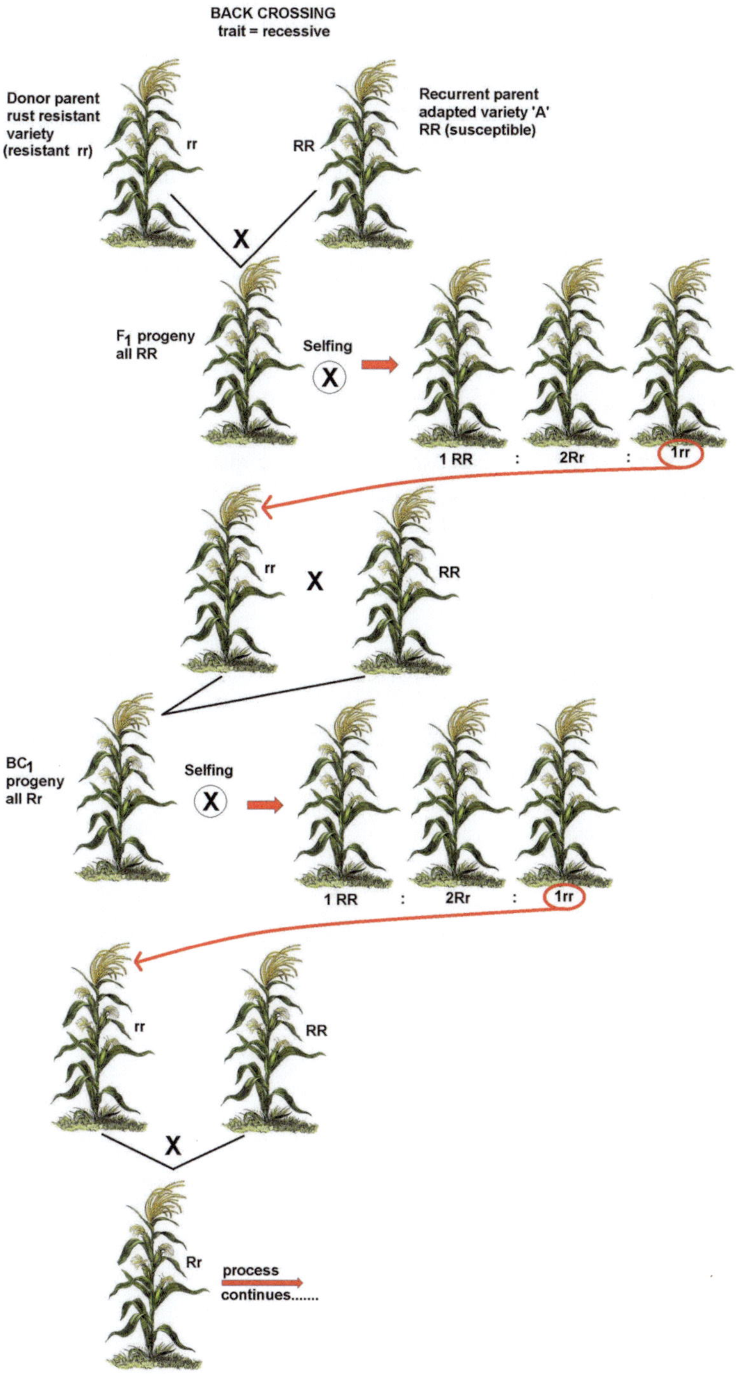

Fig. 10.3 Procedure to transfer resistance governed by recessive gene

Backcrossing accommodates traits, genes or even anonymous loci or chromosome segments. Backcrosses ensure that the proportion of genome from the donor parent shall be zero after successive generations, except for the trait of interest. If selection is applied for the desired characteristic only, then the proportion of donor genome is expected to be reduced by one-half (50%) at each generation, except on the chromosome holding the characteristic. On this chromosome, the rate of decrease is slower resulting in *linkage drag*. Obviously, if selection can also be applied against the donor genome proportion, then its rate of decrease can become faster. Phenotypic resemblance to the recurrent parent is the attribute used by breeders for quite long. Molecular marker alleles can also be used for selection. Historically, this was among the first suggested uses of molecular markers to assist breeding programmes. Reduction of linkage drag is the most difficult goal to achieve. Marker-assisted selection (MAS) is advantageous at this juncture.

Since the process of backcrossing isolates a gene, or chromosomal region, in a different genetic background, it is useful to delineate quantitative traits. In fact, it is one of the few reliable methods to validate the additive effect of a quantitative trait locus (QTL) or a candidate gene. In addition, backcrossing could be used for QTL detection to increase the precision of QTL mapping.

10.2 Recovery Rate of RP Genes

The extent of recovery of trait is dependent on the number of backcrosses done and the number of loci that differ between the recurrent parent (RP) and the donor. In the absence of genetic linkage, the average recovery of RP genes increases each backcross by one-half the percentage of the donor parent (DP) present in the previous backcross. This is demonstrated in Table 10.1, and the general equation is:

$$(1/2)^{n+1} = \%\mathrm{RP}$$

where

n equals the number of backcrosses that have been completed.

Table 10.1 Average recovery of RP genes per round of backcrossing assuming no gene linkage

No. of backcrosses	Recurrent parent (%)	Donor parent (%)
F_1	50.00	50.00
1	75.00	25.00
2	87.50	12.50
3	93.75	6.25
4	96.88	3.13
5	98.44	1.56

Table 10.2 Average recovery of RP genome when recurrent and donor parents have different alleles at multiple loci

	Backcross numbers					
Number of loci	1 (%)	2 (%)	3 (%)	4 (%)	5 (%)	6 (%)
1	50.00	75.00	87.50	93.75	96.88	98.44
2	25.00	56.25	76.56	87.89	93.85	96.90
3	12.50	42.19	66.99	82.40	90.91	95.39
4	6.25	31.64	58.62	77.25	88.07	93.89
5	3.13	23.73	51.29	72.42	85.32	92.43
10	0.10	5.63	26.31	52.45	72.80	85.43

If both parents have different alleles at multiple loci, then the number of backcrossing needed is expected to increase, as shown in Table 10.2, and the general equation by Allard in 1960 is:

$$(2^n - 1/2^n)^m = \%\mathrm{RP}$$

where

m is the number of backcrosses and m is the number of loci that differ between the RP and DP.

If DP and RP have different alleles at ten loci, only 85% of the BC_6 F_1 plants will have homozygous for all ten alleles of RP. In contrast, 98% of the BC_6 F_1 plants will be homozygous for the trait in question, if only one locus is different. If DP is closely related to RP, the number of backcross generations can be reduced.

In breeding for leaf rust resistance, the aim of backcrossing is to increase the recurrent parent's genes except for the gene for resistance. The amount of remaining genetic information (the non-target genes), on the average, from DP is reduced by 50% with each backcross.

The calculation for this data is:

$$\text{Percentage of non-target genes from donor parent} = (1/2)^{n+1}$$

where

n = number of backcrosses.

Genes getting eliminated during backcrossing are influenced by linkage of genes. Linked genes stay together and unlinked genes independently assort. When genes are far apart or on different chromosomes, they are unlinked. If they are nearer, they are inherited together. Linkage is measured by the recombination frequency/map distance (see inset of Fig. 10.1). For example, if an undesirable allele d for dwarfing is linked to R (rust resistant), and selection is only for R, d tends to be brought along in the F_1. However, when reintroducing R in each backcross, the number of opportunities for crossing over between the R and d loci occur. Therefore, the probability of eliminating d is:

$$1 - (1 - p)^n$$

where

n = number of backcrosses
p = recombination frequency between loci

It should be noted that if d and R are very close together (small map distance), it will be very hard to select R and is eliminated.

10.3 Molecular Marker-Assisted Backcrossing

In order to improve the efficiency of introgression (movement of a gene through repeated backcrossing), use of molecular markers has been investigated. This includes various aspects of the use of molecular markers for controlling the target genes, accelerating the recovery of recurrent genome or reducing linkage drag. Use of markers can gain time equivalent to about two backcross generations. Even with the largest population sizes, it is not possible to introgress more than four or five QTLs (*see chapter on "Molecular Breeding" for further details on MAS*).

Marker-assisted backcrossing for a single gene: Two types of selection are recognized:

(a) *Foreground selection*: Plants having the marker allele of the donor parent at the target locus are selected by the breeder. This is to maintain the target locus in a heterozygous state (one donor allele and one recurrent parent allele) until the final backcross is completed. After this, selected genotypes are self-pollinated. The progeny plants that are homozygous for the donor allele are selected.
(b) *Background selection*: Here, the target locus is selected based on phenotype. The breeder selects for recurrent parent marker alleles in all genomic regions except the target locus. The elimination of potential deleterious genes introduced from the donor is vital. The inheritance of unwanted donor alleles is difficult to overcome with conventional backcrossing, but can be done with markers.

Both foreground and background selections can be done by the same backcross breeding programme. They can be done either simultaneously or sequentially. A programme on combined use of foreground and background selection is illustrated in Fig. 10.4. Factors like population size of each backcross generation, distance of markers from the target locus and number of background markers used are governing this process of selection. When foreground and background selections

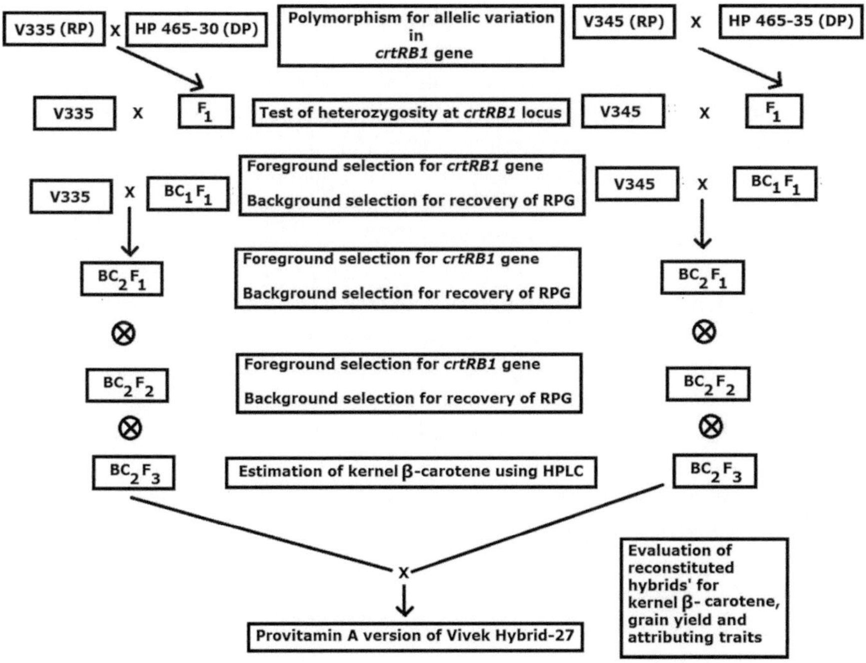

Fig. 10.4 Marker assisted backcross breeding scheme adapted from the introgression allele 1 of the crtRB1 3'TE gene into elite parent (V335 and V345) of the maize hybrid Vivek Hybrid-27 (RP: recurrent parent; DP: donor parent)

Table 10.3 Expected results of a typical marker-assisted backcrossing programme, based on simulations of 1000 replicates

		% homozygosity of recurrent parent alleles at selected markers		% recurrent parent genome	
Backcross generation	Number of individuals	Chromosome with target locus	All other chromosomes	Marker-assisted backcross	Conventional backcross
BC_1	70	38.4	60.6	79.0	75.0
BC_2	100	73.6	87.4	92.2	87.5
BC_3	150	93.0	98.8	98.0	93.7
BC_4	300	100.0	100.0	99.0	96.9

In each backcross generation, heterozygotes were selected at the target locus. Recurrent parent alleles were selected at markers flanking the target locus (2 cM on either side) and at three markers on each non-target chromosome

are combined with MAS, recovery of the recurrent parent genome is faster (Table 10.3). When the target locus is on the same chromosome, the recurrent parent genome is recovered more slowly because of the difficulty in breaking linkage with the target donor allele.

Examples from maize, barley and soybean are:

(a) In maize, the introgressions of *Bt* insect resistance transgene were accomplished. Even though the target gene could be detected phenotypically, markers are used to select for the recurrent parent genome. This process has avoided two backcross generations for the recovery of recipient genome.
(b) In barley, a marker linked to the *Yd2* gene for resistance to barley yellow dwarf virus was successfully used to select for resistance in a barley backcross breeding scheme. BC_2 F_2-derived lines containing the marker exhibited lesser leaf symptoms and higher grain yield, compared to the lines lacking the marker.
(c) In soybean, a yield QTL from a wild accession was introgressed into commercial varieties that increased yield. Even though the yield increment occurred only two of six genetic backgrounds, the process has potential to incorporate wild alleles with the assistance of markers.

Marker-assisted selection for multiple genes: Some suggestions for using markers to select for multiple genes are as follows:

(a) Number of genes undergoing selection may be limited to 3 or 4 (if they are QTLs selected on the basis of linked markers). If they are known loci, directly limit the genes to five or six.
(b) QTLs that have medium to large effects may be targeted so that their consistency can be detected in a range of environments.
(c) As illustrated in Fig. 10.5, examine the QTL analysis carefully to decide which markers to select.
(d) Stepwise backcrossing procedure may be considered. Say if four target genes are to be introgressed into the same genetic background, two parallel backcross schemes, each incorporating two target genes, can be considered. Selected individuals from each scheme are then crossed so as to have plants with all four targets genes. This procedure gives ample chance to undertake background selection in recurrent parent genome rather than selecting for all four targets simultaneously.
(e) Strategies, like F_2 enrichment, backcrossing and inbreeding, may be considered. This would allow reduction in population size (reduction in size up to 90%).

Examples from maize and tomato are:

(a) In maize, QTLs had previously been identified for second-generation European corn borer (ECB) resistance in one population and for rind penetrometer resistance (RPR), an indicator of stalk strength, in three populations. For each trait and population, selection was carried out as indicated in Fig. 10.6, with the 10 highest or 10 lowest families selected in each fraction. Each of the five selected sub-populations was recombined by random mating the selected families, followed by evaluation in field trials.
(b) In some cases, MAS was effective in moving the population in one direction (e.g. ECB susceptibility), but not in the other. Logistically, MAS was considered more advantageous for ECB resistance than for RPR, because of the greater time and expense required for ECB resistance evaluation.

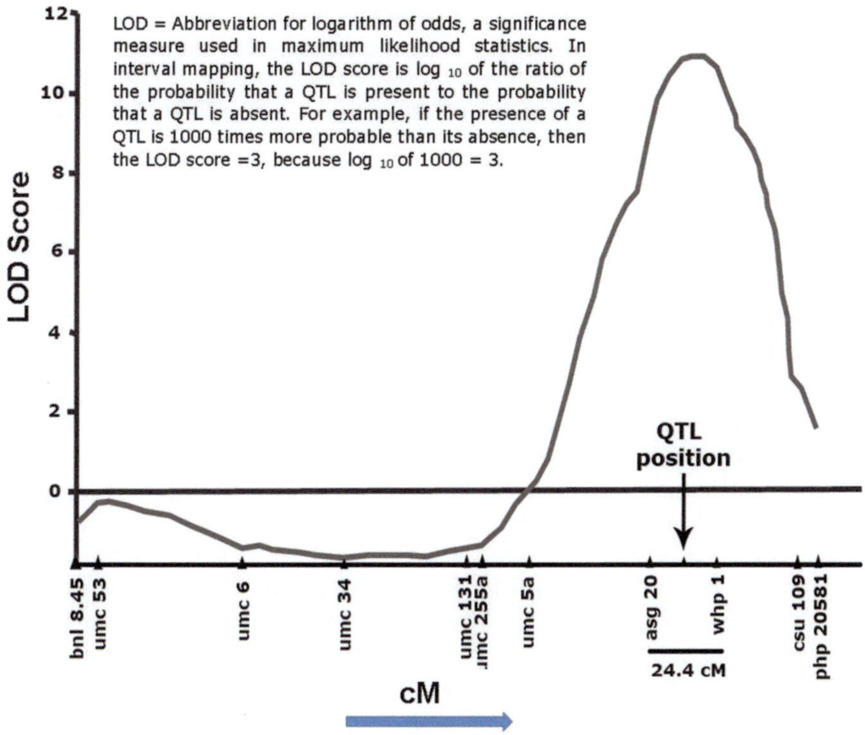

Fig. 10.5 LOD curve from a QTL analysis, indicating the most likely QTL position (peak of the curve) is in the middle of 24 cM marker interval. To select for the favourable allele at the QTL, selection on the basis of both flanking markers (asg20 and whp1) is advisable

Fig. 10.6 Selection scheme for comparing MAS with phenotypic selection for rind penetrometer resistance (RPR) in maize

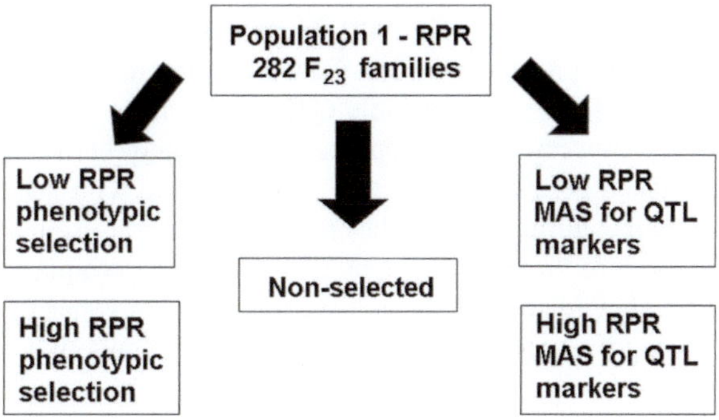

(c) An advantage of MAS is its ability to pyramid multiple resistance genes in the same variety. Combining qualitative and quantitative resistance genes and improved resistance levels are an advantage of MAS. This is done in the presence of a virulent race of the pathogen.

(d) In tomato, an MAS study for black mould resistance demonstrated the value of alleles from wild relatives. Five QTL alleles for resistance, previously detected in wild *Lycopersicon cheesmanii*, were backcrossed into a cultivated tomato background and the backcross progenies were evaluated. Three of the five alleles were effective in reducing disease severity. However, only one of the effective alleles was not associated with negative traits (see Chap. 23 for details on QTL analysis).

10.3.1 Recurrent Selection in Backcross

Backcross breeding facilitates selection on a quantitative trait to isolate genes through repeated backcrossing and selection. This is recurrent selection backcross (RSB). With many markers surrounding the fixed QTL segment, the near isogeneic lines (NIL– lines which differ from a single favourable QTL) obtained could then be used for fine mapping the QTL (see Box 10.1). Depending on the intensity of selection and the number of generations of backcrossing, a QTL remains segregating as a function of its effect on the trait. This method works best for QTL of large effects. However, to fix QTL with smaller effect, inter se mating between each generation of backcrossing (RSBI for RSB intercross) can be applied for stronger selection. RSB/RSBI does not use the same information as interval mapping and may be relevant to quantitative traits of a different genetic architecture. This is useful for exploiting very dense marker coverage around the QTL. RSB still has some advantages over interval mapping (see Chap. 23 for interval mapping).

10.4 Transfer of Quantitative Characters

Transfer of QTLs through backcross is otherwise known as advanced backcross QTL analysis: AB-QTL. AB-QTL analysis was proposed by Tanksley and Nelson in 1996. This process integrates QTL analysis with variety development, by identifying and transferring the valuable QTL alleles from wild to cultivated germplasm in a single process. In this approach, QTL and marker analyses are performed in advanced generations, like BC_2 or BC_3. The wild ancestors of crops are available in their natural habitats that represent precious source of genetic variation. But, majority of the genetic potential preserved in germplasm repositories are unutilized. Though wild germplasm are used as sources of genes for biotic resistance, its use for the improvement of polygenically inherited traits like yield, nutritional quality and stress tolerance is rather limited.

Generations beyond the BC_3 are likely to have low statistical power to detect most QTLs. Tanksley and Nelson proposed two factors that are important in determining this: (a) maximum size (in centimorgans) of the donor segment designating the QTL and (b) the amount of residual donor genome (unlinked to the targeted QTL) still present in the genome. In this way, the backcross populations get skewed towards recurrent parent alleles which make them superior over selfing generations. The number of additional generations of backcrossing required and the number of individuals need to be sampled must be well planned. This is to attain the lines having segment of the donor chromosome with the valuable QTL in the background of recurrent parent genome. Those lines are referred as QTL/nearly isogenic lines or QTL-NILs.

QTL-NILs can be derived from BC_1- or BC_2-derived populations, but for this, screening of large number of individuals (around 5000 or 10,000) is required, respectively. However, selection can be exerted to eliminate non-targeted donor segments by screening a smaller number of individuals over two sequential generations (e.g. a backcross followed by a selfing). Thus, in contrast to the BC_1 and BC_2, QTL-NILs can be derived directly from BC_3 to BC_5 selections from a comparatively small number of individuals. In other words, we can say that the more advanced the backcross population, the simpler it will be to derive a desired QTL-NILs.

10.4.1 AB-QTL in Self-Pollinated Crops

In this scheme, single elite inbred variety is initially crossed to an unrelated donor line to generate BC_1 progeny (around 100 plants). Plants selected in BC_1 are crossed again with the recurrent parent to produce BC_2 progeny of around 200 plants. The BC_2/BC_3 generation plants are evaluated in replicated trials and genotyped for marker-trait loci and selfed to produce BC_2S_1/BC_2S_2 progeny. The genotypic and phenotypic data are subjected to QTL analysis to identify donor genome regions containing favourable QTL alleles. BC_2S_1 or BC_2S_2 families assist in the detection of some recessive QTL donor alleles in addition to the expected dominant and additive donor QTL alleles. Ultimately, QTL/NILs are extracted from the superior BC_2S_1/BC_2S_2 which is used to confirm the findings from the QTL mapping or to fine map the detected QTLs. The outperforming QTL-NILs can be used as parent in future breeding programme or as new varieties (see Fig. 10.7).

10.4.2 AB-QTL in Cross-Pollinated Crops

AB-QTL analysis can be applied to cross-pollinated crops through a slight modification. The elite inbred parent, say inbred A, is used as the recurrent parent in a single cross A x B. Hybridization and backcrossing of donor with the recurrent parent are performed to produce the BC_2 population. The selected plants from the BC_2/BC_3 are genotyped for marker loci. Instead of selfing the BC_2/BC_3, they are crossed with the inbred B to produce BC_2F_1 families, and later the phenotyping is performed. The marker and the phenotypic data are used for the QTL analysis. On the basis of the QTL analysis of this data, favourable QTLs from the donor parent are identified; eventually, the QTL-NILs could be generated. Using this technique, QTL analysis has been conducted in crops like tomato, wheat, barley, rice and cotton. Some examples of AB-QTL analysis conducted in crop genetics and breeding have been listed in Table 10.4.

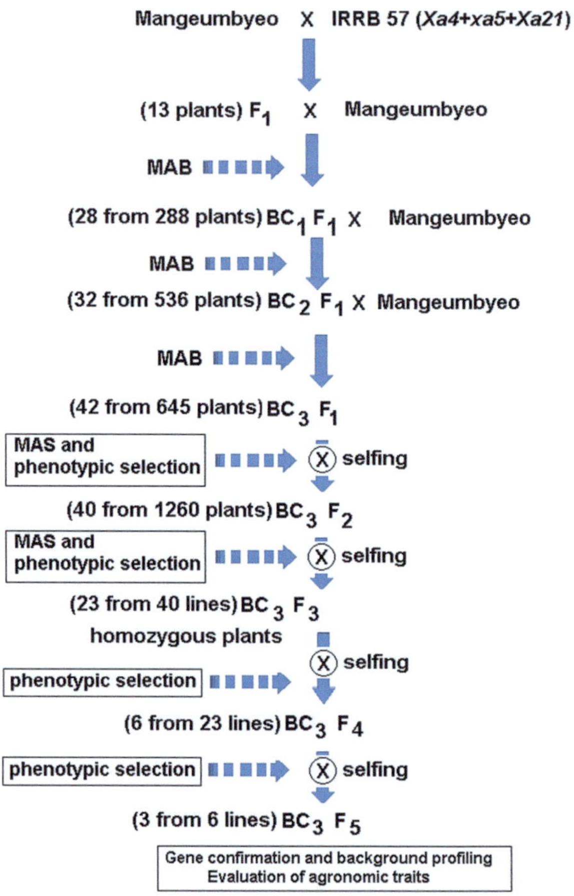

Fig. 10.7 AB-QTL and trait specific inbred line analysis for gene/QTL discovery and development of new rice cultivars or near isogenic lines by MAS. This is an example *of Xa4,xa5 and Xa 21* gene-pyramided backcross breeding lines using marker assisted foreground and background selection for bacterial blight resistance in rice. (Courtesy: Dr. Jena, IRRI; Springer)

10.4.3 Merits and Demerits of AB-QTL Method

Only smaller number of genes from donor parent will be present in BC_2 or BC_3. So, the undesirable effect of wild species on improved variety is reduced. Hence, the effect of individual QTL is measured more precisely. Since the phenotypic selection is delayed for advanced generation, the frequency of deleterious or undesirable alleles from the donor is further reduced. Therefore, the deleterious effects which are associated with balanced population (F_2, BC_1 or RILs) are minimized.

MAS performed in advanced generation is more effective than in F_2 or BC_1 as accumulation of the donor alleles is minimized in advanced generation due to breaking of assembly of favourable epistatic gene combination through recombination. In this way, AB population is skewed more towards the recurrent parent genome. QTL-NILs can be created by one or two additional backcrossing.

In some of the cases, effortless application of this method is limited. AB-QTL analysis is not likely to be useful in crops with relatively longer generation time (>2 years). The longer generation time hinders production of inbreds. In highly

Table 10.4 Some examples of AB-QTL analysis in crop plants

Crop	Wild/donor	Traits studied
Wheat	Synthetic wheat line (W&984)	Yield and yield components
	Synthetic wheat line (xx86)	Agronomic traits
	Synthetic wheat line (TA-4152-4)	Yield and yield components
	Synthetic wheat 6 x lines (Syn 022, Syn 086)	Baking quality traits
	Synthetic wheat accessions (Syn 022)	Leaf rust resistance
	Synthetic wheat accessions (Syn 084)	Drought resistance
Rice	*Oryza rufipogon* (IRGC 105491)	Agronomic traits
	Oryza rufipogon (IRGC 105491)	Yield
	Oryza rufipogon (IRGC 105491)	Yield and morphological traits
	Oryza sativa spp. *japonica* koshihikau	Grain shape
Maize	RD 3013	Grain yield and height
	Dan 232	Grain yield components
	Zea nicaraguensis	Root aerenchyma formation

heterozygous crops also, where inbred lines are not commonly employed (alfalfa, potato), application of AB-QTL is difficult.

10.4.4 Marker-Assisted Gene Pyramiding

Gene pyramiding was proposed by Nelson in 1978 for bringing together a few to several oligogenes resistant to a pathogen. This is for developing durable resistance to diseases. Pyramiding is the stacking of two or more genes controlling a single trait in a single variety. This is a straightforward process by which the same donor parent contributes all the genes. A relatively different strategy is used for gene pyramiding when two or more donor parents are to be used (Fig. 10.8). To achieve durable resistance against one or more diseases in a single cultivar, marker-assisted gene pyramiding can be successfully used to introgress oligogenes or oligogenes with QTLs.

10.4.5 Modifications of Backcross Method

Several modifications have been suggested for backcross method. They are as follows:

In the modified backcross, F_2 and F_3 generations are produced after the first and the third backcrosses. A confirmed selection for the trait is done in the F_2 and F_3 generations. Selection need not be done either for the trait being transferred or for the trait of the recurrent parent in backcross progenies. The fourth, fifth and sixth backcrosses are made in succession. In sixth backcross, a relatively larger number of progeny are used. This is useful to transfer of both dominant and recessive genes. Effective selection in F_2 and F_3 generations is equivalent to one or two additional backcrosses.

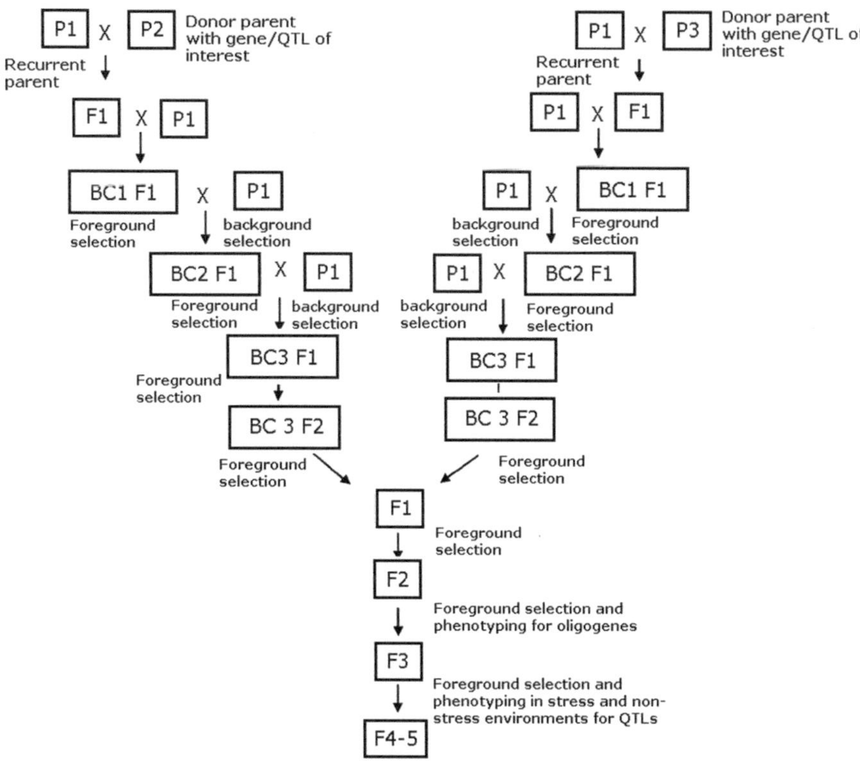

Fig. 10.8 Pyramiding of R-genes using MABC

Another scheme is backcross-pedigree method. Here, the hybrid is backcrossed one or two times to the recurrent parent to ensure transfer of majority of superior genes from the recurrent parent. Subsequently, the backcross progenies are handled according to the pedigree method. This scheme is desirable when one of the parents is superior to the other in several traits but the non-recurrent parent is agronomically weak. Superior parent is used as the recurrent parent. This ensures enough heterozygosity for transgressive segregants to appear. The varieties developed by this scheme are evaluated for yield as in pedigree method. See Table 10.5 for a comparison of pedigree and backcross methods.

10.4.6 Merits and Demerits of Backcross Breeding

The following are the merits:

(a) The newly developed genotype is nearly identical with that of the recurrent parent, except for the genes transferred. So, the outcome of a backcross programme is known beforehand which can be reproduced again.

Table 10.5 Comparison between pedigree and backcross methods

	Pedigree	Backcross
1	F_1 and subsequent generations are allowed to self-pollinate	F_1 and subsequent generations are crossed to the recurrent parent
2	New variety developed differs from the parents in traits	Differs in only one trait in question (trait transferred)
3	New variety to be extensively tested before release	Extensive testing not a prerequisite for release
4	Aim is to improve the yielding ability and other traits	Aims at improving specific trait of a well-adapted, popular variety
5	Useful in improving both qualitative and quantitative traits	Useful for the transfer of both quantitative and qualitative characters with high heritability
6	Not suitable for gene transfer from related species and for producing substitution or addition lines	Only useful for gene transfers from related species and for producing addition and substitution lines
7	Hybridization is limited to the production of F_1 generation	Hybridization with the recurrent parent is necessary for producing every backcross generation
8	F_2 and the subsequent generations are much larger than those in the backcross method	Backcross generations are small and usually consist of 20–100 plants/generation
9	Procedure here is the same for both dominant and recessive genes	Procedures are different for transfer of dominant and recessive genes

(b) Extensive field trials are not necessary since the performance of recurrent parent is already known. In annual crops, this saves up to 5 years.
(c) Since backcross programme is not dependent on environment (except for that done for abiotic stress resistance), off-season nurseries and greenhouses can be used to grow 2–3 generations each year. This reduces the time required to develop a new variety.
(d) Compared to pedigree method, smaller population is needed in the backcross method.
(e) Traits like susceptibility to disease of a well-adapted variety can be removed without affecting its performance and adaptability. Farmers will prefer such a variety since they know the performance of recurrent variety (parent) well.
(f) Backcross is the only conventional method for interspecific gene transfers.
(g) Since transgressive segregation may occur for quantitative traits, backcross can be modified.

The demerits are:

(a) A new variety cannot be superior to the recurrent parent except for the character transfer from donor parents.
(b) There is a likely chance that undesirable genes may also be transferred to the new variety.

(c) Exercise of hybridization for each backcross consumes time (6–8 backcrosses).
(d) Backcross does not permit combination of genes from more than two parents.

Box 10.1: Near-Isogenic Lines

Near-isogenic lines (NILs) are genotypes that differ at one or a few genetic loci. NILs are useful for quantitative trait locus (QTL) analysis. NILs can be used to characterize contrasting chromosomal segments on a uniform genetic background. NILs are produced by transferring ("introgressing") one or more chromosomal segments from a resistant genotype into the genetic background of a susceptible line. There are different crossing strategies to produce various kinds of NILs: (a) a single locus can be introgressed by backcrossing; (b) a large number of loci can be introgressed that span an entire region or chromosome; and (c) lines near the end of the inbreeding process that harbour residual heterozygous regions can be self-pollinated to produce NILs contrasting at those regions. Also, transgenic lines, genome-edited lines and mutants can be considered as NILs.

Analyses of chromosomal segments carrying resistance loci can be done with NILs. For example, NILs can be used to study the resistance spectra of R genes. Sets of maize NILs carrying introgressions from resistant lines into susceptible genotypes were used to identify quantitative resistant loci (QRLs) for single and multiple diseases.

Dissection of resistance components can be done with NILs. For example, in barley stripe rust, individual QRLs varied in their relative effects on different resistance components. Since most NILs are created through a few generations of backcrossing, several linked genes are likely to have been introgressed from the donor line. This is important when analysis is done of possible pleiotropic effects associated with a resistance locus. If the resistant NIL shows low yield, the genes conferring resistance are the same as the yield-reducing genes.

Further Reading

Grandillo S, Tanksley SD (2005) Advanced backcross QTL analysis: results and perspectives. In: Tuberosa R, Phillips RL, Gale M (eds) Proceedings of the International Congress "In the Wake of the Double Helix: From the Green Revolution to the Gene Revolution", Italy. Avenue Media, Bologna, pp 115–132

Kearsey MJ (2002) QTL analysis: problems and (possible) solutions. In: Kang MS (ed) Quantitative genetics, genomics, and plant breeding. CABI Publication, New York, pp 45–58

Ortiz RR (2015) Plant breeding in the omics era. Springer, New York

Paterson AH (2002) What has QTL mapping taught us about plant domestication? New Phytol 154:591–608
Remington DL, Purugganan MD (2003) Candidate genes, quantitative trait loci, and functional trait evolution in plants. Int J Plant Sci 164(3 Suppl):S7–S20
Vogel KE (2009) Backcross breeding. Methods Mol Biol 526:161–169
Zeng Z-B (1994) Precision mapping of quantitative trait loci. Genetics 136:1457–1468

Breeding Self-Pollinated Crops

11

Keywords
Pure-lines · Open-pollinated cultivars · Homozygous and homogeneous · Heterozygous and homogeneous · Homozygous and heterogeneous · Heterozygous and heterogeneous · Mass selection · Pure-line selection · Hybridization and pedigree selection · Special backcross procedures · Multiline breeding and cultivar blends · Breeding composites and recurrent selection · Hybrid varieties

As a matter of fact, breeding procedures and schemes differ with the breeding behaviour of a particular species (see Table 11.1). At the beginning of each breeding programme, the breeder should decide on the type of cultivar to breed for release to farmers. The breeding method used depends on the type of cultivar to be produced. There are basic types of cultivars, *viz.,* inbred pure lines, open-pollinated populations, hybrids and clones.

Pure-Line Cultivars Pure-line cultivars are developed in highly self-pollinated species. These are homogeneous and homozygous, attained through series of self-pollinations. Pure lines are often used as parents for the production of other hybrids. Pure-line cultivars have a narrow genetic base. They are desired for regions where uniformity is in great demand.

Open-Pollinated Cultivars Open-pollinated cultivars are developed for species that are naturally cross-pollinated. They are genetically heterogeneous and heterozygous. Two basic types are available. The first is developed by improving the general population by recurrent (or repeated) selection or bulking and increasing

P. M. Priyadarshan, *PLANT BREEDING: Classical to Modern*,
https://doi.org/10.1007/978-981-13-7095-3_11

Table 11.1 Classification of crop plants based on mode of pollination and mode of reproduction

Mode of pollination and reproduction	Examples of crop plants
Self-pollinated crops	Rice, wheat, barley, oats, chickpea, pea, cowpea, lentil, green gram, black gram, soybean, common bean, moth bean, linseed, sesame, khesari, sunhemp, chilli, eggplant (brinjal) tomato, okra, peanut, potato, etc.
Cross-pollinated crops	Corn, pearl millet, rye, alfalfa, radish, cabbage, sunflower, sugar beet, castor, red clover, white clover, safflower, spinach, onion, garlic, turnip, squash, muskmelon, watermelon, cucumber, pumpkin, kenaf, oil palm, carrot, coconut, papaya, sugarcane, coffee, cocoa, tea, apple, pears, peaches, cherries, grapes, almond strawberries, pine apple, banana, cashew, Irish, cassava, taro, rubber, etc.
Often cross-pollinated crops	Sorghum, cotton, triticale, pigeon pea, tobacco

material from selected superior inbred lines. The second type, synthetic cultivars, is derived from planned matings involving selected genotypes. Open-pollinated cultivars are with a broader genetic base.

Hybrid Cultivars They are produced by crossing inbred lines. Hybrids with hybrid vigour (or heterosis) produce superior yields. Heterosis is vital in cross-pollinated species. Hybrid cultivars are homogeneous but highly heterozygous. Since human intervention was required for artificial pollination, hybrid seed production was expensive. Male sterility is exploited to facilitate hybrid production. The natural reproductive mechanisms (e.g. cross-fertilization, cytoplasmic male sterility) are more readily economically exploitable in cross-pollinated species.

Clones Seeds are used to reproduce most crops. However, a number of species are propagated by using stems and roots. As such, the plants produced will be identical and homogeneous. However, they are highly heterozygous. Some plant species sexually reproduce but are propagated clonally (vegetatively) by choice. Clones are not only identical to each other but also identical to the parent. Such species are improved through hybridization, so that when hybrid vigour exists it can be fixed (i.e. the vigour is retained from one generation to another), and then the improved cultivars are propagated asexually. In seed-propagated hybrids, hybrid vigour is highest in the F_1, but is reduced by 50% in each subsequent generation. Clonally propagated hybrid cultivars may be harvested and used for planting the next season's crop without adverse effects. Hybrid seeds in sexually propagated species must always obtain a new supply of seeds.

Genetically, a population shall be (a) homozygous and homogeneous, (b) heterozygous and homogeneous, (c) homozygous and heterogeneous and (d) heterozygous and heterogeneous.

Homozygous and Homogeneous Cultivars Cultivars that are genetically homozygous shall produce homogeneous phenotypes. Self-pollinated species are naturally inbred and are homozygous. Breeding strategies in these species will be to obtain cultivars that are homozygous. Here, the farmer may save seeds from the current season's crop for planting the next season. Developed economies have well-established commercial seed production systems. Under such circumstances, intellectual property rights prohibit the reuse of commercial seed for planting the next season's crop. So, such a system calls for seasonal purchase of seed by the farmer from seed companies.

Heterozygous and Homogeneous Cultivars A cultivar may be genetically heterozygous yet phenotypically homogeneous. An example is hybrid cultivar production. Hybrid seeds are widely used for the production of outcrossing species like corn. Hybrid cultivar is heterozygous F_1 product derived from a cross of highly inbred (repeatedly selfed, homozygous) parents. Since F_1 is the cultivar, all plants are uniformly heterozygous and homogeneous. The F_2 seed obtained from F_1 will be heterozygous with maximum heterogeneity. The current season's seed cannot be used for planting next season, since the genes may segregate. Heterozygous for some of the genes, but keeps uniform heterozygosity in the population.

Homozygous and Heterogeneous Cultivars The component genotypes are homozygous, where large amount of diverse genotypes are included so the overall population is not uniform. Homozygous for some of the genes.

Heterozygous and Heterogeneous Cultivars The population will be heterozygous for several genes. Synthetic and composite breeding genotypes are included in this category. Here, the farmer can save seed for further planting. Composite cultivars are suited to production in developing countries, while synthetic cultivars are common in forage production all over the world. Population will not be uniform.

11.1 Self-Pollinated Crops: Methods

Self-pollinated cultivars are derived either from a single plant or from a mixture of plants. Cultivars derived from single plants are homozygous and homogeneous. However, cultivars derived from plant mixtures may appear homogeneous but may become heterozygous later since individual plants are different genotypes. The methods of breeding self-pollinated species may be divided into two broad groups – those preceded by hybridization and those not preceded by hybridization. Plant breeders use a variety of methods and techniques to develop pure lines, open-pollinated populations, hybrids and clones.

11.1.1 Mass Selection

In mass selection, seeds are collected from (usually a few dozen to a few hundred) desirable appearing individuals in a population, and the next generation is sown from the stock of mixed seed. This is often referred to as phenotypic selection since it is based on how each individual looks. It is used widely to improve old "land" varieties. Old land varieties are those that are passed down from one generation of farmers to the next over long periods. An alternate approach that has no doubt been practised for thousands of years is simply to eliminate undesirable types by destroying them in the field. No matter whether superior plants are saved or inferior plants are eliminated, the result is the same. Seeds of the selected plants make the planting stock for the next season. The Danish botanist, Wilhelm Johannsen, in 1903 developed the scheme of mass selection. This is the oldest method of breeding self-pollinated species that is widely practised.

Population improvement through increasing the frequencies of desirable genes is the purpose of mass selection. Selection is based on plant phenotype. Mass selection is imposed either once or multiple times (recurrent mass selection). However, improvement is limited to pre-existing genetic variability and no new variability is generated. Mass selection aims at improving average performance of base population. The general procedure in mass selection is to rogue out off-types, often called negative mass selection. Some breeders may rather select and advance a large number of plants that are desirable and uniform for the trait(s) of interest. This is positive mass selection. Where applicable, single pods from each plant may be picked and bulked for planting. For cereal species, the heads may be picked and bulked. The breeder plants the heterogeneous population in the field and looks for off-types to remove and discard them (Fig. 11.1). During year 1, the objective is to purify an established cultivar. Seeds from selected plants are planted in a row to confirm the purity prior to bulking. The original cultivar needs to be planted alongside for comparison. During year 2, evaluation of composite seed in a replicated trial is done, using the original cultivar as check. This evaluation is done at multi-locations for several years. The advantages of mass selection are as follows: It is rapid, simple and straightforward. Even though it is a mixture of pure lines, it is inexpensive. The cultivar produced is phenotypically fairly uniform. They are genetically broad-based, adaptable and stable. The disadvantages are as follows: Optimal selection is achieved if it is conducted in a uniform environment. The selected heterozygotes will segregate in the next generation if progeny testing is not done.

A modern refinement of mass selection is to harvest the best plants separately and to grow and compare their progenies. The poorer progenies are discarded and the seeds of best genotypes are harvested. Selection is based on both the appearance of the parent plants and the appearance and performance of their progeny. Progeny selection is usually more effective than phenotypic selection when dealing with quantitative characters of low heritability. Here, progeny testing requires an extra generation.

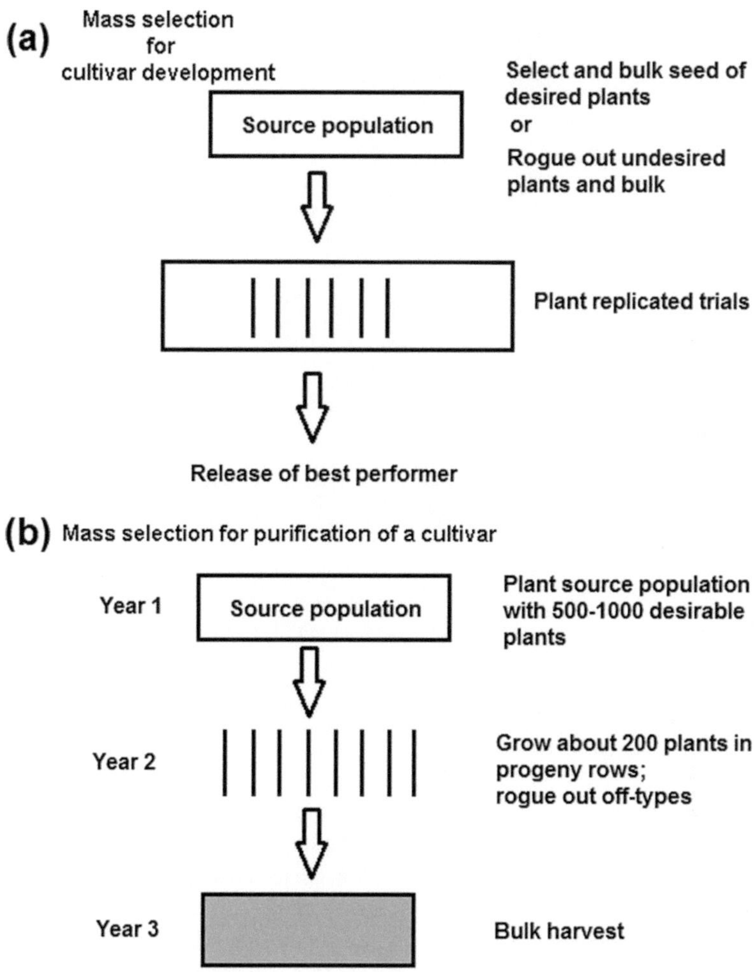

Fig. 11.1 Generalized steps in mass selection for (**a**) cultivar development and (**b**) purification of a given cultivar

11.1.2 Pure-Line Selection

The theory of the pure line was developed in 1903 by the Danish botanist Johannsen. He could demonstrate that a mixed population of self-pollinated species could be sorted out into genetically pure lines in beans (*Phaseolus vulgaris*) when he considered seed weight as a trait. Selection does not create variation, but is a passive process that eliminates variation. The pure-line theory has following attributes:

(a) Lines that are genetically different may be successfully isolated from within a population of mixed genetic types.

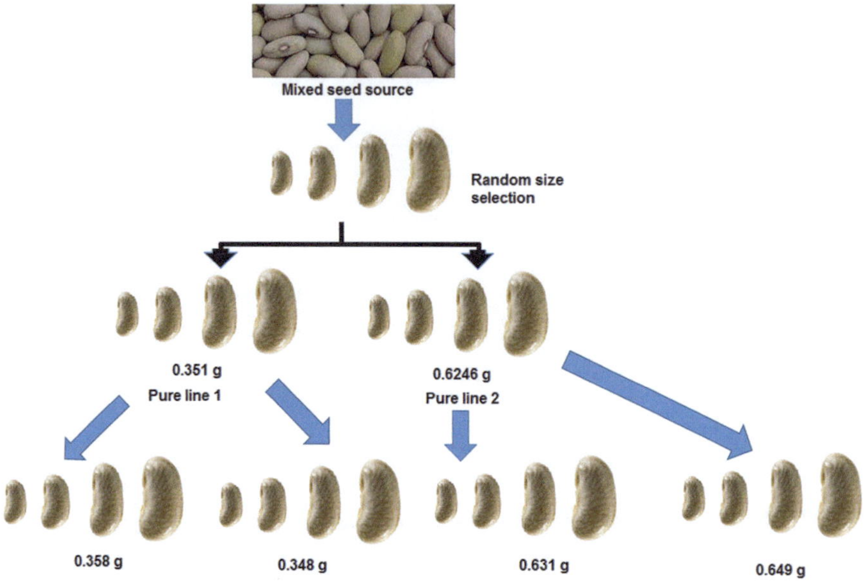

Fig. 11.2 Development of pure-line theory by Johannsen (figures of seeds are representative)

(b) Any variation that occurs within a pure line is not heritable, but variation is due to environmental factors only.

Consequently, as Johannsen's bean study showed, further selection within the line is not effective (Fig. 11.2). He could get the seeds from the market that consisted of a mixture of larger and smaller seeds. He then selected seeds of different sizes and grew them individually. Progenies of larger seeds produced larger seeds and progenies from smaller seeds produced small seeds. This showed that the variation is with a genetic basis. Nineteen lines were studied and the lot was a mixture of pure lines. Variation observed within a pure line is due to environment. Confirmatory evidence was obtained in three ways. One of the lines (line 13) had 450 mg of seed weight; he divided the seeds on weight basis. He divided the line into seeds having 200, 300, 400 and 500 mg weights and studied the progenies. The ultimate result was seeds with weight ranging from 458 to 475. The conclusion was that the variation is due to environment. The second evidence came in the form of ineffective selection within a pure line. Within the pure line with seeds of 840 mg, selection was made for large and small seeds. After six generations of selection, the progeny was with seeds of 680–690 mg. So, it was demonstrated that selection within a pure line is ineffective. The third evidence was that when parent-offspring regression was worked out in line 13, the result was zero indicating thereby that the variation observed is non-heritable.

Pure-line selection follows three distinct steps: (a) from a genetically variable population, numerous superior plants are selected; (b) progenies of the individual plant selections are grown and evaluated; and (c) extensive trials are undertaken

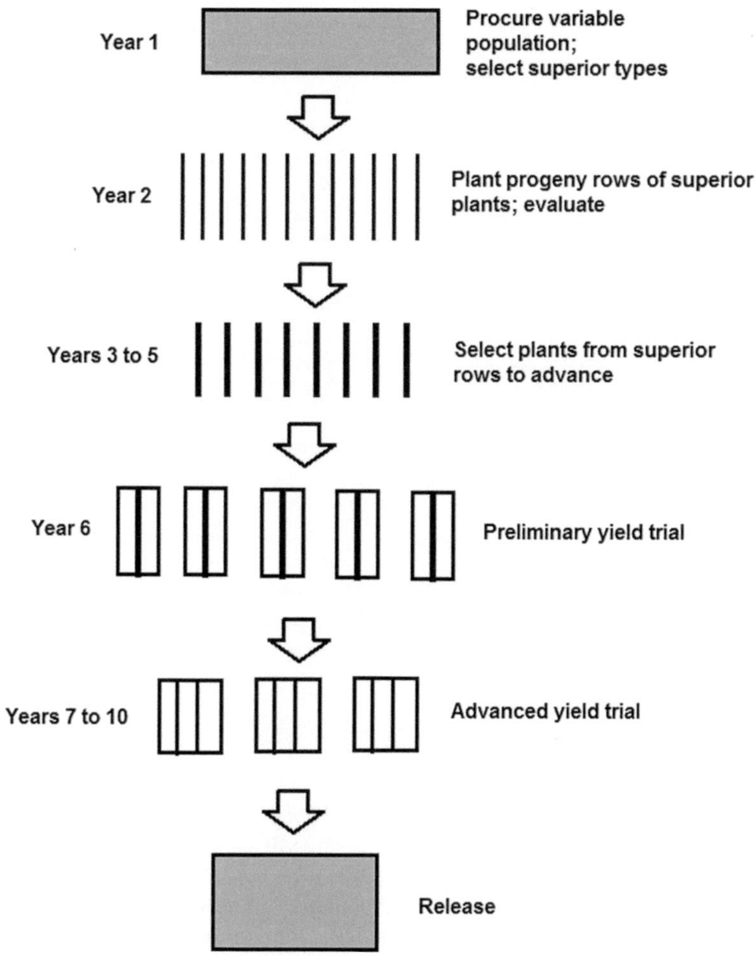

Fig. 11.3 Steps in breeding for pure-line selection

when selection can no longer be made on the basis of observation alone. The remaining selections were evaluated for superiority in yielding ability and other attributes (Fig. 11.3). Any progeny superior to an existing variety is then released as a new "pure-line" variety. During the early 1900s, existence of genetically variable land varieties that were unexploited led to the success of this method. Such variability worked as a source of superior pure-line varieties. So, this method is applicable only in genetically resourceful species.

A different pure-line selection method is the selection of single-chance variants, mutations or "sports" in the original variety. Varieties that differ in traits like colour, lack of thorns or barbs, dwarfness and disease resistance originated in this way. Please see Table 11.2 for differences between pure-line and mass selection procedures.

Table 11.2 Difference between pure-line and mass selection

Pure-line selection	Mass selection
The variety developed as a pure line	The variety is a mixture of several pure lines
It is not practised by farmers	It is practised by the farmers unknowingly
It is practised in self-pollinated crops	Practised in self- as well as cross-pollinated crops
The varieties developed are highly uniform and the variation is purely environmental	The variety is heterozygous, hence not uniform and having genetic variation
The selected plants are subjected to progeny test	Progeny test is not carried out
The variety is best pure line present in the original population	The variety is inferior to the best pure line
Varieties are having narrower adaptability and stability in performance than mixture of pure lines	The varieties developed have wider adaptability and greater stability than pure-line varieties
Pollination is controlled	Pollination is not controlled
The variety developed is homozygous and uniform in quality	The variety developed in a mixture of several types hence heterozygous
About 9–10 years is required for developing variety	About 5–7 years period is required to develop variety
Once developed variety is maintained easily	It is repeated every year to maintain purity
The variety is easily identified in seed certification programme	The variety developed is relatively difficult to identify in seed certification programme

11.1.3 Hybridization and Pedigree Selection

During the twentieth century, hybridization between selected parents was predominant in breeding of self-pollinated species. This is to combine desirable genes from two or more different varieties and to produce pure lines superior in many respects compared to parents. Genotypes are a combination of genes. The challenge of the plant breeder is to manage the innumerable number of genotypes that occur generations after generations following hybridization. Hypothetically, a cross between wheat varieties that differ by only 21 genes can produce more than 10,000,000,000 different genotypes in the second generation. At spacing normally used by farmers, more than 50,000,000 acres would be required to grow such a population to permit every genotype to occur in its expected frequency. These genotypes are hybrid (heterozygous) for one or more traits. Statistically 2,097,152 different pure-breeding (homozygous) genotypes are possible, each potentially a new pure line. These numbers call for efficient techniques in managing hybrid populations. Pedigree method is most widely used to manage such populations.

After deriving a hybrid, the breeder makes several selfed generations like F_1, F_2, F_3, etc. and keeps the ancestry record of the cultivar. Pedigree was first described by H.H. Lowe in 1927. If the two parents do not provide all desired traits, a third parent can be added by crossing it to one of the hybrids of F_1. Documentation of the pedigree enables breeders to trace parent-progeny back to an individual F_2 plant from any subsequent generation. In a segregating population, the breeder should be able to select plants with desirable traits on the basis of a single phenotype. Breeder exercises a selection among them. Plants are reselected in each subsequent generation. This is continued until a desirable level of homozygosity is attained. When homozygosity is attained, plants will be phenotypically homogeneous.

The F_2 generation offers the first chance for selection in pedigree programmes. The emphasis is on the rejection of plants with undesirable major genes. As a result of natural self-pollination, the succeeding generations offer way to pure breeding, and families derived from different F_2 plants begin to display their unique character. One or two superior plants are selected within each superior family in these generations. Emphasis shifts to selection between families by F_5 generation where pure-breeding condition (homozygosity) will be very extensive. While making these eliminations, the pedigree record will be useful. Each selected family is usually harvested in mass to obtain the larger amounts of seed needed to evaluate families for quantitative characters. This evaluation is usually carried out simulating commercial planting practices. Precise evaluation for performance and quality begins by F_7 or F_8 generation when the number of families has been reduced to manageable proportions by visual selection. The final evaluation of promising strains involves (a) observation on the number of years and locations, to detect environment-induced variations, (b) precise yield testing and (c) quality testing. Usually such tests will be conducted for 5 years at five representative locations before releasing a new variety for commercial production.

The generation-wise procedures are:

F_1 generation: F_1 leads to F_2 for selection. F_1 seed is planted for maximum seed production. Recently, plant breeders started using genetic markers in crossing programmes.

F_2 generation: F_2 generation is with the maximum genetic variation and selection starts here. If the parents differ by a larger number of genes, the rate of segregation will be higher. A large F_2 population is planted (2000–5000) usually. Selection intensity should be moderate (about 10%) since 50% of the genotypes in the F_2 are heterozygous. Selection with high heritability will be more effective, requiring lower numbers than for traits with low heritability. The F_2 is also usually space planted to allow individual plants to be evaluated for selection. In pedigree selection, each selected F_2 plant is documented.

F_3 generation: Progeny from individual plants is sown in a row that can allow homozygous and heterozygous genotypes to be distinguished. Homozygosity in F_3 will be 50% less than F_2. The heterozygotes will segregate in the rows. The F_3 generation is the beginning of line formation. Selection is based on performance against check cultivars.

F_4 generation: F_4 plants are grown as in F_3 generation. The progenies will be 87.5% homozygous. Selection in F_4 will be based on progeny rather than individual plants.

F_5 generation: Selections made in F_4 are grown in preliminary yield trials (PYTs). F_5 plants are 93.8% homozygous. PYTs are with at least two replications. This can be increased depending on the amount of seed available. The seeding rate shall be

comparable to the commercial rate with all recommended agricultural practices with check cultivars. This can include quality traits and disease resistance. Selected lines are advanced to the next generation.

F_6 generation: Superior selections from F_5 are further evaluated in competitive yield trials or advanced yield trials (AYTs), with a check (local reference variety).

F_7 and subsequent generations: Superior lines from F_6 are evaluated in AYTs for several years, at multi-locations and in different seasons as desirable. Eventually, after F_8, the most outstanding entry is released as a commercial cultivar.

There are several advantages and disadvantages for pedigree selection. Advantages are as follows: (a) unlike other methods, record keeping gives genetic information of the cultivar unavailable; (b) selection is based both on phenotype and genotype (progeny row) for selecting superior lines from segregates; (c) with the help of progeny records, only the progeny lines with target genes are carried forward; and (d) genetic purity with high degree is ensured in the cultivar. This is an advantage where certification is a prerequisite for certain markets. Disadvantages are as follows:

(a) Record keeping is slow, tedious, time-consuming and expensive.
(b) If only one growing season is possible per year, pedigree selection takes time, demanding about 10–12 years or even more.
(c) Suited for qualitative rather than for quantitative disease resistance breeding. Pedigree is not effective for accumulating the number of minor genes governing horizontal resistance.
(d) Selecting F_2 plants for quantitative traits (such as yield) may not be effective. One needs to wait till F_3 (Fig. 11.4).

Bulk Population Breeding The bulk population method of breeding differs from the pedigree method primarily in the handling of generations following hybridization. H. Nilsson-Ehle developed the procedure. Additional theoretical foundation for this was provided by H.V. Harlan and colleagues through their work on barley breeding in the 1940s. F_5 generation is sown as per commercial planting procedures in a larger plot. The crop is harvested in mass at maturity and the generation is advanced. No record of ancestry is kept. Plants having poor survival value will be naturally eliminated during the period of bulk propagation. Artificial selection applied are as follows: (a) destruction of plants that carry undesirable major genes and (b) when only part of the seeds are mature, mass selection techniques are practised, to select for early-maturing plants. The same technique can be applied to select for increased seed size. Further, as in the pedigree method of breeding, single plant selections are exercised and evaluated. Bulk population method allows the breeder to handle very large numbers of individuals inexpensively (Fig. 11.5).

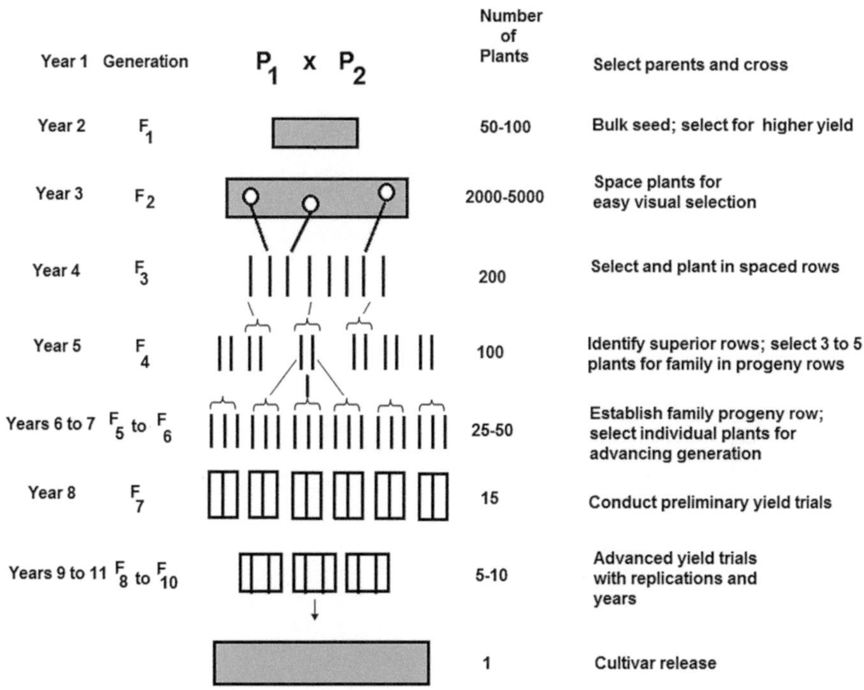

Fig. 11.4 Steps in breeding for pedigree selection

Single-Seed Descent Method This concept was first proposed by C.H. Goulden in 1941. He attained the F_6 generation in 2 years while conducting multiple plantings per year, using the greenhouse and off-season planting. In this method, F_1 population is fairly large to ensure adequate recombination among parental chromosomes. A single seed per plant is advanced in each subsequent generation until the desired level of inbreeding is attained. Selection is usually practised in F_5 or F_6. Then, each plant is used to establish a family to help breeders in selection and to increase seed for subsequent yield trials. The following are the steps:

Year 1: Selected parents are crossed to generate sizeable F_1 for the production of a large F_2 population.

Year 2: About 50–100 F_1 plants are grown in a greenhouse. They may also be grown in the field. Harvest identical F1 crosses and bulk.

Year 3: About 2000–3000 F_2 plants are grown. A single seed per plant is harvested and bulked for planting F_3.

Years 4–6: Single pods per plant are harvested to be planted as F_4. The F_5 is space planted in the field, harvesting seed from only superior plants to grow progeny rows in the F_6 generation.

Year 7: Superior rows are harvested to grow preliminary yield trials in the F_7.

Year 8 and later: Yield trials are conducted in the F_8–F_{10} generations. The most superior line is increased in the F_{11} and F_{12} as a new cultivar.

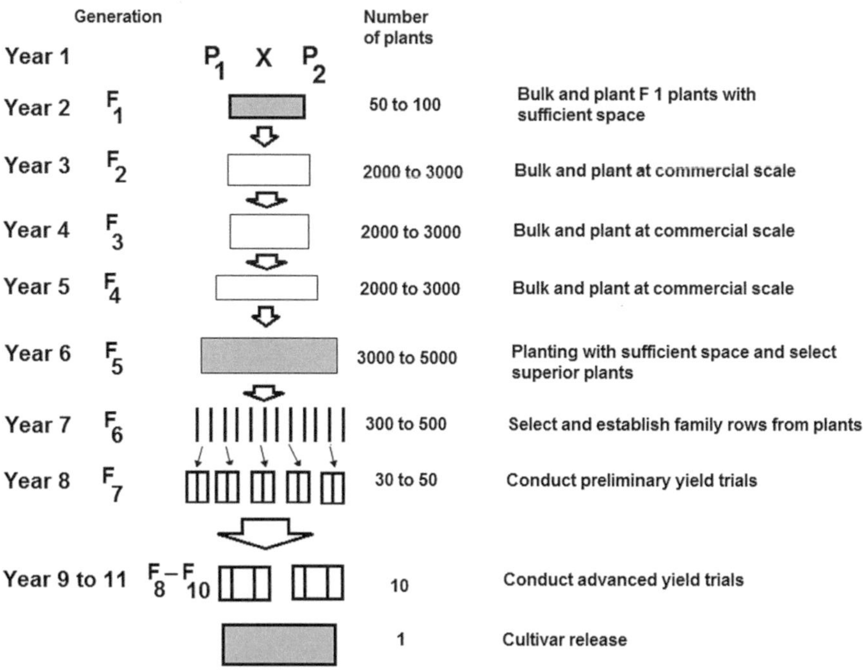

Fig. 11.5 Steps in breeding by bulk selection

The advantages of this method are as follows: (a) easy and rapid way to attain homozygosity (2–3 generations per year); (b) limited space is required in early generations (e.g. can be conducted in a greenhouse); (c) natural selection has no effect; (d) the duration of the programme can be reduced by several years by using single-seed descent; and (e) every plant originates from a different F_2 plant, resulting in greater genetic diversity. The disadvantages are as follows: (a) natural selection has no effect; (b) plants are selected based on individual phenotype not based on progeny performance; (c) inability of seed to germinate or a plant to set seed may prohibit every F_2 plant from being represented in the subsequent generation; and (d) the number of plants in the F_2 is equal to the number of plants in the F_4. Selecting a single seed per plant has a greater chance of losing desirable genes. The assumption is that the single seed represents the genetic base of each F_2. It may not be correct always that a single seed represents the genetic base of each F_2.

Backcross Breeding H.V. Harlan and M.N. Pope proposed backcross breeding in 1922. Backcross breeding is meant to substitute gene(s) rather than to improve the genotype. It is to replace an undesirable gene with a desirable one while preserving all other qualities (adaptation, productivity, etc.) (see Chap. 10). F_1 is repeatedly crossed with the desirable parent to incorporate the desirable gene. The adapted and highly desirable parent is called the *recurrent parent.* The source of the desirable gene is called the *donor*. An inferior recurrent parent will be inferior after the gene transfer, and hence, the donor should not be significantly deficient in other desirable traits.

Backcross breeding is most effective when the trait to be transferred is qualitative and dominant. It must also express in the hybrid. Quantitative traits are more difficult to breed by this method. Cytoplasmic male sterile (CMS) genotypes that are capable of hybrid production in species like corn, onion and wheat are desirable. The donor (of the chromosomes) is crossed with the recurrent parent as male again and again until all donor chromosomes are recovered in the cytoplasm of the recurrent parent. Backcrossing is also used for the introgression of genes via wide crosses. This would be a lengthy process since wild plant species possess a large number of undesirable traits. Backcross breeding can also be used to develop isogenic lines (genotypes that differ only in alleles at a specific locus) for traits (e.g. disease resistance, plant height). This is effective when the expression of a trait depends mainly on one pair of genes.

Steps for dominant gene transfer:

Year 1: Select the donor (RR) and recurrent parent (rr) and make 10–20 crosses. Harvest the F_1 seed.

Year 2: Grow F_1 plants and backcross them with the recurrent parent to obtain the first backcross (BC_1).

Years 3–7: Grow BC_1 to BC_5 progeny and backcross them to the recurrent parent as female. Select about 30–50 heterozygous backcrossed individuals that are similar to the recurrent parent that can be used in the next backcross. After each backcross, the recessive genotypes are discarded using appropriate screening techniques. For disease resistance breeding, artificial epiphytotic conditions shall be created. BC_5 progeny should very closely resemble the recurrent parent with the donor trait. In advanced generations, most plants would look like the adapted cultivar.

Year 8: Grow BC_5F_1 plants and self-fertilize them. Select several hundreds of desirable plants (300–400) and harvest them individually.

Year 9: Grow BC_5F_2 progeny rows. Select about 100 desirable non-segregating progenies and bulk.

Year 10: Yield tests involving backcrossed individuals with the recurrent parent must be conducted to determine equivalence before releasing (Fig. 11.6).

Steps for recessive gene transfer:

Years 1–2: These are the same as for dominant gene transfer. The donor parent has the recessive desirable gene (Fig. 11.7).

Year 3: Grow BC_1F_1 plants; self, harvest and bulk the BC_1F_2 seed. In disease resistance breeding, all BC_1s will be susceptible.

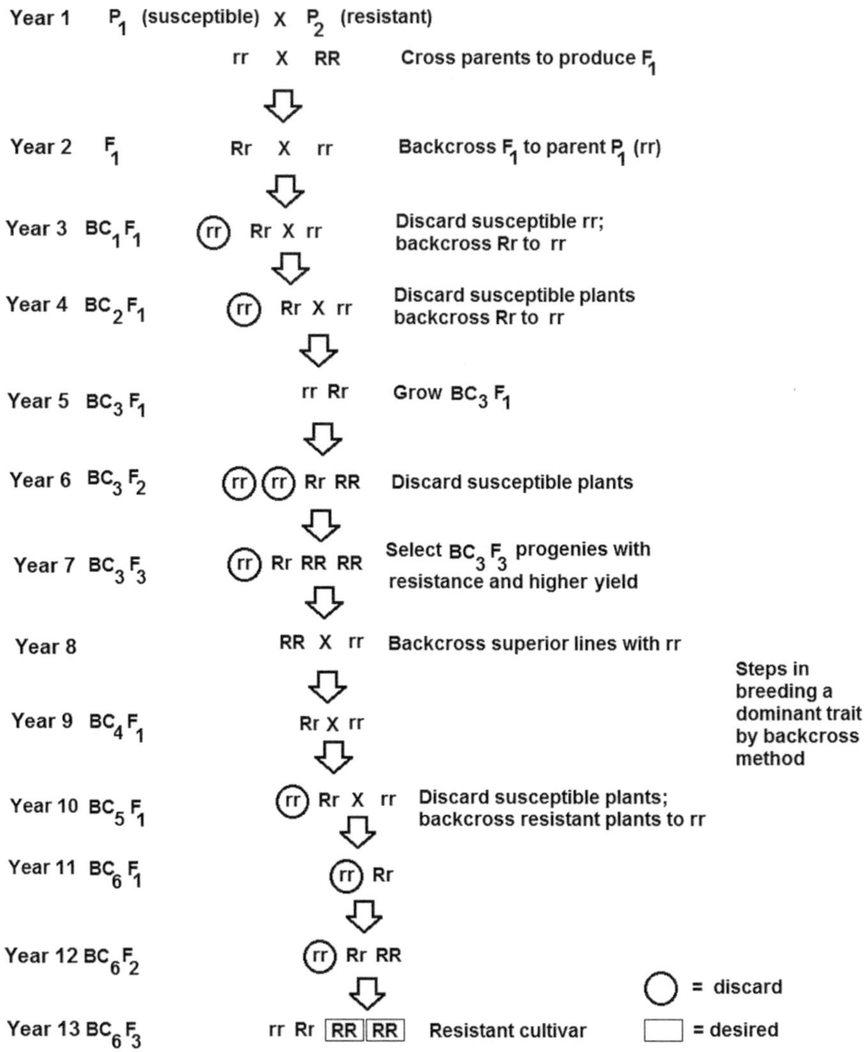

Fig. 11.6 Backcross method for transferring dominant trait

Year 4: Grow BC_1F_2 plants and screen for desirable plants. Backcross 10 to 20 plants to the recurrent parent to obtain BC_2F_2 seed.

Year 5: Grow BC_2 plants. Select 10 to 20 plants that resemble the recurrent parent and cross with the recurrent parent.

Year 6: Grow BC_3 plants; harvest and bulk the BC_3F_2 seed.

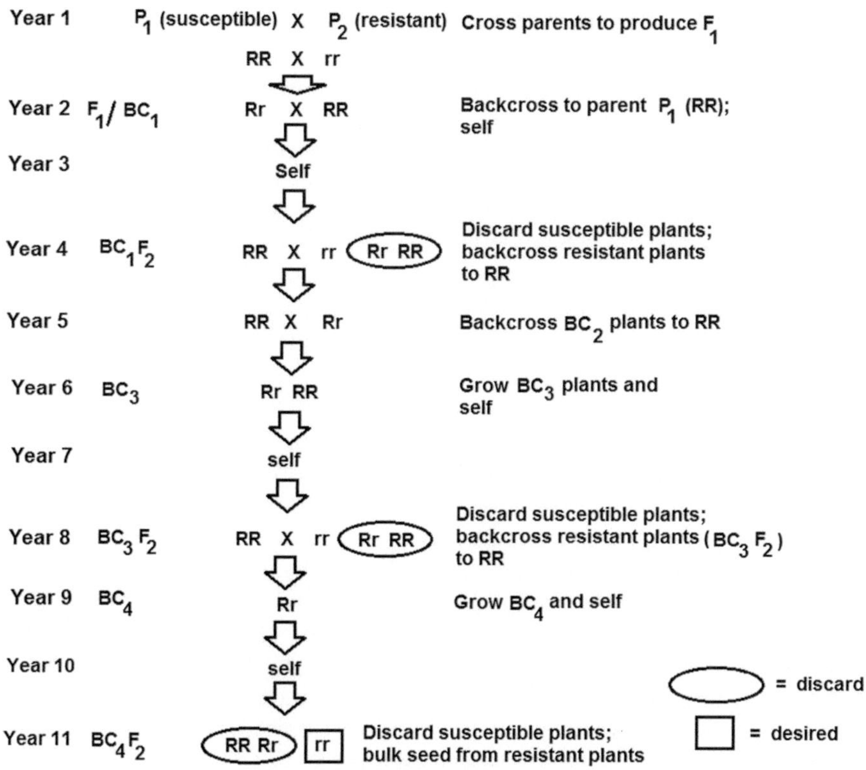

Fig. 11.7 Backcross method for transferring recessive trait

Year 7: Grow BC_3F_2 plants, screen, and select the desirable plants. Backcross 10 to 20 plants with the recurrent parent.

Year 8: Grow BC_4 plants, harvest, and bulk the BC_4F_2 seed.

Year 9: Grow BC_4F_2 plants, screen, and select the desirable plants. Backcross 10 to 20 plants with the recurrent parent.

Year 10: Grow BC_5 plants, harvest, and bulk the BC_5F_2 seed.

Year 11: Grow BC_5F_2 plants, screen, and backcross.

Year 12: Grow BC_6 plants, harvest, and bulk the BC_6F_2 seed.

Year 13: Grow BC_6F_2 plants and screen; select 400 to 500 plants and harvest separately for growing progeny rows.

Year 14: Grow progenies of selected plants, screen, and select about 100 to 200 uniform progenies; harvest and bulk the seed.

Years 15–16: Follow the procedure as in breeding for a dominant gene (Fig. 11.7).

11.2 Special Backcross Procedures

Two special backcross procedures are *congruency backcross* and *advanced backcross QTLs* (quantitative trait loci). The congruency backcross technique is a modification of the standard backcross procedure whereby multiple backcrosses, alternating between the two parents in the cross (instead of restricted to the recurrent parent), are used. The technique has been used to overcome the *interspecific hybridization* barrier of hybrid sterility, genotypic incompatibility and embryo abortion that occurs in simple interspecific crosses. The advanced backcross quantitative trait loci (QTL) method developed by S.D. Tanksley and J.C. Nelson in 1996 allows breeders to transfer QTLs from unadapted germplasm into an adapted cultivar (see Chap. 10).

11.3 Multiline Breeding and Cultivar Blends

Multilines are more expensive because each component line must be developed by a separate backcross. N.F. Jensen used this technique first to breed for more lasting form of disease resistance in oats in 1952. A multiline or blend is multiple pure lines in which each component constitutes at least 5% of the whole mixture. These pure lines are phenotypically uniform for agronomic traits (e.g. height, maturity, photoperiod), in addition to genetic resistance for a specific disease. These lines are grown separately, followed by compositing in a predetermined ratio. Multilines are mixtures involving isolines or near-isogenic lines (lines that are genetically identical except for the alleles at one locus). Mixing genotypes is to increase heterogeneity. This would decrease the risk of total crop loss from the infection of one race of the pathogen or some other biotic or abiotic factor. The component genotypes are designed to respond to different races of a pathogen.

In multiline breeding, the agronomically superior line is the recurrent parent, while the source of disease resistance constitutes the donor parent. To develop multilines by isolines, the first step is to derive a series of backcross-derived isolines or near-isogenic lines. Such a process is practised since true isolines are illusive because of linkage between genes of interest and other genes influencing other traits (Fig. 11.8). Two cultivars with contrasting features for a specific trait is the result.

11.4 Breeding Composites and Recurrent Selection

A composite cultivar is also a mixture of different genotypes. The difference between multiline and composite lies primarily in the genetic distance between the components of the mixture. While a multiline is constituted of closely related lines (isolines), a composite consists of inbred lines, hybrids, populations and other less similar genotypes.

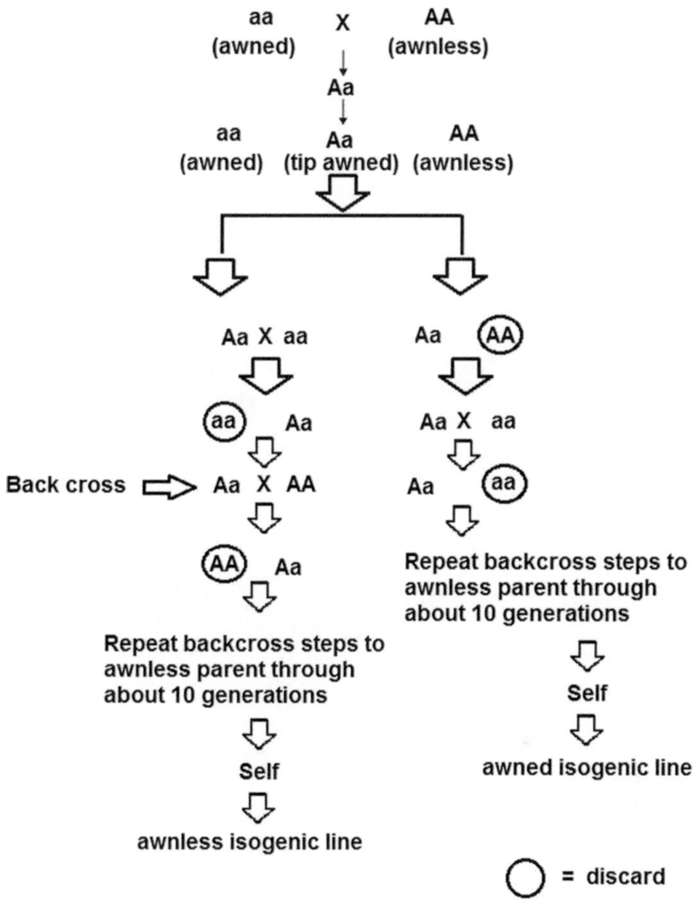

Fig. 11.8 Breeding multiline cultivars

Recurrent selection is a cyclical improvement technique aimed at gradually concentrating desirable alleles in a population. This was first developed for improving cross-pollinated species like maize. Recurrent selection ensures repeated intermating after first cross, something not available in pedigree selection. It is effective for improving quantitative traits (see Chap. 12 for a detailed account of composite breeding and recurrent selection).

11.4.1 Hybrid Varieties

The F_1 hybrid is often much more vigorous than its parents. This hybrid vigour, or heterosis, can be manifested in many ways, including increased rate of growth, greater uniformity, earlier flowering and increased yield, the last being of greatest

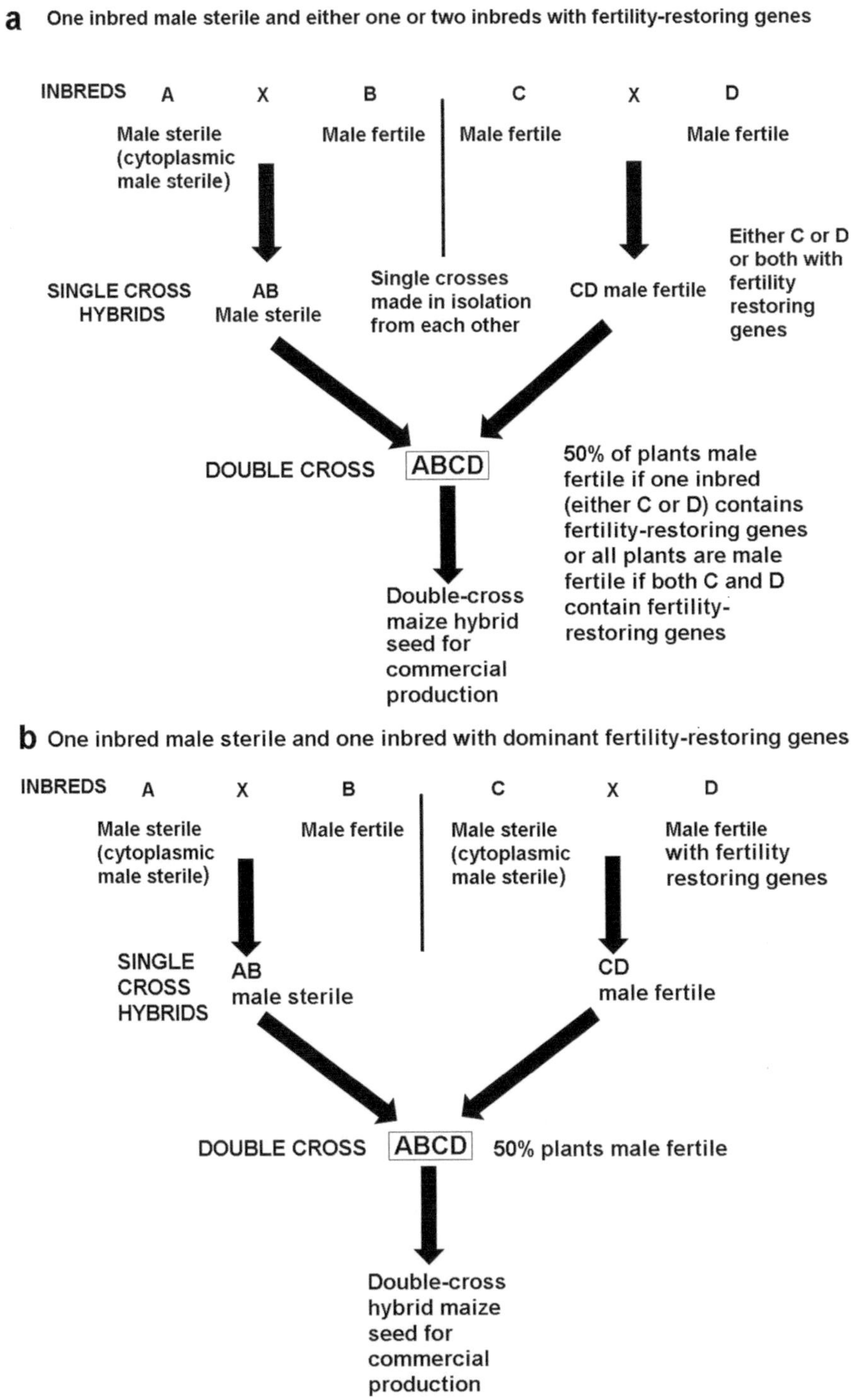

Fig. 11.9 Two methods of producing double-cross hybrid maize seeds using cytoplasmic male sterility and fertility restorer genes

importance in agriculture. Maize is an example for exploitation of heterosis. Hybrid corn production involves the following steps:

(a) Selection of superior plants.
(b) Selfing for several generations to produce a series of inbred lines. They are pure breeding and highly uniform.
(c) Crossing selected inbred lines.
(d) Select those single crosses exhibiting the highest combining ability for the character(s) to be improved for use in the double-cross hybrids.
(e) Produce double-cross hybrids from the best-performing single crosses.

Inbreds were produced and crossed in pairs. Those crosses giving superior F_1 were chosen for commercial production of hybrid seed. Single-cross hybrids did not significantly surpass the yield of open-pollinated varieties. Then came the use of the double crosses, a hybrid between two F_1s of four parents:

$$(A \times B)\,F_1 \times (C \times D)F_1$$

Double cross was more successful than single cross. The single-cross parents of the double cross were much more vigorous and higher yielding than the inbred parents of the single cross, and the hybrid seed was more vigorous and viable than the single-cross seed. For both single cross and double cross, cytoplasmic male sterility (CMS) can be used to evade labour-intensive de-tasselling (emasculating) female parents. Fertility-restoring genes are also used (see Chap. 6 on sterility) (see Fig. 11.9). As distinct from government-funded or public-good breeders, commercial breeders prefer hybrid varieties. This preference is due to the fact that heterosis breaks down in the F_2 and in later generations due to segregation. Farmers do not have any other option but to buy new F_1 planting seed from the breeder (or the licenced seed producer) each season. Hybrid varieties have been a great deal of success in maize, sunflowers, sorghum and many vegetable crops in many countries like Australia and the USA.

Further Reading

Araus JL, Cairns JE (2014) Field high-throughput phenotyping: the new crop breeding frontier. Trends Plant Sci 19(1):52–61. https://doi.org/10.1016/j.tplants.2013.09.008

Kempe K, Gils M (2011) Pollination control technologies for hybrid breeding. Mol Breed 27:417–437

Kim Y, Zhang D (2018) Molecular Control of Male Fertility for Crop Hybrid Breeding. Trends Plant Sci 23:53–65

Ramalho MAP, de Araújo LCA (2011) Breeding self-pollinated plants. Crop Breed Appl Biotechnol S1:1–7

Stamp P, Visser R (2011) The twenty-first century, the century of plant breeding. Euphytica 186:585–591

Wright SI, Kalisz S, Slotte T (2013) Evolutionary consequences of self-fertilization in plants. Proc R Soc B 280:20130133. https://doi.org/10.1098/rspb.2013.0133

Zhao et al (2014) Genomic selection in hybrid breeding. Plant Breed. https://doi.org/10.1111/pbr.12231

Breeding Cross-Pollinated Crops

12

Keywords
Selection of cross-pollinated crops · Mass selection · Recurrent selection · Intra-population improvement methods · Individual plant selection methods · Family selection methods

While methods for improving self-pollinated species tend to focus on improving individual plants, improving cross-pollinated species, on the other hand, tends to focus on improving a population of plants. A population is a large group of interbreeding individuals. The principles of population genetics are applied to effect changes in the genetic structure of a population. The change is such that only desirable genotypes predominate in the population. In this process of changing gene frequencies, new genotypes will arise. This genetic variability must be maintained so that they can be utilized for further improvements in the future.

In the breeding of cross-pollinated species, the heterozygous nature of individual plants is exploited. Individual plants within a cultivar will be heterozygous, and the cultivar will be more heterogeneous than cultivars in self-pollinated species. Here, the focus of the breeder is on improving populations instead of selecting superior individual plants. Also, more emphasis is given to quantitative inheritance in breeding systems than in self-pollinated crops.

In order to evaluate the genetics of a heterozygous mother plant, one needs to cross it with known testers, which may be either inbred or a relative to the mother plant. This gives an idea of the genetic value of a mother plant – known as *combining ability*. Combining ability is the capacity of an individual to transmit superior performance to its offspring. Combining ability is of two types: general and specific. General combining ability (GCA) is the average or overall performance of a genotype in a large series of crosses. Specific combining ability (SCA) is the performance of an individual plant in combination with another individual plant or strain. Breeding procedures in cross-pollinated crops are based largely on population

P. M. Priyadarshan, *PLANT BREEDING: Classical to Modern*,
https://doi.org/10.1007/978-981-13-7095-3_12

improvement principles, i.e. improving the frequency of genes in the population for the desired breeding objective. Some of the features promoting cross-pollination are:

Monoecy: Separation of staminate and pistillate flowers on same plant like corn (*Zea mays*) and rubber (*Hevea brasiliensis*).

Dioecy: Production of staminate and pistillate flowers on different plants like papaya and date palm.

Self-incompatibility: It is the failure to become fertilized and seed set following self-pollination.

Male or female sterility: Both inhibit seed formation. Female sterility is less common. Male sterility promotes cross-pollination.

Floral devices: Maturity of stamen and pistil at different times.

Breeding methods followed in cross-pollinated species are introduction and selection. In introduction, it is the collection of germplasm, and in selection, it is mass selection and recurrent selection. Single plant selection is not a useful breeding method in cross-pollination crops because it is prone to segregation.

12.1 Selection in Cross-Pollinated Crops

Cross-fertilizing populations of crops are characterized by a high degree of heterozygosity and heterogeneity. They have characteristic reproductive features and population structure. Existence of self-sterility, self-incompatibility, imperfect flowers and mechanical obstructions make the plant dependent upon foreign pollen for normal seed set. Each plant receives a blend of pollen from a large number of individuals each having different genetic set up. Such populations are characterized by a high degree of heterozygosity with tremendous free and potential genetic variation, which is maintained in a steady state by free gene flow among individuals within the populations. It is inappropriate, and could be rather hazardous, to take one or a few individuals to investigate or improve these populations. The enhanced fitness of heterozygotes over homozygotes of cross-pollinated crops has been manipulated in the form of two different breeding approaches, namely, population improvement and hybrid breeding in such crops.

In the development of hybrid varieties, the aim is to identify the most productive heterozygote from the population, which then is produced with the exclusion of other members of the population. In contrast, the population improvement envisages a stepwise elimination of deleterious and less productive alleles through repeated cycles of selective mating of genotypes that are more productive. Population improvement is slow, steady and a long-term programme, whereas the production of hybrids is aimed to maximize the genetic gains in much less time. Both of these breeding approaches are complementary rather than mutually exclusive and are based on sound genetic theory. The different selection methods can be summarized as follows.

Fig. 12.1 Mass selection

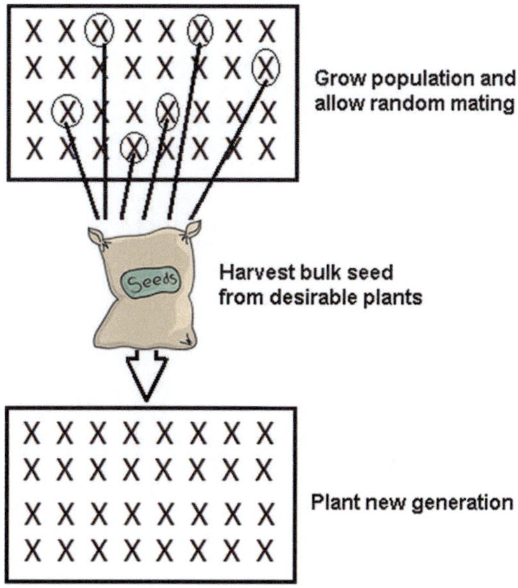

12.1.1 Mass Selection

It is the simplest, easiest and oldest method of selection where individual plants are selected based on their phenotypic performance, and bulk seed is used to produce the next generation (Fig. 12.1). Mass election proved to be quite effective in maize improvement at the initial stages, but its efficacy, especially for improvement of yield, soon came under severe criticism that culminated in the refinement of the method of mass selection. The selection after pollination does not provide any control over the pollen parent, as result of which, effective selection is limited only to female parents. The heritability estimates are reduced by half, since only parents are used to harvest seed, whereas the pollen source is not known after the cross-pollination has taken place.

12.1.2 Recurrent Selection

Plant breeders generally assemble germplasm, evaluate selected selfed plants, cross the progenies of the selected selfed plants in all possible combinations and bulk and develop inbred lines from the populations. In cross-pollinated crops, a cyclical selection approach, called *recurrent selection*, is often used for inter-mating. The cyclical selection is capable of increasing the frequency of favourable genes for quantitative traits. The classification of population improvement is several, according to the unit of selection – either individual plants or family of plants. The method can also be grouped according to the populations undergoing selection as either intra-population or inter-population. In intra-population improvement, the end

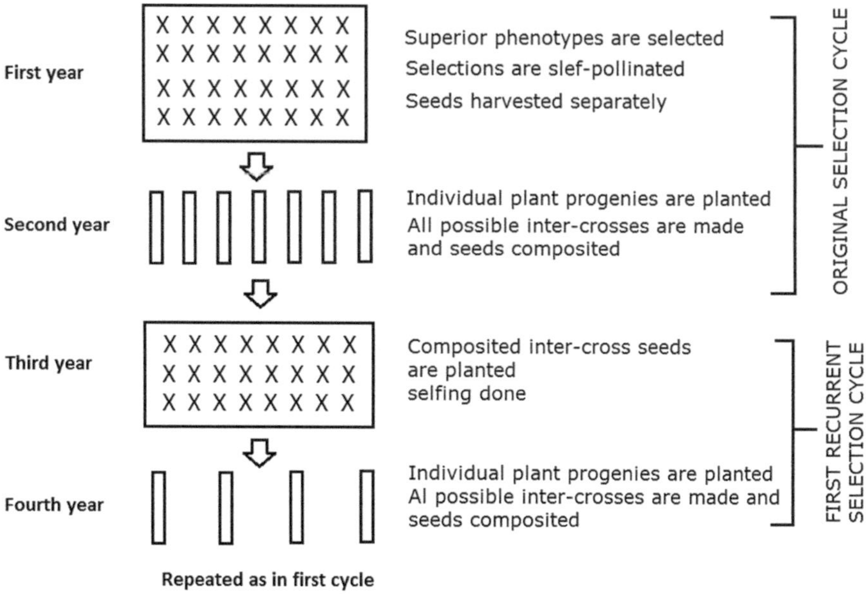

Fig. 12.2 Simple recurrent selection

product will be a population or synthetic cultivar, and it may end up elite pure lines for hybrid production. Or, it can also be used for developing mixed genotype cultivars (in self-pollinated crops). Inter-population improvement deals with the selection on the basis of the performance of a cross between two populations. The final product will be a hybrid cultivar with heterosis.

The cyclical selection is a systematic technique to isolate genotypes with desirable genes mated to form a new population (Fig. 12.2). Subsequently, this cycle is repeated. This is to improve one or more traits so that a new population that is superior to the original population is achieved. The source material may be random mating populations, synthetic cultivars and single-cross or double-cross plants. The improved population may be released as a new cultivar or used as a breeding material (parent) in other breeding programmes. Improvement of population without reduction in genetic variability is the advantage of recurrent selection. The parents should not be closely related and should have high performance regarding the traits of interest which would maximize genetic diversity. It is advisable to include as many parents as possible in the initial crossing to increase genetic diversity. The breeder is expected to decide on the number of generations of inter-mating that is appropriate for a breeding programme. Recurrent selection cycle has three main phases, viz. (a) the parents are crossed in all possible combinations and individual families are created for evaluation, (b) the families are evaluated and a new set of parents are selected, and (c) the selected parents are inter-mated to produce the population for the next cycle of selection. The aforesaid cycle is repeated several times (3–5 times). The original cycle is labelled C_0 and is called the base population. The subsequent cycles are named as C_1, C_2, . . ., C_n, etc.

Types of gene action exploited by recurrent selection range from additive partial dominance to dominance and overdominance. However, this scheme is effective only for traits of high heritability in the absence of testers (as in simple recurrent selection). So, only additive gene action is exploited in the selection for the trait in question. Selections for general combining ability (GCA) and specific combining ability (SCA) are applicable where testers are used, permitting use of other gene effects. When additive gene effects are more important, recurrent selection for GCA is more effective than other schemes. When overdominance gene effects are more important, recurrent selection for SCA is more effective than other selection schemes. Reciprocal recurrent selection is more effective than others when both additive and overdominance gene effects are more important. When additive with partial to complete dominance effects prevail, all three schemes are equally effective. The expected genetic advance may be obtained as:

$$\varDelta G = (C\, i\, VA)/y\ \sigma_{\mathrm{p}}$$

where:

ΔG = expected genetic advance
C = measure of parental control ($C = 0.5$ if selection is based on one parent and equals 1 when both parents are involved)
i = selection intensity
VA = additive genetic variance among the units of selection
y = number of years per cycle
σ_{p} = *p*henotypic standard deviation among the units of selection

Increasing selection intensity will increase selection gains. This can happen if the population advanced is not reduced to a size where genetic drift and loss of genetic variance can occur. Genetic advance per cycle can be increased by including selection for both male and female parents, maximizing available additive genetic variance, and management of environmental variance among selection units. The breeder can control genetic gain through selecting appropriate parents in a breeding programme.

There are four types of recurrent selection schemes:

(a) *Simple recurrent selection*: This is similar to mass selection with 1 or 2 years per cycle which does not involve a tester. Phenotypic scores are the basis for selection. This is otherwise called phenotypic recurrent selection.
(b) *Recurrent selection for general combining ability*: This is a half-sib progeny (only one parent known) test procedure where a wide genetic-based cultivar is used as a tester. The testcross performance is evaluated in replicated trials prior to selection.

(c) *Recurrent selection for specific combining ability*: An inbred line (narrow genetic base) is used as a tester. The testcross performance is evaluated in replicated trails before selection.
(d) *Reciprocal recurrent selection*: This scheme is capable of exploiting both general and specific combining ability. Two heterozygous populations are involved, each serving as a tester for the other.

12.2 Intra-population Improvement Methods

Commonly used intra-population improvement methods are mass selection, ear-to-row selection and recurrent selection. Intra-population methods may be based on single plants as the unit of selection (e.g. as in mass selection) or family selection (e.g. as in various recurrent selection methods).

12.2.1 Individual Plant Selection Methods

Intra-population improvement *via* mass selection is different from mass selection for self-pollinated crops. Mass selection for population improvement aims at improving the general population performance by selecting and bulking superior genotypes that already exist in the population. Here, the selection units are individual plants and based on better phenotype. Seeds from selected plants (pollinated by the population at large) are bulked to start the next generation. No crosses are made, but a progeny test is conducted. The process is repeated until a desirable level of improvement is observed.

Year-wise procedure shall be:

Year 1: Source population is planted (local variety, synthetic variety, bulk population, etc.). Undesirable plants are rogued out before flowering. Select several hundreds of plants on the basis of phenotype. Harvest and bulk.
Year 2: Process of year 1 is repeated. Bulked seeds are grown in a preliminary yield trial. Check shall be the original unselected population if the goal of the mass selection is to improve the population.
Year 3: Process of year 2 is repeated.
Year 4: Conduct advanced yield trials.

Since selection is solely on the phenotype, heritability of the trait plays a pivotal role in its effectiveness. Where additive gene action operates, the selection is most effective. Effectiveness of mass selection also depends on the number of genes involved in the control of the trait of interest. As more additive genes are involved, the greater shall be the efficiency of mass selection. The expected genetic advance through mass selection is given by the following (for one sex – female):

$$\Delta G_m = \left[(1/2)\, i\sigma^2{}_A\right] \sigma_p = \left[(1/2)\, i\sigma^2{}_A\right] / \left[\sigma^2{}_A + \sigma^2{}_D + \sigma^2{}_{AE} + \sigma^2{}_{DE} + \sigma^2{}_e + \sigma^2{}_{me}\right.$$

where

σ_p is phenotypic standard deviation in the population, $\sigma^2{}_A$ is additive variance, $\sigma^2{}_D$ is dominance variance and the other factors are interaction variances. ΔG_m doubles with both sexes. This large denominator makes mass selection inefficient for low heritability traits. Selection is limited to only the female parents since there is no control over pollination.

There are two modifications for planting the progeny that are to be evaluated. They are stratified or grid system and honeycomb design. In stratified or grid system, as proposed by C.O. Gardener, the field is divided into small grids (or sub-plots) with little environmental variance. An equal number of superior plants are selected from each grid for harvesting and bulking. On the other hand, in the honeycomb designs, as proposed by Fasoulas and Fasoula in 1995, each single plant is at the centre of a regular hexagon, with six equidistant plants, and is compared to the other six equidistant plants (Fig. 12.3) or to additional equidistant plants, depending on the intensity of selection the breeder wishes to apply. All plants grow at wide distances to exclude any interplant interference with the equal sharing of resources. As shown in the figure, this replicated R-31 honeycomb design evaluates 31 lines. Plants are placed in ascending order in horizontal rows, and the number set is repeated regularly. A notable and essential property of all honeycomb designs is the ability to form complete and moving replicates in any spot in the field and with any of the evaluated entries. Further, the designs have the ability to form moving triangular grids across the field and secure comparable conditions of evaluation for all plants. Thus, the breeder can select with equal success in both fertile and less fertile field areas, and selection takes place within and among the evaluated lines.

Crucial for the formation of moving replicates is that the starting number is different in each row and derived from simple equations by Fasoulas and Fasoula in 1995. This unique arrangement allows using the *plant yield index* to express the individual plant yields as a ratio to a common denominator, i.e. to the average of a complete moving replicate, facilitating removal of confounding effect of soil heterogeneity on single plant yields. Plants are ranked according to their yielding capacity avoiding the bias of the visual evaluation, commonly known as the "breeder's eye". The arrangement and the practically unlimited number of replications (>30) afforded by all honeycomb selection designs offer unbiased and precise estimations of crop yielding potential, although the evaluation concerns individual plants, because of the component analysis of *crop yield potential* as stated by Fasoula and Fasoula in 2002. The relevant statistical script for the analysis can be had in Fasoula et al. 2019 (see other references for further reading).

12.2.2 Family Selection Methods

Family selection methods are characterized by three general steps: (a) creation of a family structure, (b) evaluation of families and selection of superior ones by progeny

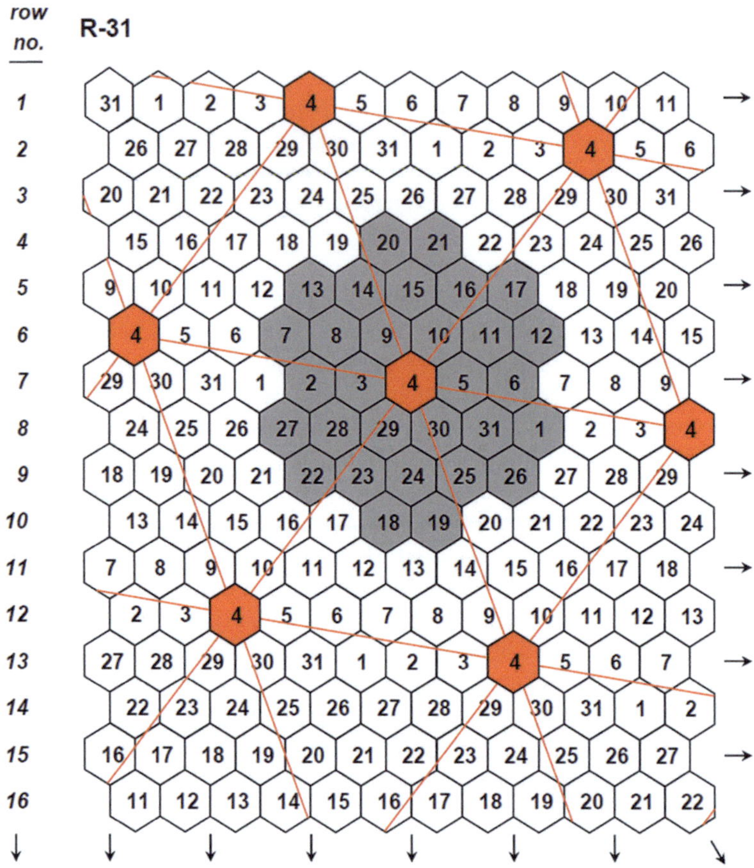

Fig. 12.3 A replicated R-31 honeycomb design for evaluating 31 lines. The complete moving replicate and the triangular grid are illustrated for plants of line 4. (Courtesy Dr. D.A. Fasoula)

testing and (c) recombination of selected families or plants within families to create a new base population for the next cycle of selection. The basic feature of this group of methods is that half-sib families are created for evaluation and recombination, both steps occurring in one generation. The populations are created by random pollination of selected female plants in generation 1. The seeds from generation 1 families are evaluated in replicated trials and in different environments for selection. There are different kinds of half-sib family selection methods like ear-to-row selection and modified half-sib selection. Ear-to-row selection is the simplest scheme of half-sib selection for cross-pollinated species.

In *ear-to-row selection*, the following procedures are followed:

Season 1: Grow the source population (heterozygous) and select desirable plants (C_0) based on the traits of interest. Harvest plants individually. Keep remnant seed of each plant.

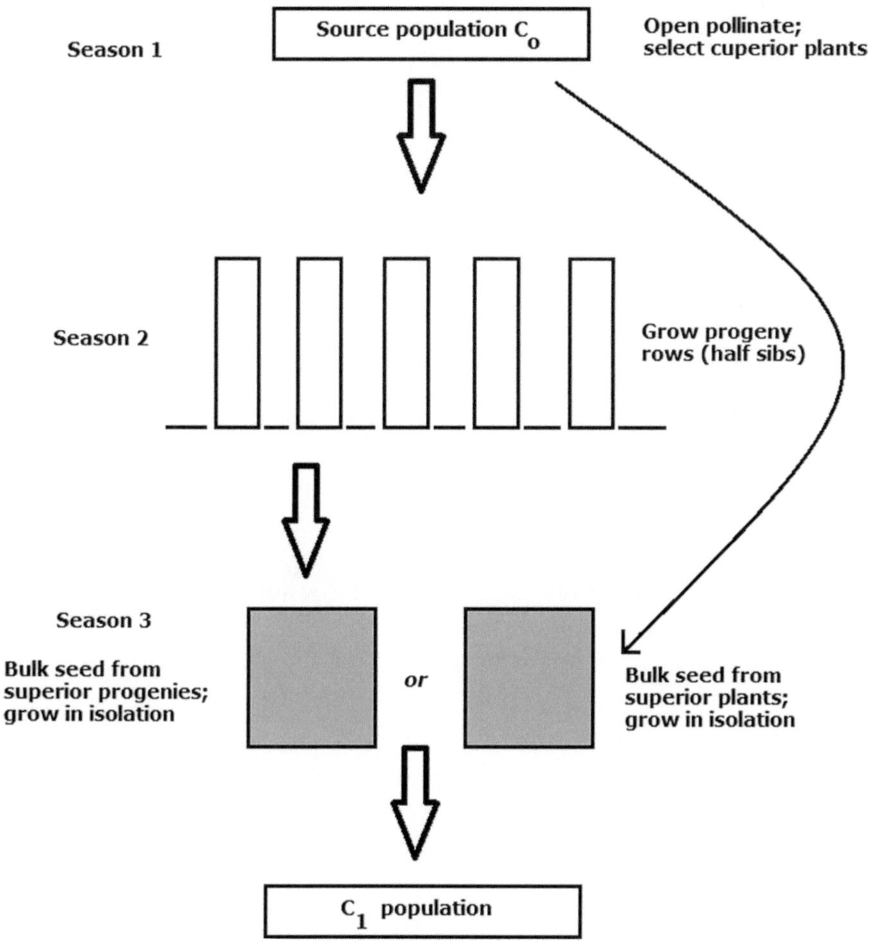

Fig. 12.4 Generalized steps in ear-to-row selection

Season 2: Grow replicated half-sib progenies (C_0 × tester) from selected individuals in one environment (yield trial). Select best progenies and bulk to create progenies for the next cycle. The bulk is grown in isolation (crossing block) and random mated.

Season 3: The seed is harvested and used to grow the next cycle (see Fig. 12.4).

In *modified half-sib selection*, the following procedures are followed:

Season 1: Select desirable plants from source population. Harvest these open-pollinated (half-sibs) individually.

Season 2: Grow progeny rows of selected plants at multiple locations and evaluate for yield performance. Plant female rows with seed from individual half-sib

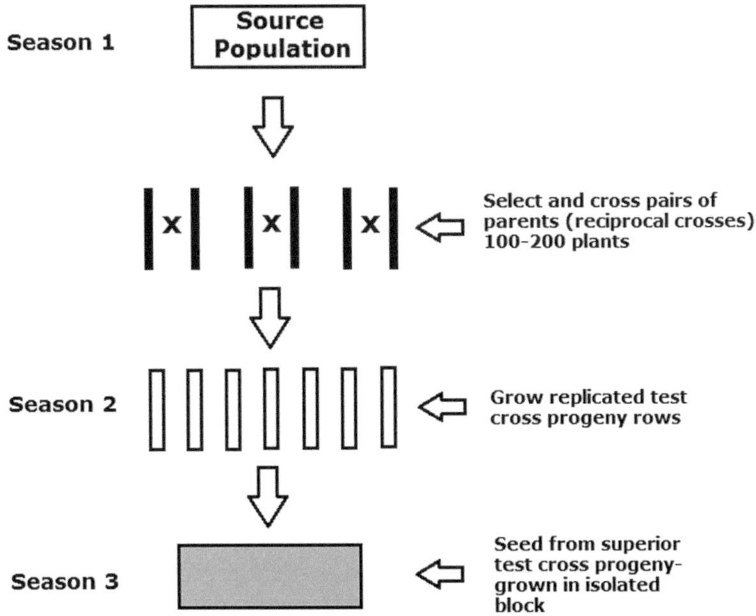

Fig. 12.5 Generalized steps in breeding by full-sib method

families, alternating with male rows (pollinators) planted with bulked seed from the entire population. Select desirable plants (based on average performance over locations) from each progeny separately. Bulk the seed to start the next cycle.

12.2.2.1 Full-Sib Family Selection

Full sibs are derived from crosses of parents from the base population. The families are evaluated in a replicated trial to identify and select superior full-sib families, which are then recombined to initiate the next cycle.

Applications Full-sib family selection has been used for maize improvement. The steps are:

Season 1: Select random pairs of plants from the base population and inter-mate, pollinating one with the other (reciprocal pollination). Make between 100 and 200 biparental crosses. Save the remnant seed of each full-sib cross (Fig. 12.5).

Season 2: Evaluate full-sib progenies in multiple location replicated trails. Select the promising half-sibs (20–30).

Season 3: Recombine the selected full sibs.

Selfed (S_1 or S_2) Family Selection

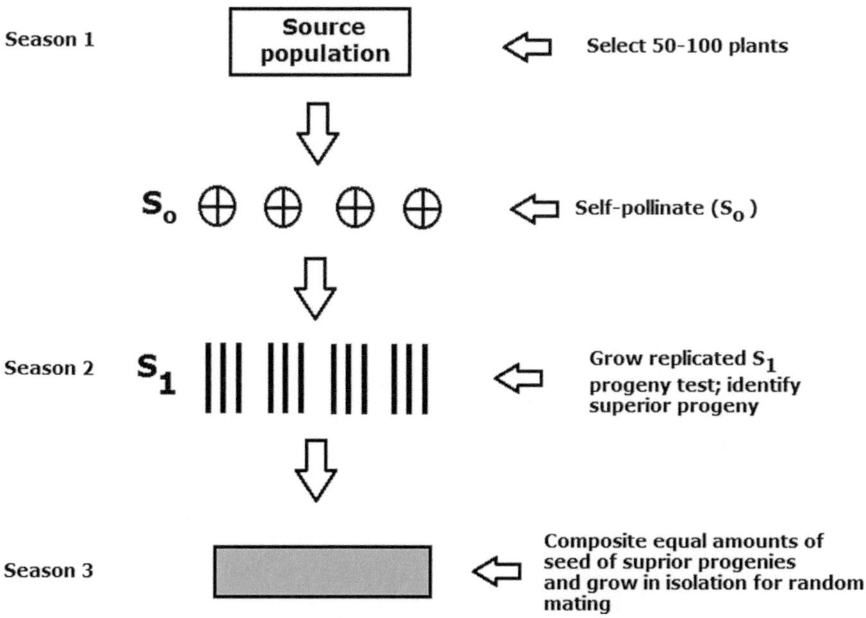

Fig. 12.6 Generalized steps in breeding based on S_1/S_2 progeny performance

An S_1 is a selfed plant from the base population. The key features are the generation of S_1 or S_2 families, evaluating them in replicated multi-environment trials, followed by recombination of remnant seed from selected families (Fig. 12.6).

Applications The S_1 appears to be best suited for self-pollinated species (e.g. wheat, soybean). It has been used in maize breeding. One cycle is completed in three seasons in S_1 and four seasons in S_2. A genetic gain per cycle of 3.3% has been recorded.

Procedure

Season 1: Self-pollinate about 300 selected S_0 plants. Harvest the selfed seed and keep the remnant seed of each S_1.
Season 2: Evaluate S_1 progeny rows to identify superior progenies.
Season 3: Random mate selected S_1 progenies to form a C_1 cycle population.

12.2.2.2 Half-Sib Selection with Progeny Test

Half-sib or half-sib family selection is called such, because only one parent in the cross is known. C.G. Hopkins in 1899 first used this procedure to alter the chemical composition of corn by growing progeny rows from corn ears picked from desirable plants. Superior rows were harvested and increased as a new cultivar.

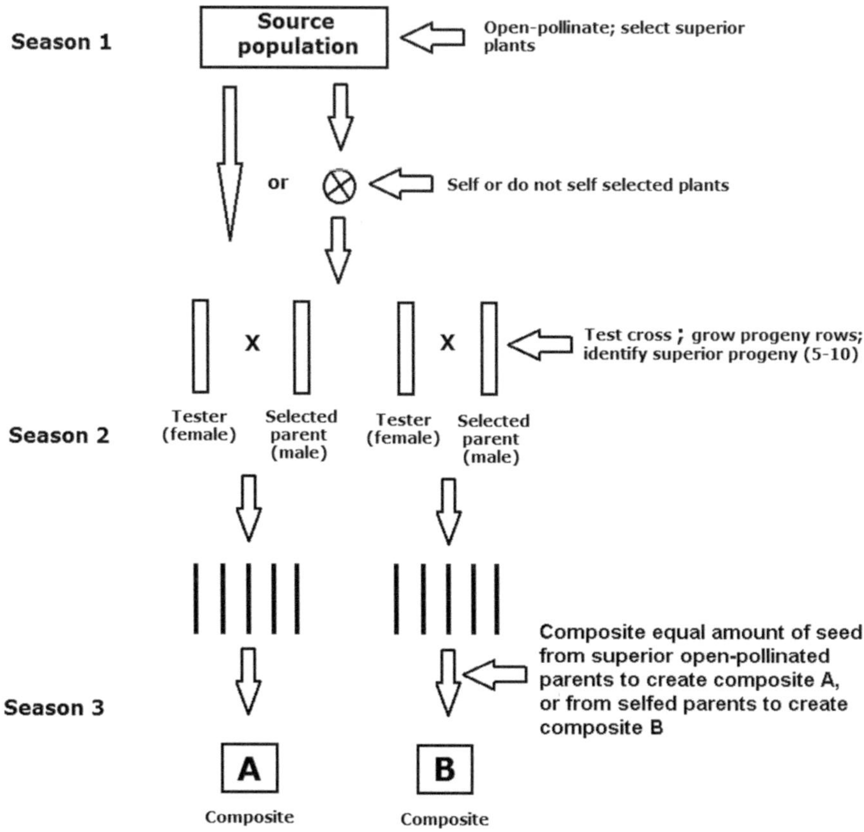

Fig. 12.7 Generalized steps in breeding by half-sib selection with progeny test

Key Features There are various half-sib progeny tests, such as the topcross progeny test, open-pollinated progeny test and polycross progeny test. A half-sib is a plant (or family of plants) with a common parent or pollen source. Individuals in a half-sib selection are evaluated based on their half-sib progeny. Unlike mass selection, in which individuals are selected solely on phenotypic basis, the half-sibs are selected based on the performance of their progenies. In this case, the pollen sources are not known.

Applications Recurrent half-sib selection has been used to improve agronomic traits as well as seed composition traits in corn. It is suited for improving traits with high heritability and in species that can produce sufficient seed per plant to grow a yield trial. Species with self-incompatibility (no self-fertilization) or some other constraint of sexual biology (e.g. male sterile) are also suited to this method of breeding.

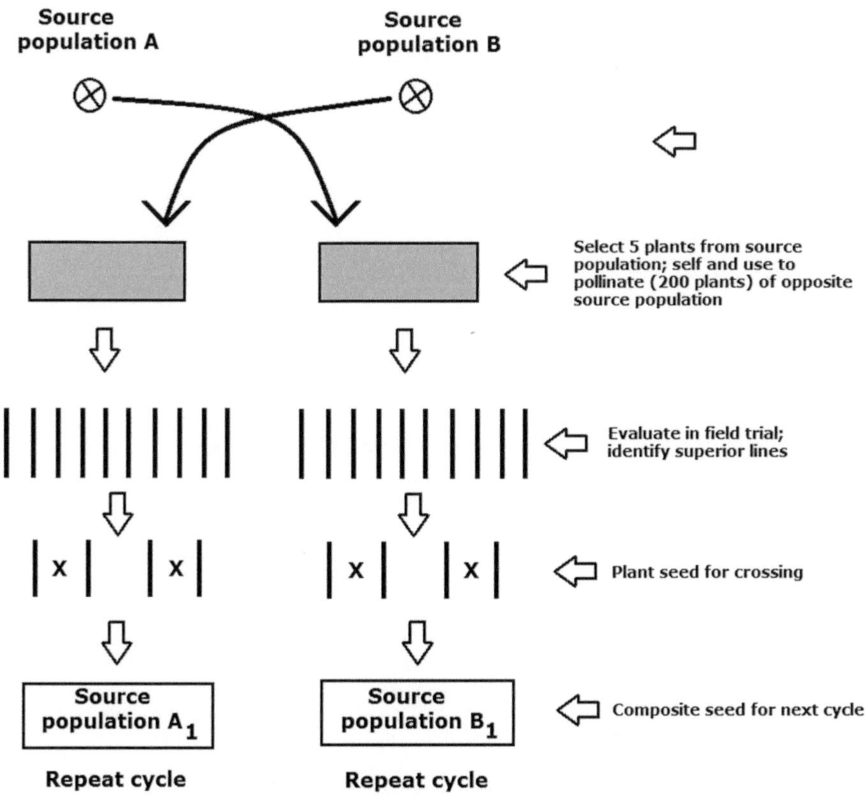

Fig. 12.8 Generalized steps in breeding by half-sib selection with a testcross

Procedure A typical cycle of half-sib selection entails three activities – crossing the plants to be evaluated to a common tester, evaluating the half-sib progeny from each plant and intercrossing the selected individuals to form a new population. In the second season, each separate seed pack is used to plant a progeny row in an isolated area (Fig. 12.7). The remnant seed is saved. In season 3, 5–10 superior progenies are selected, and the seed is harvested and composited; alternatively, the same is done with the remnant seed. The composites are grown in an isolation block for open pollination. Seed is harvested as a new open-pollinated cultivar or used to start a new population.

The advantages are as follows: (a) the procedure is rapid to conduct and (b) progeny testing increases the success of selection. The disadvantages are as follows: (a) the trait of interest should have high heritability for success; (b) it is not readily applicable to species that cannot produce enough seed per plant to conduct a yield trial; and (c) lack of pollen control reduces heritability by half.

12.2.2.3 Half-Sib Selection with a Testcross

A testcross can also be conducted to evaluate composited genotypes. This variation of half-sib selection allows the breeder to more precisely evaluate the genotype of the selected plant by choosing the most suitable testcross parent (Fig. 12.8). The half-sib lines to be composited are selected based on a testcross evaluation and not based on progeny performance. The tester may be inbred, in which case all the progeny lines will have a common parental gamete. Like half-sib selection with a progeny test, this procedure is applicable to cross-pollinated species in which sufficient seeds can be produced by crossing. However, in procedures in which self-pollination is required, the method cannot be applied to species with self-incompatibility.

Further Reading

Hoyos-Villegas et al (2018) QuLinePlus: extending plant breeding strategy and genetic model simulation to cross-pollinated populations—case studies in forage breeding. Heredity. https://doi.org/10.1038/s41437-018-0156-0

Fasoulas AC, Fasoula VA (1995) The honeycomb selection designs. In: Janick J (ed) Plant breeding reviews, vol 13. Wiley, New York, pp 87–139

Fasoula, Fasoula (2002) Principles underlying genetic improvement for high and stable crop yield potential. Field Crop Res 75:191–209

Fasoula DA, Tokatlidis IS (2012) Development of crop cultivars by honeycomb breeding. Agron Sustain Dev 32:161–180. https://doi.org/10.1007/s13593-011-0034-0

Fasoula DA (2012) nonstop selection for high and stable crop yield by two prognostic equations to reduce yield losses. Agriculture 2:211–227. https://doi.org/10.3390/agriculture2030211

Fasoula VA (2013) Prognostic breeding: a new paradigm for crop improvement. In: Janick J (ed) Plant breeding reviews, vol 37. Wiley, New York, pp 297–347

Fasoula VA, Thompson KC, Mauromoustakos A (2019) The prognostic breeding application JMP Add-In Program. Agronomy 9(1):25. https://doi.org/10.3390/agronomy9010025

Ceccarelli S (2014) Efficiency of Plant breeding. Crop Sci 55:87–97

Zhao et al (2015) Genomic selection in hybrid breeding. Plant Breed 134:1–10

Stoddard FL (2017) Climate change can affect crop pollination in unexpected ways. J Exp Bot 68:1819–1821

Wu Y et al (2016) Development of a novel recessive genetic male sterility system for hybrid seed production in maize and other cross-pollinating crops. Plant Biotechnol J 14:1046–1054

Recombinant Inbred Lines 13

Keywords
Inbred line development in cross-pollinated crops · Methods adopted for RILs · Doubled haploid breeding · Reverse breeding

13.1 Inbred Line Development in Cross-Pollinated Crops

Breeding cross-pollinated species is a challenge to the plant breeder. In plant breeding, *inbred lines* are used as stocks for the creation of hybrid *lines* to exploit heterosis. Inbred lines can be developed from a heterozygous natural population or from F_2 progeny. Inbreds are derived through repeated self-pollination. Usually, repeated self-pollinations up to 6–10 generations (i.e. 3–5 years when two seasons per year can be accomplished) are necessary to achieve homozygous inbred lines. Development of inbred parents can follow different breeding methods such as pedigree breeding, backcrossing, bulking, single-seed descent, doubled haploids.

RILs can be used for studying genetic loci underlying phenotypic traits. Since meiotic crossover events create a mosaic of parent genomes in each RIL, they are derived from crosses of divergent parents (Fig. 13.1). The mapping of QTL relies on markers, genotyped in each RIL, falling close enough to the causal loci (i.e. in linkage disequilibrium) to show a non-random association with the phenotype. There are several steps being followed for the production of RILs: selection of parent strains, selection of construction design, parent cross and F_1 cross, advanced intercross and inbreeding. These steps will be briefly discussed here.

P. M. Priyadarshan, *PLANT BREEDING: Classical to Modern*,
https://doi.org/10.1007/978-981-13-7095-3_13

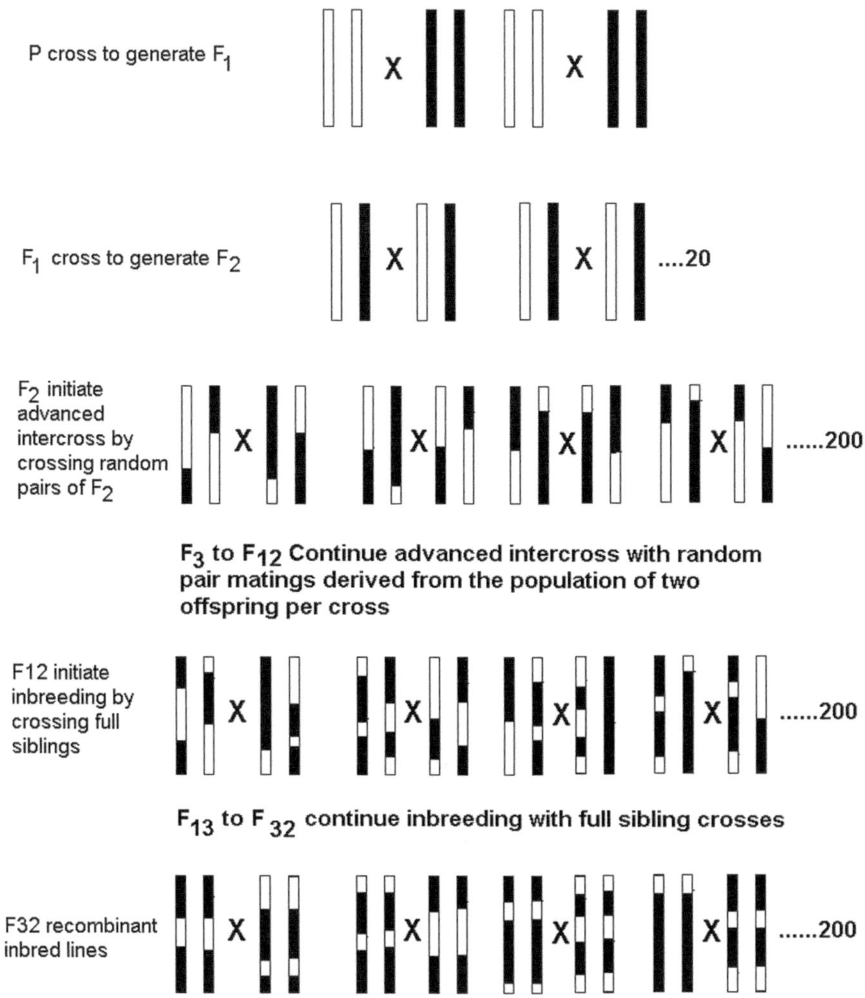

Fig. 13.1 Example of a RIL construction design. Two replicate parent crosses produce 40 F_1. Twenty F_1 crosses produce 400 F_2. Two hundred random F_2 crosses initiate the advanced intercross. Two hundred random pair matings of offspring (two from each cross) in each generation are performed for ten generations of intercrossing. Inbreeding of full siblings in all 200 lines begins at F_{12} and continues for 20 generations to F_{32}. Individuals are represented by a set of diploid chromosomes. Each parent genotype is represented by either white or black. (Courtesy: Springer Science)

13.2 Methods Adopted for RILs

13.2.1 Selection of Parent Strains

Parent strains are to be with significant phenotypic divergence. Strains with sufficient marker density need to be selected. First calculate the expected linkage map length resulting from your RIL construction design (linkage map length is the genetic distance spanned by all the chromosomes – a value that increases with increased recombination). Inbreeding to isogenicity through crosses of sibling expands the F_2 linkage map to fourfold, but selfing of siblings results in approximately twofold expansion. Intercrossing for t generations adds an additional map expansion of approximately $t/2 + 1$. In a linkage map of length L, the number of randomly placed markers needed (n) to have fraction p loci within m map units of a random marker is:

$$n = \frac{\ln(1-p)}{\ln\left(\frac{1-2m}{L}\right)}$$

Plotting the number of markers (n) vs. m for different values of p and L can give an intuitive feeling for the relationship of these variables. Once the target number of markers is established, one can confirm that potential parent pairs have sufficient genotypic divergence for this marker density. Prior to RIL construction, the full set of markers should be selected and tested on the parents for accuracy and ease of genotyping. Parents with incompatibilities are not desirable since that may result in loss of some recombinants leading to allele frequency distortions.

13.2.2 Selection of Construction Design

Factors influencing selection of design are number of RILs produced, how many generations they are inbred and how many generations they are intercrossed past the F_2 generation. Larger RIL populations are preferred that reduces the influence of drift on allele frequencies and increases the number of crossing over events. Inbreeding removes heterozygosity and generates crossover events. After t generations of full-sibling inbreeding, an initial level of heterozygosity, h_0, is approximately reduced to:

$$h_t = h_0 \times 1.17\,(0.809^t)$$

For selfing species, the expected homozygosity after t generations is $h_0/2^t$. In full-sibling inbreeding, h_0 is reduced by 86% in 10 generations and 98.3% after 20 generations. In selfed inbreeding, h_0 is reduced by 99.9% in just 10 generations. Under normal situations, 10 generations of selfed inbreeding and 20 generations of full-sibling inbreeding shall be sufficient to achieve RILs.

13.2.3 Parent Cross and F_1 Cross

One has to ensure there are a sufficient number of parent crosses. Crosses are to be replicated to generate the desired RIL population. For an average family size of B, equal sex ratios and monogamous outcrossing, the construction of a RIL population of size N will require a minimum $4N/B^2$ replicated parent crosses (see Fig. 13.1). A minimum of 2 parent crosses are needed to construct a RIL population of 200 for a species with average family size of 20.

A minimum of $2N/B$ F_1 crosses are required to generate the desired F_2 population (see Fig. 13.1). From the example above ($N = 200$, $B = 20$), 20 F_1 crosses are needed to generate an F_2 population of 400 from which 200 inbreeding lines can be set up. As with the parent crosses, it is always recommended to set up more crosses than the minimum required to guarantee sufficient numbers of F_2s.

13.2.4 Advanced Intercross

Intercross may be initiated among F_2 population. More crosses are to be set up than your desired population size since all crosses may not produce offspring out of intercrossing and inbreeding. Note that many cross designs assume an even population size. Terminology followed is very crucial. As an example, mating 84 in the F_3 generation is a cross of mating 1 from F_2 and mating 128 from the F_2 generation can be represented as: $M_1F_2 \times M_{128}F_2 = M_{84}F_3$ (M=mating scheme).

13.2.5 Inbreeding

One has to initiate inbreeding from an F_2 population onwards that involves the random pairing of F_2 individuals. A unique name has to be assigned to each inbreeding line. If it is from an advanced intercross, the details of the cross from which this advanced line is derived have to be recorded. The inbreeding needs to be continued till the desired number of generations is reached.

13.3 Doubled Haploid Breeding

Doubled haploids are generated by doubling chromosomes of haploid plants raised from either egg or sperm cells. Three widely used methods to produce DHs are (a) culture of sperm cells, microspores and anthers; (b) gynogenesis, using ovary or ovule culture; and (c) through chromosome elimination where the target species is crossed to a distant related relative and the embryos produced are cultured or rescued in vitro (Fig. 13.2). Chromosomes of the distant relative are eliminated, and the resultant plants will be with chromosomes of target species. Chromosomes of such haploid plantlets are doubled by chemical means. Such process is being successfully used in barley (*Hordeum vulgare*) crossed with *Hordeum bulbosum* and wheat (*Triticum aestivum*) crossed with maize (*Zea mays*) (see also Box 13.1).

Box 13.1: Centromere Mediated Chromosome Elimination
Chromosomes are either with paternal or maternal inheritance. Haploids can be generated either from cultured gametophyte cells that can be regenerated into haploid plants or can be induced from rare interspecific crosses, in which one parental genome is eliminated after fertilization. Centromeres from the two parent species interact unequally with the mitotic spindle, causing selective loss of chromosomes. In *Arabidopsis thaliana*, the centromere-specific histone CENH3 is manipulated to disrupt spindle fibre attachment and haploids plants are generated. When CENH3 mutants expressing altered CENH3 proteins are crossed to wild type, chromosomes from the mutant are eliminated, producing haploid progeny. In hybrids, in the early embryonic mitotic divisions, the chromosomes marked by the defective CENH3 are lost. This results in haploid plants with nuclear genome is derived from the wild-type parent. Haploids are spontaneously converted into fertile diploids through meiotic non-reduction of chromosomes (formation of $2n$ gametes resulting from failure of *reduction* during meiosis) (see Fig. 13.3).

DH achieves complete homozygosity in one generation that enables significant shortening of time to the production of pure lines. This allows more precise phenotyping and allows accurate gene-trait association in genetic mapping and gene function studies. DH technology has been successfully used in barley, wheat, maize rice, oats, rye, *Brassica* spp., legumes and fruit crops. Cotton and many legume species are not amenable to DH technology. DH only allows one or two chances of recombination, as DH lines are usually generated from F_1 or sometimes F_2 plants, limiting the diversity of the DH lines. They are ideal for estimating QTL × environment interactions as complete homozygosity with two identical sets of chromosomes allows better estimates of trait. They have only one recombination opportunity in the first generation. To increase recombination, sometime F_2 pollen/egg is used. Yet another system for the development of haploid plants is fast generation cycling system (FGCS) (see Box 13.2).

Fig. 13.2 Doubled haploid (DH) technology. (**a**) Comparison between conventional breeding and DH technology. (**b**) Diagram of three major DH technologies adopted in crop breeding: anther culture, microspore culture and chromosome elimination. *CD* = chromosome doubling with chemical treatment

Box 13.2: Fast Generation Cycling System (FGCS)

FGCS is a process to reduce generation time. It involves two steps in each generation: a) plants are grown in a controlled environment where vegetative growth and flower differentiation are accelerated through irrigation and nutrient management; b) *in vitro* culture of young embryos is undertaken to reduce the time required for seed maturity. At this step, endosperm is removed. This promotes embryo germination as it can absorb the readily available sucrose in the medium. Immature embryo culture can be carried out without waiting for full seed development. Single seed descent (SSD) is usually adopted in FGCS for developing RIL through continuous selfing from the F_2 generation until the desired level of homozygosity is reached.

FGCS is remunerative in species where DH lines are difficult to derive. Successful application of FGCS were reported in crops like barley, wheat, maize rice, oats, rye *Brassica* spp., legumes and fruit crops, where significant shortening of generation time is made possible with 6–9 generations per year, where only 1–3 generations per year would only be possible through conventional means.

The advantages are: While it takes time to derive a variety from crossing to release, DHs reduces time to develop for RILs; number of meiotic events where recombination occurs are not reduced; and selection can be exercised in any generation and Near Isogenic Lines (NILs) can be developed using the heterogeneous inbred family (HIF) selection.

13.4 Reverse Breeding

Since it is difficult to predict which parental lines will give the best progeny, hybrid breeding depends on a trial and error approach. Many pairs of parents are to be crossed and their progenies are to be tested. Reverse breeding involves production of superior hybrids and selection of parental lines. In conventional breeding, recombination of chromosome pairs results in rearrangements of genetic material, and the unique combination of genetic variation will be lost. In reverse breeding, an elected heterozygote is crossed with itself, while chromosomal recombination is suppressed by a transgene resulting in lines with homozygous chromosome pairs. For hybrid variety production, parental lines in which the genetic variation of the chromosome pairs that complements each other are selected from the reverse-breeding programme. Crossing such lines will result in uniform offspring hybrid plants which are genetically similar to the plant with which the reverse breeding was started (Fig. 13.4).

Fixation of non-recombinant chromosomes in homozygous doubled haploid lines (DHs) is accomplished by the knockdown of meiotic crossovers. The chromosome structure shall be intact. *Arabidopsis* gene ASY1 and the rice ASY1 homologue

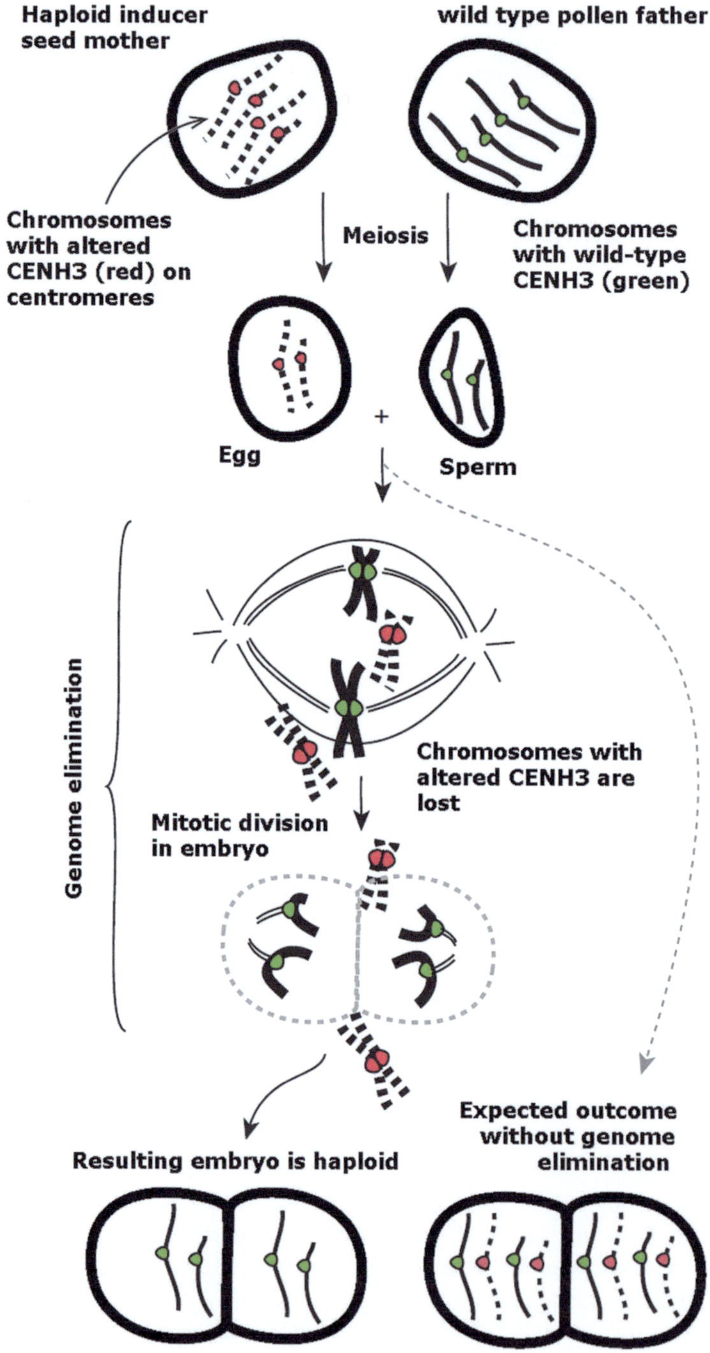

Fig. 13.3 Genome elimination induced by modification of centromeric histone H3 (CENH3). An *Arabidopsis* plant becomes a haploid inducer if the native CENH3 gene is knocked out and complemented with one encoding an altered *CENH3*. While the chromosomes of the haploid inducer are inherited efficiently upon self-crosses, they are unstable in crosses to a wild-type

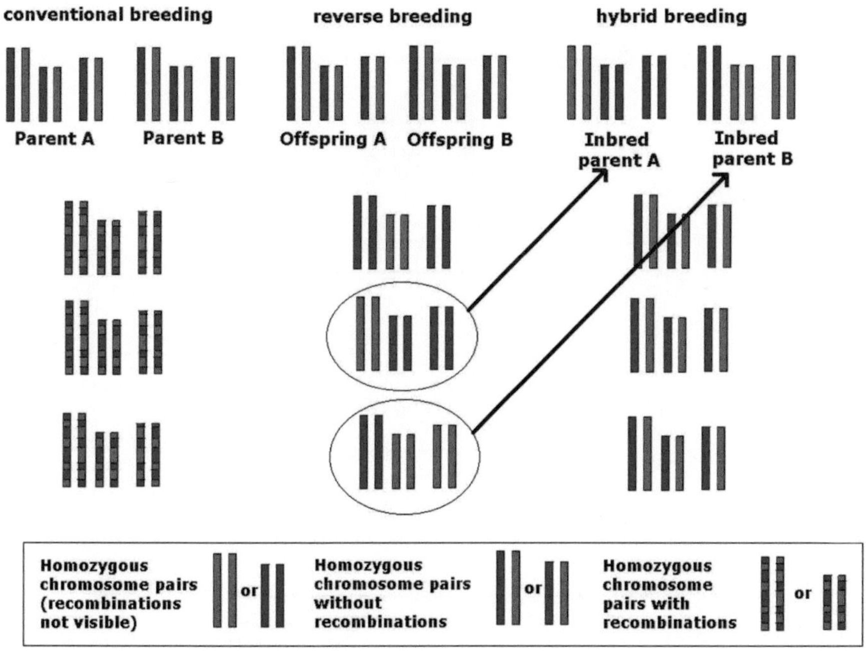

Fig. 13.4 Overview of the outcomes of different breeding programmes

PAIR2 are the examples. Such mutants display univalents at metaphase I. Gene expression is knocked out using RNA interference (RNAi) or siRNAs that result in post-transcriptional gene silencing (PTGS) (Fig. 13.5).

Reverse breeding generates homozygous parental lines and starts with a heterozygote in which meiotic recombination can be suppressed (Fig. 13.6a). The result is the production of random wild-type doubled haploids in which non-recombinant chromosomes are present (Fig. 13.6b). Also available are different genotypes with no crossovers from among reverse-breeding doubled haploids (Fig. 13.6c).

Fig. 13.3 (continued) plant. In the early embryonic mitotic divisions of a hybrid derived from this cross, the chromosomes marked by the defective CENH3 (red) are lost, resulting in a haploid plant of which the nuclear genome derives from the wild-type parent. Diploidization ensues spontaneously or after treatment with spindle inhibitors to produce a fertile dihaploid plant, which is characterized by complete homozygosity. In the lower right, the diploid hybrid produced without genome elimination is depicted. Not shown is the relatively simple step entailing the spontaneous or induced diploidization of the haploid. (Figure courtesy: *PLoS Biology*)

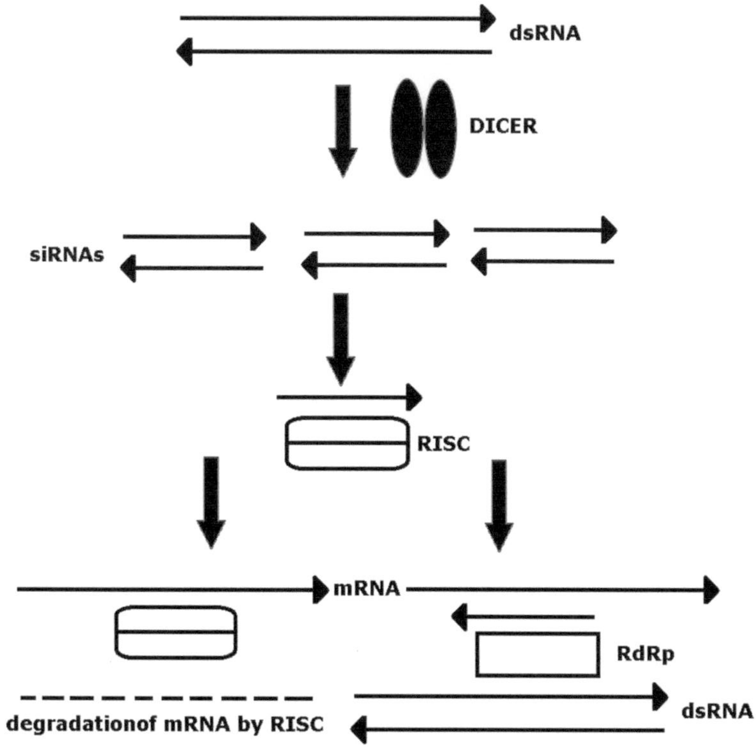

Fig. 13.5 RNAi mechanism. The cellular enzyme Dicer cleaves intracellularly synthesized or exogenously administered dsRNA into 21–25 nucleotide siRNAs. The siRNAs are incorporated into the RNA-induced silencing complex (RISC), which uses the antisense strand of the siRNA to find and destroy the target mRNA. The siRNAs can also be used as primers for the generation of new dsRNA by RNA-dependent RNA polymerase (RdRp)

13.4.1 Marker-Assisted Reverse Breeding (MARB)

MARB is being used in maize breeding. It will revert any maize hybrid into inbred lines with any level of required similarity to its original parent lines. Pericarp DNA of a hybrid is from the maternal parent, and one-half of the embryo DNA is from the maternal parent and the other half from the paternal parent. DNA from both seed embryo and pericarp (embryo represents both male and female and pericarp represents only female) can be extracted separately and high-density single-nucleotide polymorphism (SNP) chips analysed that are derived from the two parental genotypes (Fig. 13.7). Marker-assisted selection can be performed based on an Illumina low-density SNP chip designed with SNPs polymorphic between the 2 parental genotypes, which were uniformly distributed on 10 maize chromosomes. This method has the advantages of fast speed, fixed heterotic mode and quick recovery of beneficial parental genotypes compared to traditional pedigree breeding using elite hybrids.

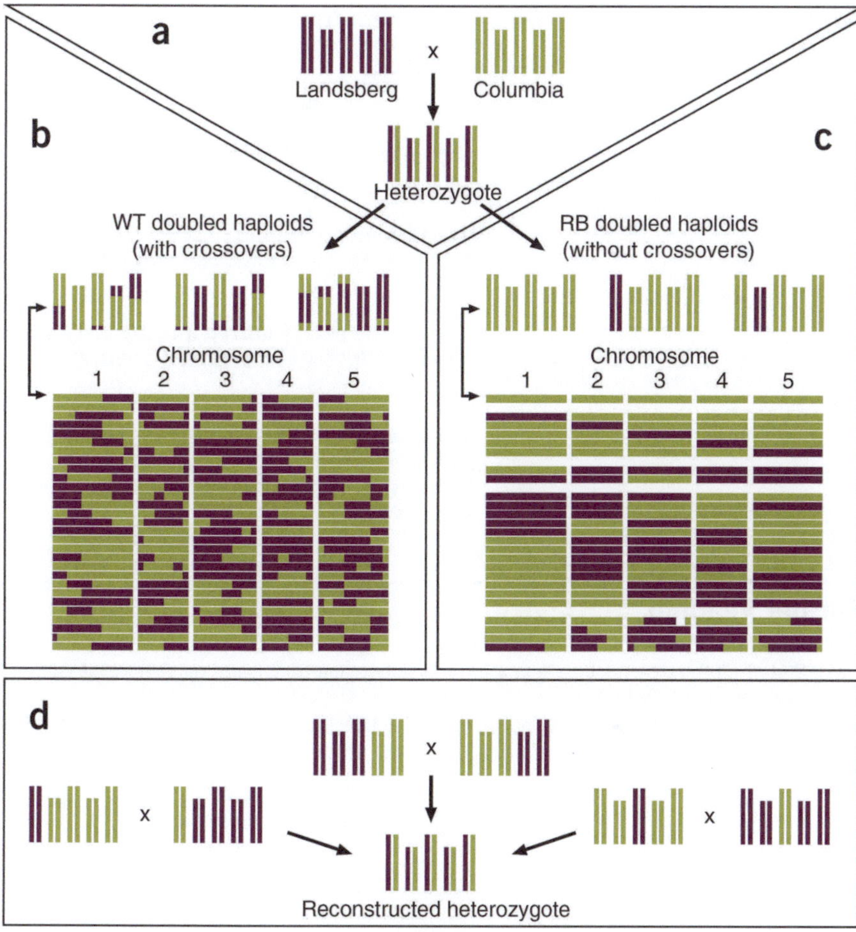

Fig. 13.6 Reverse-breeding strategy and genotypes of wild-type (WT) and reverse-breeding (RB) doubled haploid offspring in *Arabidopsis thaliana*. (**a**) Reverse breeding starts with a heterozygote in which meiotic recombination can be suppressed. (**b**) Genotype of 29 randomly selected wild-type doubled haploids. Three individuals are shown with 'classic' vertical chromosomes, but others as horizontal lines only. Each line represents chromosomes 1–5 for an individual plant. Note the presence of non-recombinant chromosomes. (**c**) 21 different genotypes are recovered, in which no crossovers occurred from among 36 reverse-breeding doubled haploids. The first row represents the genotype of one of the recovered original parents; the next seven genotypes represent chromosome substitution lines and the remainder are mosaics of Col and L*er* chromosomes. The last four represent genotypes of haploid offspring that showed crossovers. (**d**) Three pairs of reverse-breeding doubled haploids were crossed to recreate the initial hybrid; they have the RNAi transgene. (Figure courtesy: Erik Wijnker, Wageningen University; *Nature Genetics*. Figures are diagrammatic and representative)

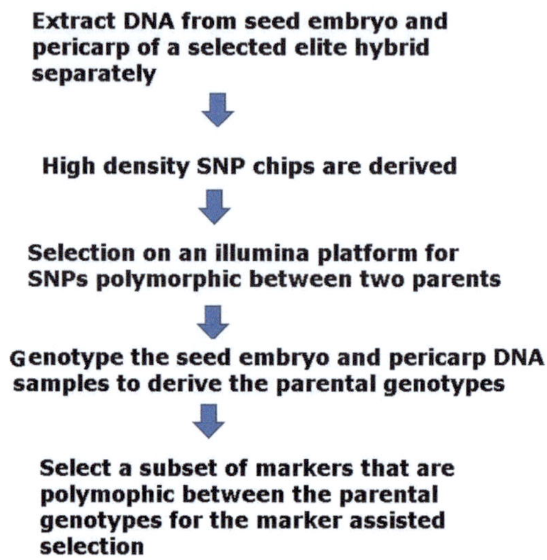

Fig. 13.7 General protocol of marker-assisted reverse breeding

Further Reading

Dirks R et al (2009) Reverse breeding: a novel breeding approach based on engineered meiosis. Plant Biotechnol J 7:837–845

Shuro AR et al (2017) Review paper on approaches in developing inbred lines in cross-pollinated crops. Biochem Mol Biol 2:40–45

Quantitative Genetics

14

Keywords
Multiple-factor hypothesis (Nilsson-Ehle) · Models, Assumptions and predictions · Partition of variance components · Linearity · The infinitesimal model · Types of gene action · Quantifying gene action · Population mean · Phenotypic variance · Breeding value · Heritability · Estimating additive variance and heritability · Models for combining ability analysis · Biparental progenies (BIP) · Polycross · Topcross · North Carolina designs · Diallels · Multiple regression analysis · Stability analysis · Regression approaches · Genetic architecture of quantitative traits

14.1 Principles of Biometrical Genetics

Most of the traits improved through breeding like yield, height, drought resistance, disease resistance in many species, etc. are quantitative. They are also called polygenic, continuous, multifactorial or complex traits. Quantitative traits are the result of cumulative action of many genes and their interactions with the environment. Thus, it can create a range of individuals that vary among themselves with continuous distribution of phenotypes. A quantitative trait is assumed to be controlled by the cumulative effect of numerous genes, known as quantitative trait loci (QTLs), as per multiple-factor hypothesis by Nilsson-Ehle (a Swedish geneticist in 1909) and East (an American in 1916). Hence, a single phenotypic trait is regulated by several QTLs.

14.1.1 Multiple-Factor Hypothesis (Nilsson-Ehle)

Nilsson-Ehle concluded kernel colour in wheat as a quantitative character. True-breeding red kernel wheat (RR) was crossed with true breeding white (rr) and the F_1

P. M. Priyadarshan, *PLANT BREEDING: Classical to Modern*,
https://doi.org/10.1007/978-981-13-7095-3_14

Table 14.1 F_2 ratio in wheat

Genotype	Genotypic ratio	Phenotype
$R_1R_1R_2R_2$	1	Dark red
$R_1R_1R_2r_2$	1	Medium dark red
$R_1r_1R_2R_2$	2	Medium dark red
$R_1r_1R_2r_2$	4	Medium red
$R_1R_1r_2r_2$	1	Medium red
$R_1r_1R_2R_2$	1	Medium red
$R_1r_2r_2r_2$	2	Light red
$R_1r_1R_2r_2$	2	Light red
$R_1r_1r_2r_2$	1	White

was red (Rr). The F_2 segregated for red and white in 3:1 ratio indicates the dominance of red over white. However, red colour among the red colour progenies indicated variation. F_1 red was not as intense as the parent. In F_2, a range of red colour was observed. In some crosses, a ratio of 15 red:1 white was found in F_2 indicating that there are two pairs of genes for red colour and that either or both of these can produce red kernels (Fig. 14.1). The intensity of colour decreased from dark red to white. The F_2 showed red shades and white as follows:

Dark red	:	1	
Medium dark red	:	4	
Medium red	:	6	15 = total red
Light red	:		4
White	:		1
Total	:		16

Two duplicate dominant alleles (R1 and R2) *cumulatively decided* the intensity of red colour

(a) Both R1 and R2 are in completely dominant over white.
(b) The high intensity of red colour depends on the number.

The F_2 ratio is available Table 14.1. If two parents differ for two genes, the segregation was 1:4:6:4:1. If three genes are involved, then F_2 segregation would be 1:6:15:20:15:6:1.

Thus, Nilsson-Ehle's multiple factor states that:

(a) Quantitative trait could be governed by several genes with independent segregation, but had cumulative effect on phenotype.
(b) There is incomplete dominance.
(c) Each gene influences expression of trait.

East (1916) reported his studies on the inheritance of corolla length in *Nicotiana longiflora*, a self-pollinated species of tobacco. This trait is governed by multiple genes. He crossed a variety, the corolla which had an average length of 52 mm, to a variety with corolla of 70 mm. Both these varieties had long been inbred and

♀ \ ♂	*R1R2*	*R1r2*	*r1R2*	*r1r2*
R1R2	*R1R1R2R2* Dark red	*R1R1R2r2* Medium red	*R1r1R2R2* Medium red	*R1r1R2r2* Intermediate red
R1r2	*R1R1R2r2* Medium red	*R1R1r2r2* Intermediate red	*R1r1R2r2* Intermediate red	*R1r1r2r2* Light red
r1R2	*R1r1R2R2* Medium red	*R1r1R2r2* Intermediate red	*r1r1R2R2* Intermediate red	*r1r1R2r2* Light red
r1r2	*R1r1R2r2* Intermediate red	*R1r1r2r2* Light red	*r1r1R2r2* Light red	*r1r1r2r2* White

Fig. 14.1 Nilsson-Ehle carefully categorized the colours of kernels in wheat in the F_2 generation and discovered that they followed a 1:4:6:4:1 ratio. This occurs because the contributions of the red alleles are additive. In this example, two genes, with two alleles each (red and white), govern kernel colour. Offspring can display a range of colours, depending on how many copies of the red allele they inherit. If an offspring is homozygous for the red allele of both genes, it will have very dark red kernels. By comparison, if it carries three red alleles and one white allele, it will be medium red (which is not quite as deep in colour). In this way, this polygenic trait can exhibit a range of phenotypes from dark red to white

therefore were homozygous. The marked differences in corolla lengths were heritable pointing out that they are controlled by genes rather than environment. East found that F_1 was intermediate with mean corolla length of 61 mm. In F_2, a much larger variation for corolla length than F_1 was observed (Table 14.2; Fig. 14.2). The variation was continuous as well. East raised 444 F_2 plants and failed to get even a single plant like either of the parents. This pointed out that more than four pairs of genes are involved in determining the length of corolla in *Nicotiana longiflora*.

Quantitative inheritance is based on the following facts:

(a) Continuous variation.
(b) A marked effect of the environment on their expression.
(c) Governed by multiple or polygenes.
(d) Each gene produces unit or individual effect. The effects of genes are additive or cumulative.
(e) Dominance is absent or partial. F_1 hybrids show blending in characters, or in other words, the F_1 hybrid is intermediate.
(f) Segregation and independent assortment of genes in F_2 is according to Mendelian inheritance, but the phenotype is in continuous range between the extreme

Table 14.2 F_2 Generation in the experiment of East

Genotype	Frequency	Number of dominant factors	Length (mm)	Frequency			
AA BB CC	1	6	70	1			
AA BB Cc	2						
AA Bb CC	2	5	67	6			
Aa BB CC	2						
AA Bb Cc	4						
Aa BB Cc	4						
Aa Bb CC	4						
AA BB cc	1	4	64	15			
AA bb CC	1						
Aa BB CC	1						
Aa Bb Cc	8						
AA bb cc	2						
Aa BB cc	2						
AA bb Cc	2	3	61	20			
Aa bb Cc	2						
Aa BB Cc	2						
Aa Bb CC	2						
Aa Bb cc	4						
Aa Bb Cc	4						
Aa Bb Cc	4						
AA bb cc	1	2	58	15			
Aa BB cc	1						
Aa bb CC	1						
Aa bb cc	2						
aa Bb cc	2	1	55	6			
aa bb Cc	2						
aa bb cc	1	0	52	1			
Number of active alleles	0	1	2	3	4	5	6
Length (mm)	52	55	58	61	64	67	70
Frequency (phenotypic ratio)	1	6	15	20	15	6	1

limits of the parents. The phenotypic proportion of F_2 is modified according to the number and nature of genes.

(g) Sometimes polygenic characters are governed by single gene too. That is, single-gene mutation may have the same effect as changes in many cumulative genes. For example, in sweet peas tallness is controlled by polygenes. Variations in the size of tall plants are partly environmental and partly polygenic, but single mutation as well can result in dwarf plants.

(h) For statistical analysis of polygenic inheritance, we owe a great deal to Mather, Haldane, Fisher, etc. Biological samples are infinite, and therefore, statistical

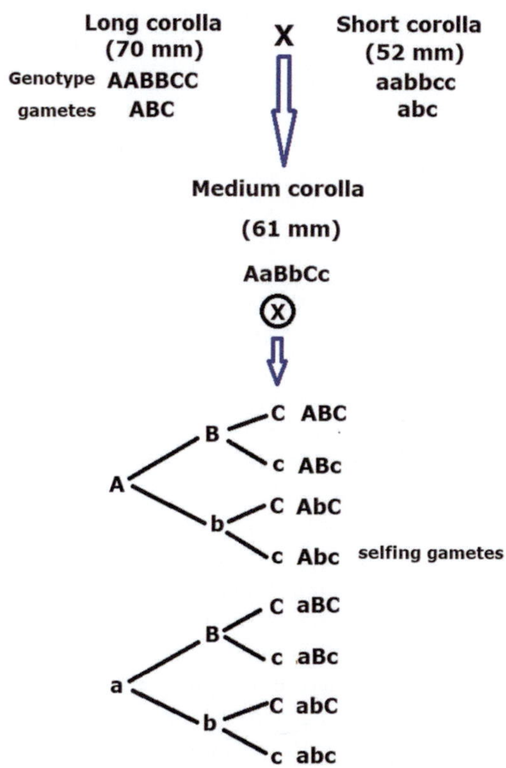

Fig. 14.2 F_2 segregation in *Nicotiana longiflora*

Table 14.3 Major differences between qualitative and quantitative genetics

Qualitative	Quantitative
Deals with the inheritance of traits of kind, viz. form, structure, colour, etc.	Deals with the inheritance of traits of degree, viz. heights of length, weight, number, etc.
Discrete phenotypic classes occur which display discontinuous variations	A spectrum of phenotypic classes occur which contain continuous variations
Each qualitative trait is governed by two or many alleles of a single gene	Each quantitative trait is governed by many non-allelic genes or polygenes
Phenotypic expression of a gene is not influenced by environment	Environmental conditions effect the phenotypic expression of polygenes variously
Concerns with individual matings and their progeny	Concerns with a population of organisms consisting of all possible kinds of matings
Analysis is made by counts and ratios	Analysis is made by statistical methods

parameters are not well defined. Sampling is essential and this can lead us only near the truth but never to the truth or reality.

Major differences between qualitative and quantitative characters are available in Table 14.3.

Polygenic traits do not follow patterns of Mendelian inheritance (qualitative traits) and are unlike monogenic traits. Instead, their phenotypes exhibit spectrum depicted by a bell curve (see Chap. 7 on basic statistics). For instance, in fruit size (controlled by a single gene with alleles "s" for small and "S" for large), the progeny would segregate into 3:1 ratio. Hence, one can infer the "genotype" (SS or Ss versus ss) by observing the "phenotype" (large or small). On the other hand, quantitative traits are complex because:

(a) Quantitative traits are controlled by multiple genes or QTLs and same phenotype can be carrier of different alleles at each QTL.
(b) Genotypes with identical QTL can exhibit different phenotypes when grown under different environments.
(c) One QTL can influence the allelic constitution of other QTL. So, inferring a genotype from the phenotype is difficult. Specialized genetic stocks must be constructed to be grown under precisely controlled environments.

QTLs include two groups of genes: (a) highly heritable traits governed by major genes with very large effects, each gene explaining a large portion of the total trait variation in a mapping population, and (b) QTLs under the regulation of many genes, each controlling small portion of the total trait variation. Most quantitative traits are controlled by a small number of major genes or QTLs. Both types of genes with moderate and minor effects also influence quantitative traits. Major genes can be analysed via segregation analysis or evolutionary and selection history. However, numerous genes with small effects cannot be investigated individually.

14.2 Models, Assumptions and Predictions

14.2.1 Partition of Variance Components

A model for partition of variance components was developed by Fisher in 1918 and further developed by Cockerham (in 1954) and Kemthrone (in 1969). In this, variances and covariances among relatives are described in terms of the variances in additive genetic effects or breeding values (V_A) and interactions of effects between alleles within loci (dominance, V_D) and among loci (epistasis, V_{AA}, V_{AD}, etc.). Such partitions are dependent on assumptions like:

(a) Genotypes follow Hardy-Weinberg equilibrium, random mating (i.e. no inbred individuals)
(b) Linkage equilibrium prevails (which requires many generations to achieve for tightly linked genes)
(c) No selection pressure

An elegant formalization for the variance-covariance matrix V of phenotypic values of a group of individuals for a single trait would be:

$$V = \mathrm{A}V_{\mathrm{A}} + \mathrm{D}V_{\mathrm{D}} + A\#\mathrm{A}V_{\mathrm{AA}} + A\#\mathrm{D}V_{\mathrm{AD}} + \ldots\ldots + \mathrm{I}V_{\mathrm{E}},$$

where A is the numerator relation matrix of individuals, D defines dominance relationships and V_E is the environmental variance. For the epistatic terms, # denotes element-by-element multiplication, but applies only for unlinked loci. Many more terms like maternal genetic effects and genotype × environment interaction may be included. This model addresses complexity elegantly. This is the strength of the model, but requirement of large data sets to allow partitioning into only very few components is its weakness.

14.2.2 Linearity

The regression of offspring phenotype on that of parent for the trait in question is usually assumed to be linear. The regression of response on selection differential will also be linear. This important assumption holds under multivariate normality of phenotypic and genotypic values and thus the central limit theorem assuming multifactorial inheritance. However, some traits like litter size or lifespan do not follow normal distribution. But adequate transformations can be invoked or departures ignored.

14.2.3 The Infinitesimal Model

Response to the first generation of selection can be predicted from the breeder's equation *Response* = h^2 *x selection differential.* Selection changes gene frequencies and genetic variance. In subsequent generations, to predict response, knowledge of individual gene effects and frequencies is a prerequisite. Fisher's "infinitesimal model", formalized by Bulmer in 1980, provides a practical but biologically unrealistic resolution such as many unlinked genes with infinitesimally small additive effect influence on selection that produces negligible changes in gene frequency and variance at each locus. Only inbreeding can change the within-family or Mendelian segregation variance. The change in between-family variance (the "Bulmer effect") depends only on the intensity and accuracy of selection practised. Hence, the selection response in successive generations can be predicted from estimable base population parameters such as heritability and phenotypic variance, selection practised and inbreeding.

14.3 Types of Gene Action

Total genetic variance is partitioned into three types – *additive, dominance* and *epistatic* variance. Adding up of the effects of each allele is additive genetic variance. Hypothetical examples of additive gene action are available in Fig. 14.3. Note that petal length in those examples is determined simply by the number of capital letter

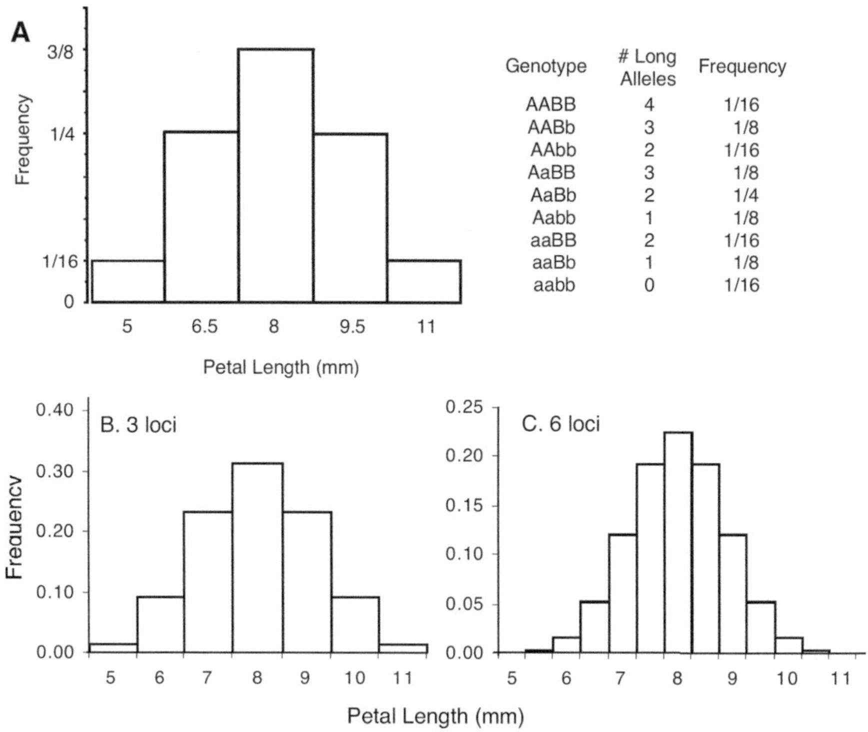

Genotype	# Long Alleles	Frequency
AABB	4	1/16
AABb	3	1/8
AAbb	2	1/16
AaBB	3	1/8
AaBb	2	1/4
Aabb	1	1/8
aaBB	2	1/16
aaBb	1	1/8
aabb	0	1/16

Fig. 14.3 A hypothetical example (based on the real petal length data in Fig. 14.2) showing genotypic values (along the *x*-axes). The three graphs show how increasing numbers of loci affecting a trait make the trait distribution more continuous in the absence of environmental deviations. In A, there are two loci with two alleles each, which is the simplest case for a trait affected by more than one locus. The loci act additively (no dominance or epistasis), so each capital letter allele adds 1.5 mm of petal length over the *aabb* genotype, which has petals with 5 mm. The frequency of each genotype is with $p = q = 0.5$ for both loci, and the graph shows the phenotypic distribution that results. B and C show the phenotypic distribution with 3 and 6 loci respectively

alleles present in the two-locus genotypes. Effect of each allele is not affected by the effect of other allele of the same locus. On the other hand, it is also not affected by the effect of other alleles of the other loci. It may be noted that additivity is not equal effects of all alleles at a locus. Dominance is the interaction between alleles of the same locus, and epistasis is characterized by interactions between alleles of different loci (Table 14.4).

Genes acting in a dominant fashion means interaction between alleles at one locus. The diploid genotype at each locus needs to be considered as a whole to determine the phenotypic effect. It is specific for a given locus. It is also specific for a given phenotypic trait. In a phenotypic hierarchy, the degree of dominance or epistasis for a given locus can vary across traits at different levels.

Table 14.4 Summary of how interactions among alleles at different levels (within or between loci) causes different types of gene action

	Interactions among alleles?	
	No interaction	Interaction
Within locus	Additive	Dominance
Between loci	Additive	Epistasis

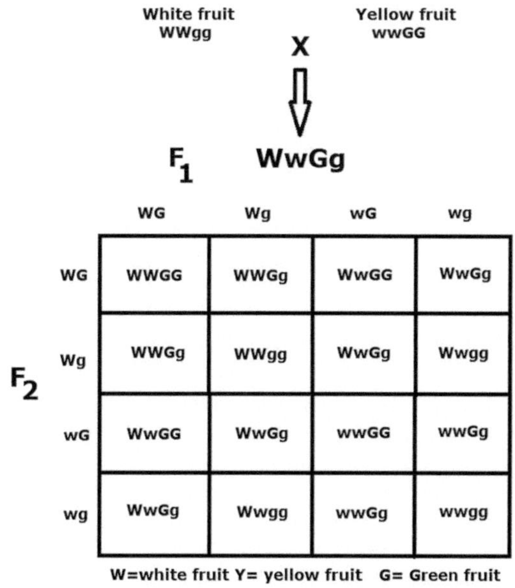

Fig. 14.4 Dominant epistasis for fruit colour in summer squash (*Cucurbita pepo*). The normal dihybrid ratio modified into 12:3:1 in F_2 generation

Epistasis is the interaction between genes. Either genes can mask each other so that one is considered "dominant," or they can combine to produce a new trait. It is the conditional relationship between two genes that can determine a single phenotype of some traits. At each locus there are two alleles that govern phenotypes. They can affect one another in such a way that, regardless of the allele of one gene, it is recessive to one dominant allele of the other (Fig. 14.4).

14.3.1 Quantifying Gene Action

The magnitude of additive and dominant action at a locus can be quantified as *a* and *d*, respectively (Fig. 14.5). Here, the midpoint between the two homozygotes is set to zero, G for the two homozygotes are $+a$ and $-a$, and for the heterozygote is *d*. A shows the additive case, B complete dominance and C partial dominance. From this we can see that the degree of dominance can be expressed as d/a, which equals 0, 1 and 1/4 in these three cases, respectively. Note that the absolute value of *d* is the same in C and D, but since *a* is smaller in D, the degree of dominance d/a is greater in

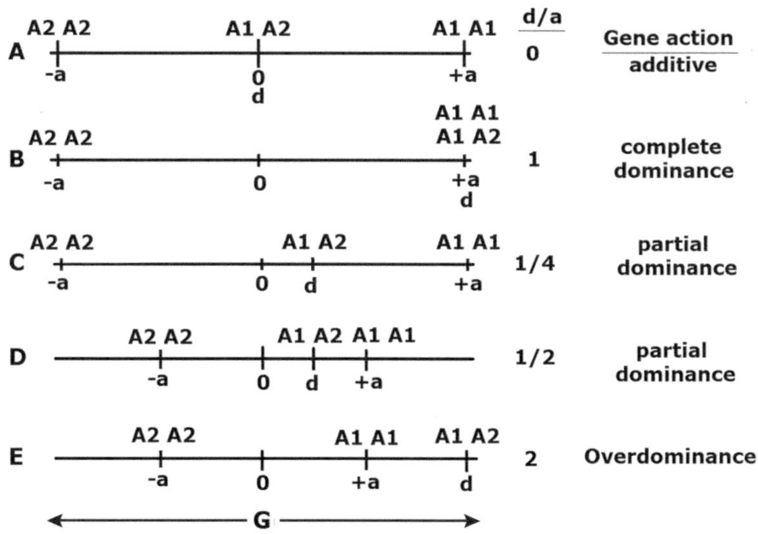

Fig. 14.5 Gene action quantified using *a* and *d*. The horizontal scale represents genotypic values

Table 14.5 Derivation of the equation for genotypic mean. To simplify the sum of the products, note that $p^2 - q^2 = (p + q)(p - q) = p - q$ because $p + q = 1$

Genotype	Frequency	Genotypic value	Product
AA	p^2	$+a$	p^2a
Aa	$2pq$	d	$2pqd$
aa	q^2	$-a$	$-q^2a$
		Sum of products =	$a(p-q) + 2pqd$

D (1/2) due to the smaller overall effect of the locus in D. E represents a locus with overdominance where $d/a > 1$.

14.3.2 Population Mean

The results from each locus can be summed to give the effects of all loci for a phenotypic trait in the absence of epistasis. Calculation of a mean is done by totalling values and dividing it by the number of individuals (Table 14.5). In this method, the value for each class (the three genotypes here) is multiplied by its frequency. After this, these products are totalled to work out mean. The frequencies are the Hardy-Weinberg values, while the values are expressed in terms of *a* and *d*. The summation gives the equation for the mean:

$$\bar{G} = \bar{P} = a(p - q) + d2pq \tag{14.1}$$

The magnitude of the additive effect and the degree of dominance is expressed in this equation. It also shows how population means are determined by the allele frequencies. The first term represents the effects of the homozygotes and shows that as a increases, the mean increases if $p > q$ and decreases if $p < q$ (recall that G for the aa homozygote is $-a$) (see also Chap. 7). The second term is the effect of heterozygotes. Again, in the absence of epistasis, these terms can just be summed over all loci affecting the trait.

14.3.3 Phenotypic Variance

Variation is the raw material for evolutionary change. Variance is absolutely vital because it is fundamental measure of variation in statistics:

$$V_x = \frac{\sum_{i=1}^{n} (X_i - \bar{X})^2}{n - 1} \tag{14.2}$$

If the phenotypic values in a population are used in the aforesaid equation, it is the phenotypic variance (V_P) for that trait. The numerator of this formula is the "sum of squares" (SS) or the sum over all individuals of the squared deviations from the mean. If there are lots of individuals with values far from the mean in a population (i.e. curves A and C in Fig. 14.6), the sum of the deviations and variance will be large. If most individuals have values close to the mean (e.g. curve B in Fig. 14.6), then the deviations and variance will be small. The denominator is the number of individuals minus one (the degrees of freedom). This makes the variance an average squared deviation from the mean. Variance is sometimes called a mean square (MS) because of this attribute.

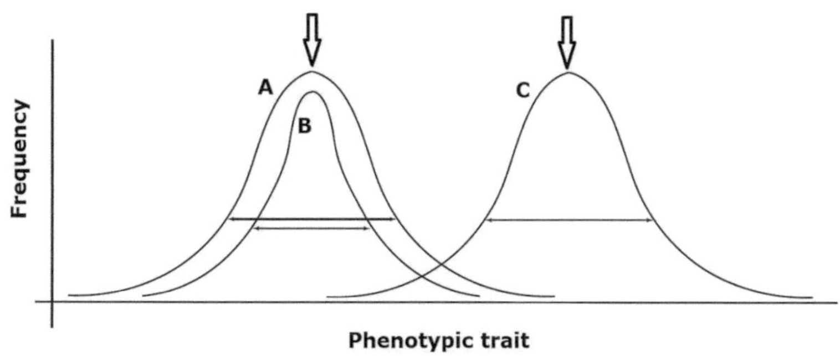

Fig. 14.6 Three normal distributions illustrating mean and variance. The mean (single-headed arrows) is just the average phenotype in the population, and the variance (double-headed arrows) is a measure of how variable the population is, in other words the width of the distribution. Populations A and B have the same mean but different variances, while A and C have different means but the same variances

Assuming that there is no correlation or interaction between the genotypes and the environment, the total variance in the population can be partitioned into additive components. The simplest partition is:

$$V_{\mathrm{P}} = V_{\mathrm{G}} + V_{\mathrm{E}}$$

V_{G} is the genotypic variance and V_{E} is the environmental variance. This partitioning is most useful for clonal or highly self-pollinated organisms. However, since they pass their diploid genotypes onto their offspring, it is less useful for cross-pollinated species. Here the genotypes are created anew in each offspring by a random combination of an allele from each parent at each locus. Therefore, for cross-pollinated species we need to further partition V_{G}:

$$V_{\mathrm{G}} = V_{\mathrm{A}} + V_{\mathrm{D}} + V_{\mathrm{I}} \quad (14.3)$$

where V_{A} is the additive genetic variance, V_{D} is the dominance variance and V_{I} is the interaction or epistatic variance (the latter two are collectively referred to as non-additive genetic variance).

and the total phenotypic variance can be rewritten as:

$$V_{\mathrm{P}} = V_{\mathrm{A}} + V_{\mathrm{D}} + V_{\mathrm{I}} + V_{\mathrm{E}} + V_{\mathrm{GE}} \quad (14.4)$$

In sexually reproducing species, additive genetic variance is the most important, because only the additive effects of genes are passed on directly from parents to offspring. Only one allele at each locus is transferred from each parent to create new dominance relationships in dominance and epistatic effects. This happens in offspring of sexually reproducing species. Similarly, independent assortment of alleles at different loci creates new epistatic effects.

Additive variance in terms of allele frequencies and gene action is:

$$V_{\mathrm{A}} = 2pq\,[a + d\,(q - p)]^2 \quad (14.5)$$

V_{A} is most important in determining changes in mean phenotypic value across generations. V_{A} is measured on the basis of resemblance between relatives, and primarily this resemblance is caused by additive variation. On the other hand, dominance and epistasis exert influence on the offspring not to look like the average of their parents. For instance, let us consider a hypothetical cross between two rice genotypes (Table 14.6). One is with BBCC genotype and the other with bbcc genotype (for plant height). If additivity is complete, the offspring (all BbCc) is expected to have genotypic values equal to the average of the parents, i.e. (41.91 + 41.62)/2 = 41.76. However, due to both under-dominance and epistasis in this case, the double heterozygote offspring have only $G = 40.81$.

The equation for additive variance in terms of allele frequencies and gene action is:

$$V_{\mathrm{A}} = 2pq\,[a + d\,(q - p)]^2 \quad (14.6)$$

Table 14.6 Hypothetical example of plant height in rice (cm). Genotypes at two loci (sample length in parentheses). The B locus exhibits complete dominance. Note that these are estimates of genotypic values, because they are the averages of a number of individuals of the same genotype

	BB and Bb	bb
CC	41.91 (46)	40.96 (119)
Cc	40.81 (113)	42.13 (32)
cc	40.94 (150)	41.62 (21)

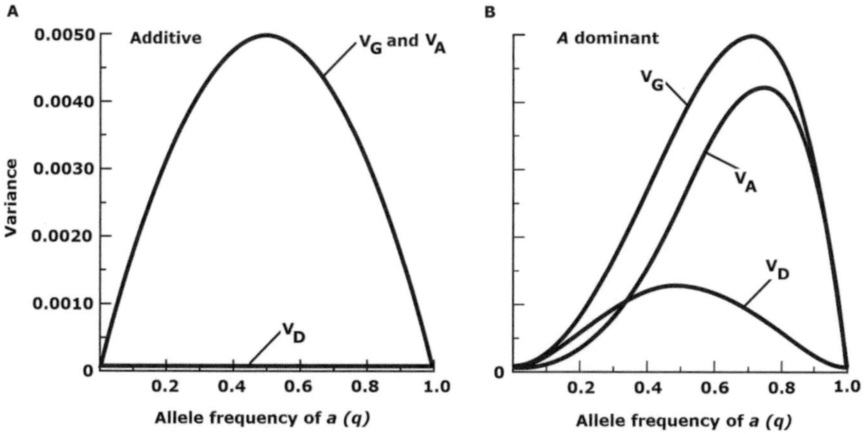

Fig. 14.7 Genotypic variance V_G, additive genetic variance V_A and dominance variance V_D for a single locus with two alleles in a hypothetical population. Note that the *x*-axis is the frequency of the *a* allele, which is recessive in panel B. Because this is a single locus, there is no epistatic variance. (A) A completely additive locus, $a = 0.1$, $d = 0$. (B) Complete dominance, $a = d = 0.0707$. From Eqs. 14.6, 14.7 and 14.8

2pq is maximum at $p = q = 0.5$, indicating thereby that genetic variance is high at intermediate allele frequencies. Such a situation also confirms that if one allele is rare, most individuals are homozygous for the other alleles. Hence, there will be little variance in the population. If there is no dominance, $d = 0$, then the equation reduces to:

$$V_A = 2pqa^2 \tag{14.7}$$

This means that additive variance is maximized at $p = q = 0.5$ (Fig. 14.7a). When there is dominance, the maximum variance occurs when the recessive allele is more common ($q = 0.75$), making the $d\,(q\text{-}p)$ term large and positive (Fig. 14.7b). This is because with dominance and equal allele frequencies, 75% of the individuals in the population have the dominant phenotype. As q becomes larger than 0.75, additive variance drops because the first $2pq$ term drops faster and then the $d(q-p)$ term increases. Note that the dominance variance does peak at $p = q = 0.5$ (Fig. 14.7b) and this is because the equation for dominance variance is similar to Eq. 14.7 in that the allele frequencies are only in the $2pq$ term:

$$V_{\mathrm{D}} = (2pqd)^2 \tag{14.8}$$

Variance is defined as the squared deviation from the mean (Eq. 14.2) because all these equations for variance have a squared term. Negative variability is meaningless because variability cannot be negative. Estimates of variances can be negative because of experimental error.

14.3.4 Breeding Value

Genotypes are not passed on from parents to offspring, but are created afresh because of the combination of alleles from each parent at each locus. The effect of an individual's genes on the value of the trait is the breeding value. This is caused by additive effect of genes. Otherwise known as "additive genotype", the variance of these breeding values is V_{A}. So, breeding values are prominent than G in sexually reproducing species. While assisting estimation of genetic correlations, breeding values may reduce bias in measuring selection. Best linear unbiased prediction (BLUP) is the method of estimating breeding values.

14.3.5 Heritability

Phenotype evolves in response to artificial or natural selection. This is determined by heritability. Heritability is the proportion of the total phenotypic variance that is due to genetic causes. In other words, heritability is a statistic used to measure the degree of variation in a phenotypic trait that is due to genetic variation between individuals in that population (see Box 14.1). There are two kinds of heritability: broad sense and narrow sense. Broad-sense heritability is based on genotypic variance:

$$h_{\mathrm{B}}^2 = \frac{V_{\mathrm{G}}}{V_{\mathrm{P}}} \tag{14.9}$$

Box 14.1: History and Misconceptions of Heritability

Since Sewall Wright used h (for heredity) to denote the correlation between genotype and phenotype in his path coefficient mode, it has become standard to use the symbol h^2 for heritability. h^2 is the proportion of variation in the phenotype that is attributable to the path from genotype to phenotype. Ronald Fisher in 1918 explained the relationship between relative resemblance in terms of correlation and regression coefficients. He also gave example of percentage of the total variance in stature in humans that can be ascribed to genotypes and to 'essential genotypes'. Such percentages are nowadays called broad-sense and narrow-sense heritability. It is thought that J. L. Lush, an

(continued)

Box 14.1 (continued)
animal breeder, was the first to formally use the term 'heritability' in 1940 to describe the proportion of variation that is due to hereditary factors.

There is a misconception that heritability is the proportion of a phenotype that is passed on to the next generation. Genes are only passed on and not the phenotypes. However, narrow-sense heritability is the variation because of additive genetic effects. Half of these effects are passed on from each parent, but the actual half is unique to each offspring.

High heritability is caused by variation in genotypes. That means in a population, phenotype is the good predictor of a genotype. However, it does not mean that the phenotype is not determined by the genotype alone and because environment manipulates the phenotype.

A low heritability means that of all observed variation, a small proportion is caused by variation in genotypes. But in no way the additive genetic variance is small. This difference matters because the response to natural or artificial selection depends on the amount of genetic variation in the population. Many phenotypes relating to fitness in natural populations have a large amount of additive genetic variation relative to the mean.

There is a belief that heritability is informative about the nature of differences between groups. This misconception comes in two forms. The first misconception is that when the heritability is high, groups that differ greatly in the mean of the trait in question must do so because of genetic differences. The second misconception is that the observation of a shift in the mean of a character over time for a trait with high heritability is a paradox. This is due to Flynn effect, because for IQ, a large increase in the mean has been observed in numerous populations. Heritability should not be used to make predictions about changes in mean in the population over time. Also, predictions on the differences between groups based on heritability will be erroneous. This is because in each individual calculation, the heritability is defined for a particular population. Populations are to be dealt differently while calculating heritability. An example comes from the White males born in the United States. They were the tallest in the world in the mid-nineteenth century and about 9 cm taller than Dutch males. Towards the end of the twentieth century, although the height of males in the United States had increased, many European countries had overtaken them and Dutch males are now approximately 5 cm taller than white US males, a trend that is likely to be environmental rather than genetic in origin.

The probability of detecting a gene with large effect increases with heritability in many gene mapping experiments. This never indicates that there is a relationship between heritability and the number or size of genes affecting that trait.

Broad-sense heritability estimates how phenotypic variation is determined by genotypic variation. It includes dominance and epistatic variances and is most useful in clonal or highly self-pollinated species where genotypes are passed from parents to offspring in an intact fashion. Narrow-sense heritability is applicable to outbreeding species. It is calculated as proportion of total phenotypic variance that is determined by additive variance:

$$h_{\mathrm{N}}^2 = \frac{V_{\mathrm{A}}}{V_{\mathrm{P}}} \tag{14.10}$$

Since V_{E} is part of V_{P} (Eq. 14.4), heritability can differ among environments. This is more evident when the equation for heritability is rewritten with the components of V_{P}:

$$h^2 = \frac{V_{\mathrm{A}}}{V_{\mathrm{A}} + V_{\mathrm{D}} + V_{\mathrm{I}} + V_{\mathrm{E}}} \tag{14.11}$$

Therefore, as V_{E} increases, heritability decreases, because less of the phenotypic variance is additive genetic. $\mathrm{V_I}$ = epistatic variance.

For example, heritability of wing width in male *Drosophila melanogaster* was much greater under control ($h^2 = 0.69$) as compared to stressful conditions ($h^2 = 0.09$). This lower heritability was caused by a much greater environmental variance under stress ($V_{\mathrm{E}} = 9.2$). V_{E} was only 0.9 under control conditions. Since expression of genetic variance can be affected by the environment, the numerator of heritability can also be affected by the environment. Such an effect is called genotype-by-environment interaction (see Chap. 20).

Heritability has the following uses: (a) predicts the effectiveness of selection; (b) chooses breeding methods for effective selection; (c) gives leads on the response of various traits to selection pressure; (d) gives predictions on the performance under vivid intensity of selection; (e) assists in determination of selection index; and (f) works as a guide to estimate the proportion of variation that is due to genotypic or additive effects.

14.3.6 Estimating Additive Variance and Heritability

Additive genetic variance is responsible for creating resemblance among relatives compared to the resemblance among unrelated members of the population. Quantitative genetics uses this fact to separate V_{A} from non-additive variance and V_{E}. V_{G} cannot be separated from V_{E} through raising the organisms in a controlled laboratory environment (for eliminating V_{E}). This is because V_{E} is in the denominator of heritability (Eq. 14.11). One overestimates heritability in the field by reducing V_{E} in the lab. In fact, in the presence of GE interaction, the lab estimate of V_{A} does not accurately reflect V_{A} in the field.

Offspring-parent regression: For reasonably precise estimates of heritability using offspring-parent regression, 30 to 50 pairs of parents are usually necessary.

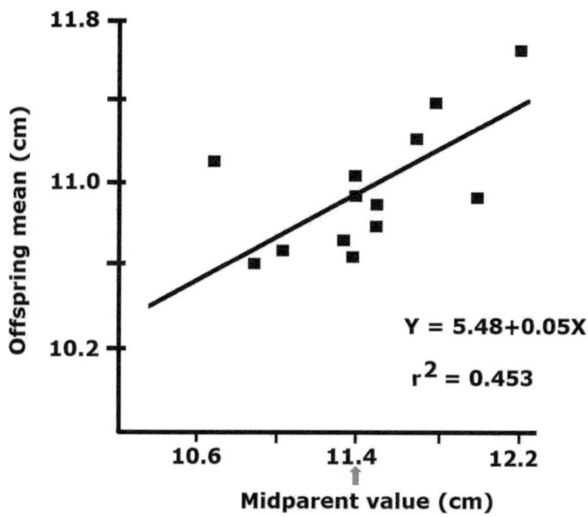

Fig. 14.8 Offspring-parent regression of awn length in wheat (hypothetical)

The procedure followed is to measure the trait(s) of interest on one or typically both the parents and raise their offspring. This offspring average is then regressed on the measurements of the male parents and female parents and/or the average of the two parents (called the mid-parent; Fig. 14.8). In this offspring-parent regression, each family is represented as one point. Therefore, each of the 13 points represents the average awn length of the two parents on the *x*-axis and the average awn length of all the offspring of those two parents on the *y*-axis. Linear regression gives the best-fitting straight line through these points, which produces an equation for the line:

$$Y = a + bX \tag{14.12}$$

The estimate of heritability is the slope of offspring on mid-parent. If the slope is steeper, then the offspring resemble more to their parents. Additive genetic variation that is responsible for higher proportion of phenotypic variance is passed on from parents to offspring. For example, if the slope is one, then for an increase of one unit of phenotypic value of the parents, you get an increase of one unit in the offspring. Or in other words, the phenotypic value of average offspring will be exactly the same value as that of average parent. A slope of one also means that the spread of points along the *x*-axis is the same as the spread of points along the *y*-axis. Since all phenotypic variance is additive genetic, such variance in the parents is passed onto the offspring.

14.4 Models for Combining Ability Analysis

Plant breeder considers two principal objects in most breeding programmes: (a) identification of genotypes for commercial release and (b) promising lines to be used as parents in future crosses. Lines for commercial release are selected based on multi-environment trial data. The selection of promising parents can be done following mating designs like biparental progenies (BIP), polycross, topcross, North Carolina (I, III, III), diallels (I, II, III, IV) and Line x tester design. Through following such designs, the genetic influences of a line can be partitioned into additive and non-additive components. Combining ability or productivity in crosses is vital in plant breeding programmes. It is the ability to combine desirable genes or traits during hybridization so that traits are transmitted to their progenies.

General combining ability (GCA) and specific combining ability (SCA) play a pivotal role in inbred line evaluation and population development in crop breeding. GCA is the average performance of a genotype in a series of hybrid combinations. But certain hybrid combinations perform better or poorer than expected on the basis of the average performance of parents. Such a phenomenon is called SCA. Parents exhibiting high average combining ability are believed to have good GCA. On the other hand, if their ability to combine well is confined to a particular cross, they are expected to be with high SCA. From a statistical angle, GCA is main effect and SCA is interaction effect. GCA is governed by additive and additive × additive gene interactions. SCA is regarded as an indication of loci with dominance variance (non-additive effects) and all the three types of epistatic interaction components if epistasis were present. They include additive × dominance and dominance × dominance interactions. Here, we will discuss mating designs used for combining ability analysis such as biparental progenies (BIP), polycross, topcross, North Carolina (I, II, III), diallels (I, II, III, IV) and Line X tester design.

14.4.1 Biparental Progenies (BIP)

This is the simplest mating design proposed by Comstock and Robinson in 1952. It is otherwise known as paired crossing design. A large number of plants (n) are selected at random and are crossed in pairs to produce 1/2 n full-sib families. Their progeny is tested and the observed variation partitioned by straightforward analysis of variance into between and within families. If r plants per family are evaluated, the variation within (w) and between (b) families may be analysed following details as given in Table 14.7. Even though simple, it is not sufficient enough to yield information to estimate all parameters required. Since the progeny are either full-sib or unrelated, only two statistics are available for estimating V_A, V_D, V_{EW} and V_{EC}. Dominance is assumed to be absent ($V_D = 0$), and individuals from the same family do not share the same environment ($V_{EW} = 0$), and there is a chance that the analysis will lead to an overestimation of the genetic component relative to the environmental component.

Table 14.7 Analysis of variance for BIP design

Source of variation	df	MS	EMS
Between families	$\{\frac{a}{b}n\} - 1$	MS_1	$\sigma^2 w + r\sigma^2 b$
Within families	$\frac{a}{b}n(r-1)$	MS_2	$\sigma^2 w$
Total	$\frac{a}{b}nr - 1$		

where *n* and *r* refer to the number of parents and plant samples within each cross respectively; $\sigma^2 b$ is the covariance of full sibs; ($\sigma^2 b$ = Cov FS = ½ V_A +*1/4* V_D+V_{EC} =*1/r* (MS_1 – Ms_2)) and σ^2w = {σ^2_G – Cov FS} + σ^2_{EV} = *1/2*V_A +*3/4* V_D + V_{Ew}= MS_2; *V_{Ew} is the environmental source of variation for variance within the crosses. When you assume that dominance is zero, then σ^2b = ½ V_A and σ^2w= ½ V_A+V_{Ew}*

Table 14.8 ANOVA table of polycross design with many replications

Source	df	MS	EMS	Variance components
Progenies	$g-1$	M_1	$\sigma^2_e + r\sigma^2_{prog}$	$\sigma^2_{prog} = \text{Cov(HS)} = \frac{1+F}{4}\sigma^2_A$
Blocks	$r-1$	M_2	–	–
error	$(g-1)\,(r-1)$	M_3	σ^2_e	$\sigma^2_e - \sigma^2$

14.4.2 Polycross

This is for inter-mating of a group of cultivars through natural crossing in isolated block. The term polycross was coined by Tysdal, Kiesselbach and Westover in 1942. Terminology was to indicate progeny from seed of a line that was subject to outcrossing with other selected lines growing in the same block. This design is suitable for obligate cross-pollinators like forage grasses and legumes, sugarcane and sweet potato. To ensure equal chance for each individual to cross with all other individuals, a proper design in the polycross block is critical. When less than ten genotypes are used, Latin square experimental design is suggested as most appropriate so as to ensure equal chance of random inter-mating in the polycross nursery. However, one has to ensure that synchronous flowering happens in all the individuals to have equal chances of cross-pollination. This design is used to produce synthetic cultivars, select families in recurrent breeding or evaluate the GCA of entries. Here, progenies from individual plants that are half-sib families are tested. The covariance within families is:

$$\text{Cov (HS)} = \frac{1 + F\sigma 2\,A}{4}$$

where F is the inbreeding coefficient of the genotypes being tested. ANOVA is in Table 14.8. The variance component σ^2_{prog} is an estimate of $\frac{1+F\sigma 2\,A}{4}$ when the parents are non-inbred, $F = 0$. A comparison of the coefficients with the corresponding coefficients in case of parent-offspring covariance indicates that the precision of the estimate of σ^2A is lower for the topcross or polycross than for the covariance between parents and offspring. Polycross is suitable for identifying mother plants

Table 14.9 Skeleton of ANOVA for half-sib family test by topcross

Source	df	MS	EMS	Variance components
Progenies	$g-1$	M_1	$\sigma^2_{e} + r\sigma^2_{prog}$	$\sigma^2_{prog} = \text{Cov(HS)} = \frac{1+F}{4}\sigma^2_A$
Blocks	$r-1$	M_2	–	–
error	$(g-1)\,(r-1)$	M_3	σ^2_{e}	$\sigma^2_{c} = \sigma^2$

with superior genotypes based on the performance of general combining ability of progeny.

14.4.3 Topcross

Topcross is crossing between a selection, a line, and a clone with a common pollen parent. Jenkins and Brunsen in 1932 proposed this method for testing inbred lines of maize. Later, this method was renamed as topcross by Tysdal and Grandall in 1948. Topcross progenies provide information about only GCA. Progenies from individual plants are tested that are half-sib families. The covariance within the families is:

$$\text{Cov (HS)} = \frac{1+F}{4}\sigma^2{}_A$$

where F is the inbreeding coefficient of the genotypes tested (Table 14.9).

The variance component $\sigma^2{}_{prog}$ is an estimate of $1 + F/4\ \sigma^2{}_A$ calculated from:

$$\sigma^2_{prog} = V(m_1) + (m_2)$$

Shortfalls of this design are as follows: (a) a single tester may not be sufficient enough to offer wide genetic background for testing the inbred stocks and (b) if the test inbreds are more, then the number of crosses become too many.

14.4.4 North Carolina Designs

Design I is widely used for both theoretical and practical plant breeding purposes (Fig. 14.9). This design is to estimate additive and dominance variances and for evaluation of full- and half-sib recurrent selection. It demands larger quantity of seed for replicated evaluation trials. So, this method is not of use in breeding species that are not capable of producing larger quantity of seed. However, NC design I can be used for both self- and cross-pollinated species that produces larger quantity of seeds. As a nested design, each member of a group of parents used as males is mated to a different group of parents. NC design I is a hierarchical design with non-common parents nested in common parents. The total variance is partitioned as given in Table 14.10.

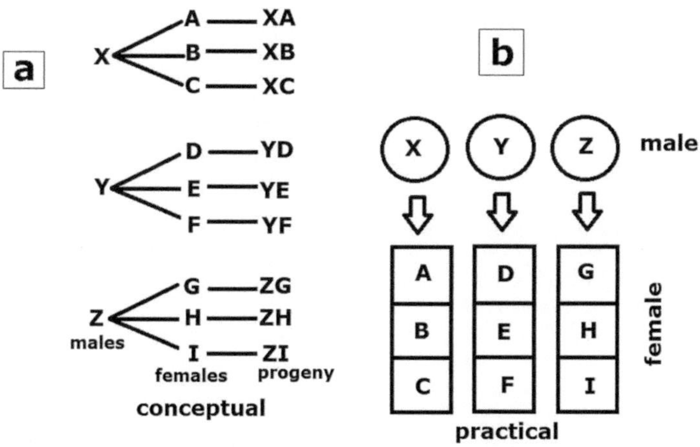

Fig. 14.9 North Carolina design I. (**a**) This design is a nested arrangement of genotypes for crossing in which no male is involved in more than one cross. (**b**) A practical layout of the field

Table 14.10 Partition of total variance

Source	df	MS	EMS
Males	$n-1$	MS_1	$\sigma^2_w + r\sigma^2_{mf} + rf\sigma^2_m$
Females	$n_1 (n_2-1)$	Ms_2	$\sigma^2_w + r\sigma^2_{mf}$
Within progenies	$n_1 n_2 (r-1)$	MS_3	σ^2_w

$\sigma^2_m = \{MS_1 - Ms_2\}/rn_2 = ¼\ V_A$
$r\sigma^2_{mf} = \{MS_1 - M_3\}/r = (1/4)\ V_A + (1/4)\ V_D$
$\sigma^2_w = MS_3 = (1/2)\ V_A + (3/4)\ V_D + E$

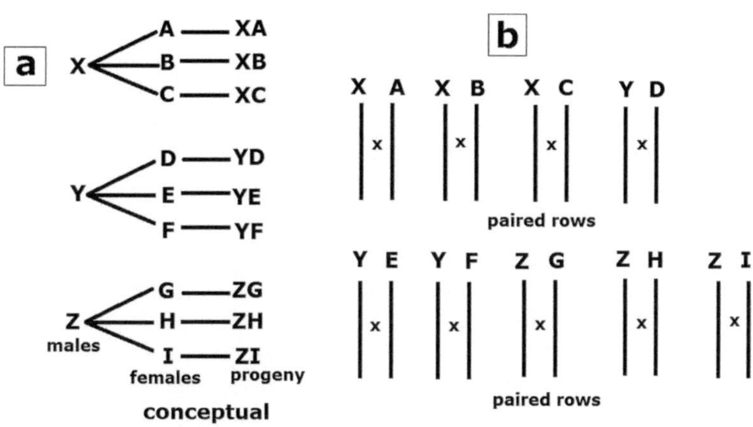

Fig. 14.10 North Carolina design II. (**a**) This is a factorial design. (**b**) Paired rows may be used in the nursery for factorial mating of plants

In NC design II, each member of a group of parents that are used as males is mated to each member of another group of parents used as females. Design II is similar to design I but is a factorial mating scheme (Fig. 14.10). It is used to evaluate

Table 14.11 ANOVA for GCA and SCA

Source	df	MS	EMS
Males	n_1-1	MS_1	$\sigma^2_w + r\sigma^2_{mf} + rn\sigma^2_m$
Females	n_2-1	MS_2	$\sigma^2_w + r\sigma^2_{mf} + rn_1\sigma^2_f$
Males × females	$(n_1-1)(n_2-1)$	MS_3	$\sigma^2_w + r\sigma^2_{mf}$
Within progenies	$n_1n_2(r-1)$	MS_4	σ^2_w

$\sigma^2_m = \{MS_1 - Ms_3\}/rn_2 = (¼)\ V_A$
$r\sigma^2_f = \{MS_2 - M_3\}/rn_1 = (1/4)\ V_A$
$r\sigma^2_{mf} = \{MS_3 - M_4\}/r = (1/4)\ V_D$
$\sigma^2_w = MS_4 = (1/2)\ V_A + (3/4)\ V_D + E$

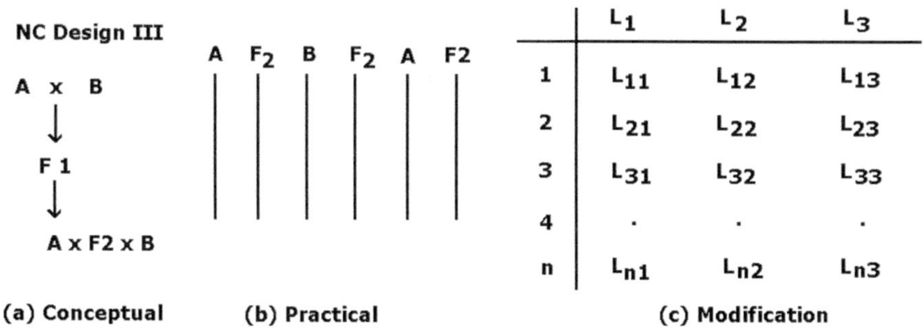

Fig. 14.11 North Carolina design III. (**a**) The conceptual form, (**b**) the practical layout, (**c**) the modifications

Table 14.12 Skeleton of NC III ANOVA

Source of variation	df	MS	Expected mean squares
Testers, p	1	M_4	$\sigma^2 + r\sigma^2_{np} + rmK^2p$
Males (F_2),m	$m-1$	M_3	$\sigma^2 + 2r\sigma^2_n$
Testers x parents	$m-1$	M_2	$\sigma^2 + r\sigma^2_{np}$
Within FS families/error	$(r-1)(2m-1)$	M_1	σ^2
Total	$2mr-1$		

inbred lines for combining ability. The design is successful in species with multiple flowers where each plant can be used repeatedly as both male and female. Crossing involving a single group of males to a single group of females is kept intact as a unit through blocking. It follows a two-way ANOVA where variation is partitioned into difference between males and females and their interactions. This design allows breeder to measure both GCA and SCA. ANOVA is in Table 14.11.

In NC design III, a random sample of F_2 plants is backcrossed to the two inbred lines from which the F_2 descended. NC III is most powerful among all three NC designs. Kearsey and Jinks in 1968 by adding a third tester (not just the two inbreds) made the design more powerful (Fig. 14.11). Their modified version is called triple test cross. NC III is capable of testing for non-allelic (epistatic) interactions which other designs are incapable of. It can also estimate additive and dominance variance (Table 14.12).

Table 14.13 Skeleton of ANOVA for method I diallel

				Expected mean squares	
Source	df	SS	SS	Model I	Model II
GCA	p–1	S_g	M_g	$\sigma 2 + 2p\left(\frac{1}{p-1}\right)\Sigma g^2 i$	$\sigma^2\left(\frac{2(p-1)}{p}\right)\sigma^2 g + 2p\sigma^2 g$
SCA	p(p–1)/2	S_s	M_s	$\sigma^2 + \left(\frac{2}{p(p-1)}\right)\Sigma\Sigma Sij^2$	$\sigma^2\left(\frac{2(p^2-p+1)}{p^2}\right)\sigma s^2$
Reciprocal eff.	p(p–1)/2	S_r	M_r	$\sigma^2 + 2\left(\frac{2}{p(p-1)}\right)\sum_{kj}\Sigma r_{\mathrm{ii}}^2$	$\sigma^2 + 2\sigma^2 r$
Error	m	S_e	M_e	σ^2	

14.4.5 Diallels

In diallel mating, the parental lines are crossed in all possible combinations (both direct and reciprocal crosses) to recognize parents as best or poor general combiners by GCA and the specific cross combinations by SCA. It may become impractical sometimes to conduct an experiment using a complete diallel cross design. Under such circumstances, a subset of crosses (partial diallel) can be used.

The most frequently used methods in the diallel analysis are Griffing's diallel procedures, where Griffing suggested four different diallel methods for use in plants: (a) Method 1 (full diallel), parents, F_1 and reciprocals; (b) Method 2 (half diallel), parents and F_1s; (c) Method 3, F_1s and reciprocals; and (d) Method 4, F_1s. These four methods have been widely used to study the patterns of inheritance of different traits in many crops. These diallel methods of Griffing are generally used for 1 year or one location trials (Table 14.13).

Estimates of variation are partitioned into sources due to GCA and SCA in all diallel types. The reciprocal crosses estimate the variation due to maternal effects, which are expected for some traits. A relatively larger GCA/SCA variance ratio demonstrates importance of additive genetic effects, and a lower ratio indicates predominance of dominance and/or epistatic gene effects. As per overall analysis, if mean squares for GCA and SCA are significant, then only GCA and SCA effects for individual lines are calculated.

14.5 Multiple Regression Analysis

Multiple regression analysis analyses the straight-line relationships among two or more variables. Multiple regression estimates the βs in the equation:

$$y_j = \beta_0 + \beta_1 x_{1j} + \beta_2 x2_j \ldots\ldots\ldots + \beta_p x_{pi} + \varepsilon_j$$

The xs are the *independent variables* (IVs), and y is the *dependent variable* (DV). The subscript j represents the observation (row) number. The βs are the unknown *regression coefficients*. Their estimates are represented by bs. Each β represents the original unknown (population) parameter, while b is an estimate of this β. The ε_j is the error (residual) of observation j.

Multiple regression analysis studies the relationship between a *dependent* (response) *variable* and p *independent variables* (predictors, regressors, IVs). The sample multiple regression equation is:

$$\hat{y}_j = b_0 + b_1x_{1j} + b_2x2_j \ldots\ldots\ldots + b_px_{pi}$$

If $p = 1$, the model is *simple linear regression.*

The intercept, b_0, is the point at which the regression plane intersects the y-axis. The b_is are the slopes of the regression plane in the direction of x_i. These coefficients are called the partial regression coefficients. Each partial regression coefficient represents the net effect the ith variable has on the dependent variable, holding the remaining xs in the equation constant.

A large part of a regression analysis consists of analysing the sample *residuals*, e_j, defined as:

$$e_j = y_i - \hat{y}_j$$

Once the βs have been estimated, various indices are studied to determine the reliability of these estimates. One of the most popular of these reliability indices is the correlation coefficient. The correlation coefficient ranges from −1 to 1. When the value is near zero, there is no linear relationship. As the correlation gets closer to plus or minus one, the relationship is stronger (see Chap. 7). The regression equation is only capable of measuring linear, or straight-line, relationships.

14.5.1 Regression Models

The basic regression model is:

$$y = \beta_0 + \beta_1x_1 + \beta_2x2_j \ldots\ldots\ldots + \beta_px_p + \varepsilon$$

This expression represents the relationship between the dependent variable (DV) and the independent variables (IVs) as a weighted average in which the regression coefficients (βs) are the weights. Unlike the usual weights in a weighted average, it is possible for the regression coefficients to be negative.

A fundamental assumption in this model is that the effect of each IV is additive. Now, no one really believes that the true relationship is actually additive. Rather, they believe that this model is a reasonable first approximation to the true model. To add validity to this approximation, you might consider this additive model to be a Taylor series expansion of the true model. However, this appeal to the Taylor series expansion usually ignores the "local neighbourhood" assumption. Another assumption is that the relationship of the DV with each IV is linear (straight line). Here again, no one really believes that the relationship is a straight line. However, this is a reasonable first approximation. In order to obtain better approximations, methods have been developed to allow regression models to approximate curvilinear relationships as well as non-additivity.

14.6 Stability Analysis

A successful new variety must have higher yield and other essential agronomic attributes. This superiority over other varieties needs to be proven under a wide range of environments. The differences in performance among genotypes in their yielding potential are due to genotype-environment (GE) interactions. While the genotypic composition of the variety remains stable, variations in yield are often termed "phenotypic stability" to refer to fluctuations in the phenotypic expression of yield. There are two concepts in stability analysis: static and dynamic. In static concept, a stable genotype exhibits an unchanged performance irrespective of any variation in the environment. This means its variance among environments is zero.

In the dynamic concept of stability, genotypic response to environmental conditions varies significantly. The estimated or predicted level agrees with the level of performance actually measured when defining stability. However, Becker in 1981 termed this type of stability as the agronomic concept that separates it from the biological concept of stability. Such an observation makes this concept equivalent to the static concept. Univariate parametric stability statistics measure uncertainty in the respective biometrical analysis. In addition, univariate non-parametric stability statistics have been proposed, which is based on rank orders of genotypes and which do not need any assumptions about distribution of observed values. Multivariate techniques have also been introduced for stability analysis.

To present stability analysis, a two-way linear model is assumed for convenience as follows:

$$X_{ij} = \mu + e_j + g_i + (\mathrm{ge})_{ij} + \varepsilon_{ij}$$

where X_{ij} is the observed phenotypic mean value of genotype i ($i = 1, \ldots,$ G) in environment j ($j = 1, \ldots,$ E) and μ, e_j, g_i, ge_{ij} and ε_{ij} represent the overall population mean, the effect of the jth environment, the effect of the ith genotype, the effect of the interaction between the ith genotype and the jth environment and the mean random error of the ith genotype in the jth environment, respectively, with $\bar{X}_i$, $\bar{X}_j$ and $\bar{X}$ denoting the marginal means of genotype i environment j and the overall mean respectively.

14.6.1 Static Concept

Early in 1917, Roemer measured phenotypic stability using variance of a genotype over a wide range of environments. The environmental variance was measured as:

$$s_{xi}^2 = \sum_i \frac{(X_{ij} - \bar{X}_i)^2}{E - 1}$$

This environmental variance of genotypes detects all deviations from the genotypic mean. The assessment of genotypes can be done though significance tests for comparing variances. As per this static concept, a desirable genotype will not react at all in changing environmental conditions. This would be useful for quality traits like resistance to diseases and traits like winter hardiness. While considering yield, breeder's objective shall be to select genotypes that are stable and high yielding. Stability evaluated through static concept shall be poor yielders. So, for studying yield stability, dynamic concept is recommended.

14.6.2 Dynamic Concept

Most genotypes react similarly to favourable and unfavourable environments when yield or other quantitative traits are considered.

Wricke in 1962 proposed ecovalence as stability measure to denote the GE interaction effects for each genotype, squared and summed across all environments. This may be estimated as follows:

$$W_i^2 = \sum_i (X_{ij} - \bar{X}_{i.} - \bar{X}_{.j} + \bar{X}_{..})^2$$

where X_{ij} is the mean performance of the ith genotype in the jth environment and X_i and $X_{.j}$ are the genotype and environment mean deviations, respectively. X is the overall mean. For this reason, genotypes with a low $W^2{}_i$ value have smaller deviations from the mean across environments and are therefore more stable. A genotype with $W^2{}_i = 0$ is considered stable.

Shukla in 1972 further proposed the variance component of each genotype across environments as another relevant measure of phenotypic stability. It measures stability rather than performance. According to Shukla's stability variance ($\sigma^2{}_i$) $G \times E$ sum of squares is partitioned into components, one corresponding to each genotype and estimated as:

$$\sigma_{\mathrm{i}}^2 = \frac{1}{(G-1)(G-2)(E-1)} \left\{ G(G-1) \sum_j (X_{ij} - \bar{X}_{i.} - \bar{X}_{.j} + \bar{X}_{..})^2 - \sum_i \sum_j (X_{ij} - \bar{X}_{i.} - \bar{X}_{.j} + \bar{X}_{..})^2 \right\}$$

where G is the number of genotypes, E is the number of environments, X_{ij} is the mean yield of the ith genotype in the jth environment, $X_{i\cdot}$ is the mean of the ith genotype in all environments, $X_{.j}$ is the mean of all genotypes in jth environments and $X_{..}$ is the overall mean.

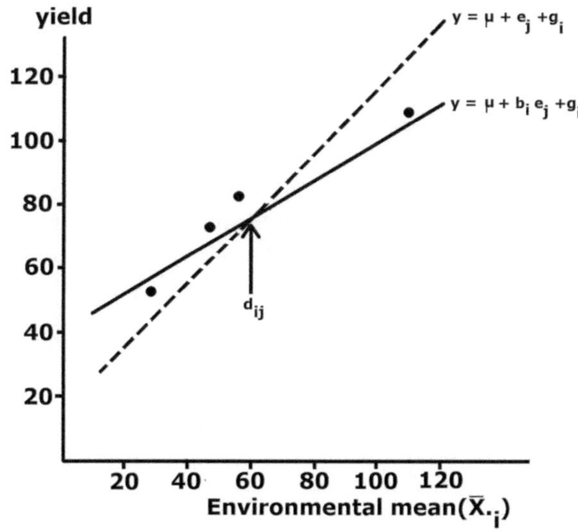

Fig. 14.12 Graphical representation of the regression approach

A genotype is identified as stable if the stability variance of a genotype was equal to the environmental variance ($\sigma^2_{i} = 0$). Significant σ^2_i value shows that a genotype's performance throughout the environments is unstable. Genotypes with a non-significant or negative σ^2_i would be regarded stable throughout the environments.

14.6.3 Regression Approaches

When we use usual biometrical model, the assumption is that no covariance exists between environments and of GE interactions. Comstock and Moll in 1963 stated that when we consider each genotype separately, this covariance differ from zero. The standardized description of this covariance is regression coefficient. The linear regression coefficient of genotypes in response to varying environments was calculated first by Stringfield and Salter in 1934. Yates and Cochran in 1938, Finlay and Wilkinson in 1963, Eberhart and Russell in 1966 and Perkins and Jinks in 1968 all further elaborated this technique.

The deviations between actual and predicted values normally decrease by the amount of covariance between environmental and GE interaction effects. The straight line $Y = \mu + b_i\, e_j + g_i$ fits the data better than $Y = \mu + e_j + g_i$ (Fig. 14.12). The effects of GE interaction may be expressed as:

$$(\mathrm{ge})_{ij} = \beta_i e_j + d_{ij}$$

where β_i is the linear regression coefficient for genotype i and d_{ij}, a deviation. Two slightly different regression techniques are proposed to explain part of GE interactions. Either GE interaction effects may be regressed on environmental effects (β_i of Perkins and Jinks), or X_{ij} values may be regressed on means of environments (b_i of Finlay and Wilkinson). Both these statics are equivalent.

$$b_i = 1 + \frac{\sum_i \left(X_{ij} - \bar{X}i. - \bar{X}_{.j} + \bar{X}..\right)\left(\bar{X}.j - \bar{X}..\right)}{\sum \left(\bar{X}.j - \bar{X}..\right)^2}$$

where X_{ij} is the performance of the *i*th genotype in the *j*th environment, $X_{i.}$ is the mean performance of the *i*th genotype and $X_{.j}$ is the mean performance of the *j*th environment. *X*.. is the overall mean. The regression coefficient (b_i) mainly indicates adaptation of a genotype to several environments. It also describes the linear response between environments which is also described by b_i.

As it could be seen in Fig. 14.12**,** a genotype with regression line above that of overall mean performance is regarded as stable. It can adapt to all environments. When the regression line crosses overall mean performance, the genotype is considered to be with specific adaptation to an environment. If its regression line is placed below that for the overall mean performance, the genotype is having an average performance. High-yielding genotypes will have larger values for b_i as they are particularly adapted to favourable environments. Such genotypes when cultivated in poor environments would exhibit a lesser than optimal performance. When cultivated under optimal environments, they could achieve maximum performance.

In addition to the coefficient of regression, the deviation mean squares ($s^2{}_{\mathrm{di}}$) describe the contribution of genotype i to GE interactions as explained by Eberhart and Russell:

$$s^2{}_{\mathrm{di}} = \frac{1}{E-2}\left[\sum_i \left(X_{ij} - \bar{X}_{i.} - \bar{X}_{.j} + \bar{X}..\right)^2 - (b_i - 1)\sum_i \left(\bar{X}j. - \bar{X}..\right)^2\right]$$

As per Eberhart and Russell model, genotypes are grouped based on their variance of the regression deviation. While a genotype with variance in regression deviation equal to zero is highly predictable, a genotype with regression deviation more than zero is less predictable. Both methods of Finlay and Wilkinson and Eberhart and Russell (b_{i} and $s^2{}_{\mathrm{di}}$) are used in different ways to assess the reaction of genotypes to varying environmental conditions. While the coefficient of regression b_i characterizes the specific response of genotypes to environmental effects and may be regarded as response parameter, $s^2{}_{\mathrm{di}}$ is strongly related to the remaining unpredictable part of variability of any genotype and therefore is considered as a stability parameter. Genotypes with zero b_{i} values would be stable according to the static concept. Genotypes with average performance have the value of one (Fig. 14.13).

For a more comprehensive account of QTL mapping, readers may refer to Chap. 23 on molecular breeding.

14.7 Genetic Architecture of Quantitative Traits

Quantitative traits exhibit continuous patterns of variation determined by the combined effect of genes and the environment. This genetic variation is the raw material for adaptation and evolution. Last hundred years witnessed continuous efforts to

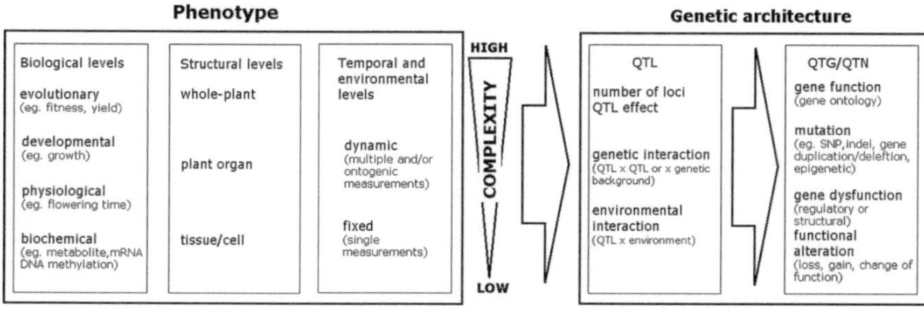

Fig. 14.13 Phenotypic levels and genetic architecture components of quantitative traits. Diagram depicts the different analytical phenotypic levels of quantitative traits depending on biological organization, plant structure or temporal and environmental scales. Given the phenotypic hierarchies of organisms at biological and structural (modular) levels, complex whole-plant traits that are affected by a large number of small effect loci (e.g. plant growth or yield) can be fractionated in several lower-level components (at molecular or cellular levels) with simpler genetic bases. In addition, quantitative traits can be analysed at different temporal and/or environmental levels differing in complexity. The architecture of quantitative traits is first determined at genetic (QTL) level and subsequently at the DNA (QTG/QTN) level. QTL, QTG and QTN: quantitative trait locus, gene and nucleotide, respectively. (Figure courtesy: Elsevier)

define genetic and molecular basis of quantitative traits. This is to determine and estimate the additive/dominance effect of genes, the pleiotropic relationships and their interactions with the environment. The genetic basis of quantitative traits ranges between simple oligogenic (few QTL with large effect) to complex polygenic (many QTL with small effect) governance. Quantitative trait genes and nucleotides (QTGs and QTNs, respectively) have been characterized in several plant species during the last decade. Model traits, such as flowering time, growth or plant defence, highlight a broader evolutionary perspective across plant kingdom.

Two-way statistical analyses detected digenic epistasis as a significant component of quantitative variation. Similarly, interactions between nuclear and chloroplast genes have impact on plant defence and growth traits. Epistasis among natural alleles has been addressed in detail. Differential pleiotropic effects on branching and flowering have been demonstrated in multiple segregating populations of *A. thaliana* with two-gene to four-gene interactions. Standard two-way tests may not work with while analysing transgenic genotypes. Understanding the molecular bases of such complex interactions will give light to the evolution of gene networks accounting for quantitative variation. In environments differentiated by biotic or abiotic factors, analysis of individual QTL/QTGs/QTNs can reveal genetic causes that determine phenotypic plasticity. A set of such genes for flowering time is known to interact with temperature and photoperiod suggesting importance of climatic adaptation. Such studies indicate considerable environmentally governed pleiotropy.

Currently the genetic architecture of quantitative traits are studies under three heads: (a) small effect QTL that are often masked by large effect loci but uncovered by multi-trait and multi-level analyses, (b) range of small effect and large effect mutations and (c) pleiotropy dependent on genetic and environmental interactions.

Plant adaptation by quantitative trait variations can be explained by comprehensive studies on nuclear, chloroplastic and mitochondrial networks.

Further Reading

Bazakos C et al (2017) New strategies and tools in quantitative genetics: how to go from the Phenotype to the Genotype. Annu Rev Plant Biol 68:435–455

Barrett RDH et al (2005) Experimental evolution of *Pseudomonas fluorescens* in simple and complex environments. Am Nat 166:470–480

Etterson JR (2004) Evolutionary potential of *Chamaecrista fasciculata* in relation to climate change. I. Clinical patterns of selection along an environmental gradient in the Great Plains. Evolution 58:1446–1458

Falconer DS, Mackay TCF (1966) Introduction to quantitative genetics. Longman, London

Fisher K et al (2004) Genetic and environmental sources of egg size variation in the butterfly *Bicyclus anynana*. Heredity 92:163–169

Gienapp P et al (2008) Climate change and evolution: disentangling environmental and genetic responses. Mol Ecol 17:167–178

Lynch M, Walsh B (1998) Genetics and analysis of quantitative traits. Sinauer Associates, Sunderland

Merilä J et al (2004) Variation in the degree and costs of adaptive phenotypic plasticity among *Rana temporaria* populations. J Evol Biol 17:1132–1140

Mousseau TA, Fox CW (eds) (1998) Maternal effects as adaptations. Oxford University Press, New York

Saastamoinen M (2008) Heritability of dispersal rate and other life history traits in the Glanville fritillary butterfly. Heredity 100:39–46

Via S, Hawthorne DJ (2005) Back to the future: genetic correlations, adaptation and speciation. Genetica 123:147–156

Waldmann P (2001) Additive and non-additive genetic architecture of two different-sized populations of *Scabiosa canescens*. Heredity 86:648–657

Charmantier A, Garant D (2005) Environmental quality and evolutionary potential: lessons from wild populations. Proc R Soc Biol Sci 272:1415–1425

Falconer DS, Mackay TFC (1996) *Introduction to quantitative genetics*. Longman, Harlow

Hill WG et al (2008) Data and theory point to mainly additive genetic variance for complex traits. PLoS Genet 4:e1000008

Macgregor S et al (2006) Bias, precision and heritability of self-reported and clinically measured height in Australian twins. Hum Genet 120:571–580

Visscher PM et al (2006) Assumption-free estimation of heritability from genome-wide identity-by-descent sharing between full siblings. Public Libr Sci Genet 2:e41

Visscher PM, Hill WG, Wray NR (2008) Heritability in the genomics era—concepts and misconceptions. Nat Rev Genet 9:255–266

Part IV

Specialized Breeding

Heterosis

15

Keywords

Historical aspects · Dominance hypothesis · Over-dominance hypothesis · Heterosis and epistasis · Epigenetic component to heterosis · Physiological basis · Molecular basis · Inbreeding depression · Prediction of heterosis · Phenotypic data-based prediction of heterosis · Molecular marker-based prediction of heterosis · Achievements by heterosis · Heterosis breeding in wheat, rice and maize

There are many definitions for heterosis:

> Heterosis or hybrid vigour is the superiority of a hybrid offspring over the average of both its genetically distinct parents
>
> or
>
> hybrid vigour is the increased vigour or other superior qualities arising from the crossbreeding of genetically different plants
>
> or
>
> Heterosis is superiority of F_1 in one or more characters over its better parental or mid parental value
>
> or
>
> heterosis is that progeny of diverse varieties exhibit greater biomass, speed of development, and fertility than both parents.
>
> or
>
> Heterosis is the phenomenon observed when the F_1 progeny of a cross exhibit improved or transgressive values traits over their parents.

P. M. Priyadarshan, *PLANT BREEDING: Classical to Modern*,
https://doi.org/10.1007/978-981-13-7095-3_15

15.1 Historical Aspects

Joseph Koelreuter (1733–1806) was the first to record heterosis in tobacco hybrids. G.H. Shull in 1914 proposed the term heterosis to replace the older term heterozygosis. Heterosis can also be defined as the tendency of a crossbred organism to have qualities superior to those of either parent (Fig. 15.1). Heterosis is opposite to inbreeding depression. When a hybrid inherits traits from its parents that makes them unfit for survival, the result is referred to as outbreeding depression. Heterosis is a multigenic complex trait and is the sum total of many physiological and phenotypic traits including magnitude and rate of vegetative growth, flowering time, yield and resistance to biotic and abiotic environmental stresses.

Heterosis can be either positive (yield, quality, disease resistance) or negative (plant height, maturity duration). It is predominant in cross-pollinated species than in self-pollinated. Heterosis confines to F_1 generation only, and due to segregation and recombination, it declines in subsequent generations. It is governed mostly by nuclear genes or by the interaction between nuclear and cytoplasmic genes. Heterosis can be either fully exploited in hybrids or partially exploited as in synthetic and composite varieties.

Performance of hybrids relative to their parents can be described as:

(a) Better-parent heterosis will have best values for the trait in question. Mid-parent heterosis is more than average of its two parents. Mid-parent has limited agronomic relevance.
(b) A phenotype can be either additive (not significantly different from the average of the two parents) or non-additive. Based on the phenotypes of two parents,

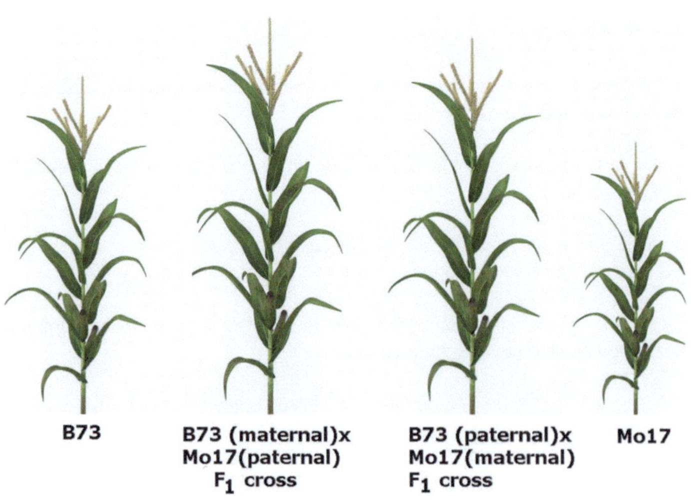

Fig. 15.1 Phenotypic manifestation of heterosis in maize. On the left is an average B73 genotype, and on the right is Mo17 phenotype. The central two are B73 (maternal) × Mo17 (paternal) F_1 cross and the reciprocal cross (diagrammatic)

non-additive phenotypes can be further classified. They are either partially dominant (differs from mid-parent but does not reach parental levels), or dominant (not significantly different from one parent), or over/under dominant (substantially outside the range of the parental phenotypes) (Fig. 15.2).

In agriculture, heterosis is a multibillion-dollar business. In various crops, yield enhancement through heterosis has been tremendous. In the USA, 44.8 million hectares (111 million acres) were required to produce 51 million metric tons of

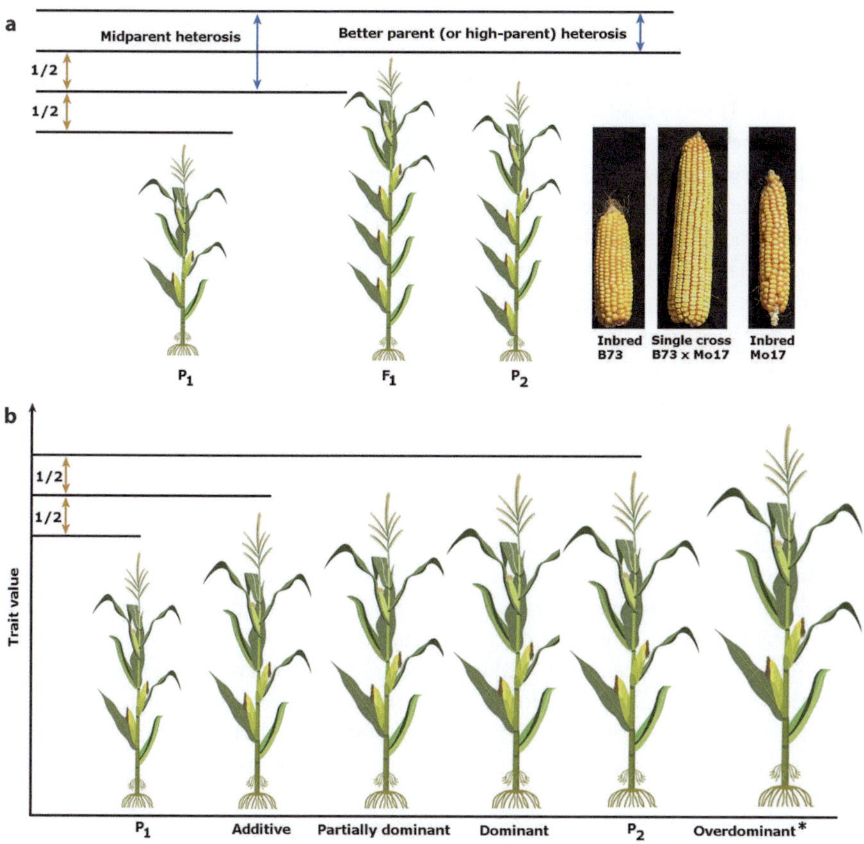

Fig. 15.2 Types of heterosis as judged through phenotypic level of a trait. (**a**) Better-parent heterosis describes the trait-specific performance of a hybrid relative to its parent having the best value for that trait. Mid-parent heterosis describes the performance of a hybrid relative to the average of its two parents. Although mid-parent heterosis is an intriguing biological phenomenon, it has limited agronomic relevance. (**b**) The phenotypic level for any trait in a hybrid can be described using several terms. Any phenotype can be described as additive (not significantly different from the average of the two parents) or non-additive (*asterisks*). Quite often, terms like mid-parent, high/low parent-like, or above high parent/below low parent are used to describe molecular patterns in hybrids rather than the terms additive, dominant and overdominant (diagrammatic)

maize grain in 1932, with a mean yield of 1.66 metric tons/ha. In 1994, it took only 32 million ha to produce 280 million metric tons of grain, with a mean yield of 8.69 tons/ha. Again in the USA, in 1996, 21 vegetable crops occupied 1,576,494 ha (3.9 million acres), with a mean of 63% of the crop in hybrids. Without any increase in land use, heterosis saved around 220,337 ha of land per year, feeding 18% more people. At the International Rice Research Institute, Manila, the best rice hybrids yielded 17% more rice over the best inbred-rice varieties between 1986 and 1995.

In China, 15–20% yield increment was achieved in hybrid rice varieties with heterosis. Hybrid rice are planted in 17 million hectares that comprises 58% of the total national rice area. This success in China has encouraged others like India, Vietnam, the Philippines, Indonesia and Bangladesh to follow popularizing hybrid rice technology since the 1990s. China derived "super hybrid rice" that yields more than 13 tons/ha, and their national average rice grain production increased from 6.21 t/ha in 1996 to 6.89 t/ha in 2015.

Maize yields increased by nearly 2% a year through popularizing heterotic F_1 hybrids during 1930–1940 in the USA. Simultaneously, improved use of farm machinery and fertilizers was augmented. Also, adoption of systems like double haploids to achieve inbred lines in a speedy way compared to conventional methods was made. The fact that farmers were willing to purchase F_1 hybrids each year from breeding companies also augmented research on heterosis.

15.2 Types of Heterosis

Heterosis is seldom called euheterosis or true heterosis, mutational heterosis, balanced heterosis and pseudo-heterosis or luxuriance. If types of estimation are considered, heterosis can be average or relative heterosis, heterobeltiosis, useful or standard or economic heterosis. Mutational heterosis is the simplest of all types. All non-lethal, dominant and adaptively superior alleles eliminate recessive and unfavourable alleles. This is termed as mutational heterosis. Balanced heterosis is with gene combinations more adaptive to environmental conditions. Pseudo-heterosis or luxuriance is superiority over parents in vegetative growth, but not in yield and adaptation. Such progenies are sterile or with low fertility. When heterosis is estimated over mid-parental value (i.e. average of two parents), average heterosis is:

$$= [(F_1 - MP)/MP] \times 100$$

where F_1 = value of F_1 and MP = mean value of two parents.

Heterobeltiosis is a performance over the better parent.

$$= [F_1 - BP]$$

where F_1 = value of F_1 and BP = value of better parent.

If heterosis is estimated over standard commercial hybrid, it is standard heterosis or useful or economic heterosis.

$$= [(F_1 - SH)/SH] \times 100.$$

Various causes of heterosis can be listed as genetic basis (dominance hypothesis, overdominance hypothesis, epistasis), physiological basis, cytoplasmic basis and biochemical basis.

15.2.1 Dominance Hypothesis

The dominance hypothesis was proposed by Davenport in 1908 and also by Bruce, Keeble and Pellew in 1910. As the most widely accepted hypothesis, it postulates that heterosis is the result of the superiority of dominant alleles, when recessive alleles are deleterious. Deleterious recessive genes are hidden, and the hybrid exhibits heterosis. Both the parents differ for dominant genes. Imagine genetic constitution of parents as AABBccdd and aabbCCDD. Heterosis will be proportional to the number of dominant genes contributed by each parent.

$$\begin{array}{ccccc} \text{AABBccdd} & \times & \text{aabbCCDD} & \rightarrow & \text{AaBbCcDd} \\ \text{Parent 1} & & \text{Parent 2} & & \text{Hybrid} \end{array}$$

The dominance model postulates due to complementation by the superior parent alleles on slightly deleterious alleles of other parent line, the F_1 generation display heterotic characteristics. This can lead to F_1 offspring that exceed the trait values observed in either parent. If slightly deleterious alleles ("a" and "b") are present in the genomes of parental lines P_1 and P_2, which have genotypes aa,BB and AA,bb, respectively (Fig. 15.3a), on hybridization, the F_1 offspring will be heterozygous at both loci, i.e. genotype Aa, Bb. The deleterious alleles at both loci can thus be complemented, leading to increased fitness or enhanced values of other traits observed. Due to independent segregation, the heterotic of F_1 progeny is not stably inherited in subsequent generations.

15.2.2 Overdominance Hypothesis

The overdominance hypothesis was independently developed by East and Shull in 1908 and supported by Hull in 1945. According to this hypothesis, due to complementation between divergent alleles, superiority of heterozygote over parents is achieved. East in 1936 further explained that a series of alleles a1, a2, a3, a4, etc. with gradual increment in divergence results in heterosis. Higher will be heterosis with more divergent alleles. A combination of a1, a4 will be with higher heterosis compared to other combinations. Also, synergistic allelic interaction at specific heterozygous loci will be superior. In Fig. 15.3b, $*$B is an allele variant of B (irrespective of dominance in this case). F_1 hybrids inherit both alleles and act synergistically to cause a heterotic effect. If $*$B is not inherited, the F_1 progeny exhibit no heterotic effect.

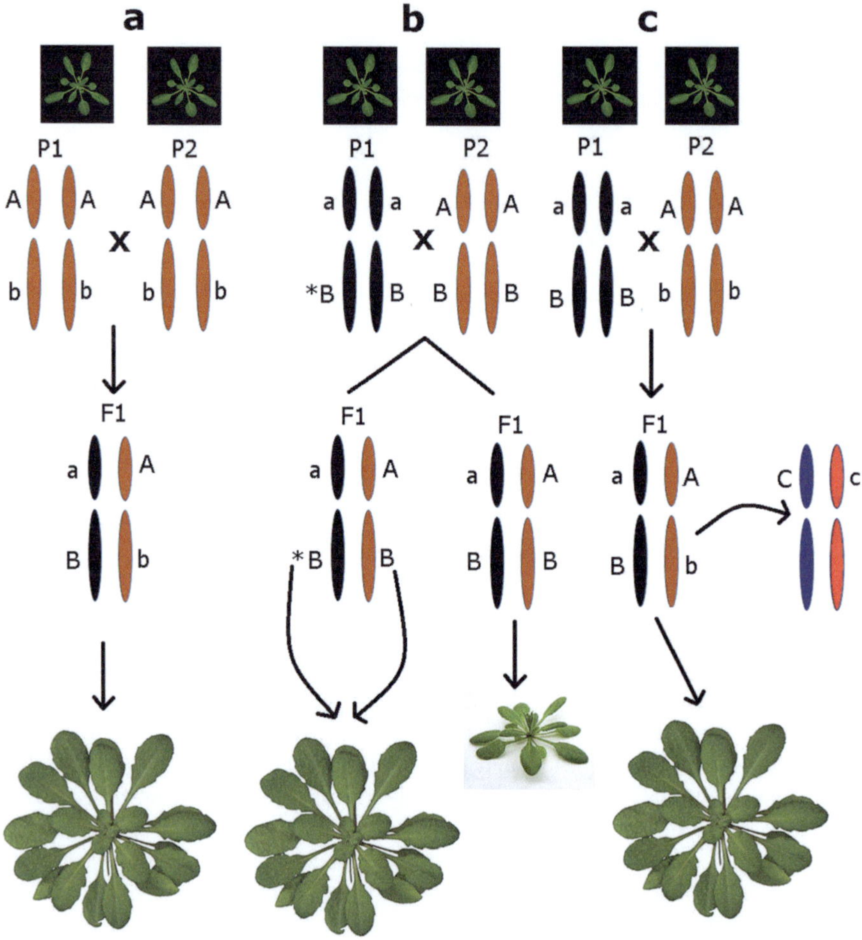

Fig. 15.3 Schematic representation of genetic models for explaining heterosis in *Arabidopsis thaliana*. (**a**) Dominance model; (**b**) overdominance model; (**c**) epistasis (Courtesy: Springer International)

15.2.3 Heterosis and Epistasis

Epistasis refers to interaction between alleles of two or more different loci. Otherwise known as non-allelic interaction, it involves dominance effects (dominance × dominance) as seen in cotton and maize. Epistasis can be detected or estimated by various biometrical models. Many heterotic epistatic relationships could in principle occur in F_1 hybrids when one allele is complemented and its gene product affects the function of one or more products of other genes. The gene product of dominant allele "A" has an epistatic interaction with the gene product of "C", in an unlinked locus (Fig. 15.3c). This interaction can cause heterotic effects in the F_1. An allele having an epistatic relationship with the allele of another locus *in trans* can mimic an overdominant heterotic QTL. The molecular basis of heterosis is expected to be complex and multigenic. It must also be reminded that any single mechanism cannot explain heterosis.

15.2.4 Epigenetic Component to Heterosis

Though aforesaid models are acceptable, a comprehensive understanding of heterosis is not available. Every aspect of heterosis cannot be fully explained by the sum total of all genetic interactions in a hybrid F_1 genome. Whether non-genetic mechanisms governing heterosis can exist is the question. Epigenetic effects following non-Mendelian inheritance can regulate heterosis. Due to differential modification of the epigenetic state, same genotype can display diverse phenotypes. There are epialleles at loci with identical DNA sequences but display vivid epigenetic states that can influence a variety of phenotypes. This is a deviation from Mendelian inheritance. DNA methylation, histone modifications and chromatin remodelling and the RNAi pathway (including RNA-directed DNA methylation, RdDM) are some of the most studied epigenetic mechanisms. Such mechanisms can epigenetically modify DNA sequences. Epigenetic variation can cause gene expression to spatio-temporally change throughout the development of an organism and during gametogenesis and sexual reproduction. Such epigenetic changes are briefly explained here (Box 15.1).

Box 15.1 Epigenetic Changes in Hybrids

Molecular properties are common among the different hybrid systems even though the basis of heterosis may differ in different crops. Increased leaf area results in a greater total chlorophyll content and a greater production of photosynthate. This can lead to greater biomass and seed yield. In rice, genes involved in photosynthesis are differentially expressed presuming that supply of photosynthate is critical. A small number of genes could be generating hybrid vigour exclusively since the vigour gets reduced over generations.

An epigenetic distance between parents is provided by DNA methylation. This is provided by the interaction between DNA methylation and the gene activities responsible for hybrid vigour. Mutations in these genes could provide direct evidence for the role of epigenetics in hybrid vigour. This is seen in *Arabidopsis,* maize and rice. Genes with altered expression include loci involved in responses to hormones and to biotic and abiotic stress.

DNA methylation also interacts with covalent modifications of the histone octamers that "pack" the DNA into nucleosomes and then to chromatin. This modification leads to covalent change in histone proteins, usually on their N-terminal tails. Such a change causes nucleosome rearrangement, chromatin

(continued)

Box 15.1 (continued)
remodelling and altered transcriptional potential. The overexpression or knocking out histone deacetylase genes can lead to non-additive gene expression in hybrids at some loci, which could in principle lead to overdominance for a trait controlled by the locus. It is likely that heterosis could be associated with alterations of epigenetic histone modifications. Small RNAs (sRNAs) can also govern regulation of heterosis. Prominent among such RNAs are microRNAs (miRNA) and small interfering RNAs (siRNAs). DNA methylation associated with 24-nucleotide small interfering RNAs exhibit transallelic effects in hybrids. Some of the transmethylation changes are inherited, and some affect gene expression. sRNA levels show substantial variation between parental inbred lines and their F_1 hybrid or allopolyploid offspring in several taxa. These sRNAs can work through RNA-directed DNA methylation (RdDM). During RdDM, double-stranded RNAs (dsRNAs) are modified into 21–24 nucleotide small interfering RNAs (siRNAs) that regulate methylation of homologous DNA loci.

DNA Methylation This is an epigenetic mechanism that governs gene expression. This epigenetic signalling can fix genes in the “off” position. DNA is a combination of four nucleotides: cytosine, guanine, thymine and adenine. Addition of a methyl (CH_3) group to the fifth carbon atom of a cytosine ring makes DNA methylation. This conversion of cytosine to 5-methylcytosine is catalysed by DNA methyltransferases (DNMTs). These modified cytosine residues usually lie next to a guanine base (CpG methylation), and the result is two methylated cytosines positioned diagonally to each other on opposite strands of DNA. Many studies have indicated that cytosine methylation (mC) may be involved in heterotic expression. In maize, mC patterns differ in heterotic F_1 in relation to their parents. In rice, mC patterns in inbred lines result in transcript level changes. Such changes are with differentially methylated regions (DMRs) in the F_1 hybrids.

Heterosis and Histone Modifications DNA is packed into nucleosomes and then to chromatin with the aid of histone octamers. The covalent modification of histone proteins, usually on their N-terminal tails, causes nucleosome rearrangement. Such nucleosome rearrangement causes chromatin remodelling and altered transcriptional potential. There is a possible link between histone modifications and heterosis. In *A. thaliana*, altered histone modifications regulated the genes involved in the circadian clock that underwent transcriptional changes in both diploid and allotetraploid F_1 hybrids. Starch biosynthesis and growth rate are governed by circadian clock. When the internal circadian rhythm matches with that of the environment, such plants are seen to be more vigorous than plants that do not have such a matching.

sRNAs and Heterosis Epigenetic control may also involve small RNA molecules (of 20–27 nucleotide long). These are non-coding RNAs. Such sRNAs can induce immune system to counteract against deleterious foreign viral RNA or transposons.

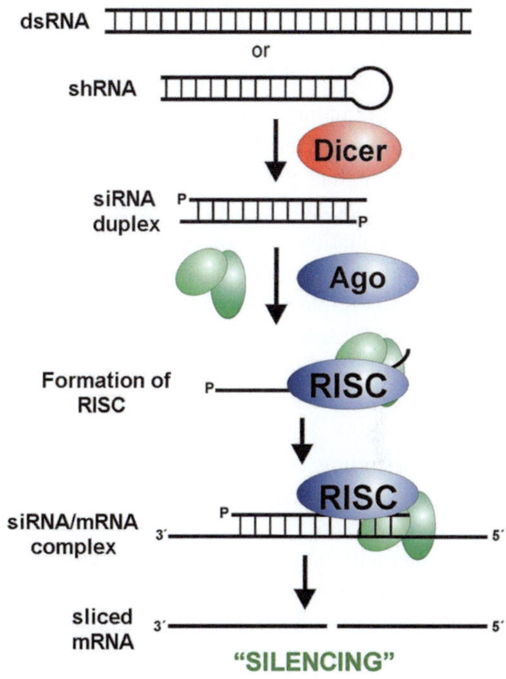

Fig. 15.4 Major steps of siRNA biogenesis and siRNA-mediated gene silencing

Such mechanisms involve transcriptional gene silencing (TGS) and post-transcriptional gene silencing (PTGS). There are two major classes of sRNAs: microRNAs (miRNA) and small interfering RNAs (siRNAs) (Fig. 15.4). miRNA precursors are transcribed from MIR genes (microRNA genes) by RNA POLYMERASE II (RNA Pol II) and are then cleaved ("diced") to a length of 20–27 nucleotide long DICER-LIKE 1 (DCL1). The mature miRNAs are then loaded into the RNA Induced Silencing Complex (RISC), accompanied by the ARGONAUTE 1 (AGO1) endonuclease. The loaded complex is then guided to messenger RNAs with sequence similarity to the mature miRNAs in order to cleave the mRNA transcripts and/or inhibit translation. sRNA-mediated pathways might be necessary for heterosis. HUA ENHANCER 1 (HEN1) is an *A. thaliana* methyltransferase that methylates mature sRNAs of both siRNA and miRNA classes to increase their stability. This indicates that the association between sRNAs and some heterotic traits is important in governing heterosis.

15.3 Physiological Basis

Heterosis is expressed as various metabolic and physiological traits. Physiological explanations ranging from hybrid enzymes to energy efficiency have been put forth to explain heterosis. The typical and general heterotic plant phenotype is large in size (i.e. "hybrid vigour"), as compared to its parents or common open-pollinated varieties. This greater size shall be due to greater biomass achieved during the growth duration as the parent materials. Physiological or molecular logic and evidence must first explain this large phenotype. Explanations like "hybrid enzymes", "mitochondrial metabolism", "metabolic flux" or "metabolic balance" will remain premature without link between physiological or molecular logic and the larger phenotype. Heterotic large phenotype is attained mainly via a greater cell number rather than greater cell size. A greater rate of cell division is set in early embryo development. This is followed by a compounding effect in cell division and organ differentiation, towards a luxuriant plant.

A partial explanation of heterosis can be increased to assimilate partitioning. Increased partitioning can also lead to greater grain number. Photosynthesis and the availability of a carbohydrate pool must be considered as crucial in this respect. The central role of the sink-source relationship in regulating the grain yield of crop plants is conspicuous. The sink regulates source activity by signals which is not well understood. Sink demand can even kill the source. Source and sink automatically adjusts depending upon the demand for assimilates. Breeders look forward in deriving genotypes with very large sink that are expected to yield more. For this, one has to increase the source effectiveness. A classical case is the uniculm Gigas wheat lines with large spike carrying a large number of florets. However, yield per unit area of the Gigas genotype was lower than that of standard wheat due to floret abortion. The reason for abortion is the normal rate of photosynthesis in the Gigas plant. This indicates that it is unreasonable to say that a genotype with large sink can realize higher yield without increased photosynthesis over its parents. So, the current knowledge of photosynthesis must be revamped to explain a large hybrid sink.

15.4 Molecular Basis

Studies on transcriptomes, metabolomes and proteomes have provided some details on the molecular basis of heterosis. Transcriptomes (a set of all RNA molecules, including mRNA, rRNA, tRNA and other non-coding RNA produced in one or a population of cells), proteomes (entire set of proteins expressed at a time) and metabolomes (collection of all metabolites) have provided molecular insights into regulatory networks of hybrid vigour (Fig. 15.5a).

Transcriptomic changes are complex but the trends are:

(a) Additive and non-additive gene expression changes are more correlated with genetic distance than with genome dosage, and non-additive gene expression is more common in interspecific hybrids than in intraspecific hybrids. A well-known example of non-additive gene expression is nucleolar dominance, which refers to epigenetic silencing of the ribosomal RNA genes from one parent in interspecific hybrids of plants. For example, in *A. thaliana* rRNA genes are silenced in *Arabidopsis* allopolyploids that are formed in a cross between *A. thaliana* and *A. arenosa*. rRNA genes from one parent are silenced by mechanisms including DNA methylation, histone modifications and small RNAs. *A. arenosa* genes are dominant over *A. thaliana* genes in *Arabidopsis*

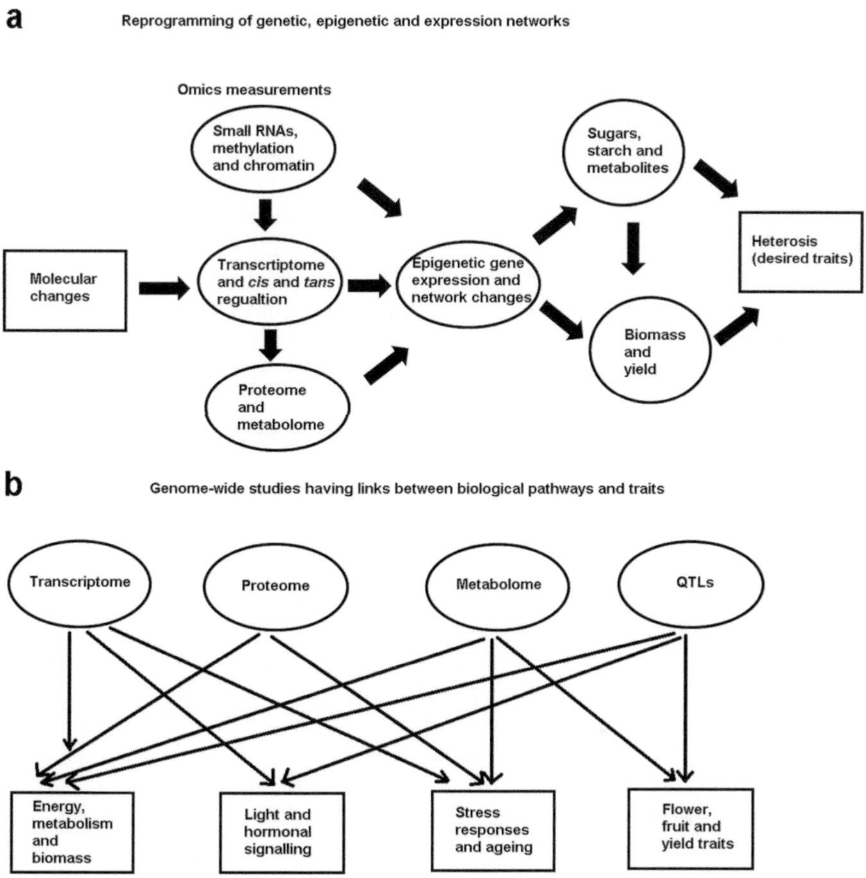

Fig. 15.5 Molecular changes at epigenetic, genomic, proteomic and metabolic levels lead to heterosis traits. (**a**) Changes in the epigenome (including chromatin modifications and DNA methylation), small RNAs, the transcriptome and the proteome result in epigenetic gene expression and regulatory network changes, some of which are associated with quantitative trait loci (QTLs). These changes can cause heterosis in traits such as metabolism, growth and yield. Note that vigour components of physiology and metabolism (e.g. sugar and starch levels) are connected to heterosis in biomass and yield. (**b**) Genome-wide studies of transcriptomes, proteomes, metabolomes and QTLs identify collective changes in biological pathways and phenotypic traits in hybrids, which include energy, metabolism and biomass, light and hormonal signalling, stress responses and ageing, and flowering, fruiting and yield. The arrows represent connections that have been shown in studies to date, and the numbers indicate references to these studies. Many of these pathways and traits are under the control of "master regulators" (such as the circadian clock). These traits are also interconnected and may affect one another and exert feedback effects on the regulators (Courtesy: Nature Reviews Genetics)

allotetraploids. Such expression of dominance is also found in cotton allotetraploids.

(b) Gene expression changes correspond to alterations of biological networks (Fig. 15.5b) In *Arabidopsis* allotetraploids, non-additively expressed genes are enriched in the gene ontology classes of energy, metabolism, stress response and phytohormone signalling. In *A. thaliana* hybrids, gene expression changes also correlate with an increased capacity for photosynthesis. These findings are consistent with increased photosynthetic and metabolic activities that correlate with heterosis in *Arabidopsis* hybrids and allopolyploids.

(c) Genome-wide changes in gene expression in interspecific hybrids and allopolyploids can result from *cis*- and *trans*-regulatory divergence between hybridizing species (*cis*-regulatory genes are typically located on the same DNA strand opposed to *trans*, which refers to the effects on genes not located on the same strand or farther away, such as transcription factors). In *Arabidopsis* F_1 allotetraploids and their progenitors, overall there are more genes that have *cis*-regulatory changes than *trans*-regulatory changes. Some genes with enhancing *cis* and *trans* changes are associated with stress responses, thus promoting growth and adaptation; some other genes with compensating *cis* and *trans* changes are related to biosynthetic and metabolic processes, which maintain growth, developmental stability and vigour in allotetraploids.

Proteomics Additive and non-additive proteomic patterns have been found in the embryos, in the roots and in the nuclei and mitochondria of the ear of maize hybrids, in mature embryos of rice hybrids and in the leaves of *Arabidopsis* autopolyploids and allopolyploids. Isoforms or allelic variants exist in maize hybrids with high or low levels of heterosis than in their parents, thus suggesting transgressive effects. Some of these isoforms are known to respond to stresses. Although transcriptomic and proteomic studies both reveal non-additive changes, non-additively accumulated proteins or peptides do not necessarily match non-additively expressed genes. This suggests that there are changes in post-transcriptional and translational regulation in hybrids and polyploids.

Metabolomics Biomass heterosis is correlated with increased levels of metabolic activity, which depends on the maternal parent. In recombinant inbred lines (RILs), biomass is significantly correlated with specific combination of metabolites. In tomato, 14–20 metabolites were sufficient to predict freezing tolerance among different F1 hybrids. Genotypes that contain genetic loci from wild species, approximately 50% of all metabolic loci tested were associated with QTLs for whole-plant yield traits. In maize, for 26 metabolites in leaves, single-nucleotide polymorphisms (SNPs) have been identified that explain 32% of genetic variation in these metabolites among inbred lines. A limited number of particular metabolites provide useful “biomarkers” for the prediction of heterosis.

15.5 Inbreeding Depression

Inbreeding depression is the reduction of fitness in the progeny of related individuals compared to the progeny of unrelated individuals. The conceptual opposite of heterosis is inbreeding depression (see Table 15.1 for differences between heterosis

Table 15.1 Differences between heterosis and inbreeding depression

Nature of the difference	Inbreeding depression	Heterosis
Genetic variation	Must be present within the species or population	Can appear in F_1 individuals between genetically uniform populations or strains
Effect of genetic drift in small populations	Lowers inbreeding depression due to mildly deleterious mutations in small populations	Heterosis due to mildly deleterious mutations is highest for small populations or highly inbreeding populations
Likelihood of outbreeding depression and its consequences	Unlikely without strong isolation or local adaptation, and therefore unlikely to affect the magnitude of inbreeding depression within a population	May lower the magnitude of heterosis
Complementary interactions between different deleterious recessive mutations	Can cause inbreeding depression if loci are linked, so homozygosity for the genome region lowers fitness (pseudo-overdominance)	Can cause heterosis even if loci are unlinked and even if heterozygous alleles at the loci cause phenotypes that are between those of the homozygotes

and inbreeding depression). Prolonged inbreeding in cross-pollinated species like maize leads to progressive accumulation of deleterious traits such as slow growth, low fertility and diseases. The molecular basis of this mechanism is not clear. A widespread genetic hypothesis is that inbreeding opens deleterious recessive mutations. This contention is questionable because most recessive alleles are not detrimental. Heterosis is also likely to be governed by non-defective alleles. The expression and/or function of heterozygous, non-defective alleles lead to advantageous performance in hybrids relative to inbred individuals. If the genetic variation within a population is higher, it is less likely that the population could suffer from inbreeding depression. So, in molecular terms, inbreeding depression and heterosis are not absolutely opposites. They are also governed by genetic and epigenetic interactions of non-defective alleles. Since linked deleterious mutations and a single heterozygous locus cannot be distinguished, it is difficult to quantify the different genetic contributions to inbreeding depression or heterosis. The main genetic hypotheses for inbreeding depression fall into single locus and multilocus.

Single-locus hypothesis says that since homozygotes are rare except after inbreeding, recessive mutant alleles present at low frequencies in populations can contribute to inbreeding depression. This hypothesis (Fig. 15.6 top row) is often called "the dominance model". Heterozygotes for a loss-of-function often have the same level of function as wild-type homozygotes ("directional dominance"). Mildly deleterious mutations that are partially recessive lead to heterozygotes, and their fitness is only approximately 5–25% higher than the homozygote average. Such mutations in aggregate contribute to larger effects. Overdominant alleles (Fig. 15.6 bottom row) are maintained by balancing selection. Balancing selection also sometimes maintains chromosomal inversion polymorphisms and polymorphisms for other large genome regions with suppressed recombination. When homozygous,

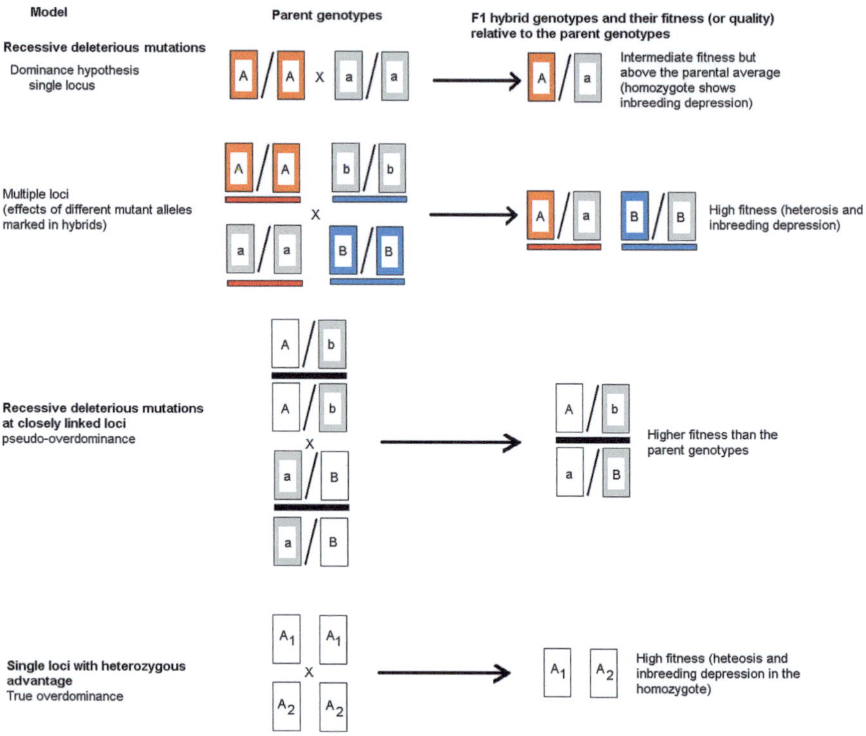

Fig. 15.6 Summary of the main genetic hypotheses for inbreeding depression. These hypotheses were developed by maize geneticists early in the twentieth century but have proved difficult to test (see text). The increased homozygosity of inbred individuals can lower fitness either because of deleterious mutations with recessive effects, which cause homozygotes to have lower survival or fertility (top and middle rows), or because loci exist with different alleles that result in the higher fitness of heterozygotes (overdominance, bottom row). For the dominance and pseudo-overdominance (mutational) hypotheses, the figure shows how the higher homozygote frequencies for recessive deleterious mutant alleles (indicated as *a* and *b*) among inbred individuals will cause lower fitness than in more heterozygous outbred individuals or hybrids. In the overdominance hypothesis, inbred individuals are less likely to be heterozygous for the two alleles (*A*1/*A*2) than outbred individuals or hybrids and therefore have lower fitness. (Courtesy: Nature Reviews Genetics)

such mutations are with lower fitness making the region overdominant. In some cases, polymorphic chromosomal rearrangements are responsible for inbreeding depression for male fertility. The recessive alleles with harmful effects on a trait may have beneficial effects on other traits. However, it is unlikely that dominant alleles always give higher fitness.

In two or more loci, pseudo-overdominance may govern the inbreeding depression and heterosis. Complementation happen between unlinked deleterious alleles in a hybrid, producing heterosis (Fig. 15.6 top row). Also, a genome region could contain two or more closely linked genes in repulsion phase (Fig. 15.6 middle row).

Even though two distinct loci are involved, homozygotes for the chromosomal region may lead to reduced performance thus ending with overdominant factors in QTL studies.

If many deleterious alleles are present in an outbred population with multiplicative and non-multiplicative interactions, homozygous alleles in a genotype will determine its fitness. Homozygosity acts multiplicatively towards fitness reducing effects and will occur when the traits are independently affected by mutations. Multiplicative effects result in a linear decline on a logarithmic scale. If mutations reduce fitness more than additively, synergism can occur. Completely additive alleles (no dominance) might not lead to inbreeding depression, but two or more such loci can cause heterosis. The multiplicative combination of component traits can influence yield.

15.6 Prediction of Heterosis

Over the years, several methods were employed to predict heterosis, such as per se performance of parental lines, mitochondrial complementation, combining ability and genetic diversity estimated from geographical origin, coefficient of parentage, multivariate analysis of morphological traits and isozyme and molecular marker analysis. Among these methods, mitochondrial complementation-based heterosis prediction is unpopular since the results were not reproducible. Hence, this method will not be discussed here. Apart from these methods, gene expression is being used in recent studies to predict heterosis.

15.6.1 Phenotypic Data-Based Prediction of Heterosis

Heterosis prediction can be done through per se performance of parental lines, combining ability and genetic diversity studies using the phenotypic data collected from field evaluation of genotypes. There are contrasting conclusions regarding the effectiveness of per se performance in the prediction of heterosis. Studies in maize and sugarcane concluded that there was no association between per se performance of parental lines and heterosis in F_1 hybrids. Same is the case with many other crops. Therefore, it can be concluded that heterosis prediction based on per se performance of parents may not be a reliable indicator of heterosis.

Identification of superior parental lines for developing heterotic hybrids was generally done by employing combining ability tests such as top-cross test, polycross test, single-cross test, diallel mating and line × tester analysis, though with variable levels of success. In general, selection of parental lines with high general combining ability (GCA) effects resulted in the development of heterotic hybrids in any crop. However, heterotic combinations could also be derived from parents exhibiting low GCA effects as noticed in rice and such combinations could not be derived from parents with high GCA. It is important to note that the strong relationship between the mean performance and GCA of inbred lines may be due to the presence of additive genetic variance. The non-additive genetic variance or specific combining ability (SCA) has to be given due consideration since it has a direct impact on heterosis. The extent of genetic diversity between the two parents has been proposed as a possible indicator for the prediction of heterosis. But the extent of correlation varied widely from one trait to another and from one data set to another. Due to lack of consistency in the prediction of heterosis based on genetic diversity and combining ability among parental lines through field evaluation, there is an immense need for the prediction of heterosis based on molecular marker polymorphism without field evaluation. There are heterotic groups that refer to genetically diverse groups of genotypes/parental lines, and crosses among them may result in heterotic hybrids.

15.6.2 Molecular Marker-Based Prediction of Heterosis

As discussed earlier, genetic divergence of parental lines is thought to be related to heterosis. Thus, biochemical and molecular marker-assayed genetic variation of the parental lines may potentially be useful for predicting heterosis. Prior to molecular markers, isozymes were the commonly used biochemical markers for the prediction of heterosis. Isozymes were considered to be unpopular for heterosis predictions since they could sample only a limited number of loci and it is unlikely that these loci have direct effect on the phenotypic expression of the targeted trait. The use of molecular markers led heterosis prediction into a new phase. High positive correlation of yield heterosis with genetic distance based on RAPD and SSR markers was reported for indica x indica and japonica x japonica crosses but not for indica x japonica crosses. Table 15.2 summarizes the studies conducted on the prediction of heterosis using molecular markers in different crops and their conclusions.

With the popularity of single-locus markers like microsatellite markers, several efforts were made in rice in assessing the utility of SSR markers for heterosis prediction based on the relationship between molecular diversity of parental lines. Prediction based on functional markers, especially the markers associated with genes controlling heterosis for yield traits, might be more powerful than that based on anonymous markers. Molecular markers would be useful for predicting hybrid performance only when a significant portion (>50%) of the selected markers is linked to QTL. Once such informative markers are identified, they should be tested in different populations of parental lines varying in their genetic background to ascertain their consistency in the prediction of heterosis. If successful, such prediction methods may lead to the selection of the limited number of parental combinations for synthesizing experimental hybrids for field evaluation for the identification of highly heterotic hybrids, thus increasing the efficiency of hybrid development.

Besides molecular marker data, the transcriptomic and metabolomic data also have the potential for the prediction of heterosis. One logic is that the differentially expressed genes (DEGs) are related to heterosis. Microarrays have been more popularly used to study such expression. Analysis of the differential expression of genes at different developmental stages of hybrids and their parental lines have

Table 15.2 Heterosis prediction in different crops using molecular markers

Crop	Marker type	Plant material	Conclusions
Rice	Pedigree record, quantitative traits and SSR	37 maintainers, 43 restorers and 34 hybrids	Prediction is difficult through SSR and pedigree-based diversity of complex traits
	RAPD and SSR	41 hybrids from a half diallel with 10 *japonica* cultivars	Proposed the role of "key" DNA markers in the prediction of heterosis
	SSR	13 CMS lines, 19 restorers and 151 hybrids	Prediction of heterosis based on effect-increasing loci was more effective
	SSR and EST-SSR	Nine CMS lines, 32 restorers and 20 hybrids	Prediction of heterosis is better using EST-SSRs
Maize	Morphological data and RAPD	28 open-pollinated varieties in a diallel scheme and 378 hybrids	Low and positive correlation was observed between RAPD-based GD and SCA for yield
	RAPD	13 inbred lines and 78 hybrids	RAPDs are not suitable for the prediction of yield performance of hybrids
	AFLP and SSR	18 S_3 inbred lines in a partial diallel mating design	Single-cross performance can be predicted through AFLP-based GD
	SSR	15 elite inbred lines and 105 hybrids	Prediction of yield heterosis using SSR markers is difficult
Wheat	RFLP and RAPD	Eight-parent diallel cross and top cross (4 males × 25 females)	A weak correlation was observed between parental diversity and hybrid performance
	RAPD	10 CMS lines, 10 restorers and 41 hybrids	No significant correlation was observed between RAPD marker-based GD with heterosis
	RAPD	18 parental lines and 76 F_2 hybrids	GDs between parents can be a potential predictor of hybrid performance for selected traits
Cotton	RAPD and SSR	Three CMS lines, 10 restorers and 22 hybrids	The relationship between SSR marker heterozygosity and hybrid performance can be used to predict the fibre length during interspecific hybrid cotton breeding

Abbreviations: *RFLP* restriction fragment length polymorphism, *RAPD* randomly amplified polymorphic DNA, *AFLP* amplified fragment length polymorphism, *SSR* simple sequence repeats, *EST* expressed sequence tag, *GD* genetic distance, *QTL* quantitative trait loci, *SCA* specific combining ability

proven to be a useful methodology to identify the genes associated with heterosis. In maize, it was concluded that the prediction of hybrid performance was more precise with transcriptome-based distances using selected markers than earlier prediction models involving DNA markers or the estimates of general combining ability. Recently, whole-genome prediction (WGP) was suggested to be a powerful complementary approach in hybrid breeding for highly polygenic traits with prediction accuracies in the range of 0.72–0.81 for SNPs and 0.60–0.80 for metabolites. Since gene expression-based approaches are expensive and demand sophisticated infrastructure, they may not be suitable for the routine screening of the large number of parental lines. So, there is an immense need for the development of easy, cheap, rapid and routinely usable assays that will help those involved in hybrid development to predict heterosis in different crops. So, the use of PCR-based markers targeting the sequence polymorphism responsible for the differential gene expression shall be a better for the prediction of heterosis.

15.7 Achievements by Heterosis

Heterosis was first exploited in rice. Some of the rice varieties developed with the use of heterosis in India are listed in Table 15.3. Agriculture got benefited by heterosis for over 100 years. Many crop and vegetable F_{1s} are cultivated over large areas. This has augmented agricultural practices and seed industry business. Given its economic importance and scientific interest, researchers have used quantitative genetics, physiology and molecular approaches in an effort to understand the basis of heterosis.

15.7.1 Heterosis Breeding in Wheat

The main goal of hybrid breeding in wheat is to systematically exploit heterosis. For this, grouping of lines into genetically divergent pools is a prerequisite to exploit heterosis. Because of intensive exchange of elite lines, divergent groups in wheat may not exist in a given environment. For making genetic diversity among pools, collection of elite lines from vivid target environments is a method that can be practised. However, this approach is complicated by the different requirements for vernalization, photoperiod, quality and frost tolerance. Heterosis in wheat can be explained as (a) the joint action of multiple loci with the favourable allele either partially or completely dominant, (b) overdominant gene action at many loci and (c) epistatic interactions between non-allelic genes. Several classical quantitative genetic experiments were undertaken to explain gene actions underlying heterosis. Since the parameters reflect the net contribution of gene effects at all loci, such studies are of limited use.

To elucidate the genetic basis of heterosis, two prominent experimental designs have been applied: North Carolina Design III (NC III) and the triple testcross design (TTC) (Fig. 15.7). In NC III, hybrids from a cross between two inbred lines are backcrossed to its parents. The TTC is an extension of NC III, where the segregating population is backcrossed to the F_1s. NC III enables the identification of loci contributing to heterosis. Contribution of a particular gene to heterosis is a function of its dominance and its cumulative effects with all other loci in the genome. NC III never enables partitioning of main and interaction components, but TTC allows estimation of interaction effects to an extent.

Table 15.3 Rice varieties derived through heterosis

Rice hybrids	Year of release	Duration (days)	Yield (t/ha)	Developed by	Recommended for the sates of
APHR 1	1994	130–135	7.14	APRRI, Maruteru (ANGRAU), Hyderabad	Andhra Pradesh
APHR 2	1994	120–125	7.52	APRRI, Maruteru (ANGRAU), Hyderabad	Andhra Pradesh
CNRH 3	1995	125–130	7.49	RRS, Chinsurah (W.B.)	West Bengal
DRRH 1	1996	125–130	7.30	DRR, Hyderabad	Andhra Pradesh
KRH 2	1996	130–135	7.40	VC Farm, Mandya, UAS, Bangalore	Bihar, Karnataka, Tamil Nadu, Tripura, Maharashtra, Haryana, Uttarakhand, Orrisa, West Bengal, Pondicherry, Rajasthan
PHB 71	1997	130–135	7.86	Pioneer Overseas Corporation, Hyderabad	Haryana, Uttar Pradesh, Tamil Nadu, Andhra Pradesh, Karnataka
ADTRH 1	1999	115–120	7.10	TNRRI, Aduthurai (TNAU)	Tamil Nadu
Sahyadri 3	2005	125–130	7.5	RARS, Karjat (BSKKV)	Maharashtra
HKRH-1	2006	139	9.41	RARS, Karnal (CCSHAU)	Haryana
Haryana Shankar Dhan-1 (HKRH-1)	2006	139	9.40	HAU, Haryana RARS, Kaul (CCS, HAU)	Haryana
JRH-4	2007	110–115	7.50	JNKVV, Jabalpur	Madhya Pradesh
JRH-5	2007	105–108	7.50	JNKVV, Jabalpur	Madhya Pradesh
Indira Sona	2007	120–125	7.0	IGKKV, Raipur	Chhattisgarh
JRH- 8	2008	105–110	7.50	JNKVV, Jabalpur	Madhya Pradesh
DRH - 775	2009	97	7.70	Methelix Life Sciences, Pvt. Ltd., Hyderabad	Bihar, Chhattisgarh, Jharkhand, Madhya Pradesh, Uttar Pradesh, Uttarakhand, West Bengal
27P31 (IET 21415)	2012	125–130	8–9	PHI Seeds Pvt. Ltd., Hyderabad- 82	Jharkhand, Maharashtra, Karnataka, Tamil Nadu, Uttar Pradesh, Bihar, Chhattisgarh
27P61 (IET 21447)	2012	132	6.70	PHI Seeds Pvt. Ltd., Hyderabad- 82	Chhattisgarh, Gujarat, Andhra Pradesh, Karnataka, Tamil Nadu

(continued)

Table 15.3 (continued)

Rice hybrids	Year of release	Duration (days)	Yield (t/ha)	Developed by	Recommended for the sates of
25P25 (IET 21401)	2012	110	6.70	PHI Seeds Pvt. Ltd., Hyderabad- 82	Uttarakhand, Jharkhand, Karnataka
Arize Tej (HRI 169) (IET 21411)	2012	125	7.0	Bayer Bio Science Pvt. Ltd., Hyderabad – 81	Bihar, Chhattisgarh, Gujarat, Andhra Pradesh, Tamil Nadu
PNPH 24 (IET 21406)	2012	120–130	5.8–6.9	Nuziveedu Seeds Limited, Medchal Mandal, Ranga Reddy- 501,401 (A.P.)	Bihar, West Bengal, Odisha
VNR 2245 (IET 20716)	2012	120–125	7.0–7.2	VNR Seeds Pvt. Ltd., Raipur 492,099	Chhattisgarh, Tamil Nadu

India is the second largest wheat-producing nation (11.9% share) after China (with 16.9% share). India and China together with Russian Federation, the USA and Canada contribute to more than half of the global wheat production. Wheat is grown on more land area than any other food crop (220.4 million hectares in 2014). In 2016, world production of wheat was 749 million tons, making it the second most-produced cereal after maize. Since 1960, world production of wheat and other grain crops has tripled and is expected to grow further. Seedling vigour, improved root system, resistance to insects/diseases, adaptability, increased yield and improved milling and baking characteristics are the six possible factors to heterosis in wheat. It is possible for heterosis to be expressed by an F_1 hybrid in any part of the plant into which the products of photosynthesis are channelled. Heterosis in grain yield must arise from an increase in the production of one or more of the plant's yield components. The weight of grain produced from a single plant is the product of the number of fertile tillers/plant, grains/ear and the weight of an individual grain. One of the underlying differences between the tillers and the number and weight of grain is the period of growth at which they are formed. The establishment of potential tillers begins at the four-leaf stage. Grain weight is largely determined in post-anthesis stage. Grains/ear is of course the product of number of spikelets/ear and grains/spikelet.

There is a need to have parental lines with better yield components that can be accumulated for harnessing heterosis at commercial level. Such parental lines can be developed by pre-breeding activities or diversification through utilization of diverse germplasm lines. In order to widen the genetic base of bread wheat, the emphasis has been laid on introgressing genes from unexploited *buitre* types, synthetic hexaploids

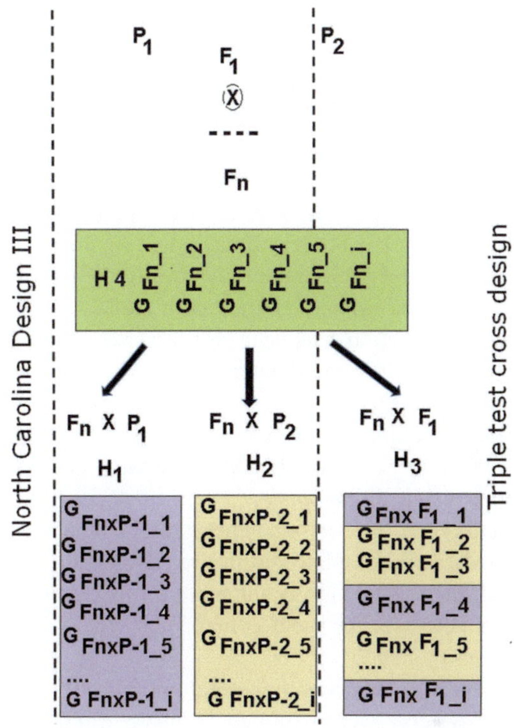

Fig. 15.7 Experimental designs for determining the genetic basis of heterosis. Both NC III and TTC designs begin with an F_2 segregating population having *i* plant individuals, created from a cross between two parental inbred lines ($P\text{-}_1$ and $P\text{-}_2$) that differ in the trait of interest. Instead of selfing the F_2 to produce $F_2{:}_3$ progeny, in the NC III scheme, all F_2 individuals are backcrossed as female parents with pollen from each parental line: $P\text{-}_1$ and $P\text{-}_2$. The individuals in the two resulting lines, denoted by GFnxP-1_i and GFnxP-$_2$_i, are then scored for studied phenotypes. In the TTC scheme, the F_2 individuals are further backcrossed to F_1 to generate the third line GFnxF_1_i. The third line provides additional information to distinguish dominant effects. (See heterosis breeding in wheat)

and Chinese sub-compactoid ear germplasm. The *buitre* lines have robust stem, long spikes, more spikelets, more grains/spike, large leaf area and broad leaves. The synthetic hexaploids developed at CIMMYT (International Maize and Wheat Improvement Center :Spanish acronym: *Centro Internacional de Mejoramiento de Maíz y Trigo*) were endowed with genetic richness for high grain weight, delayed senescence (stay green), high molecular weight (HMW) glutenins, resistance to Karnal bunt and yellow rust. Similarly, Chinese germplasm lines have robust stem, more grains/spike and new sources of yellow rust resistance. The desirable attributes from buitre types, synthetic hexaploids and Chinese germplasm were introgressed into "PBW 343" and "WH 542" background. The advanced bulks developed through utilization of diverse material have shown wide range of variability. The introgression for 1000-grain weight (herbicide tolerant lines) was also observed from the Chinese germplasm lines, and a number of transgressive segregants were obtained having 1000-grain weight of more than 65 g.

The work on development of hybrid wheat started in 1962 at global level in many countries. Ing. Riccardo Rodriguez initiated the research efforts at CIMMYT in 1962. The elite CIMMYT lines were transferred with *T. timopheevii* cytoplasm, the fertility restorer (Rf) genetic stocks were developed, and the experimental hybrids were produced. However, with the advent of semi-dwarf high-yielding wheat varieties, the emphasis got further strengthened only for popularization and genetic improvement of pure-line varieties, and as a result, the research efforts on hybrid wheat got distracted. The work was discontinued as no significant results of heterosis were observed for commercial exploitation. The research efforts were readdressed at CIMMYT during 1997–2002 in collaboration with the Monsanto Co. to develop a practical hybrid wheat production scheme in Northern Mexico and to identify spring hybrid bread wheat with superior yield potential, leaf-rust resistance and acceptable quality, under optimal conditions. In India, under Directorate of Wheat Research at Karnal, hybrid wheat development through CMS and CHA approach in network mode commenced from 1995. Through CMS approach, cytoplasmic male sterile lines were developed using *T. timopheevii, T. araraticum*, *Ae. caudata* and *Ae. speltoides* as source parents. Two exotic genetic stocks registered as "PWR 4099" and "PWR 4101" indicated complete fertility restoration in *T. timopheevii*-based CMS lines. Although there is no significant result for heterosis for yield in totality, few hybrids showed heterosis for yield components, viz. spikelet number, spike length and tillers/plant.

The insufficient levels of heterosis, low seed multiplication rate and complexity of the hybridization systems were explored as major limiting factors for hybrid wheat development. The discovery of an effective cytoplasmic male sterility and pollen fertility restoration systems in wheat using *Aegilops caudata* cytoplasm opened up new avenues, but the stability of male sterility across the locations is another bottleneck. *T. timopheevii* seems to be the most suitable one for commercial production of hybrid seed. The inclusion of yield potential in the bread wheat is also an important issue. As wheat is allohexaploid, the transfer of donor traits from related species takes in more negative traits than the positive components. Table 15.4 summarizes events related to hybrid wheat development.

15.7.2 Heterosis Breeding in Rice

China and India are the largest rice producers. Compared to India, China's rice production is greater since all its rice area is irrigated, while India has less than half of its area irrigated. Further, Indonesia, Bangladesh, Vietnam and Thailand are in the order of hierarchy. These seven countries all had average production of more than 30 million tons of paddy and together account for more than 80% of world production (estimates of 2006–2008). Rice is the third highest produced agricultural commodity with a world production of 759.6 million tons in 2017.

Chinese Professor Yuan Longping is popularly known as the "Father of Hybrid Rice". He developed genetically inherited male sterility in rice enabling only cross-pollination. This mechanism is widely being used worldwide to develop hybrid rice. China initiated research on hybrid rice in 1964 and became the first country to

Table 15.4 Events related to hybrid wheat development

1919	Heterosis was first reported in wheat for plant height (Freeman)
1934	Heterosis first reported in wheat for yield
1951	Cytoplasmic male sterility introduced into wheat using *Aegilops caudata* cytoplasm (Kihara)
1957	The USA is the first country to plan hybrid wheat production
1958	CMS research started on wheat in Kansas
1959	Nuclear male sterility reported in Wheat
1961	Fertility restorers found in adapted wheat varieties
1961	DeKalb Agricultural Association begins the first commercial hybrid wheat breeding programme
1961:	The variety "Gaines" becomes the first semi-dwarf wheat to be released in the USA
1961:	Fertility restorers found in adapted wheat varieties
1962:	Source of CMS found in *Triticum timopheevii*
1962:	First commercially feasible CMS system proposed
1966:	McDaniel and Sarkissian proposed the theory of mitochondrial heterosis
1971:	de Vries commences publishing papers dealing with the suitability of wheat for cross-fertilization
1972:	"XYZ" system proposed for the utilization of nuclear male sterility (NMS) (Driscoll)
1974:	First commercial CMS hybrid wheat released in the USA
1981:	Hybrid wheat varieties released by Cargill in the USA and by DeKalb in Australia
1982:	Monsanto starts HW (Hybrid Wheat) programme in the USA and Europe based on CHA Genesis
1982:	New CMS wheat hybrids make an impact on the US market
1984:	OECD begins work on international certification scheme for hybrid wheat
1984:	Hybrid wheat varieties enter registration trials in the UK
1986:	Hybrid wheat varieties released in Argentina by Cargill
1990:	Cargill cease production and sale of hybrid wheat in the USA but continue commercialization in Australia and Argentina
1995:	ICAR (DWR) initiated work on hybrid wheat in a network mode through Chemical Hybridizing Agent (CHA) and CMS approach
2000:	Monsanto Co. stops GENESIS-based hybrid production and HW activities in the US and Europe
2002:	DuPont/Hybrinova stops Croisor-based hybrid production and HW activities in Europe
2003:	DWR and NCL Pune got the US Patent (US2003/0192070A1) for chemical composition for complete male sterility, its process for preparation and use
2007:	ICAR (DWR) discontinued work on hybrid wheat through CHA approach
2009:	ICAR initiated network project on hybrid wheat using CMS approach

produce hybrid rice commercially. Hybrid rice breeding has been based on using cytoplasmic male sterility (CMS) or photo-thermogenetic male sterility (P-TGMS). A breeding system using three lines (a CMS line, CMS maintainer and CMS restorer lines) was established in 1973. A two-line hybrid rice system using P-TGMS was established in the 1980s, and two-line hybrid rice was widely used by 1998. First three hybrid rice varieties were released by China in 1974, and by 1976, commercial hybrid rice cultivation began. Rice scientists succeeded to overcome negative traits like inferior grain quality and susceptibility to diseases which derived strains superior than inbred counterparts. Hybrid rice has been widely adopted in China – the world's biggest producer of rice – with around 56% of the rice planted in China made up of hybrid rice. In 2009, hybrid rice yielded around 6.6 tons per hectare – well above the world average of 4.2 tons. In 2011, Indonesia, Vietnam, Myanmar, Bangladesh, India, Sri Lanka, Brazil, the USA and the Philippines followed the success story of China. IRRI was actively involved in hybrid rice research since 1979. Research at IRRI focuses on producing hybrid rice with consistently high-yield heterosis (hybrid vigour), good grain quality, tolerance to key environmental stresses, multiple resistances to insect pests and diseases, and high seed production yield. Hybrid Rice Development Consortium (HRDC) by IRRI in 2008 to collaborate more closely with partners to develop new hybrid rice.

In China, hybrid varieties could obtain about 30% grain yield advantage over inbred (pure-line) varieties. In the first 20 years of cultivation, hybrid rice could be extended to about 50% of the area that helped China to increase rice yield from 5.0 t/ha of conventional rice to 6.6 t/ha, reaching consistently 7.5 t/ha in the Sichuan province (see Fig. 15.8). Hybrid rice has now become a commercial success in several Asian countries, such as Vietnam, India, the Philippines and Bangladesh. If hybrid rice were not developed, an estimated 6 million ha of extra area should have been required. In the last few decades, the USA, Brazil and other South American countries have also begun the commercial production of hybrid rice. Improved hybrid rice, with resistance genes to many diseases, were derived through both normal breeding and genetic engineering.

The use of *indica* x *japonica* crosses has long been considered a promising approach to broaden the genetic diversity and to enhance the heterosis of rice. However, F_1 semi-sterility has generally been encountered in inter-subspecies crosses of rice, making it meaningless for direct use in hybrid rice breeding. In addition, distant crosses do not always increase F_1 yield, and this is particularly true when the parental lines belong to different subspecies.

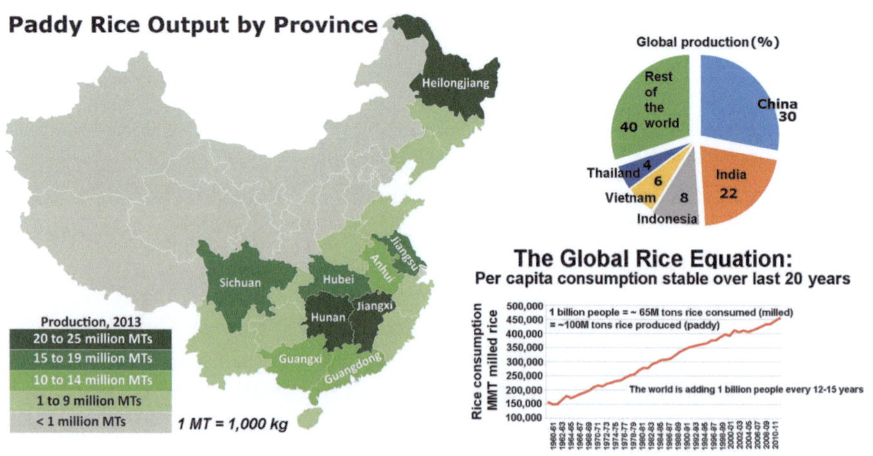

Fig. 15.8 Rice production in China compared against global production

It is now considered that *indica*-inclined or japonica-inclined lines are generally advantageous for a higher F_1 yield. In recent decades, China could integrate *japonica* component into *indica* breeding programmes. In regions where japonica is grown, they have integrated *indica* components into japonica background. By this, a series of *indica*-inclined or *japonica*-inclined rice lines have been derived and used as parental lines to develop super rice varieties.

Super Hybrid Rice Chinese Ministry of Agriculture initiated a programme with the aim to achieve very high yields (target: 10 tons/ha in the majority of Chinese rice-growing areas and up to 12 tons/ha in large field trials). Super hybrid rice involves heterosis achieved through hybridization between indica and japonica rice varieties (inter-subspecific) as well as pyramiding of heterosis genes for different rice ecotypes and the incorporation of useful genes (including genes for anti-herbivore resistance) from near and distant relatives. Some of these new-generation hybrids (i.e. Liangyoupeijiu and Liangyou 293) have demonstrated high yields in field trials. Some of the super hybrid rice varieties developed by China between 2005 and 2016 are available in Table 15.5.

In 1981, the Ministry of Agriculture, Forestry and Fisheries of Japan launched large-scale collaborative research projects to develop super-high-yielding rice with improved agro-techniques. Over 15 years, release of some super-high-yield varieties that produced brown rice with 10 t/ha, an increase by 50% compared to the control variety Akihikari was achieved. By the late 1980s, the grain yield of Chenxing, Aoyu 326 and Beilu 130 was close to 10 t/ha. However, these super-high-yield varieties could not gain popularity among farmers due to low seed setting rate, poor quality and limited adaptability.

In 1989, the International Rice Research Institute (IRRI) launched a plan to breed for the new plant type (NPT) rice, with a goal of 20% yield increment compared to the existing high-yielding varieties or producing an yield of 15 t/ha. In 1994, IRRI announced that its NPT rice reached 12.5 t/ha, a 20% increase against control variety. But these NPT rice had a low rate of seed setting, poor grain filling and weak resistance against brown plant hopper.

India began a relatively small programme of the Indian Council of Agricultural Research (ICAR), focusing on hybrids for irrigated cultivation in 1989. United Nations Industrial Development Organization (UNIDO) and Food and Agriculture Organization (FAO), Mahyco Research Foundation, the Asian Development Bank (ADB), IRRI and the National Agricultural Technology Project (funded by the World Bank) and India's Ministry of Agriculture altogether funded $8 million. Despite these investments and efforts, hybrid rice in India faced several challenges that delayed the government's goal of achieving hybrid rice cultivation in 25% of rice area by 2015. But the proportion of area under hybrid rice grew at a rate of about 40% per year since 2005, contributed by the states of Jharkhand, Bihar, Uttar Pradesh and Uttarakhand. Currently, efforts by the private sector to promote hybrid rice in eastern India are significant. Yield of inbred varieties in these states are fairly low (approx. 2.5 tons/ha), and hybrid rice could contribute more.

Table 15.5 Super rice varieties certified by the Ministry of Agriculture of China (2005–2016)

Year	Number of varieties	Super rice varieties
2016	10	Jijing 511, Nanjing 52, Huiliangyou 996, Shenliangyou 870, Deyou 4727, Fengtianyou 553, Wuyou 662, Jiyou 225, Wufengyou 286, Wuyouhang 1573
2015	11	Yangyujing 2, Nanjing 9108, Diandao 18, Huahang 31, Hliangyou 991, Nliangyou 2, Yixiangyou 2115, Shenyou 1029, Yongyou 538, Chunyou 84, Zheyou 18
2014	18	Longjing 39, Liandao 1, Changbai 25, Nanjing 5055, Nanjing 49, Wuyunjing 27, Yliangyou 2, Yliangyou 5867, Liangyou 038, Cliangyouhuazhan, Guangliangyou 272, Liangyou 6, Liangyou 616, Wufengyou 615, Shentaiyou 722, Nei5you 8015, Rongyou 225, Fyou 498
2013	12	Longjing 31, Songjing 15, Diandao 11, Yangjing 4227, Ningjing 4, Zhongzao 39, Yliangyou 087, Tianyou 3618, Tianyouhuazhan, Zhong9you 8012, Hyou 518, Yongyou 15
2012	13	Chujing 28, Lianjing 7, Zhongzao 35, Jinnongsimiao, Zhunliangyou 608, Shenliangyou 5814, Guangliangyouxiang 66, Jinyou 785, Dexiang 4103, Qyou 8, Tianyouhuazhan, Yiyou 673, Shenyou 9516
2011	9	Shennong 9816, Nanjing 45, Wuyunjing 24, Yongyou 12, Lingliangyou 268, Zhunliangyou 1141, Huiliangyou 6, 03you 66, Teyou 582
2010	12	Xindao 18, Yangjing 4038, Ningjing 3, Nanjing 44, Zhongjiazao 17, Hemeizhan, Guiliangyou 2, Peiliangyou 3076, Wuyou 308, WufengyouT 025, Xinfengyou 22, Tianyou 3301
2009	10	Longjing 21, Huaidao 11, Zhongjiazao 32, Yangliangyou 6, Luliangyou 819, Fengliangyouxiang 1, Luoyou 8, Rongyou 3, Jinyou 458, Chunguang 1
2007	12	Ningjing 1, Huaidao 9, Qianzhonglang 2, Liaoxing 1, Chujing 27, Longjing 18, Yuxiangyouzhan, Xinliangyou 6380, Fengliangyou 4, Nei2you6, Ganxin 688, IIyouhang 2
2006	21	Tianyou 122, Yifeng 8, Jinyou 527, Dyou 202, Qyou 6, Qianliangyou 2058, Yyou 1, Zhuliangyou 819, Liangyou 287, Peizataifeng, Xinliangyou 6, Yongyou 6, Zhongzao 22, Guinongzhan, Wujing 15, Tiejing 7, Jijing 102, Songjing 9, Longjing 5, Longjing 14, Kenjing 14
2005	28	Xieyou 9308, Guodao 1, Guodao 3, Zhongzheyou 1, Fengyou 299, Jinyou 299, IIyouming 86, IIyouhang 1, Teyouhang 1, Dyou 527, Xieyou 527, IIyou 162, IIyou 7, IIyou 602, Tianyou 998, IIyou084, IIyou 7954, Liangyoupeijiu, Zhunliangyou 527, Liaoyou 5218, Liaoyou 1052, IIIyou 98, Shengtai 1, Shennong 265, Shennong 606, Shennong 016, Jijing 88, Jijing 83

15.7.3 Heterosis Breeding in Maize

Maize (*Zea mays* L.) is a versatile C_4 crop grown under a range of agroclimatic zones and considered as queen of cereals with high production levels. Among resource poor communities of tropical and subtropical regions, maize is the major source of nutritional security. George Harrison Shull first reported heterosis in maize in 1908. The total area under maize cultivation in tropical countries is 100 million hectares, and it yields 9 t/ha in temperate zones. Maize has the longest history of breeding for yield and other agronomic traits under stressed environments through traditional breeding methods. Hybrid breeding, especially the double-cross hybrids of 1960s, has been widely adopted to improve tropical maize productivity.

D.F. Jones in 1918 was the first to invent the double-cross hybrid. A double-cross is created by making two single-cross hybrids (A × B) and (C × D) and then crossing the two hybrids of single crosses. Seeds from the second cross are sold to farmers. Such hybrid seeds geared up corn cultivation in the USA. However, for the first 30 years of twentieth century, the US agricultural economy was in recession. When New Deal farm policies were implemented, the farmers were willing to invest procurement of hybrid seed. Double-cross hybrids were replaced by three-way hybrids and further by single crosses in the 1970s. A three-way cross uses three inbred lines, (A × B) × C. Single crosses only contain two A × B. Single-cross hybrids are the most sought after with higher yield Corn Belt.

Molecular breeding and doubled haploid (DH) technologies are the two major technologies of the twentieth century that have made positive impact on maize productivity. Studies using SSR markers revealed (done at International Maize and Wheat Improvement Centre – CIMMYT) higher heterozygosity and lesser genetic purity in inbreds derived from tropical germplasm. SSR markers for abiotic stress were utilized in breeding programmes. The genome structure of maize reveals 80% repetitive and 32% sequences that diverged within maize (paralogous sequences) with numerous transposons (sequence that can move to new position within the genome of a single cell). Paradoxically, it is presumed that the extent of nucleotide diversity between any two maize lines is higher than the genetic distance between a chimpanzee and human.

Linkage analysis and association studies are the two major techniques to dissect genetic architecture of complex traits. Linkage analysis is the traditional method used to detect the co-segregation of a small genomic region (QTL) governing a trait of interest in families or pedigrees of known ancestry using RFLPs and SSRs. Using linkage mapping, hundreds of marker-trait associations were proved in tropical maize research. But, only very few of this could be utilized in commercial breeding programmes. One of the reasons could be that the QTLs detected in biparental population using interval mapping are relevant only for programmes that involve parents to detect the QTL. High interference of G x E interactions and low heritability are probable demerits of linkage mapping of traits. On the contrary, association study is a precision and high-resolution method for mapping genes (or loci) underlying complex traits based on linkage disequilibrium (LD) in populations.

Association study broadly falls into two classes: "candidate-gene studies" and "whole-genome studies". The "candidate-gene"-based association study is hypothesis-based analysis. The "candidate genes" are selected for association mapping, either by their location in a genomic region that has been roughly identified via linkage analysis. Alternatively, whole-genome association study, also called genome-wide association study (GWAS), is an approach for establishing marker-trait associations, and most important of this include the use of natural genetic resources, i.e. germplasm lines, instead of segregating mapping population that saves time and occurrence of historical recombinations (selections) that allows multiple alleles per locus, making increased map resolution. GWAS is a powerful NGS tool, used to dissect complex traits.

Doubled haploid (DH) technology through in vivo haploid induction has been largely adopted by commercial breeding programmes. This is a well acclaimed technique for reducing time taken for a breeding cycle and to generate parental lines (see Chap. 13 for details account on doubled haploids).

Allelic Variation and Heterosis One of the most common approaches towards documenting allelic diversity is to compare the sequence of genic regions (including coding regions, introns, untranslated regions and single copy DNA surrounding genes) from multiple strains or varieties in order to identify variation. This variation can then be used for mapping or association studies. On average, indel polymorphisms (insertion/deletion polymorphism) occur every 309 bp, and SNPs occur every 79 bp. The analysis of 300–500 bp amplicons (a piece of DNA or RNA that is source of amplification or replication events) found that 44% of the sequences contained at least one polymorphism in maize variety B73 relative to variety Mo17. In general, it is estimated that there is one polymorphism in every 100 bp in any two randomly chosen maize inbred lines. Maize has a relatively high level of sequence polymorphism compared to many other species. Structural genome diversity involves large-scale chromosomal differences, altered location of genes or differences in the presence of sequences. Large-scale genome differences between different maize inbred lines were first identified by Barbara McClintock who analysed heterochromatic knob content and size to characterize genome variation in maize. Recent studies have documented differences in the content for several classes of repetitive DNA between maize inbreds at the chromosomal level.

Further Reading

Birchler JA et al (2010) Heterosis. Plant Cell 22:2105–2112

Birchler JA (2015) Heterosis: the genetic basis of hybrid vigour. Nat Plants 1:15020

Fu D et al (2015) What is crop heterosis: new insights into an old topic. J Appl Genetics 56:1–13

Herbst RH et al (2017) Heterosis as a consequence of regulatory incompatibility. BMC Biol 15:38. https://doi.org/10.1186/s12915-017-0373-7

Huang X et al (2016) Genomic architecture of heterosis for yield traits in rice. Nature 537:629–633

Lauss K et al (2018) Parental DNA methylation states are associated with heterosis in epigenetic hybrids. Plant Physiol 176:1627–1645

Xing J et al (2016) Proteomic patterns associated with heterosis. Biochim Biophys Acta (BBA) – Proteins Proteomics 1864:908–915

Induced Mutations and Polyploidy Breeding

16

Keywords

Mutation Breeding: · History · Mutagenic agents · Physical mutagenesis · Chemical mutagenesis · Types of mutations · Practical considerations · Mutation breeding strategy · In Vitro Mutagenesis · Gamma gardens or atomic gardens · Factors affecting radiation effects · Direct and indirect effects · Molecular mutation breeding · TILLING and EcoTILLING · Site-directed mutagenesis · MutMap · FAO/IAEA joint venture for nuclear agriculture · Mutation breeding in different countries · **Polyploidy Breeding:** · Types of changes in chromosome bumber · Methods for inducing polyploidy · Mechanisms of polyploidy formation · Molecular consequences of polyploidy · Molecular tools for exploring polyploidy genomes

16.1 Mutation Breeding

Mutation is a sudden heritable change that occurs in the genetic information of an organism not caused by genetic segregation or genetic recombination; but induced by chemical, physical or biological agents. Mutation breeding follows three strategies:

(a) Induced mutagenesis: mutations occur because of irradiation (gamma rays, X-rays, ion beam, etc.) or treatment with chemical mutagens
(b) Site-directed mutagenesis: mechanism of creating mutations at a defined site in a DNA molecule
(c) Insertion mutagenesis: done through DNA insertions; by genetic transformation and insertion of T-DNA or activation of transposable elements

For crop breeding, multiple mutant alleles are the sources of genetic diversity. The vital issue in mutation breeding is the diligence to isolate and select individuals with target mutation. This process involves two major steps: mutant screening and

P. M. Priyadarshan, *PLANT BREEDING: Classical to Modern*,
https://doi.org/10.1007/978-981-13-7095-3_16

mutant confirmation. In mutant screening, the breeder fixes certain traits to be selected. This involves selection of individuals that meet specific selection criteria like early flowering, disease resistance, etc. Mutant confirmation is done through reevaluating the putative mutants under controlled and replicated environments. By this process, false mutants can be revealed. In general, mutations vital for crop improvement usually involve single bases and may or may not affect protein synthesis.

16.1.1 History

Reports on mutant crops from China were available as early as 300 BC. Towards the late nineteenth century, Hugo de Vries was the first to identify mutations while "rediscovering" Mendelian laws. He could consider such variability as heritable that was distinctive from segregation and recombination. He coined the term "mutation". Such variability was described as shock-like changes (leaps) in existing traits. After the discovery of mutagenic action of X-rays, radiation-induced mutations were used as tools for generating novel genetic variability. This was demonstrated in maize, barley and wheat by Stadler. The first commercial mutant variety was produced in tobacco in 1934. The number of commercially released varieties rose to 484 by 1995. This number sharply increased with time (Fig. 16.1). They include fruit trees, ornamentals and food crops. Agronomic traits like lodging resistance, early maturity, winter hardiness and product quality (e.g. protein and lysine content) were the most sought after traits in breeding. Mutagenesis became very popular from the 1950 as a breeding tool, and a range of crops and ornamentals were subjected to induced mutations to increase trait variation.

16.1.2 Mutagenic Agents

Agents that induce artificial mutations are called mutagens. They are grouped as chemical and physical. Planting materials are exposed to physical and chemical mutagenic agents to induce mutations. Materials like whole plants, usually seedlings, and in vitro cultured cells can be used for mutation induction. Seed is the most commonly used plant material. Plant forms as bulbs, tubers, corms and rhizomes are also used. In vegetatively propagated crops, vegetative cuttings, scions or in vitro cultured tissues like leaf and stem explants, anthers, calli, cell cultures, microspores, ovules, protoplasts, etc. are used. Gametes can be mutated through immersion of spikes, tassels, etc. Whereas chemical mutagens are preferably used to induce point mutations, physical mutagens induce gross lesions, such as chromosomal abbreviation or rearrangements. Frequency and types of mutations are direct results of dosage and rate of exposure or rather than its type. The choice of a mutagen will be based on the safety of usage, ease of use, availability of the mutagens, effectiveness in inducing certain genetic alterations, suitable tissue, cost and available infrastructure among other factors.

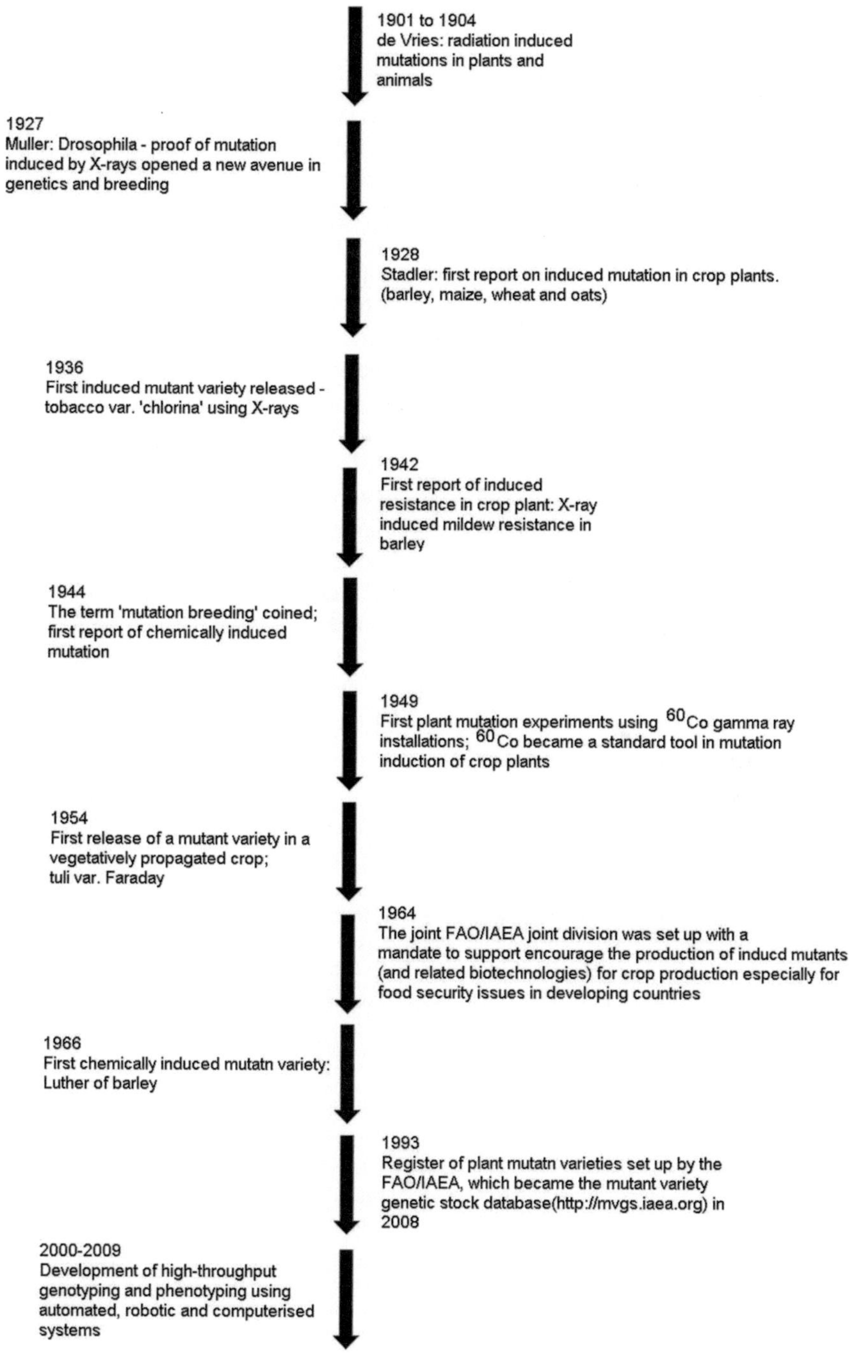

Fig. 16.1 Milestones in induced mutagenesis

16.1.3 Physical Mutagenesis

Physical mutagens, mostly ionizing radiations, have been used widely for developing more than 70% of mutant varieties for the last 80 years. Radiation is energy travelled through a distance in the form of waves or particles. Radiation is a high-energy level of electromagnetic (EM) spectrum that is capable of dislodging electrons from the nuclear orbits of the atoms. The impacted atoms, become ions, hence, the term ionizing radiation. These ionizing components of the EM include cosmic, gamma (γ) and X-rays. The most commonly used physical mutagens and their properties are shown in Tables 16.1a and 16.1b. X-rays were the first to be used to induce mutations. After this, various subatomic particles (neutrons, protons, beta particles and alpha particles) were used in nuclear generators to emit radiations. Gamma radiation from radioactive cobalt (^{60}Co) is widely used. Since it has high penetrating potential and is hazardous, gamma rays can be used for irradiating whole plants and delicate materials like pollen grains. In most cases, DNA double-strand breaks lead to mutation. Since gamma rays have shorter wavelength, they possess more energy than protons and X-rays, which gives them the strength to penetrate deeper into the tissue. Neutrons are used in dry seeds as they cause serious damage to the chromosomes. The mutagenic potential of UV rays had been confirmed in many organisms. Emission of UV light (250–290 nm) has a modest capacity to infiltrate tissues and goes deeper into the tissue and can cause a great number of variations in the chemical composition. The advantage of using physical mutagenesis over

Table 16.1a Commonly used physical mutagens

Mutagen	Source	Characteristics	Hazard
X-rays	X-ray machine	Electromagnetic radiation; penetrates tissues from a few millimetres to many centimetres	Dangerous, penetrating
Gamma rays	Radioisotopes and nuclear reaction	Electromagnetic radiation produced by radioisotopes and nuclear reactors; very penetrating into tissues; sources are 60-Co (Cobalt-60) and ^{137}Cs (Caesium-137)	Dangerous, very penetrating
Neutrons	Nuclear reactors or accelerators	There are different types (fast, slow, thermal); produced in nuclear reactors; uncharged particles; penetrate tissues to many centimetres; source is ^{235}U	Very hazardous
Beta particles	Radioactive isotopes or accelerators	Produced in particle accelerators or from radioisotopes; are electrons; ionize; shallowly penetrating; sources include ^{32}P and ^{14}C	May be dangerous
Alpha particles	Radioisotopes	Derived from radioisotopes; a helium nucleus capable of heavy ionization; very shallowly penetrating	Very dangerous
Protons	Nuclear reactors or accelerators	Produced in nuclear reactors and accelerators; derived from hydrogen nucleus; penetrate tissues up to several centimetres	Very dangerous
Ion beam	Particle accelerators	Produced positively charged ions are accelerated at a high speed (around 20–80% of the speed of light) deposit high energy on a target	Dangerous

Table 16.1b Types and properties of ionizing radiations used for plant-induced mutagenesis

	Properties		
Type of radiation	description	Energy	Penetration in plant tissue
X-rays	Electromagnetic radiation	50–300 keV	A few mm to many cm
Gamma rays	Electromagnetic radiations similar to X-rays	Up to several MeV	Through whole parts
Neutron (fast, slow and thermal)	Uncharged particle, slightly heavier than proton, observable only through interaction with nuclei	From less than 1 eV to several MeV	Many cm
Alpha particles	A helium nucleus, ionizing heavily	2–9 MeV	Small fraction of a mm
Beta particles, fast electrons or cathode rays	An electron (– or +) ionizing much less densely than alpha particles	Up to several MeV	Up to several cm
Protons or deuterons	Nucleus of hydrogen	Up to several GeV	Up to many cm
Low-energy ion beams	Ionized nucleus of various elements	Dozens of keV	A fraction of mm
High-energy ion beams	Ionized nucleus of various elements	Up to GeV	A fraction of cm

chemicals is the degree of accuracy and reproducibility. Among them, gamma rays are most sought after due to its uniform penetrating power. During the past two decades, ion beams have become more popular. They consist of particles travelling along a path that vary in mass from a simple proton to a uranium atom which is generated through particle accelerators. The positively charged ions are accelerated at a high speed (about 20%–80% of the speed of light) and form high-linear energy transfer (LET) radiation. LET radiation causes significant biological effects, such as chromosomal aberration, lethality, etc. Ion beams induce deletion of fragments of various sizes and are less repairable.

For inducing mutations, doses that lead to 50% lethality (LD_{50}) have often been chosen. It is the amount of substance required (usually per body weight) to kill 50% of the test population. Very often it is argued that LD_{50} is quite arbitrary and might lead to a high number of (mostly deleterious) mutations. LD_{50} can lead to loss of desirable mutations due to plant mortality or due to poor agronomic performance. Therefore, in self-pollinated species, a mutation rate targeting a lower LD (e.g. LD20) with a survival rate of 80% appears to be more ideal. The isotope ^{60}Co has a half-life of 5.27 years and emits radiations of energies 1.33 MeV and 1.17 MeV (mega electron volt).

Ionizing radiations break chemical bonds in the DNA molecule, deleting a nucleotide or substituting it with a new one. Radiation being applied at a proper dose depends on radiation intensity and duration of exposure. *Roentgen* (r or R) is the unit to measure dosage of radiation. Rontgen is named after Wilhem Conrad

Röntgen a German physicist, who during 1895 produced and detected electromagnetic radiation that earned him the first Nobel Prize in physics in 1901. The exposure may be chronic (continuous low dose administered for a long period) or acute (high dose over a short period). The dose rate is not necessarily positively correlated with the proportion of useful mutations. A high dose need not necessarily produce best results. The mutagen dose depends on the mutation load and the chance to find desirable mutations.

16.1.3.1 Ion Beams

Ion beams are usually generated by particle accelerators, e.g. cyclotrons, using ^{20}Ne, ^{14}N, ^{12}C, ^{7}Li, ^{40}Ar or ^{56}Fe as radioisotope sources. These ion beams are responsible for linear energy transfer (LET). Like physical mutagens, as LET increases, lethality, chromosomal aberration, etc., are also induced. The LET for gamma rays and X-rays accounts in the range of 0.2–2.0 keV/μm and hence is called low-LET radiations. However, the high-LET radiations from carbon (23 keV/μm) and iron (640 keV/μm) ion beams extend larger and wider ionization energy. High-LET ion beam radiations cause more localized, dense ionization within cells than those of low-LET radiations. The choice of ion beam depends on the characteristics of the ion with respect to electrical charge and velocity. Dose (in Gy=Gray units) is proportional to the LET (in keV/μm) and number of particles. An ideal irradiation dose provides highest mutation rate at any locus. Through applying different doses in a given time and screening the irradiated population, acceptable mutants can indicate the best dose. Scientists may consider traits like survival rate, growth rate, chlorophyll mutation, etc., as early indicators for mutation and this exercise requires sizeable investment. Advantages of ion beam mutagenesis include low dose with high survival rates, induction of high mutation rates with wide range of variation. Ion beam is an excellent tool for mutation breeding to improve horticultural and agricultural crops with high efficiency. In rice, salt-resistant lines were developed through ion beam irradiation. This was developed with 30–60 keV low-energy ion beam.

16.1.3.2 Aerospace Mutagenesis

In the recent past, plant materials were sent to aerospace to study probable mutation induction in space. The speculation is that cosmic radiation, microgravity, weak geomagnetic field, etc. contains the potential agents of mutation induction. However, much is not known on the genetics of aerospace mutagenesis. Gamma rays induce nucleotide substitutions and small deletions of 2–16 bp, and the mutation frequency is estimated to be one mutation/6.2 Mb. Fast neutrons are believed to result in kilobase-scale deletions.

More than 90% of the space radiation is composed of protons, neutrons, heavy particles, rays and microgravity. China had sent more than 400 varieties of 50 species to outer space by 8 recoverable satellites. From this exercise, more than 50 new varieties with high yield, high quality and drought tolerance have been commercialized. Though progress is made, mechanisms governing mutation induction is still under investigation.

16.1.4 Chemical Mutagenesis

Though the action is milder, the advantage with chemical mutagens is that they can be used without sophisticated machinery. However, undesirable changes are higher than in physical mutagenesis. Usually, the material is soaked in a solution of the mutagen to induce mutations. Extra care must be taken for health protection since chemical mutagens are carcinogenic. Thus, safety data sheets should be carefully read and the mutagenic agent should be appropriately inactivated before disposal. Although a large number of mutagens are available, only a small number is recognized by IAEA (International Atomic Energy Agency). Such mutagenic agents are responsible for over 80% of the registered new mutant plant varieties reported in the (IAEA) database. Of these, three compounds are significant: ethyl methanesulphonate (EMS), 1-methyl-1-nitrosourea and 1-ethyl-1-nitrosourea, which account for 64% of these varieties.

One of the most effective chemical mutagenic groups is the group of alkylating agents (these react with the DNA by alkylating the phosphate groups as well as the purine and pyrimidine). Another group is that of the base analogues (they are closely related to the DNA bases and can be wrongly incorporated during replication). Examples are 5-bromo-uracil and maleic hydrazide (Table 16.2). There is a clear advantage with the point mutations created by chemical mutagens. Point mutations have the potential to generate not only loss-of-function but also gain-of-function phenotypes. This happens when the mutation leads to a modified protein activity or affinity, like tolerance to the herbicide (glyphosate or sulphonylurea). Factors like concentration, the length of treatment and the temperature of the experiment influence the efficiency of mutagenesis. Since chemical mutagens are very reactive, it is advisable to use fresh batches of the chemical(s).

EMS reacts with guanine or thymine by adding an ethyl group which causes the DNA replication machinery to recognize the modified base as an adenine or cytosine, respectively. Chemical mutagenesis induces a high frequency of nucleotide substitutions, and a majority of the changes (70–99%) in EMS-mutated populations are GC to AT base pair transitions. Sodium azide (Az) and methylnitrosourea (MNU) are also used in combination.

All chemical mutagens are strongly carcinogenic, and extreme care should be taken while handling and disposal. EMS is an IARC group 2B carcinogen. Working with MNU can be sometimes difficult as it is unstable above 20°C. EMS solutions can be deactivated in a solution of 4% (w/v) NaOH and 0.5% (v/v) thioglycolic acid. Chemical mutagens (EMS, DES, Az) have been applied for treating banana shoot tips to produce variants for tolerance to *Fusarium* wilt. EMS has also been successful in obtaining a wide range of variations in petal colour and in salt-tolerant lines in sweet potato.

Table 16.2 Commonly used chemical mutagens

Mutagen group	Example	Mode of action
Alkylating agents	1-Methyl-1-nitrosourea (MNU); 1-ethyl-1-nitrosourea (ENU); methyl methanesulphonate (MMS); ethyl methanesulphonate (EMS); dimethyl sulphate (DMS); diethyl sulphate (DES); 1-methyl-2-nitro-1-nitrosoguanidine (MNNG); 1-ethyl-2-nitro-1-nitrosoguanidine (ENNG); N,N-dimethyl nitrous amide (NDMA); N,N-diethyl nitrous amide (NDEA)	React with bases and add methyl or ethyl groups, and depending on the affected atom, the alkylated base may then degrade to yield an abasic site, which is mutagenic and recombinogenic, or mispair to result in mutations upon DNA replication
Azide	Sodium azide	Same as alkylating agents
Hydroxylamine	Hydroxylamine	Same as alkylating agents
Antibiotics	Actinomycin D; mitomycin C; azaserine; streptonigrin	Chromosomal aberrations also reported to cause cytoplasmic male sterility
Nitrous acid	Nitrous acid	Acts through deamination, the replacement of cytosine by uracil, which can pair with adenine and thus through subsequent cycles of replication lead to transitions
Acridines	Acridine orange	Intercalate between DNA bases thereby causing a distortion of the DNA double helix and the DNA polymerase in turn recognizes this stretch as an additional base and inserts an extra base opposite this stretched (intercalated) molecule. This results in frameshift, i.e. an alteration of the reading frame
Base analogues	5-Bromouracil (5-BU); maleic hydrazide; 5-Bromodeoxyuridine; 2-aminopurine (2AP)	Incorporate into DNA in place of the normal bases during DNA replication thereby causing transitions (purine to purine or pyrimidine to pyrimidine); and tautomerization (existing in two forms which interconvert into each other, e.g. guanine can exist in keto or enol forms)

16.1.5 Types of Mutations

Mutations can be broadly divided into (a) intragenic or point mutations (occurring within a gene in the DNA sequence); (b) intergenic or structural mutations within chromosomes (inversions, translocations, duplications and deletions) and (c) mutations leading to changes in the chromosome number (polyploidy, aneuploidy and haploidy). In addition, there are nuclear and extranuclear or plasmon (chloroplast and mitochondrial) mutations. Mutational changes at the molecular

Fig. 16.2 (**a**) Transition and transversion. Transitions are interchanges of pyrimidine (C T) or purine (A G) bases. Transversions are interchanges of pyrimidine for purine bases or vice versa (**b**) Frameshift mutation: This type of mutation occurs when the addition or loss of DNA bases changes a gene's reading frame. A reading frame consists of groups of three bases that each codes for one amino acid. A frameshift mutation shifts the grouping of these bases and changes the code for amino acids. The resulting protein is usually non-functional. Insertions, deletions and duplications can all be frameshift mutations

level are accomplished through substitution of one base by the other. This happens through mispairing of bases between pyrimidines and purines.

Basically, transitions (point mutations that changes purine to another purine $A \leftrightarrow G$ or pyrimidine to another pyrimidine $C \leftrightarrow T$) and transversions (when a purine is changed to pyrimidine or vice versa) are the simplest kinds of base pair changes. However, they may result in phenotypically visible mutations (Fig. 16.2a). Another common error would be addition or deletion of a nucleotide base pair when one of the bases manages to pair with two bases or fails to pair at all. Such sequence changes in the reading frame of the gene's DNA are known as frameshift mutations. Since they can change the message of the gene starting with the point of deletion/addition, they are more prominent (Fig. 16.2b). Base sequence may be inverted because of chromosome breakage. On the other hand, reunion of the broken ends can result in different DNA molecules in a reciprocal fashion. Duplication of a DNA sequence is yet another common mechanism changing the structure of gene leading to gene mutation.

16.1.6 Practical Considerations

The dose of a mutagen that ensures optimum mutation frequency with minimum unintended damage is regarded as the optimal dose. In case of physical mutagens, tests of radiosensitivity (from radiation sensitivity) give estimates. It gives an indication of the quantity of recognizable effects of radiation exposure. Since it is a predictive value, it gives guidance on the choice of optimal exposure dosage.

Important factors influencing the outcome of chemical mutagenesis are:

(a) Inherent traits of tissue
(b) Environment
(c) Concentration of mutagen
(d) Treatment volume
(e) Treatment duration
(f) Temperature
(g) Presoaking of seeds
(h) pH (7.0)
(i) Catalytic agents (Cu^{2+} and Zn^{2+})
(j) Post-treatment handling

Factors influencing the outcome of physical mutagenesis are:

(a) Oxygen
(b) Moisture content
(c) Temperature
(d) Physical ionizing agents (electromagnetic [EM] and ionizing radiation)
(e) Dust and fibres (e.g. from asbestos)
(f) Biological and infectious agents (both viral and bacterial)

In general, the steps differ for sexually and asexually propagated crops, but common principles also exist.

The common practical considerations are:

(a) Thorough understanding of the genetic makeup of the traits to be improved. Polygenic traits have lesser chances of inducing mutations than monogenic traits.
(b) Knowledge of reproduction – sexual or asexual. For asexually propagated species, it is to be either in vitro or in vivo. If it is sexually propagated, type of fertilization (self or cross) is to be used.
(c) Determination of the material that is to be used for the propagation prior to treatment, i.e. gametes or seeds for sexually propagated crops; and stem cuttings, buds, nodal segments or twigs for asexually propagated ones.
(d) Knowledge of the karyotype, especially when there are hybridization barriers.
(e) Selection of an appropriate mutagen and dose. A pilot assay is advisable before large-scale treatment of propagules. Radiation dose is expressed in rads (radiation-absorbed dose) which is equivalent to absorption of 100 ergs/g (rad

is a unit of absorbed radiation dose, defined as 1 rad = 0.01 Gy = 0.01 joule/kg). The unit kilorad (kR which is 1000 rads) which was in use earlier is replaced by gray units (Gy). These two can be interconverted as 1 kR is equivalent to 10 Gy. A concept of LD_{50} (lethal dose 50%) is used to refer the optimum dose to be used in the experiment. By definition LD_{50} is the dose which causes 50% lethality in the organism used for irradiation in definite time. Generally, irradiated populations are generated by using an LD_{50} dose treatment and with a dose lower than LD_{50}. It can be determined by exposing different subsamples of the target plant material (seeds) to a range of doses (low to high) and monitoring survival of the plants in field (up to flowering or maturity). There are species sensitive to radiation. In such cases, doses lower than LD_{50} are also appreciable to reduce mutation load. Therefore, it is preferred to work out radiosensitivity test between LD_{25} or LD_{30} and LD_{50} to obtain desired mutation.

(f) Appropriate infrastructure (irradiation house, laboratories, screen house, fields, etc.) desired mutant selection.

(g) Isolation of chimaeras from stable mutants.

16.1.7 Mutation Breeding Strategy

The advantage of mutation breeding over other breeding strategies depends on efficient selection of superior variants in the second (M_2) or third (M_3) generation as summarized in Fig. 16.3. The generation nomenclature starts with M_0 for seed or pollen mutagenesis and M_0 V_0 for vegetative organs, where M stands for the meiotic and V for the vegetative generation. All materials are labelled with a "0" prior to mutagenesis and with a "1" after mutagenesis is performed. The first generation is unsuitable for evaluation as plants will be genotypically heterogeneous (chimaeric). The first generation suitable for selection in a seed-mutagenized material is M_2. Several cycles are needed to make a vegetatively propagated material genotypically homogeneous and to stabilize the inheritance of mutant alleles.

The first step in mutation breeding is to reduce the number of potential variants among the mutagenized seeds or other propagules of the first (M_1) plant generation to a significant level to allow close evaluation and analysis. The population size needs to be effectively managed. Population size depends on the inheritance pattern of the target gene. Hence, it is advisable to select mutagens that yield high frequency of mutations in order to reduce the population size of M_1. Genetically, M_1 mutant plants are heterozygous because only one allele is affected by one mutation. Probability of mutating both the alleles is extremely low. In M_1, dominant mutations can be identified as recessive mutation where expression is impossible. Screening for mutations in subsequent generations among segregants is the advisable option. In this way, breeder generates homozygotes for dominant or recessive alleles. M_1 population must be self-pollinated as cross-pollination that will produce new variation. Screening and selection starts in M_2 generation. Three main types of screening/ selection techniques are: physical/mechanical, visual/phenotypic and "other" methods. Physical or mechanical selection can be used efficiently to determine the shape, size, weight, density of seeds, etc. using appropriate sieving machinery.

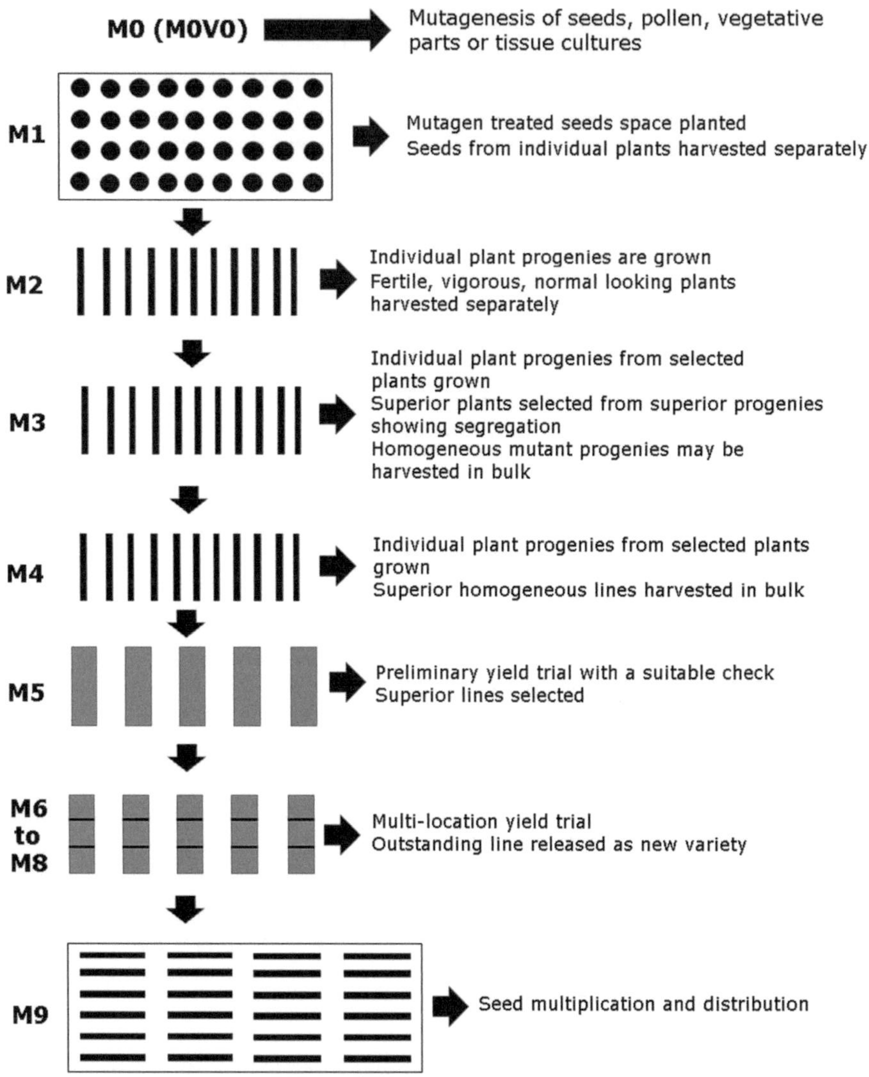

Fig. 16.3 Steps in mutation breeding. Traditional mutation breeding scheme. Each row describes the steps for a specific generation

Visual screening is the most effective and efficient method for identifying mutant phenotypes. Visual/phenotypic selection is often used in selection for plant height, adaptation to soil, growing period, disease resistance, colour changes, earliness in maturity, climate adaptation, etc. In the category of "others", physiological, biochemical, chemical and physicochemical procedures for screening may be used for selection of certain types of mutants. When a mutant line appears to possess a promising trait, the next stage is seed multiplication for extensive field trials. In this case, the mutant line, the mother cultivar and other varieties (local check) will be tested.

16.1.8 In Vitro Mutagenesis

In vitro mutagenesis is induction of mutations in explants or in vitro cultures (protoplasts, cells, tissues and organs). This is applicable to both seed-propagated and vegetatively propagated crops. In the latter, it is advantageous where a large number of uniformly growing in vitro cultures can be used. Cultured cells, organs and tissues have a developmental pattern; therefore those can be synchronized and separation of chimaeras can be done more efficiently. For seed crops, the use of haploid culture may provide additional benefits. In vitro mutagenesis involves the following steps:

(a) Selection of proper target material (explants or cultures)
(b) Mutagen selection, determination of proper dose and post-treatment handling and subcultures
(c) Regeneration of plants for mutant selection

A variety of explants are available like apical meristems, axillary buds, roots and tubers. Subcultures will determine chimaeras. In the first vegetative generation (M_1V_1), mutations are not expressive. If superior mutants are detected early, these should be monitored for stability in further generations i.e. up to M_1V_4 or M_1V_6. In banana, using recurrent irradiation in vitro, increased in vitro shoot multiplication and morphological variations were observed. Resistant plants to black sigatoka were derived through carbon ion beam irradiation of in vitro plantlets of banana (cv. Williams and Cavendish Enano).

Chimaeras can be easily isolated in in vitro culture by repetitive subculturing, normally involving about four generations (M_1V_4). In seed crops, backcross to the original line can exclude unwanted mutant genes (see Table 16.3 for details). It is feasible to exercise selection of agronomically useful and genetically determined traits in in vitro culture. Usage of culture medium added with a certain amount of herbicide, salt or aluminium or exposure of cultures to physical stress such as cold or heat can be exercised. This is to select cells/tissues with required tolerance or resistance. Such cells/tissues can be isolated, multiplied through subcultures and regenerated into plants. In vitro cultured explants provide a wider choice of controlled selection where large populations can be screened as against lower number of individuals in the case of in vivo plants.

16.1.9 Gamma Gardens or Atomic Gardens

This is a form of mutagenesis where plants are exposed to radioactive sources (cobalt-60) in order to generate mutations, some of which turned out to be useful. This resulted in the development of over 2000 new varieties of plants, most of which are now used in agricultural production. The "Todd's Mitcham" peppermint variety, resistant to verticillium wilt, produced by Brookhaven National laboratory, USA, during 1950s, is one of the first examples of variety produced by a gamma garden.

Table 16.3 In vitro mutagenesis in vegetatively propagated crops

Crop species	Treated material	Mutagen and dose (LD_{50} or applied dose)	Plant regeneration route	Selected mutants/lines
Banana (*Musa* spp.)	Shoot tips	Carbon ion beam (0.5–128 Gy)	Direct regeneration	Disease-resistant lines
Banana (*Musa* spp.)	Shoot tips	γ rays (60 Gy)	Direct regeneration	Mutant Novaria; earliness
Banana var. Lakatan, Latundan	Shoot tips	γ rays (40 Gy)	Direct regeneration	Height reduction, larger fruit size
Banana var. Latundan	Shoot tips	γ rays (25 Gy)	Direct regeneration	Mutant variety Kluai Hom Thong KU1
Pineapple var. Queen (*Ananas comosus* L.) Merr.	Crowns	γ rays	Axillary bud regeneration	Lines with reduced spines
Begonia rex	In vitro cultured leaflets	γ rays (30–40 Gy)	Adventitious bud regeneration	Leaf colour and shape mutants
Potato	Callus cultures	γ rays (30–50 Gy)	Adventitious bud regeneration	Tuber colour mutants
Sugarcane	Buds/callus cultures	γ rays (20–25 Gy)	Organogenesis/ embryogenesis	Mutants for agronomic traits
Cassava	Somatic embryos	γ rays (50 Gy)	Embryogenesis	Morphological mutants; mutants with storage root yield, altered cyanogen
Sweet potato	Embryogenic suspensions	γ rays (80Gy)	Embryogenesis	Mutants for salt tolerance
Pear	In vitro shoots	γ rays (3.5 Gy)	Microcuttings from shoots	Mutants for fruit shape and size
Chrysanthemum morifolium	Rooted cuttings	γ rays (25 Gy)	Direct shoot organogenesis	Yellow flower mutants (from white and red flower varieties)

The Rio Star grapefruit, developed at the Texas A&M Citrus Center in the 1970s, now accounts for over three quarters of the grapefruit produced in Texas is yet another example. After World War II, there was a concerted effort to find peaceful uses of atomic energy. One of the ideas was to subject plants to irradiation to produce mutations in plenty, through which disease - or cold-resistant or unusual coloured varieties can be derived. Such experiments were conducted in giant gamma gardens of the USA, Europe and the former USSR. Though modern genetic engineering replaced the need for atomic gardening, still the legacy being continued by the

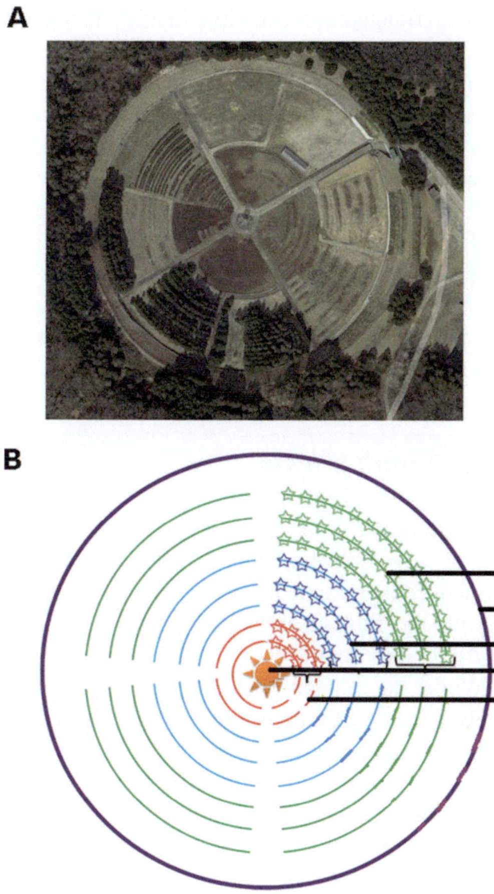

Fig. 16.4 (**a**) Aerial view of the gamma garden at the Institute of Radiation Breeding, Hitachiōmiya, Ibaraki Prefecture, Japan. (**b**) Layout of a gamma garden

Institute of Radiation Breeding in Japan that currently owns the largest and possibly the only surviving gamma garden in the world, at Hitachiōmiya in Ibaraki Prefecture (Fig. 16.4a). The circular garden measures 100 m in radius and enclosed by an 8-m high-shielding dike wall. Radiation (gamma rays) comes from a cobalt-60 source placed inside a central pole. The aim is to produce traits responsible for tolerance to fungus or consumer-friendly fruit colours. Overall development of new crop varieties with new traits is the purpose. In the words of nanotechnologist Paige Johnson of the University of Tulsa, Oklahoma, "if you think of genetic modification today as slicing the genome with a scalpel, in the 1960s they were hitting it with a hammer".

These gardens were designed to test effects of radiation on plant life. However, research gradually turned towards inducing beneficial mutations. They were typically five acres in size and were arranged in a circular pattern with a retractable radiation source in the middle (Fig. 16.4b). Plants were usually laid out like slices of a pie, stemming from the central radiation source. Radioactive bombardment will be usually for about 20 h, after which scientists wearing protective equipment would enter the garden to assess results. The plants nearest to the centre usually died, while the ones further out often featured tumours and other growth abnormalities. Plants beyond these were with a higher than usual range of mutations. These gamma gardens have continued to operate in the 1950s. Research into the potential benefits of atomic gardening has continued, most notably through a joint operation between the International Atomic Energy Agency (IAEA) and the UN's Food and Agriculture Organization (FAO). Japan's Institute of Radiation Breeding is well known for its modern-day usage of atomic gardening techniques.

16.2 Factors Affecting Radiation Effects

Ionizing radiation is energetic and penetrating, and its chemical effects in biological matter are due to initial physical energy deposition events, referred to as the *track structure*. Ionizing radiation exists in either particulate or electromagnetic types. The particulate radiation interacts with the biological tissue either by ionization or excitation. The ionizations and excitations tend to be localized, along the tracks of individual charged particles. While the photon penetrates the matter without interactions, it can be completely absorbed by depositing its energy or it can be scattered (deflected) from its original direction and deposit part of its energy as:

(a) Photoelectric interaction: a photon transfers all its energy to an electron positioned usually in the outer shell of the atom. The electron ejects from the atom and begins to pass through surrounding matter.
(b) Compton scattering: a portion of the photon energy is absorbed and the photon is scattered with reduced energy.
(c) Pair production: the photon interacts with the nucleus and an electron and a positively charged positron is produced. This only occurs with photons with energies in excess of 1.02 MeV.

16.2.1 Direct and Indirect Effects

Radiation damage causes damage to DNA molecules either directly or indirectly. In the direct action, radiation disrupts the molecular structure. This structural change either damages or kills the cell. Later, surviving damaged cells may have abnormalities. This process becomes predominant with high-LET radiations such as particles and neutrons and high radiation doses. In the indirect action, the radiation hits the water molecules, the major constituent of the cell and other organic molecules in the cell, whereby free radicals such as hydroxyl (HO) and alkoxy (R-O) are produced. Exposure of cells to ionizing radiation induces high-energy

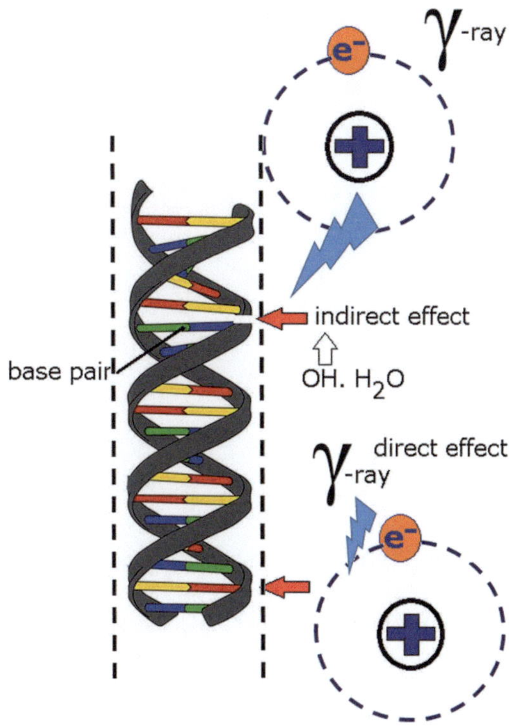

Fig. 16.5 Physical, biochemical and biological response mechanisms of radiation

radiolysis of H_2O molecules into H^+ and OH^- radicals. Such radicals are chemically reactive and in turn recombine to produce superoxide (HO_2) and peroxide (H_2O_2) that incur oxidative damage to molecules of the cell.

Free radicals are characterized by an unpaired electron and causes molecular structural damage to the DNA. Hydrogen peroxide is also toxic to the DNA. The result of indirect action on the cell is impairment of function or death. Number of free radicals produced by ionizing radiation depends on the total dose. Majority of radiation-induced damage is by indirect action since water constitutes nearly 70% of the composition. In addition to the damages caused by water radiolysis products, cellular damage may also involve reactive nitrogen species (RNS) and other species. This can occur as a result of ionization of atoms on constitutive key molecules (e.g. DNA). Either direct or indirect, the ultimate effect is the biological and physiological alterations. This may be manifested seconds or decades later. In the evolution of these alterations, genetic and epigenetic changes may be involved (Fig. 16.5).

16.2.2 Biological Effects

Biological effects are ionization of atoms of biomolecules that may cause chemical changes or eradicate its functions. The energy transmitted may act directly causing

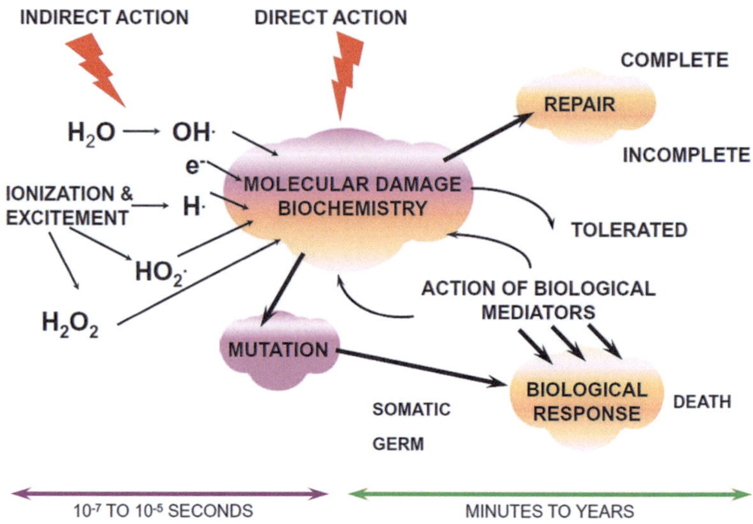

Fig. 16.6 Direct and indirect actions of radiation

ionization of the biological molecule or indirectly act through ionization of the water molecules that surround the cell (Fig. 16.6). Due to this, proteins can lose the functionality of its amino groups and thus increasing its chemical responsiveness. Enzymes would be deactivated and lipids will suffer peroxidation. Carbohydrates will get dissociated and nucleic acid chains will have ruptures/modifications. By all means, DNA is the primary target of radiation as it contains genes with information of cell functioning and reproduction. The energy deposition is a random process. Even low doses can deposit enough energy to result in cellular changes or cell death. But cells can recuperate from this damage. If the repair of DNA damage is incomplete, signalling pathways leading to cell death through apoptosis (death of cells as a normal and controlled part of an organism's growth or development) can happen. If mutation occurs, the cell will survive with modification in the DNA sequence. Mutated cells are capable of reproduction.

16.3 Molecular Mutation Breeding

Cells with damaged DNA will survive only when these damages are repaired correctly or erroneously. The result of erroneous repairs will be fixed in the genome as induced mutations. The nature and extent of DNA damage determines the molecular feature of induced mutations. For example, EMS often leads to G/C to A/T transition, while ion beam could cause deletion of DNA fragment of various sizes. While nucleotide substitution may produce a dominant allele, DNA deletions will cause recessive mutations. So, when a recessive mutation is required, irradiation may be preferred. When we need herbicide resistance (dominant mutation), the use of chemical mutagen is preferred.

Mutagenesis research has been revolutionized by advances in genomics including methods to detect genetic variation and select mutant phenotypes like:

(a) Transposon mutagenesis or transposition mutagenesis (a process that allows genes to be transferred to a host's chromosome)
(b) Insertional mutagenesis (creation of mutations of DNA by the addition of one or more base pairs. This can occur naturally or mediated by viruses or transposons)
(c) TILLING (targeting induced local lesions in genomes), ecoTILLING (ecotype TILLING) and high-resolution melting (HRM)
(d) Site-directed mutagenesis

Of these, the last two will be dealt here in some detail, since the first two are largely done in microorganisms.

16.3.1 TILLING and EcoTILLING

TILLING is a method that allows directed identification of mutations in a specific gene. TILLING was first done in *Arabidopsis thaliana* and thereafter successfully used in corn, wheat, soybean, tomato and lettuce. TILLING relies on the ability of a special enzyme to detect mismatches in normal and mutant DNA strands when they are annealed. Seed is treated with either ethyl methanesulphonate (EMS) or sodium azide to generate a population of plants with random point mutations. By selectively pooling the DNA and amplifying with unlabelled primers, mismatched heteroduplexes are generated between wild-type and mutant DNA. Heteroduplexes are incubated with the plant endonuclease CEL I (that cleaves heteroduplex mismatched sites), and the resultant products are visualized on a Fragment Analyzer. Subsequent analysis of the individual plant DNA from the pool DNA identified the plant bearing the mutation. There are 10 steps in TILLING (Fig. 16.7). This is a high-throughput process to identify single-nucleotide mutations in a gene of interest. This is also a powerful detection method that can result from chemical-induced mutagenesis. TILLING was first used by Claire McCallum in the late 1990s in *Arabidopsis.*

Outline of the basic steps for typical TILLING and EcoTILLING assays:

(a) Seeds are mutagenized with chemical mutagens. The resulting M_1 plants are self-fertilized.
(b) DNA samples are prepared from M_2 individuals for mutational screening. DNA is collected from a mutagenized population (TILLING) or a natural population (EcoTILLING).
(c) For TILLING, DNAs are pooled. Typical EcoTILLING assays do not use sample pooling, but pooling has been used to discover rare natural single-nucleotide changes.
(d) After extraction and pooling, samples are typically arrayed into a 96-well format.

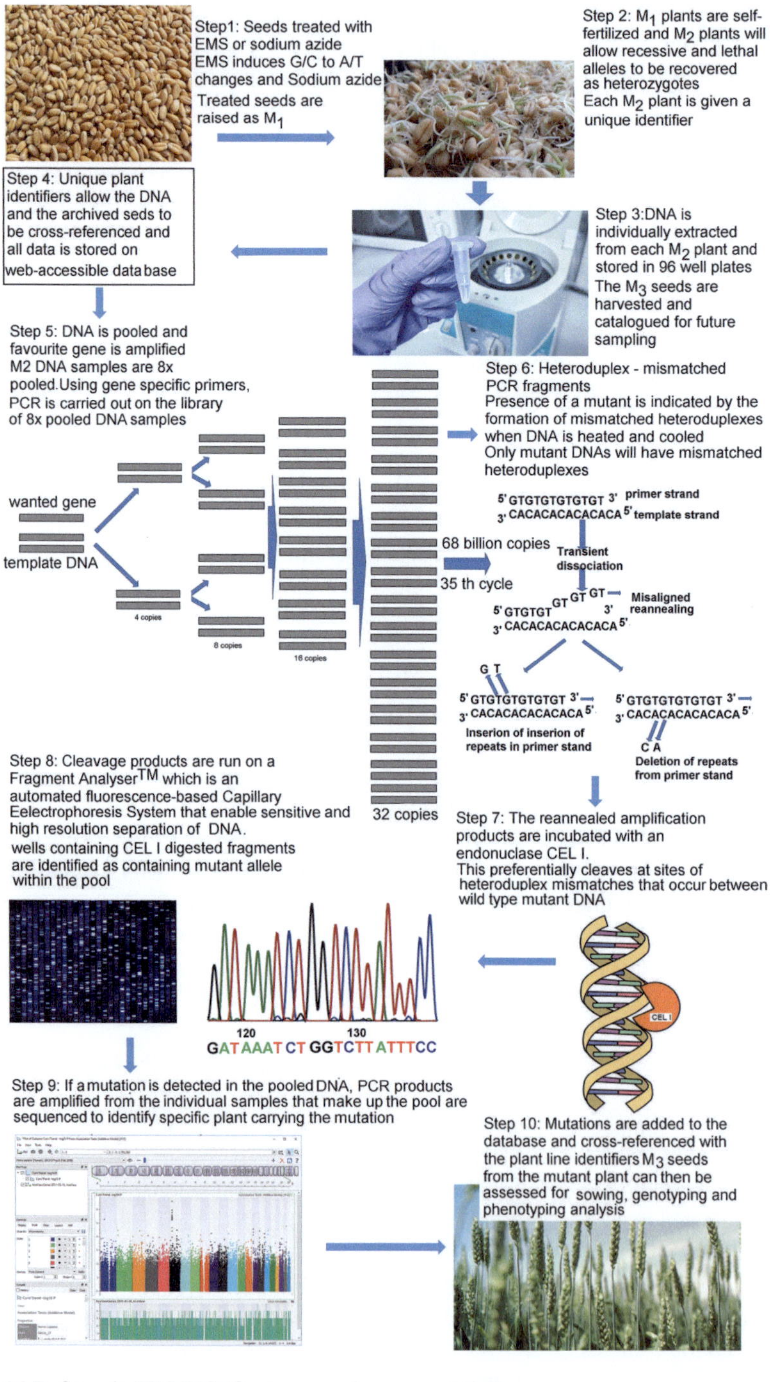

Fig. 16.7 Steps in TILLING (figures are only representative)

(e) The target region is amplified by PCR with gene-specific primers that are end-labelled with fluorescent dyes.
(f) Following PCR, samples are denatured and annealed to form heteroduplexes that become the substrate for enzymatic mismatch cleavage. Cleavage at mismatched site done by enzyme CEL I.
(g) Cleaved bands representing mutations or polymorphisms are visualized using denaturing polyacrylamide gel electrophoresis.

EcoTILLING uses TILLING techniques to look for natural mutations. DEcoTILLING is an altered method of TILLING and EcoTILLING to identify fragments. After NGS sequencing technologies were discovered, TILLING by sequencing has been developed based on Illumina sequencing of target genes amplified from multidimensionality pooled templates to identify possible single-nucleotide changes (see Chap. 24 on Genomics for details).

16.3.2 Site-Directed Mutagenesis

Site-directed mutagenesis makes specific and intentional changes to the DNA. This is otherwise known as oligonucleotide-directed mutagenesis and is used for investigating the structure of DNA, RNA and protein molecules and for protein engineering. The basic procedure requires the synthesis of a short DNA primer. This synthetic primer contains the desired mutation and is also complementary to the template DNA around the mutation site, so it can hybridize with the DNA in the gene of interest. The mutation may be a single base change (point mutation), multiple base changes, deletion or insertion. DNA polymerase is used to extend the single-strand primer that copies the rest of the gene sequence. The gene thus copied contains the mutated site and is then introduced into a host cell as a vector and cloned. DNA sequencing is undertaken to select the desired mutation.

The aforesaid method using single-strand primer extension was inefficient due to a low yield of mutants.

Some of the modified methods for site-directed mutagenesis are:

(a) *Kunkel's method*: This was introduced by Thomas Kunkel in 1985. Here, the DNA fragment to be mutated is inserted into a phagemid (DNA-based cloning vector) and is then transformed into an *E. coli* strain deficient in two enzymes, dUTPase (dut) and uracil deglycosidase (udg). Both enzymes are part of a DNA repair pathway that protects the bacterial chromosome from mutations by the spontaneous deamination of dCTP to dUTP. The dUTPase deficiency prevents the breakdown of dUTP, resulting in a high level of dUTP in the cell. The uracil deglycosidase deficiency prevents the removal of uracil from newly synthesized DNA. As the double mutant *E. coli* replicates the phage DNA, its enzymatic machinery may, therefore, mis-incorporate dUTP instead of dTTP, resulting in single-strand DNA that contains some uracils (ssUDNA). The ssUDNA thus produced is extracted from the bacteriophage that is released into the medium and then used as template for mutagenesis. An oligonucleotide containing the desired mutation is used for primer extension. The heteroduplex DNA thus formed consists of one parental non-mutated strand containing dUTP and a mutated strand containing dTTP. The DNA is then transformed into an *E. coli* strain carrying the wild-type *dut* and *udg* genes. Here, the uracil-containing parental DNA strand is degraded, so that nearly all of the resulting DNA consists of the mutated strand.

(b) *Cassette mutagenesis*: Cassette mutagenesis need not involve primer extension using DNA polymerase. Here, a fragment of DNA is synthesized and then inserted into a plasmid. It involves the cleavage by a restriction enzyme at a site in the plasmid and subsequent ligation of a pair of complementary oligonucleotides containing the mutation in the gene of interest to the plasmid. Usually, the restriction enzymes cut the plasmid permitting sticky ends of the plasmid and insert to ligate to one another. This method can generate mutants at close to 100% efficiency. The drawback with this method is that it will allow mutations only at sites that can be cleaved by the restriction enzymes.

(c) *PCR site-directed mutagenesis*: Cassette mutagenesis mutates restriction sites only. This may be overcome by using polymerase chain reaction with oligonucleotide primers so that a larger fragment may be generated, covering two convenient restriction sites. The fragment containing the desired mutation can be separated from the original by gel electrophoresis. Variations employ three or four oligonucleotides, two of which may be non-mutagenic oligonucleotides that cover two convenient restriction sites and generate a fragment that can be digested and ligated into a plasmid, whereas the mutagenic oligonucleotide may be complementary to a location within that fragment well away from any convenient restriction site. These methods require multiple steps of PCR so that the final fragment to be ligated can contain the desired mutation. The design process for generating a fragment with the desired mutation and relevant restriction sites can be cumbersome. Software tools like *SDM-Assist* can simplify the process.

16.3.3 MutMap

MutMap is a method of rapid gene isolation using a cross of a mutant to wild-type parental line. The large F_2 population will be screened to isolate mutant through SNP (single-nucleotide polymorphism) analysis (Fig. 16.8). This technique applied in rice can be explained as follows:

Use a mutagen (say EMS) to mutagenize a rice cultivar (X) that has a reference genome sequence. To make the mutated gene homozygous, plants of first mutant generation (M_1) are self-pollinated to raise M_2 and further generations. Phenotypes in the M_2 and advanced generations are screened to isolate recessive mutants with altered traits like plant height, tiller number and grain number per spike. This mutant is crossed with the cultivar used for inducing mutations (wild type). The resulting F_1 is self-pollinated, and the F_2 are grown in the field for scoring the phenotype. Since

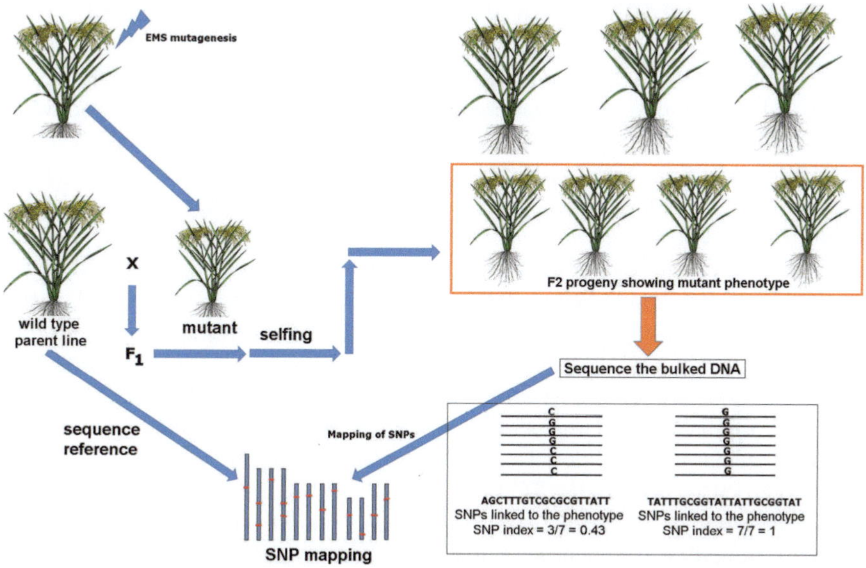

Fig. 16.8 A scheme for MutMap in rice. A rice cultivar with a reference genome sequence is mutagenized by EMS. A semi-dwarf phenotype mutant is crossed to the wild-type plant of the same cultivar used for the mutagenesis. F_2 is raised from F_1 to have both mutant and wild-type phenotypes. Crossing of the mutant to the wild-type parental line ensures detection of phenotypic differences at the F_2 generation between the mutant and wild type. DNA of F_2 displaying the mutant phenotype are bulked and subjected to whole-genome sequencing followed by alignment to the reference sequence. SNPs with sequence reads composed only of mutant sequences (SNP index of 1) are closely linked to the causal SNP for the mutant phenotype (courtesy: *Nature Biotechnology*)

F_2 progeny are derived from a cross between the mutant and its parental wild-type plant, the number of segregating loci responsible for the phenotypic change is minimal (in most cases, one). But the segregation of phenotypes in F_2 shall be prominent even if the phenotypic differences are small. It is appropriate to use SNPs to identify nucleotide changes incorporated into the mutant. They are detected as insertion-deletions (indels) between mutant and wild type. In the F_2 progeny, the majority of SNPs will segregate in a 1:1 mutant/wild type ratio. However, the SNP responsible for the change of phenotype is homozygous in the progeny showing the mutant phenotype. When DNA samples are collected from recessive mutant of F_2 progeny, and bulk sequenced, 50% mutant and 50% wild-type sequence reads are expected. However, the causal SNP and closely linked SNPs should show 100% mutant and 0% wild-type reads. On the other hand, SNPs loosely linked to the causal mutation should have >50% mutant and <50% wild-type reads. If SNP index is defined as the ratio between the number of reads of a mutant SNP and the total number of reads corresponding to the SNP, this index would equal 1 near the causal gene and 0.5 for the unlinked loci.

16.4 The FAO/IAEA Joint Venture for Nuclear Agriculture

Over the last 45 years, the Joint FAO/IAEA Programme of Nuclear Techniques in Food and Agriculture (headquartered in Vienna, Austria) supported worldwide countries' efforts to attain food security. The Plant Breeding and Genetics Section of this programme assists countries in using radiation-induced mutations, facilitated by biotechnologies, to develop superior crop varieties. The mandate of Joint FAO/IAEA Programme is constitution of field projects in developing countries, coordination of collaborative research network and a research and development laboratory arm in Seibersdorf, outside Vienna, Austria. As of now, there are a total of 86 field projects relating to the development of mutants dealing with biotic, abiotic and nutritional aspects (Tables 16.4a, 16.4b and 16.4c) (The information provided is not exhaustive). Through Technical Cooperation Projects (TCP), the technology transfer is accomplished characterized through strengthening of human and infrastructural capabilities. The irradiation facilities (majority are with cobalt-60 sources) are provided through TCP.

As per FAO/IAEA Mutant Varieties Database, more than 3222 mutant varieties are released in different countries. China, India, the former USSR, the Netherlands, Japan and the USA are the leading countries having the highest number mutant varieties. Highest proportion of mutants (>50%) is with gamma rays compared to other mutagens (Table 16.5). Crop wise, cereals stand first followed by ornamentals and legumes (see Table 16.6). Rice stands first (700 mutant varieties) in among crops followed by barley, wheat, maize, durum wheat, oat, millet, sorghum and rye (Table 16.7). As per the FAO/IAEA database, 1825 mutants (accounting to 57%) have either better agronomic or botanical traits. Of these, 577 (18%) mutants are developed for increase in yield and related traits, 321 (10%) mutants for better quality and nutritional content, 200 (6%) mutants for biotic and 125 (4%) mutants for abiotic stress tolerance. These programmes have benefited the local economies through contributing millions of dollars annually.

Table 16.4a Applications of induced mutagenesis for biotic stress resistance in plant breeding

Highlight	Crop
Resistance to bacterial wilt (*Ralstonia solanacearum*)	Tomato
Resistance to stem rot (*Sclerotinia sclerotiorum*)	Rape seed
Resistance to powdery mildew (*Podosphaera leucotricha*) and apple scab (*Venturia inaequalis*)	Apple
Resistance to *Ascochyta* blight and *Fusarium* wilt	Chick pea
Resistance to yellow mosaic virus	Mungbean
Resistance to black stem rust	Durum wheat
Resistance to stripe rust	Wheat
Resistance to blast, yellow mottle virus, bacterial leaf blight and bacterial leaf stripe	Rice
Resistance to *Myrothecium* leaf spot and yellow mosaic virus	Soybean
Resistance to bacterial blight, cotton leaf curl virus	Cotton
Resistance to Phytophthora nicotianae var. *parasitica*	Sesame
Resistance against pathogen striga (*Striga asiatica*)	Maize

Table 16.4b Applications of induced mutagenesis for abiotic stress resistance in plant breeding

Highlight	Crop
Lodging resistance, acid sulphate soil tolerance	Rice
Semi-dwarf cultivar/dwarf	Rice
	Sunflower
Early maturity	Rice
High fibre quality	Cotton
Adaptation	Rice
Acidity and drought tolerance	Lentil (*Lens culinaris* Medikus), maize
Tolerance to cold and high altitudes	Rice
Acidity and drought tolerance	Rice
Salinity tolerance	Rice, barley, sugarcane

Table 16.4c Applications of induced mutagenesis in the improvement of crop quality and nutritional traits in plant breeding

Highlight	Crop
Oil quality improvement	Soybean
	Canola
	Peanut
	Sunflower
Improvement of protein quality	Soybean, maize
High-amylose content preferred by diabetes patients because it lowers the insulin level, which prevents quick spikes in glucose contents	Cassava
Oilseed meals low in phytic acid desirable in poultry and swine feed	Soybean
Phytate (storage compund of phosphorus in seeds)	Barley
High-resistant starch in rice (RS) preferred by diabetic patients	Rice
Giant embryos (containing more plant oils); low amylose content; low protein content (for special dietary needs) rice	Rice
Dark green obovate leaf pod; increased seed size, higher yield, moderately resistant to diseases, increased oil and protein content	Groundnut

Table 16.5 Number of officially released mutant varieties

Mutagen	Number of released mutant cultivars
Gamma rays	910
X-rays	311
Gamma chronic	61
Fast neutrons	48
Thermal neutrons	22
Ethyl methanesulphonate	106
Sodium azide	11
N-Ethyl-N-nitrosourea	57
N-Ethyl-N-nitrosourea	46

Source: FAO

Table 16.6 Number of released mutant varieties in cereals and legumes

Species	Number of mutants
Cereals	
Avena sativa (oat)	23
Hordeum vulgare (barley)	304
Oryza sativa (rice)	815
Secale cereale (rye)	4
Triticum aestivum (bread wheat)	254
Triticum turgidum (durum wheat)	31
Zea mays (maize)	96
Total	1527
Legumes	
Arachis hypogea (groundnut)	72
Cajanus cajan (pigeon pea)	7
Cicer arietinum (chickpea)	21
Dolichos lablab (hyacinth bean)	1
Lathyrus sativus (grass pea)	3
Lens culinaris (lentil)	13
Glycine max (soybean)	170
Phaseolus vulgaris (French bean)	59
Pisum sativum (pea)	34
Trifolium alexandrinum (Egyptian clover)	1
T. incarnatum (crimson clover)	1
T. pratense (red clover)	1
T. subterraneum (subterranean clover)	1
Vicia faba (faba bean)	20
V. mungo (black gram)	9
V. radiata (mung bean)	36
V. unguiculata (cowpea)	12
Total	462

16.4.1 Mutation Breeding in Different Countries

Continent wise, Asia stands first in terms of mutant varieties released (Fig. 16.9). China stands first in terms of development of new varieties through induced mutagenesis. It is well ahead of other countries in number of released varieties (Fig. 16.10). Crop wise, cereals own the maximum percentage of varieties released (48%) (Fig. 16.11).

Japan used irradiation, chemical mutagenesis and somaclonal variation to release 242 mutant varieties. Due to successful efforts of Institute of Radiation Breeding, 61% of these varieties were induced by gamma rays. Some mutant cultivars of Japanese pear exhibit resistance to diseases. In addition, 228 indirect use (hybrid) mutant varieties primarily generated in rice and soybean have found value as

Table 16.7 Leading rice varieties obtained by mutation breeding

Country	Variety	Details
Pakistan	Shada Shua-92 Khushboo-95	Yield potential of 7 t/ha; fine grain quality; cultivated on over 60,000 ha; generating 21 million USD to the rural economy
	Sarshar	Yield potential of 8.5 t/ha; covers over 160,000 ha; contributing an additional 223 million USD to the rural economy
		Short stature; high yield of 5.5 t/ha; cultivated on over 200,000 ha; generating an additional 8 million USD to farmers
		Yield potential of 9.5 t/ha; cultivated on over 80,000 ha; generating an additional income of 32 million USD to farmers
Myanmar	Shwewartun	Improved grain yield, seed quality and early maturity; covered more than 800,000 ha in 1989–1993; approximately 17% of the area under rice in Myanmar
Thailand	RD6 and RD15	In 1989–1998, these two varieties yielded 42.0 million tons paddy or 26.9 million tons milled rice, which was worth USD 16.9 billion
China	Zhefu 80 Jiahezazhan and Jiafuzhan	Short life cycle (105_108 days); high-yield potential; wide adaptability; high resistance to rice blast and tolerance to cold even under infertile conditions or poor management; total area of 10.6 million ha in 1986–1994
		Early maturity; high yield and grain quality; plant hopper- and blast resistance and wide adaptability; planted on ca. 363,000 ha in Fujian province of China
Vietnam	VND_95_20 TNDB_100 and THDB	Grown on more than 300,000 ha/year; has become the top variety in southern Vietnam, both as an export variety and in terms of its growing area
		Tolerant to high salinity and acid sulphate soils; grown on over 220,000 ha in 2009
Egypt	Giza 176 and Sakha 101	Leading varieties with a potential yield of 10 t/ha
Japan	18 varieties	Income worth US$ 937 million per year
India	PNR-102 and PNR-381	Income worth US$ 1748 million per year
Costa Rica	Camago 8	Current annual planted area 30% rice-growing area in Costa Rica
Australia	Amaroo	Current annual planted area 60–70% of the rice-growing area in Australia
California, USA	Calrose 76; M-7; M-101; S-201 and M-301	Cultivated on over 220,000, 450,000, 150,000, 675,000 and 150,000 ha of land respectively

parental breeding germplasm resources in Japan. In 2005, the total cultivated area of mutant rice cultivars was 2,10,692 ha (12.4% of the total cultivated rice area). Income from mutant cultivars was estimated to be nearly 250 billion Yen (2.34 billion US dollars) in 2005.

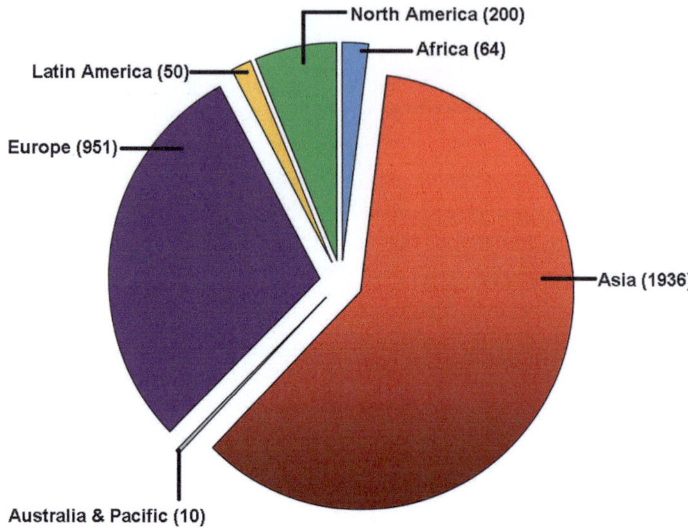

Fig. 16.9 Number and proportion of mutant cultivars released, categorized by continents (source: IAEA mutant Database)

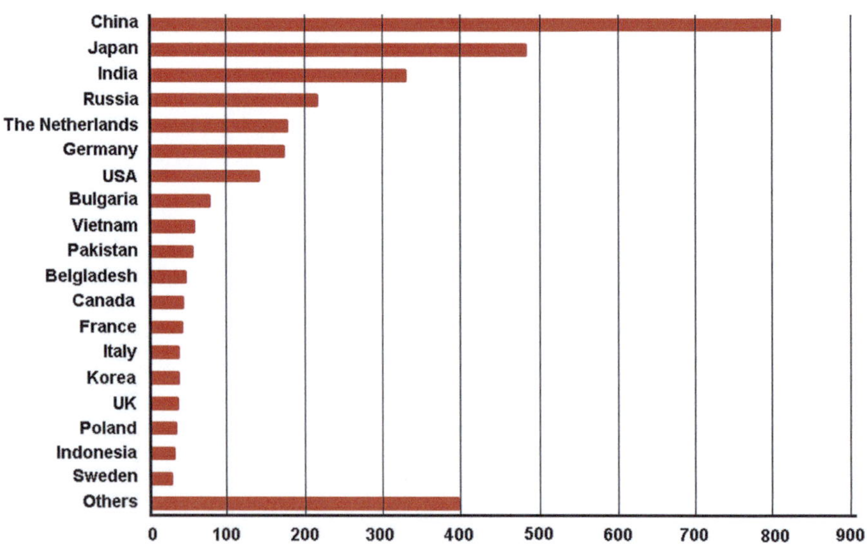

Fig. 16.10 Number of mutant cultivars released in different countries (source: FAO)

India initiated sustained efforts to use induced mutations in the late 1950s. Between 1950 and 2009, India developed about 329 mutant varieties in rice, wheat, barley, pearl millet, jute, groundnut, soybean, chickpea, mung bean, cowpea, black gram, sugarcane, chrysanthemum, tobacco and dahlia. Indian Agricultural Research Institute (IARI), Bhabha Atomic Research Centre, Tamil Nadu

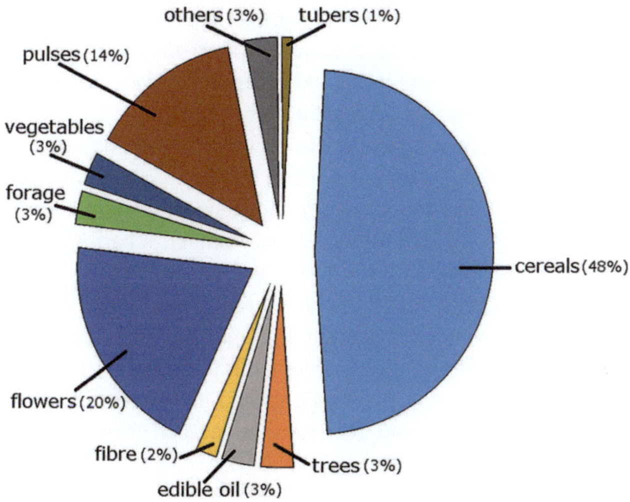

Fig. 16.11 Mutants released in various crops

Agricultural University and the National Botanical Research Institute were the prime institutions involved. Several gamma-irradiated rice mutants were released in India as high-yielding varieties under the series "PNR". Two early ripening and aromatic rice varieties, "PNR 381" and "PNR 102", are currently popular with farmers in the states of Haryana and Uttar Pradesh.

Wide use of high-yielding varieties made Vietnam the second largest exporter of rice, exporting 4.3 million tons per year. Currently, mutant varieties contribute to 15% of the annual rice production. Around 55 mutant varieties have been developed, most of which are rice. Mutant rice are planted in over 1.0 million ha, including Hatay, Bacgiang, Nghean, Vinhphuc, Hanam, Thaibinh and Hanoi of northern Vietnam, which led to poverty relief. Besides higher yield, varieties with aroma, protein and amylase content were also derived. Tolerance to salinity, cold, drought and lodging was given prime importance. Nearly 2,540,000 ha are cultivated with mutant varieties of crops with a return of 374.4 million USD.

In Thailand, the work on induced mutations in rice commenced in 1965 and was stimulated in cooperation with IAEA. Two aromatic *indica*-type varieties of rice, "RD6" and "RD15", which were developed by gamma irradiation of a popular rice variety, "KhaoDawk Mali 105" ("KDML 105") and were released in 1977 and 1978, respectively. Even after 40 years, these varieties are still popular. RD6 has glutinous endosperm and retains all of the grain characters, including the aroma of its parent variety. In contrast, RD15 is non-glutinous and aromatic, similar to the parent, but ripens 10 days earlier than the parent. According to the Bureau of Economic and Agricultural Statistics of Bangkok, during 1997–1998, RD6 was grown on 2,524,576 ha, covering 32.1% of the area under rice that produced 4,599,995 tons paddy.

In Bangladesh, more than 44 mutant varieties belonging to 12 different crop species have been released through mutation breeding. The Bangladesh Institute of

Nuclear Agriculture in Mymensingh is the prime institution for mutation breeding that released up to eight mutant rice varieties. Rice mutants, including Binasail, Iratom-24 and Binadhan-6, were all planted in a cumulative area of 795,000 ha and contributed substantially towards food security.

USA produced a semi-dwarf gene allele (sd1) in rice through gamma ray mutagenesis. This triggered the American version of the "Green Revolution" in rice. Stadler, a high-yielding wheat mutant, is another success story. Stadler is resistant to leaf rust and loose smut with lodging resistance. Luther, a barley mutant, had 20% increased yield, shorter straw, higher tillering and better lodging resistance. Luther was grown in 120,000 acres with an estimated return of 1.1 million US dollars per year. It was used extensively in crossbreeding and several mutants were released. Pennrad is yet another high-yielding winter barley mutant with winter hardiness, early ripening and better lodging resistance grown in 100,000 ha in the USA. The grapefruit varieties, Star Ruby and Rio Red, developed through thermal neutron mutagenesis are sold under the trademark "Rio Star".

In Pakistan, at the Nuclear Institute for Agriculture and Biology, crops selected for improvement include rice, chickpea, mungbean and cotton. Improvement has been sought in plant architecture, maturity period, disease resistance, etc. The primary triumph of the Nuclear Institute of Agriculture is the release of four improved varieties of rice that were obtained using induced mutagenesis (Table 16.7).

European countries have been active in mutation breeding programmes. Bulgaria released 76 new cultivars produced from induced mutagenesis of which maize has the largest number of varieties (26 varieties). Kneja 509, a maize hybrid, occupies up to 50% of the growing area. In other European countries, development of short height and high-yielding mutant cultivars of barley 'Golden Promise' and 'Diamant' have made a major impact on the brewing industry. These have also been used as parents for many leading barley cultivars across Europe, North America and Asia. Golden Promise (developed through gamma ray irradiation of malting cultivar 'Maythorpe') was released in Czechoslovakia in 1965 through gamma ray irradiation of 'Valticky'. 'Diamant' has 12% increased yield, 15 cm shorter in height, occupying 43% of the barley area. Golden Promise is popular in Ireland, Scotland and the UK for brewing. These mutants are part of the commitment of the Joint FAO/IAEA programme for global food security. Mutation breeding-derived crop varieties around the world demonstrate the potential as a flexible and practicable approach to have desirable crop varieties. There are several host institutions all over the world to conserve mutant stocks (see Table 16.8). Few of the crop varieties released through classical mutagenesis since 2010 is available in Table 16.9.

16.5 Polyploidy Breeding

Polyploids are organisms with multiple sets of chromosomes in excess of the diploid number. Polyploidy is a natural mechanism that provides adaptation and speciation. Among angiosperms, 50% to 70% of the species have undergone polyploidy during the course of evolution. Flowering plants form polyploids at a significantly high

Table 16.8 Some characterized mutant stocks of crops and the host institutions

Crop	Host institution
Maize	The Maize Genetics Cooperation Stock Centre, University of Illinois, Urbana/Champaign, IL, USA
Arabidopsis	European Arabidopsis Stock Centre (or Nottingham Arabidopsis Stock Centre, NASC), University of Nottingham, Sutton Bonington Campus, UK
	Arabidopsis Biological Resource Centre, (ABRC), Ohio State University, OH, USA
Tomato	CM Rick Tomato Genetics Resource Centre, University of California at Davis, CA, USA
Cucurbits (cucumber, melon, cucurbit and watermelon)	Cucurbit Genetics Cooperative (CGC), North Carolina State University Raleigh, NC, USA
Rice	The Oryzabase of the National BioResource Project – Rice National Institute of Genetics, Japan
	IR64 Rice Mutant Database of the International Rice Functional Genomics, International Rice Research Institute, Manila, Philippines
	Plant Functional Genomics Lab., Postech Biotech Center, San 31 Hyoja-dong, Nam-gu Pohang, Kyoungbuk, Korea
Barley and wheat	Barley mutants, Scottish Crop Research Institute, Dundee, Scotland
	Barley and Wheat Genetic Stock of the USDA-ARS, USDA-ARS Cereal Crops Research Unit, Fargo, ND, USA
	Wheat Genetics Resource Center, Kansas State University, Manhattan, KS, USA
	Wheat Genetic Resources Database of the Japanese National BioResource Project
Pea	Pea mutants, John Innes Centre, Norwich, UK

frequency of 1 in every 100,000 plants. To understand polyploidy, a few basic notations need be defined. The total number of chromosomes in a somatic cell is designated "2*n*". The total number of chromosomes in a somatic cell is twice the haploid number (*n*) in the gametes (see Fig. 16.12). There may be more polyploid species in a given genera. The haploid chromosome number of diploid species of a polyploidy series is known as the basic chromosome number (*x*). For example, in wheat, we have tetraploid and hexaploid wheat (see Fig. 16.13). The ploidy of some of the major crops in the world is represented in Table 16.10.

16.5.1 Types of Changes in Chromosome Number

Polyploids are classified as euploids or aneuploids based on their chromosomal composition. Euploids are in majority that are multiples of the complete set of chromosomes specific to a species. Based on composition of genome, euploids are either autopolyploids or allopolyploids. A common class of euploids are tetraploids (see Table 16.11).

Table 16.9 Few crop varieties released through classical mutagenesis since 2010

Name	Common name	Commercial name	Trait improved	Country	Registration year
Glycine max	Soybean	Albisoara	Drought tolerant, high protein content and high yield	Republic of Maldova	2010
Pinus avium	Cherry	ALDAMLA	Improved fruit quality	Turkey	2014
Glycine max	Soybean	Amelina	High protein content and high yield	Republic of Maldova	2010
Arachis hypogaea	Ground nut	Binachinabadam-5	Salinity tolerance	Bangladesh	2011
Oryza sativa	Rice	Bijnadhan-14	Flowering in long days, short height, long grains	Bangladesh	2013
Triticum aestivum	Wheat	Binagom-1	Salt tolerance	Bangladesh	2016
Sesamum indicum	Sesame	Birkan	Higher yield	Turkey	2011
Prunus avium	Sweet cherry	BURAK	Improved quality, yield and size	Turkey	2014
Vigna radiata	Mungbean	Chai Nut 84-1	Improved quality, yield and size	Thailand	2012
Glycine max	Soybean	Clavera	Increased yield and drought tolerant	Republic of Maldova	2010
Capsicum annum	Vegetable Pepper	F1 Orange Beauty	Improved food quality, disease resistance	Russian Federation	2011
Oryza sativa	Rice	Goldami 1ho	Improved food quality	Republic of Korea	2011
Arachis hypogaea	Ground nut	GPBD 5	Larger seed	India	2010
Triticum aestivum	Wheat	Hangmai 901	Increased yield, drought tolerant	China	2011
Carthamus tinctorious	Safflower	Inshas 10	High yield, modified quality and insect resistance	Egypt	2011
Lycopersicon esculentum	Tomato	Lanka Cherry	Easily distinguishable pear shaped fruits	Sri Lanka	2010
Triticum aestivum	Wheat	Longfumai 19	High yield, drought tolerant	China	2010

(continued)

Table 16.9 (continued)

Name	Common name	Commercial name	Trait improved	Country	Registration year
Glycine max	Soybean	Mutiara 1	High yield, high protein content and disease resistance	Indonesia	2010
Solanum tuberosum	Potato	NAHITA	Early maturity	Turkey	2016
Sorghum bicolor	Sorghum	PAHAT	Higher yield, semi-dwarf, early maturity, improved grain quality	Indonesia	2011
Oryza sativa	Rice	Pandan Putri	Higher yield, early maturity, tolerance to bacterial leaf blight	Indonesia	2010
Glycine max	Soybean	Rosa	Higher yield, biotic stress resistance	Bulgaria	2010
Hordeum vulgare	Barley	Scope	Herbicide tolerance, higher yield, early maturity	Australia	2010

Source: Joint FAO/IAEA mutant variety database

Autopolyploidy Autopolyploids are otherwise called autoploids. They are with multiple sets of basic set (x) of chromosomes of the same genome. In nature, autoploids result from union of unreduced gametes or can be induced artificially. Natural autoploids include tetraploid crops like alfafa, peanut, potato and coffee and triploid bananas. Such species occur spontaneously through chromosome doubling. In ornamentals and forages, chromosome doubling led to increased vigour. Induced autotetraploids in watermelon are utilized for producing seedless triploid hybrids. This is accomplished through treating diploids with mitotic inhibitors like *dinitroanilines* and *colchicine*. Apart from chromosome counts, ploidy status of induced polyploids can be determined through chloroplast count in guard cells; morphological features such as leaf, flower or pollen size (gigas effect) and flow cytometry.

Allopolyploidy They are also called alloploids. Alloploids are a combination of genomes of different species. Hybridization of two or more genomes followed by chromosome doubling or fusion of unreduced gametes leads to such phenomena. This process occurs in nature as a key process of speciation in angiosperms and ferns. Economically important natural alloploids are strawberry, wheat, oat, upland cotton, oilseed rape, blueberry and mustard. Each genome is designated by a

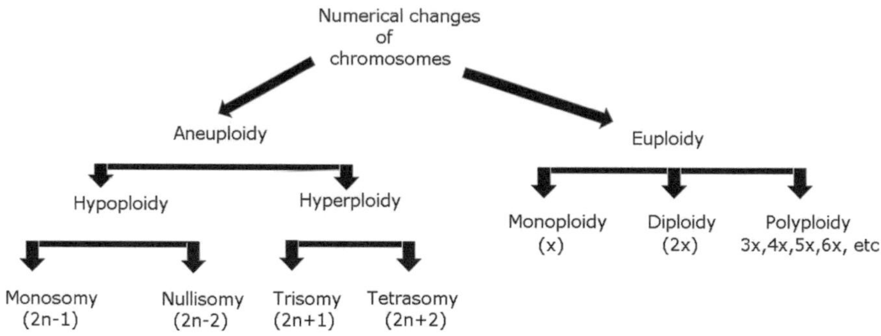

Fig. 16.12 Different kinds of changes in chromosomes (x = basic chromosome number; $2n$ = somatic chromosome number)

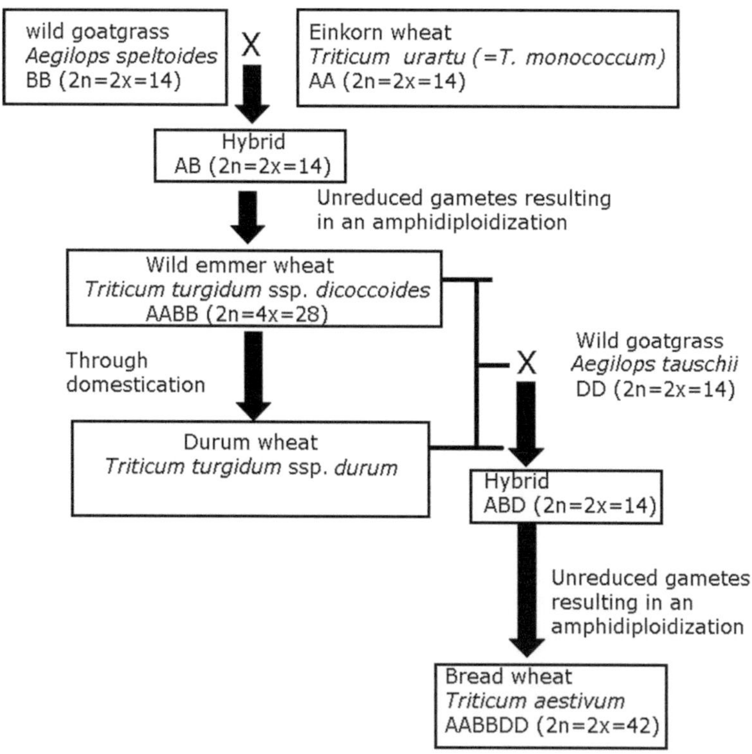

Fig. 16.13 Derivation of bread wheat

different letter to differentiate between the sources of the genomes in an alloploid. The cultivated mustards (*Brassica* spp.) can be explained in a triangle with each genome represented by a letter (Fig. 16.14a). The degree of homology between genomes differs with some being able to undergo chromosome pairing. The phenomenon becomes segmental alloploidy when only segments of chromosomes of the

Table 16.10 Examples of polyploid crops (somatic chromosome number is in brackets)

Crop	Species
Cereals	*Triticum aestivum* (6× = 42); *T. durum* (4× = 28); *Avena sativa* (6× = 42); *A. nuda* (6× = 42)
Forage grasses	*Dactylis glomerata* (4× = 28); *Festuca arundinacea* (4× = 28); *Agropyron repens* (4× = 28); *Paspalum dilatatum* (4× = 40)
Legumes	*Medicago sativa* (4× = 32); *Lupinus albus* (4× = 40); *Trifolium repens* (4× = 32); *Arachis hypogaea* (4× = 40); *Lotus corniculatus* (4× = 32); *Glycine max* (4× = 40)
Industrial plants	*Nicotiana tabacum* (4× = 48); *Coffea* spp. (4× = 44 fino a 8×); *Brassica napus* (4× = 38); *Saccharum officinale* (8× = 80); *Gossypium hirsutum* (4× = 52)
Tuber plants	*Solanum tuberosum* (4× = 48); *Ipomoea batatas* (6× = 96); *Dioscorea sativa* (6× = 60)
Fruit trees	*Prunus domestica* (6× = 48); *Musa* spp. (3× = 33; 4× = 44); *Citrus aurantifolia* (3× = 27); *Actinidia deliciosa* (4× = 116); *P. cerasus* (4× = 32)

Table 16.11 Common types of changes in chromosome number

Type	Change in chromosome number	Symbol
Heteroploid	Change from the n state	
A. Aneuploid	One of a few chromosome extra or missing from 2n	2n ± few
Nullisomic	One chromosome pair missing	2n-2
Monosomic	One chromosome missing	2n-1
Double monosomic	Two non-homologous chromosome missing	2n-1-1
Trisomic	One extra chromosome	2n + 1
Double trisomic	Two extra non-homologous chromosomes	2n + 1 + 1
Tetrasomic	One extra chromosome pair	2n + 2
B. Euploid	Number genomes different from two	
Monoploid	Only one genome present	x
Haploid	Gametic chromosome number of the concerned species present	n
C. Polyploid		
(1). Autopolyploid	More than two copies of the same genome present	
Autotriploid	Three copies of the same genome	3x
Autotetraploid	Four copies of the same genome	4x
Autopentaploid	Five copies of the same genome	5x
Autohexaploid	Six copies of the same genome	6x
Autooctaploid	Eight copies of the same genome	8x
(2). Allopolyploid	Two or more distinct genomes; each genome has two copies	
Allotetraploid	Two distinct genomes; each has two copies	$(2x_1 + 2x_2)$
Allohexaploid	Three distinct genomes; each has two copies	$(2x_1 + 2x_2 + 2x_3)$
Allooctaploid	Four distinct genomes; each has two copies	$(2x_1 + 2x_2 + 2x_3 + 2x_4)$

combining genomes differ. These chromosomes are not homologous but are homoeologous chromosomes. Homoeologous chromosomes indicate ancestral homology. Induced alloploidy is rare. Through hybridization and chromosome doubling, allotetraploid was induced in *Cucumis sativus* x *Cucumis hystrix* cross. This was done to explain the molecular mechanisms involved in diploidization (tendency of polyploids to act as diploids). Cytogenetic analysis carried out in advanced generations established molecular mechanisms involved in stabilization of newly formed allopolyploids.

A prototypic allopolyploid (allotetraploid) was synthesized by G. Karpechenko in 1928. He expected a fertile hybrid with leaves of cabbage (*Brassica*) and roots of radish (*Raphanus*). Both these species are with 18 chromosomes, and they allow intercrossing. Hybrid progeny was produced, but this hybrid was functionally sterile because chromosomes of cabbage and radish were not homologous. However, one part of the hybrid plant produced some seeds. On planting, these seeds produced fertile individuals with 36 chromosomes but were allopolyploids. They had apparently been derived from spontaneous, accidental chromosome doubling to $2n_1 + 2n_2$ in one region of the sterile hybrid which underwent normal meiosis. Thus, in $2n_1 + 2n_2$ tissue, there is a pairing partner for each chromosome, and balanced gametes of the type $n_1 + n_2$ are produced. These gametes fuse to give $2n_1 + 2n_2$ allopolyploid progeny, which also are fertile. This kind of allopolyploid is sometimes called an amphidiploid. Unfortunately for Karpechenko, amphidiploid he made had roots of cabbage and the leaves of radish. He called this *Raphanobrassica* (Fig. 16.14b). Treating a sterile hybrid with colchicine doubles chromosomes thus make them fertile. Allopolyploidy is a major force of speciation.

Aneuploidy Aneuploids contain either an addition or subtraction of one or more specific chromosome(s). Univalent and/or multivalent formation arises during meiosis. A range of 30–40% of the progeny derived from autotetraploid maize are aneuploids. Univalents arise because of unequal distribution of chromosomes during anaphase I. Similarly, multivalents are formed due to non-separation of homologous chromosomes during meiosis that leads to unequal migration of chromosomes to opposite poles. This process is called non-disjunction. Such aneuploids are with reduced vigour. Depending on the number of chromosomes gained or lost, aneuploids are classified as monosomy (2n-1), nullisomy (2n-2), trisomy (2n + 1), tetrasomy (2n + 2) and pentasomy (2n + 3).

16.5.2 Methods for Inducing Polyploidy

Colchicine first isolated in 1820 by the French chemists P. S. Pelletier and J. B. Caventou inhibits the formation of spindle fibres that temporarily arrests chromosomes at the anaphase stage. Colchicine is extracted from autumn crocus (*Colchicum autumnale*). Chromosomes have replicated during anaphase, but in the absence of cell division, polyploid cells are formed. Other mitotic inhibitors, namely,

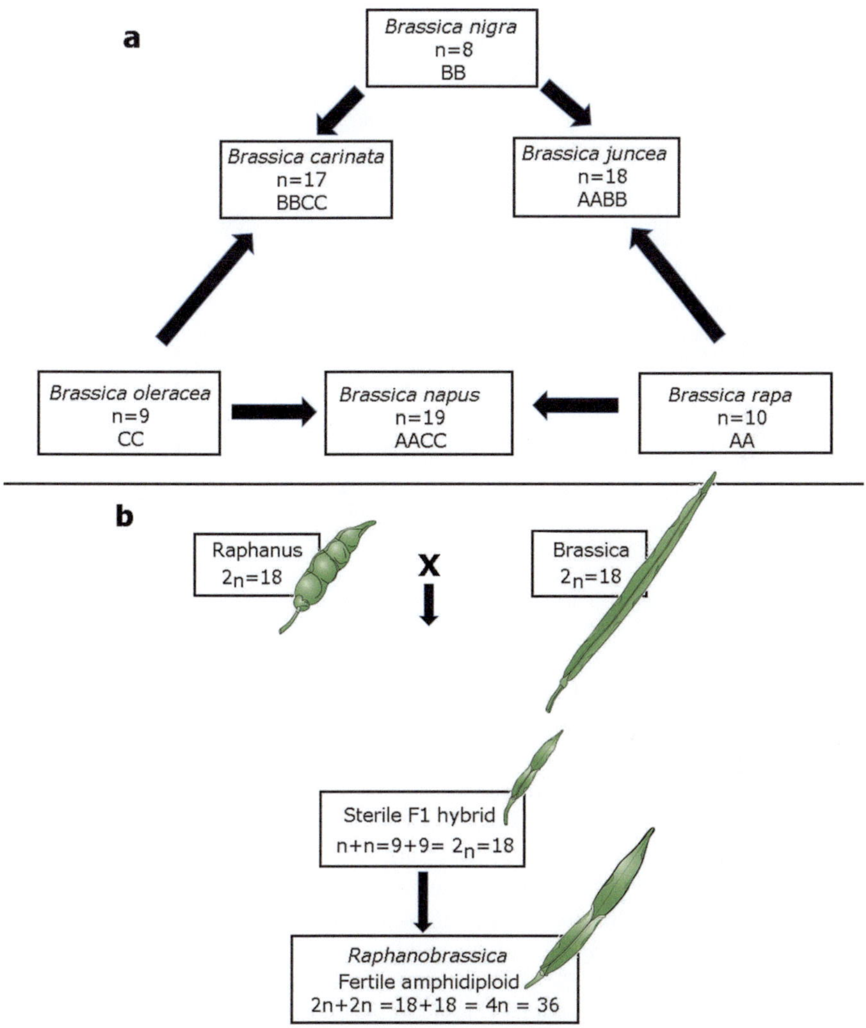

Fig. 16.14 (**a**) Triangle showing origin of cultivated mustard. (**b**) Origin of amphidiploid (*Raphanobrassica*) formed from cabbage (*Brassica*) and radish (*Raphanus*). The fertile amphidiploid arose in this case from spontaneous doubling in the $2n = 18$ sterile hybrid

dinitroanilines, oryzalin, trifluralin, amiprophos-methyl and N_2O gas, have also been identified and used as chromosome doubling agents. Seedlings with actively growing meristems are seen to be the best material to induce polyploidy. Seedlings or apical meristems can be soaked in colchicine solution. Older shoots when treated lead to cytochimaeras. Chemical solutions can be applied to buds using cotton, agar or lanolin or by dipping branch tips into a solution for a few hours or days. The efficacy can be increased by using surfactants, wetting agents and other carriers (dimethyl sulphoxide). Polyploidy in low frequencies can be induced by the use of

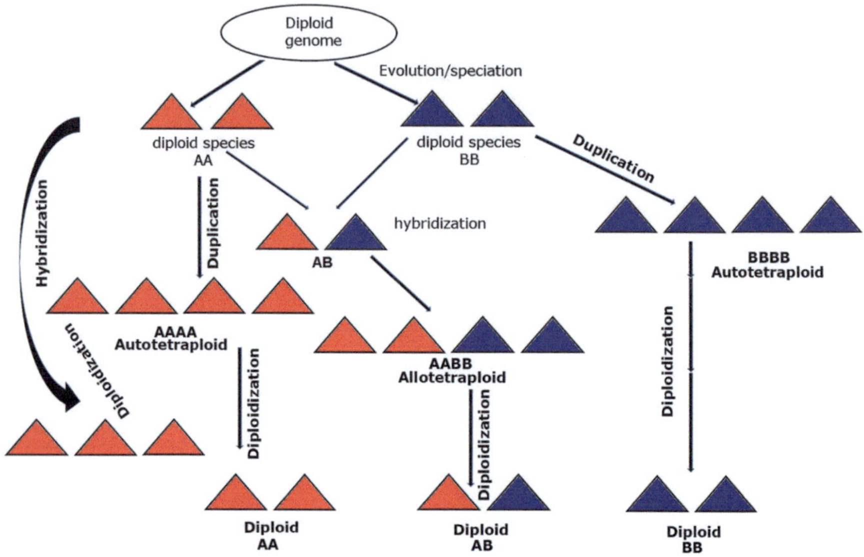

Fig. 16.15 Major pathways in the formation of polyploids

heat or cold treatment, X-ray or gamma ray irradiation. Exposure of maize plants or ears to high temperature (38–45 °C) at the time of first zygotic division produces 2–5% tetraploid progeny. Similar heat treatments are used in barley, wheat and rye to induce polyploidy.

Spontaneous induction of polyploidy in plants happens by several cytological means. Non-reduction of gametes during meiosis is one such way which is known as meiotic nuclear restitution. Such gametes are with 2n chromosomes like somatic cells. This could be due to aberrations related to spindle formation and abnormal cytokinesis. The union of non-reduced gametes form polyploids. This happens in open-pollinated diploid apples. In interspecific crosses between *Digitalis ambigua* and *Digitalis purpurea*, 90% of F_2 progenies show spontaneous allotetraploids. Autohexaploid *Beta vulgaris* (sugar beet) is another example. Alfalfa from cultivated autotetraploid varieties apparently are from the union of reduced (2x) and unreduced (4x) gametes. Polyspermy is another mechanism seen in orchids where one egg is fertilized by several male nuclei. The major pathways involved in polyploidy formation are represented in Fig. 16.15.

16.5.3 Molecular Consequences of Polyploidy

Polyploidy is widespread in flowering angiosperms and is one of the main causes behind the rapid diversification. It is a major route for the creation of new genes through gene duplication and diversification. This contention is still getting debated. Studies on molecular consequences of polyploidization commenced only recently.

Polyploids have a tendency to return to a diploidized state, a process known as diploidization. Diploidization experiences changes in chromosome organization, gene order, expression and epigenetic modification. This may involve abnormal chromosome segregation, rearrangement and breakage (Fig. 16.16a,b). In synthetic allotetraploids between doubled haploid *Brassica oleracea* (C genome) and *Brassica rapa* (A genome), abnormal chromosomal segregations led to aneuploidy in the first generation itself, with an aneuploidy rate of 24%. This aneuploidy rate rises to 95% in the 11th generation. This high rate of aneuploidy never reduces the homoelogs. The number of homeologs is maintained at four copies (i.e. the loss of chromosome 1 from the A genome is usually associated with gain of the same chromosome from the C genome, and vice versa). This is a compensating aneuploidy that indicates a dosage balance requirement. As such, the newly generated polyploids display higher rate of genome rearrangements leading to loss of chromosomal fragments (Fig. 16.16a).

Polyploidization initially results in multiplication of gene content. Genome sequencing has thrown light on gene loss in species that were subjected to polyploidization during course of evolution over several million years (Ma). Only 17% of duplicate sequences were retained in *A. thaliana* after a paleopolyplodization (β) event that took place ~50 Ma. In *Glycine max*, two rounds of whole-genome duplications took place ~59 and ~13 Ma in the paleopolyploid phase. In the more recent duplication event, 56.6% of duplicates are no longer detectable, compared to 74.1% genes lost after the older *Glycine* polyploidization. Thus, for the younger and the older duplication events, the rates of gene loss are 4.4% and 1.3% per million years (Myr), respectively. This indicates that the greater rate of gene loss in the initial phases slowed down over time. The loss of polyploidy-derived genes is *fractionation*. This is a mechanism by which removal of duplicates derived from polyploidization happens (Fig. 16.16b). Also, at the expression level, this phenomenon is reflected. Genes located on one sub-genome show higher expression than indicating genome dominance. Fractionation of genes leads to preferential gene retention. A number of distinguishing characteristics are seen in retained duplicate sequences compared to those single copy sequences. They are biased gene function, higher gene complexity (number of exons and protein domains), increased gene expression and parental genome dominance. The elevated mutation rate in polyploids reflects over increased transposable element activities. The proliferation of transposons in polyploids is due to reduced population size, masked deleterious transposon insertion and/or conflict in transposition repressors due to genome merger (Fig. 16.16c).

16.5.4 Molecular tools for Exploring Polyploid Genomes

A combination of genetic mapping, molecular cytogenetics, sequence and comparative analysis can shed light on the nature of ploidy evolution, from the base of the plant kingdom to intra- and interspecific hybridization. Some of the techniques that

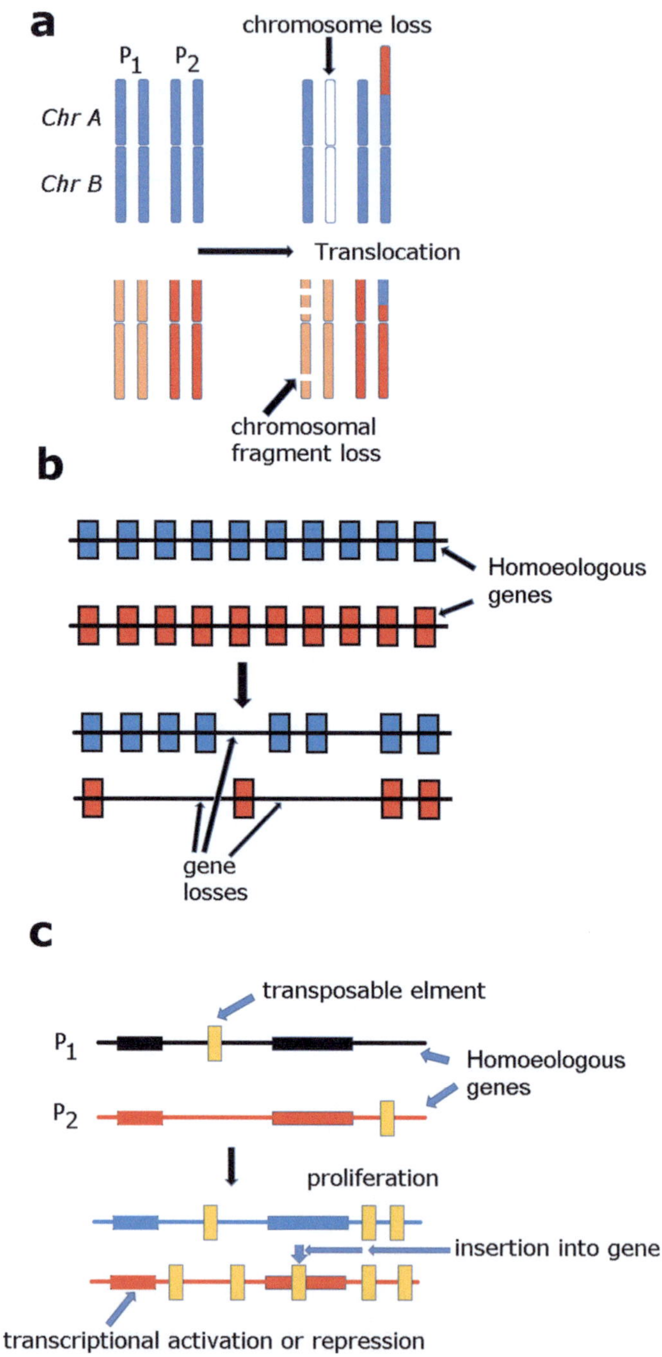

Fig. 16.16 Genomic consequences of polyploidy. (**a**) Some possible scenarios with respect to genomic rearrangements, such as chromosome loss, chromosomal translocation and chromosome

can endeavour such analysis are as follows (see Chap. on Genomics for further details on these techniques):

(a) In Situ Hybridization: In situ hybridization is a bridge between chromosomal and molecular level of genome investigations. This detects positions of unique sequences and repetitive DNAs along the chromosome(s). Fluorescent in situ hybridization (FISH) is a bit advanced, which detects fluorescent labels linked to DNA probes that can be visualized in a fluorescence microscope. Genomic in situ hybridization (GISH) is yet another advanced tool where total genomic DNA of species is hybridized as a probe on chromosomes. This leads to an analysis of whole genome discrimination rather than localization of specific sequences. There are several examples on the use of these techniques. In newly synthesized allotetraploid genotypes of *Brassica napus*, extensive genome remodelling due to homeologous pairing between the chromosomes of the A and C genomes were demonstrated. A combined GISH and FISH analysis demonstrated that in natural populations of *Tragopogon miscellus,* extensive chromosomal variation (mainly due to chromosome substitutions and homeologous rearrangements) was present up to the 40th generation following polyploidization.
(b) *Molecular Marker-Based Genetic Mapping:* Genetic mapping in polyploids is complicated compared to diploid species. The need of large populations and use of complicated statistical methods make the process more difficult to obtain reliable genetic distance estimates. A simple way is to use only single-dose markers from each parent, i.e. those segregating 1:1 in the mapping population (e.g. a population obtained from the cross Mmmm × mmmm in a tetraploid species).
(c) *Methylation-Sensitive Molecular Markers:* The use of an AFLP-like method using restriction enzymes sharing the same recognition site but having differential sensitivity to DNA methylation (isoschizomers – pairs of restriction enzymes specific to the same recognition sequence) is efficient for the determination of genome-wide DNA methylation patterns. This process otherwise known as methylation-sensitive amplified polymorphism (MSAP) is based on the use of the isoschizomers *Hpa*II and *Msp*I (both recognizing the 5'-CCGG sequence) but affected by the methylation state of the outer or inner cytosine residues. New and acceptable results were derived in newly synthesized polyploids by the use of this technique. In F_4 allotetraploids of *Arabidopsis*, frequent changes occurred when compared to the parents with increases and decreases in methylation. The change in methylation patterns equally affected both repetitive DNA sequences and low-copy DNAs.

Fig. 16.16 (continued) fragment loss, have been depicted in a simplified manner using only two chromosomes. P1, parent 1; P2, parent 2. (**b**) The process of gene loss in a parent-of-origin manner, termed *fractionation*. In the depicted scenario, the chromosomal copy from P2 loses most of the genes. (**c**) Proliferation of transposable elements over time. Such proliferation may lead to changes in gene order, gene function and gene expression

(d) *Comparative Genome Analysis:* Comparative genomics addresses several pertinent questions in genome evolution. Several phylogenetic and taxonomic studies revealed ancient polyploidy events and the evolution of novel genes that enabled adaptive processes. Recent genomic research revealed the relevance of polyploidy in angiosperm evolution and also suggested several ancient whole genome duplication (WGD) events. Transposable elements must have played a pivotal role in enhancing functional changes through genome reorganization following allopolyploidization.

(e) *High-Throughput DNA Sequencing:* High-throughput DNA sequencing coupled with computational analysis provides answers for the genetic analysis of polyploids. In *B. napus*, the polyploidy issue was done by sequencing leaf transcriptome across a mapping population. The Wheat Genome Initiative (http://www.wheatgenome.org/) individual or groups of homeologous chromosomes were analysed by flow cytometry separation. While in cultivated wheat gene duplications were predominant, wild wheat was characterized by deletions. Exon capture helped in variant discovery in polyploids that played a crucial role in the origin of new adaptations. SNPs have been utilized in the detection of variation in plant polyploidy. Illumina GoldenGate assay identifies a high number of SNPs in tetraploid and hexaploid wheat. In elite maize inbred lines, more than one million SNPs have been identified in Illumina sequencing platform.

Further Reading

Beyaz R, Ildiz M (2017) The use of gamma irradiation in plant mutation breeding. In: Jurić S (ed) Plant engineering. IntechOpen. https://doi.org/10.5772/intechopen.69974

Bourke PM (2018) Tools for genetic studies in experimental populations of polyploids. Front Plant Sci 9(513):2018. https://doi.org/10.3389/fpls.2018.00513

Ibrahim R et al (2018) Mutation breeding in ornamentals. Ornamental crops. Springer, pp 175–211

Jankowicz-Cieslak et al (2017) Biotechnologies for plant mutation breeding. Springer, Cham

Mason AS (2015) Creating new interspecific hybrid and polyploid crops. Trends Biotechnol 33:436–441

Sattler MC et al (2016) The polyploidy and its key role in plant breeding. Planta 243:281–296

Schaart JG (2016) Opportunities for products of new plant breeding techniques. Trends Plant Sci 21:438–449

Distant Hybridization

17

Keywords
Barriers in production of distant hybrids · Pre-zygotic incompatibility · Post-zygotic incompatibility · Failure of zygote formation and development · Embryonic incompatibility and embryo rescue · Transgressive segregation · Nuclear-cytoplasmic interactions

Distant or wide hybridization is the mating between individuals of different species or genera that combines diverged genomes into one nucleus. This process breaks the species barrier for gene transfer. It enables transfer of whole genome of one species to another, thus inflicting changes in genotypes and phenotypes of the progenies. Many of the day-to-day crop plants are the result of natural distant hybridization and speciation (Table 17.1). The origin of many allopolyploid species is through chromosome doubling of wide hybrids. Repeated backcrossing of wide hybrids to their parents is yet another way of gene introgression. This happens through infiltration of chromosomes or chromosome fragments from one species to another. Chromosome manipulation through wide hybridization for crop improvement can be classified into three main categories:

(a) Incorporation of single-chromosome or chromosome fragment from a wild species (also referred to as alien) into a crop to enhance genetic diversity. The resultant alien chromosome substitutions, additions or translocation lines can assist breeders to transfer desirable traits from wild and weedy plants to cultivated species.
(b) Induction of chromosome doubling to incorporate all alien chromosomes to produce amphidiploid. Amphidiploids result in a new crop. The man-made crop Triticale (X *triticosecale* Wittmack) is an amphidiploid between wheat (*Triticum turgidum* L. or *Triticum aestivum* L.) and rye (*Secale cereale* L.).

P. M. Priyadarshan, *PLANT BREEDING: Classical to Modern*,
https://doi.org/10.1007/978-981-13-7095-3_17

Table 17.1 Crop species and proposed progenitors

Common name	Family	Crop species	Proposed progenitor
Banana	Musaceae	*Musa acuminata* (AAA Group) cv Dwarf Cavendish	Several *Musa acuminata* subspecies
Barley	Poaceae	*Hordeum vulgare*	*Hordeum vulgare* subsp. *spontaneum* (synonym of *Hordeum spontaneum*)
Cassava	Euphorbiaceae	*Manihot esculenta*	*Manihot esculenta* subsp. *flabellifolia* (synonym of *Manihot esculenta*)
Chickpea	Leguminosae	*Cicer arietinum*	*Cicer reticulatum*
Maize	Poaceae	*Zea mays*	*Zea mays* subsp. *parviglumis* (synonym of *Zea mays*)
Pearl millet	Poaceae	*Pennisetum galucum*	*Pennisetum americanum* subsp. *monodii* (synonym of *Pennisetum violaceum*)
Oat	Poaceae	*Avena sativa*	*Avena sterilis*
Ground nut/peanut	Leguminosae	*Arachis hypogaea*	*Arachis monticola*
Rapeseed	Brassicaceae	*Brassica napus*	*Brassica rapa* and *Brassica oleracea*
Rice	Poaceae	*Oryza sativa*	*Oryza rufipogon*
Sesame	Pedaliaceae	*Sesamum indicum*	*Sesamum indicum* var. *malabaricum*
Sorghum	Poaceae	*Sorghum bicolor*	*Sorghum bicolor* subsp. *verticilliflorum* (synonym of *Sorghum arundinaceum)*
Soybean	Leguminosae	*Glycine max*	*Glycine soja* (synonym of *Glycine max* subsp. *soja)*
Sugarcane	Poaceae	*Saccharum officinarum*	*Saccharum robustum*
Common wheat	Poaceae	*Triticum aestivum*	*Triticum turgidum* and *Aegilops tauschii*
Durum wheat	Poaceae	*Triticum turgidum*	*Triticum turgidum subsp. dicoccoides (synonym of Triticum dicoccoides)*

(c) Production of haploids through elimination of alien chromosomes: Haploid is very useful in doubled haploid breeding, a true-breeding crop like wheat and rice can quickly fix genetic recombination and thus enhance breeding efficiency or facilitate genetic analysis (see Fig. 9.5).

Type 1 is the manipulation for single chromosome, while types 2 and 3 are the genome manipulation by the loss and the addition of alien genome, respectively. The F_1 hybrid between a crop and an alien species is the first step (se Fig. 9.5). Crossability is vital to achieve this step. Some genes or QTL for crossability have been found in tetraploid wheat (*T. turgidum* L.) and common wheat (*Triticum aestivum*). Utilization of crossable genes/QTL along with the application of techniques like embryo rescue and hormone treatment on post-pollination, successful production of F_1 hybrid can be achieved.

17.1 Barriers in Production of Distant Hybrids

Distant hybridization is dependent on the processes relating to pollination and fertilization that occur in a series of events from the germination of pollen grains to pollen tube growth and from double fertilization to zygote and endosperm development. Barriers that reduce gene flow can be divided into several categories:

(a) Pre-pollination barriers: geographic, habitat, mechanical and temporal isolation
(b) Post-pollination, pre-zygotic barriers: conspecific pollen precedence or gametic incompatibilities
(c) Intrinsic post-zygotic barriers: hybrid sterility, unviability or breakdown
(d) Extrinsic post-zygotic barriers: reductions in hybrid fitness due to the external environment

Pre-zygotic barriers provide greater contribution to speciation than post-zygotic barriers. Domestication via polyploidy is an exception norm since whole-genome duplication results in substantial post-zygotic isolation. Geographical isolation arises due to limited contact among taxa due to geological and climatic divide. Such an isolation fragments populations. Geographic isolation is the most effective barrier to gene flow. The vast majority of speciation is because of complete (allopatry) or partial (parapatry) geographic isolation.

17.1.1 Pre-zygotic Incompatibility

The incompatibility that happens before fertilization is pre-zygotic incompatibility. There are genetically predetermined pre-zygotic barriers like differences in blossoming period, and crossing may be prevented by ecological factors including a difference in habitation areas.

Genetically determined pre-zygotic types of reproductive isolation manifest a *progamic incompatibility* (during growth of pollen and pollen tubes) and *syngamic incompatibility* (in double fertilization). Indeed, the impossibility to cross *Triticum aestivum* wheat genotypes, which carried the dominant *Kr* genes, with *Secale cereale* rye is due to the inability of the pollen tubes to penetrate the embryo sac. Common wheat carries five genes responsible for this trait: *Kr1*, *Kr2*, *Kr3*, *Kr4* and *Skr* located in the *5B*, *5AL*, *5D*, *1A* and *5BS*.

Introgression of the recessive *kr1* alleles to several wheat genotypes leads to the enhancement of crossing capacity among themselves and also with rye and barley (*Hordeum vulgare*). Wheat-barley hybrids and wheat with barley chromosome introgression are notable outcomes of such exercise. However, hybrids between wheat × cultivated barley and common wheat × maize in which the *Kr* gene activity is not manifested are also found. Apparently, genetically controlling interspecific and intergeneric cross compatibility is more complicated. If the pollen tubes reach the ovary and enter the embryo sac, disorders may occur during fertilization.

Temperature and illuminance are two pertinent factors that influence the ability to cross. In vitro pollination is a renowned technique which, in combination with the cultivation of ovaries, seed buds and isolated embryos, is practised to overcome incompatibility caused by disorders of the pollen tube growth and fertilization failure. Treating plants with phytohormones before and after pollination to stimulate pollen tube growth and fertilization is a technique that can be practised. This is allowed not only to interspecific crossings but also to crossbreed species which belong to different subtribes (*H. vulgare* × *T. aestivum*), *H. geniculatum* (= *H. marinum* ssp. *gussoneanum*) ($2n = 28$) × *T. aestivum* and different tribes (*T. aestivum* × *Zea mays; T. aestivum*×*Pennisetum glaucum*).

17.1.2 Post-zygotic Incompatibility

The hybrid cells may encounter aberrations at different development periods from zygote division to the formation of the reproductive organs in the F_1 hybrids and their progeny. One of the causes for these disorders is allopolyploidy, which is the main cause that gives genomic shock to end with genetic and epigenetic changes in hybrids. Such shocks will induce selective elimination of DNA sequences, ending with reduction in genome size and gene loss. The activation of mobile elements results in chromosome rearrangements and the resulting "transcriptome shock" changes gene expression. The development or non-development depends on the rearrangements in hybrid genomes. Some of them may become reproductively isolated species, carrying heterosis for traits. Such hybrids can outperform the parental species in productivity, survivability and adaptability.

17.1.3 Failure of Zygote Formation and Development

The alternation of diploid sporophytic stage (2n) and haploid gametophytic stage (n) is the characteristic feature of angiosperms. Pollen grain (male gametophyte) carries two sperm cells (male gametes). The female gametophyte (FG), called the embryo sac, produces the female gametes and usually is enclosed within the maternal, sporophytic ovule (Fig. 17.1). Fusion of male and female gametes occurs during double fertilization. The ovules become seeds. FG development is closely regulated as it is essential for successful seed formation. FG development in flowering plants begins after meiosis, when one of four haploid daughter cells develops into the functional megaspore (FM). FM undergoes three rounds of syncytial mitotic divisions, followed by cellularization to produce seven cells belonging to four cell types, each with a defined position, morphology, and

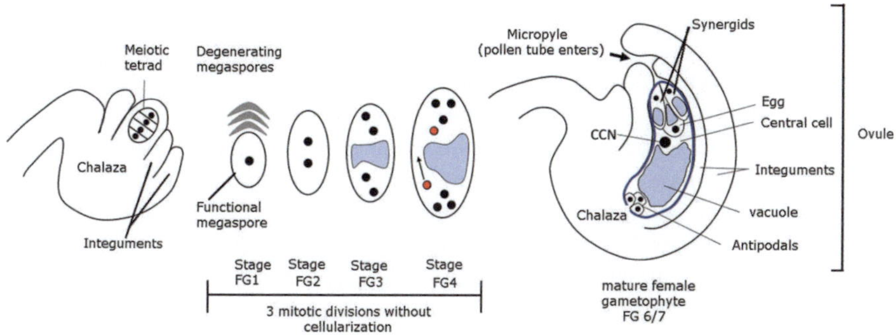

Fig. 17.1 Female gametophyte development. The progression of female gametophyte development is shown from left to right. After meiosis, a single haploid cell, usually the basal (chalazal) cell, will enlarge and form the functional megaspore while the remaining products of meiosis degenerate. This haploid megaspore will have three mitotic divisions accompanied by nuclear movement to create a defined pattern at each division. From stage FG4, the large vacuole (blue) separates the nuclei along the chalazal-micropylar axis. At FG5, the polar nuclei (red) migrate to meet each other and eventually fuse. At FG6/FG7, the mature female gametophyte has seven cells: two synergids, egg cell, central cell with large diploid nucleus (central cell nucleus, or CCN) and three antipodal cells (which are present through FG7 though much diminished)

specialized function (Fig. 17.1). Two FG cell types are gametic: the egg cell (1n) and the central cell (2n, homodiploid). These undergo double fertilization by two sperm cells of the pollen tube to produce the embryo (2n) and endosperm (3n), respectively. There are two accessory cell types called synergids and antipodals. Synergids attract pollen tube. The function of antipodals is currently unknown. These four cell types (egg cell, central cell, synergids and antipodals) are specified from the eight haploid nuclei that have descended from the FM. After the first mitotic division of the FM (stage FG_2), the two daughter nuclei are physically sequestered at either end of the embryo sac by the enlarging vacuole, creating a morphological axis (FG_3). After two further divisions (FG_5), one of the four nuclei at each end migrates around the central vacuole towards the centre. These polar nuclei will fuse, forming the central cell nucleus (FG_6). At the same time, the remaining nuclei begin to differentiate by cellularization according to their position along the distal (micropylar)-proximal (chalazal) axis. At maturity, the pollen tube enters the ovule through the micropyle. At the micropylar end of the gametophyte, the synergid cells and egg cell are in close proximity but have different morphologies, including nuclear position. The smaller synergid nuclei are oriented closer to the micropyle and egg nucleus towards the central cell.

17.1.4 Embryonic Incompatibility and Embryo Rescue

The early stages of post-zygotic development are crucial for the development of hybrid seeds. After double fertilization, incompatibility may emerge beginning from the first zygote division that can end up with disorders of endosperm development.

In vitro embryo rescue at early stages of embryo development can be a technique to overcome embryonic incompatibility. Depending on the species, the time and methods of embryo isolation can be standardized. In vitro embryo rescue was first used in lax and is now widely used in a variety of species.

The extreme incompatibility between alien genomes occurs as a total or partial chromosome elimination of one of the parents from the embryonic hybrid cells. This kind of DNA elimination is one way of getting rid of alien DNA via its destruction. This phenomenon was first noticed by Karpechenko in as early as 1920s. In barley, wheat, oat, tobacco, tomato and cabbage, single-parent chromosome elimination is typical. Single-parent genome elimination leads to haploid embryos. Partial genome elimination results in haploidy with genome of one parent supplemented with singular chromosomes of the other.

Mechanisms for single parent elimination are best studied in *H. vulgare* × *H. bulbosum* and intertribal combination of *T. aestivum* × *Pennisetum glaucum*. The process of chromosome elimination is followed by further events like spatial separation of parental genomes in the interphase nucleus, sister chromatid disjunction failure in the anaphase of the haplo-producer species, chromosome rearrangements and the formation of micronuclei, heterochromatin formation and DNA fragmentation in micronuclei and the destruction of micronuclei by endonucleases.

Inactivation of the centromere is the cause for chromosome elimination in *H. bulbosum* in the *H. vulgare* × *H. bulbosum hybrid combination*. This is determined by the fact that in contrast to active centromeres of *H. vulgare*, the inactive centromeres of *H. bulbosum* do not contain (or contain a low level) of the *CENH3* histone, which is the kinetochore complex assembly site of the normal centromere. The power to eliminate of the *H. vulgare* with respect to the genome of *H. bulbosum* emerges in combinations, in which both parents carry the same chromosome number (*H. vulgare* ($2n = 14$) × *H. bulbosum* ($2n = 14$) or *H. vulgare* ($2n = 28$) × *H. bulbosum* ($2n = 28$) – i.e. at the parental genome ratio 1: 1). The genes responsible for the elimination are located in the short arms of the second and third chromosomes of the cultivated barley.

Hybrid combinations with the single-parent chromosome elimination (*H. vulgare* × *H. bulbosum, T. aestivum* × *Z. mays* and *T. aestivum* × *P. glaucum*) are useful to obtain doubled haploid lines. The partial elimination of maize chromosomes in hybrids of *Avena sativa* × *Z. mays* is useful in mapping maize genome. Temperature has bearing on the process of chromosome elimination. An increase of temperature to 30 °C speeds up the chromosome elimination, and a temperature lower than 18 °C, inhibits this process.

17.1.5 Transgressive Segregation

When phenotypic trait value hybrids fall outside the range of parental variation, it is transgressive segregation. Transgressive segregation can produce novel genotypes with ability to adapt to a new environments. Transgressive segregation is manifested

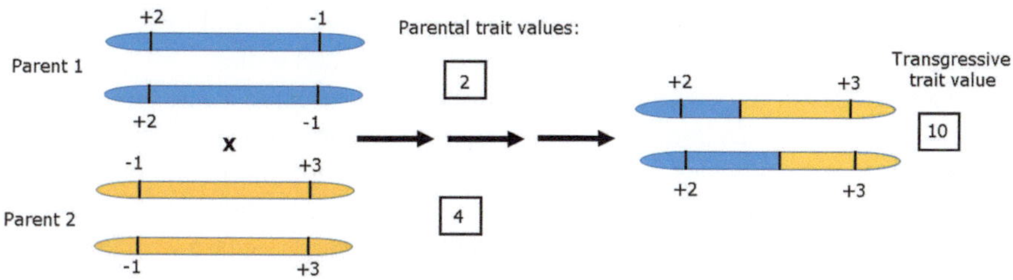

Fig. 17.2 Complementary gene action causes transgressive segregation. Complementary gene action occurs when additive alleles for a multilocus trait act in opposition to one another in both parent lineages but sort in favour of one direction of effect in segregating hybrids. Individual loci contributing to a trait are indicated along a chromosome with their additive contribution to the trait value. The total trait value for each genotype is indicated by the boxed number. One possible hybrid genotype is depicted that has acquired all + alleles and, therefore, has a transgressive trait value

in the F_2 generation and quite different from heterosis. This difference suggests possible distinct genetic mechanisms for the two phenomena. It is found that 97% of studies reporting parental and hybrid trait values include at least one transgressive trait. Like heterosis, causes of transgressive segregation are many that require serious investigation.

Complementary gene action and epistasis are the genetic mechanisms that cause transgressive segregation. The complementary gene action model entails that both parents have additive alleles of opposing sign at different loci (affecting a multilocus trait). This gene arrangement could be in favour of one direction in the segregating hybrids. As an example, one would expect that a late-generation hybrid may acquire + alleles for a trait from both parents across different loci (Fig. 17.2). This is an oppositional multiple gene system that Nilsson-Ehle in 1911 reported in wheat (*Triticum aestivum*). The epistasis model would explain non-additive interactions between loci from different parents that can cause extreme trait values in hybrids. Latest advancements in genomics suggest mechanisms involving small interfering RNAs. Epigenetic regulation and small RNA activity can also be pivotal to transgressive segregation.

17.2 Nuclear-Cytoplasmic Interactions

The genetic information is unequally distributed among the genomes of the nucleus, mitochondria and plastids. The nuclear genome controls the organelle gene expression through regulation at post-transcriptional level. This process is called *anterograde regulation*. The organelle genomes involve in retrograde regulation, activating many signalling pathways governing nuclear gene expression. Such interactions between nuclear and organelle genomes are defined as nuclear-cytoplasmic interactions. Any anomaly at such interactions can lead to nuclear-cytoplasmic conflicts. Cytoplasmic male sterility (CMS) is the result of such conflicts. This is associated with mutations in mitochondrial genes, which can influence the target nuclear genes governing production of flower's organs and pollen.

Many defects in the evolutionarily developed nuclear-cytoplasmic balance may appear in wide hybridization. In wide hybrids, two evolutionarily different genomes are combined into a nucleus and kept in the maternal cytoplasm. Reciprocal hybrids have same hybrid genome with a different cytoplasm. If the reciprocal hybrids differ, such differences are due to cytoplasmic effects or nuclear-cytoplasmic interactions. Such differential gene expression can also be mediated by small non-coding RNAs. The differences between reciprocal hybrids may also be due to parent-of-origin effects, which have a significant effect in the development period of hybrid seeds. Such effects lead to abnormal development of endosperm and the hybrid embryo.

The other models to study the role of nuclear-cytoplasmic interactions are alloplasmic lines (nuclear-cytoplasmic hybrids). Theoretically, two major events must take place in order to form an alloplasmic line: a) substitution of the maternal nuclear genome for the paternal nuclear genome in the process of recurrent crossings of hybrids with the paternal species and b) an evolutionarily fixed transfer of organelle genomes through the maternal line. In alloplasmic lines of *Triticum, Allium cepa, Brassica napus, Nicotiana tabacum,* fertility can be restored by pollinating these lines with those lines containing nuclear genes of fertility restoration on an alien cytoplasm. As an example, the restoration of the fertility of alloplasmic lines of common wheat carrying the cytoplasm of *Triticum timopheevii* (because of the development of viable pollen) is controlled by a polygenic system of the main eight nuclear Rf1–Rf8 genes (fertility restorer), which are located in the common wheat chromosomes 1A, 7D, 1B, 2DS, 6B, 6D, 7B and 6DS. It is also regulated by three less effective genes located in chromosomes 2A, 4B and 6A.

The nuclear-cytoplasmic conflict is expressed based on the phylogenetic distance between the species that contributed the nuclear and cytoplasmic genomes. In alloplasmic lines of common wheat, with cytoplasm of the *Aegilops* sp. and barley *Hodeum chilense* (wild barley), significant changes in transcription and metabolism occurred in hybrids involving *Hordeum.* This is because taxonomically, *Hordeum* is more remote from wheat than the *Aegilops* sp. It was found that wide hybridization of wheat changes the mechanism of the mtDNA transfer. The transfer takes place either through the paternal line instead of the maternal or biparental inheritance takes place.

Further Reading

Baack E et al (2015) The origins of reproductive isolation in plants. New Phytol 207:968–984

Dempewolf H et al (2017) Past and future use of wild relatives in crop breeding. Crop Sci 57:1070–1082

Goulet BE et al (2017) Hybridization in plants: old ideas, new techniques. Plant Physiol 173:65–78

Liu D et al (2014) Distant hybridization: a tool for interspecific manipulation of chromosomes. In: Pratap A, Kumar J (eds) Alien gene transfer in crop plants, volume 1: innovations, methods and risk assessment. Springer, New York

Widmer A (2009) Evolution of reproductive isolation in plants. Heredity 102:31–38

Host Plant Resistance Breeding 18

Keywords

Concepts in insect and pathogen resistance · Host defence responses to pathogen invasions · Vertical and horizontal resistance · Biochemical and molecular mechanisms · Systemic acquired resistance (SAR) · Induced systemic resistance · Qualitative and quantitative resistance · Genes for qualitative resistance · Genes for quantitative resistance · Pathogen detection and response · Signal transduction · Resistance through multiple signalling mechanisms · Classical breeding strategies · Back cross breeding · Recurrent selection · Multi-stage selection · Marker assisted breeding strategies · Monogenic vs. QTLs · Marker assisted backcross breeding (MABC) · Pyramiding resistance genes · Marker-assisted selection (MAS) · Modern approaches to biotic stress tolerance

Biotic stresses are the damage to plants caused by other living organisms such as bacteria, fungi, nematodes, insects, viruses and viroids. The resistance to biotic stresses can be defined as under:

> *Those characters that enable a plant to avoid, tolerate or recover from attacks of insects under conditions that would cause greater injury to other plants of the same species* – Painter R.H. (1951)

> *Those heritable characteristics possessed by the plant which influence the ultimate degree of damage done by the insect* – Maxwell F.G. (1972)

Some of the biotic stresses that devastated the world in the past are the potato blight in Ireland, coffee rust in Brazil, maize leaf blight in the USA. The great Bengal (India) famine in 1943 is also said to be due to crop failure. Annually, it is estimated that almost 15% of global crop yields are lost due to diseases. Since tropics and subtropics favour disease development, the extent of such losses varies with crop and the region. Chemical control was considered as an efficient method; however,

P. M. Priyadarshan, *PLANT BREEDING: Classical to Modern*,
https://doi.org/10.1007/978-981-13-7095-3_18

the use of pesticide/fungicide dramatically increased, and the overall crop loss has not decreased. This is due to the upsurge of different races of pathogens over a period of time. Breeding for host resistance offers an effective alternative to fungicides/ pesticides that can be combined with other management practices as part of an integrated programme. For example, disease-resistant crops perform better with timely planting and harvest and with crop diversification. The dynamics behind host-pathogen interactions is that virulent pathogen populations can arise and attack resistant crop varieties. Resistance breeding is therefore an ongoing process. So, wild relatives, landraces and other germplasm are being used in resistance breeding. Though resistance based on a single gene (simple resistance) shall be effective in short term, practically useful long-term resistance demands multiple scale genetic complexity. Irrespective of the fact that the resistance is short term or long term, it depends on how the breeder manipulates the systems. At the genotype level, resistance is influenced by the number of resistance genes and their specific combination in the host. So, direct or indirect effects of resistance genes on other valued traits like grain quality, adaptation to environmental conditions and yield are to be taken into account. Many important terms are involved in plant disease resistance (Table 18.1).

It is widely believed that phytopathogenic agents (insects, pests, fungi, viruses) lodge genetic polymorphism. Climatic factors can influence/modify this polymorphism. The available polymorphism can be instrumental in the production of aggressive strains that can alter the host-pathogen interaction. The vulnerability towards diseases is controlled by genetic structure of the crop (Table 18.2). Line cultivars (e.g. wheat, barley, oats, peas) that are homozygous at all loci and are homogeneous phenotypes are prone to diseases. This is true with asexually propagated clonal cultivars also (potato, strawberry, banana, fruit trees). Asexually propagated species (tuber, bulb, cutting) enable more pathogens to survive than those propagated sexually. Single-cross hybrids are also homogeneous due to the controlled crossing of two inbred lines. The segregating three-way and double-cross hybrids are with high buffering capacity due to their heterogeneous genetic structure with majority of loci heterozygous. Most crops in industrial countries are genetically uniform and are prone to disease epidemics. A list of major pests and diseases of economically important crops is available in Table 18.3.

18.1 Concepts in Insect and Pathogen Resistance

Organisms are generally classified as producers, green plants, consumers (organisms exploiting other organisms) and decomposers (organisms using dead organisms). Green plants are used by a multitude of consumers like herbivores (mammals, snails, insects) to typical parasites (insects, mites, fungi, bacteria). Plants have a range of defence mechanisms to ward off most of these consumers. These defence mechanisms are *avoidance*, *resistance* or *tolerance*. Avoidance operates before parasitic contact and decreases the frequency of incidence. After parasitic contact has been established, the host may resist the parasite by decreasing its growth or

Table 18.1 Common terms used in plant disease resistance studies

Term	Definition
Adult-plant resistance	Resistance only visible in the adult stage of a plant, i.e. at the generative phase. Adult-plant resistance can be inherited monogenically or quantitatively and need not to be durable
Aggressiveness	Degree of pathogenicity in a quantitative host-pathogen interaction; it varies quantitatively from low to highly aggressive indicating a low to high damage of the host
Avirulence (gene)	A gene (*Avr*) in a pathogen that causes the pathogen to elicit an incompatible (defence) response in a resistant host plant. Interaction of an avirulence gene product with its corresponding plant resistance (R) gene is highly specific and usually provokes a hypersensitive reaction
Broad-spectrum resistance locus	Individual locus that confers resistance to multiple races of a pathogen species or multiple taxa of pathogens
Durable resistance	Resistance that remains effective for a long period when applied on a large scale in a region that is undergoing regular epidemics of the pathogen
Epistasis	Interaction between genes at different loci
Pathogenicity	Ability of an (micro)organism to damage a healthy plant
Pathotype	Isolate with a special combination of avirulences/virulences
Pathosystem	Combination of a specific host and pathogen species or a complex of closely related pathogen species
Quantitative trait locus (QTL)	Markers linked to the genes that underlie a quantitative trait; it should be remembered that there is only a genetic linkage between markers and genes based on recombination frequencies
Race	Isolates within a pathogen species that are distinguishable by their virulence, but not by morphology. Today, races are often a complex combination of virulences, thus *pathotype* might be the better term
Qualitative resistance	Race-specific resistance inherited by single R genes, also named vertical resistance or hypersensitivity resistance following the gene-for-gene concept
Quantitative resistance	Resistance inherited by several genes with minor effects, usually non-race-specific and prone to non-genetic interactions, also named horizontal resistance
Virulence	Degree of pathogenicity in a qualitative host-pathogen interaction; low virulence indicates a virulence to a few R genes, high virulence to many R genes

tolerate its presence by suffering relatively little damage. Avoidance is mainly active against animal parasites and includes such diverse mechanisms as volatile repellents, mimicry and morphological features like hairs, thorns and resin ducts. Resistance is usually of chemical nature. Little is known of tolerance; it is very difficult to measure and is usually confounded with quantitative forms of resistance. Parasites classified as fungi, bacteria, viruses or viroids are considered as disease-inciting parasites or pathogens.

Resistance mechanisms are the most important defence mechanisms employed by crops. Avoidance and tolerance play a minor role here. In the competition between

Table 18.2 Reproductive system, type of cultivar and genetic structure of the cultivar

Reproductive system	Type of cultivar	Genetic structure (Genotype/phenotype)	Vulnerability
Sexual: Self-pollination	Line cultivar	Homozygous/homogeneous	High
Cross-pollination	Population cultivar	Heterozygous/heterogeneous	Low
Controlled crossing	Hybrid cultivar	Heterozygous/homogeneous (Assuming a single-cross hybrid)	High
Asexual: Vegetative	Clonal cultivar	Heterozygous/homogeneous	High

Table 18.3 Major pests and diseases of economically important crops

Bacterial diseases	
Beans, Rice	Blight
Cotton	Black Arm
Tomato	Canker
Potato	Ring Rot, Brown Rot
Fungal diseases	
Sugarcane	Red Rot
Bajra (Pearl Millet)	Ergot, Green Ear, Smut
Pigeon Pea, Cotton	Wilt
Ground Nut	Tikka
Rice	Blast
Paddy, Papaya	Foot Rot
Wheat	Rust, Powdery Mildew
Coffee	Rust
Potato	Late Blight
Grapes, Cabbage, Cauliflower, Bajra, Mustard	Downy Mildew
Radish, Turnip	White Rust
Viral diseases	
Potato	Leaf Roll, Mosaic
Banana	Bunchy Top
Papaya	Leaf Curl
Tobacco	Mosaic
Carrot	Red Leaf

plant and pathogen, the latter has developed widely different host ranges. Pathogens such as *Pythium* species, *Rhizoctonia solani* Kühn, and *Sclerotinia sclerotiorum* (Lib.) de Bary have a wide host range; they are non-specialized, polyphagous pathogens or generalists. *Sclerotinia* can attack hundreds of plant species belonging to at least 64 families of flowering plants and gymnosperms. A large proportion of the pathogens have a narrow host range known as monophagous pathogens or specialists. *Puccinia hordei* Otth. and *Phytophthora phaseoli*, which infect barley (*Hordeum vulgare* L.) and lima beans (*Phaseolus lunatus* L.), respectively, are the examples. There are several technical terms involved in the study of host-pathogen interactions. They are available in Box 18.1.

Box 18.1: Terms Involved in Host-Pathogen Interactions

Avirulence gene (Avr): a gene, the product of which, as defined by Flor's gene-for-gene hypothesis, is recognized by a plant R-gene and activates ETI.

Chitin elicitor binding protein (CEBiP): a plant PRR that binds the PAMP chitin.

Chitin elicitor receptor kinase 1 (CERK1): an RLK required for CEBiP-triggered PTI.

EF-Tu receptor (EFR): a plant PRR that binds the PAMP EF-Tu.

Effector-triggered immunity (ETI): plant defence responses activated following the recognition by the plant of pathogen effectors.

Flagellin sensing 2 (FLS2): a plant PRR that binds the PAMP flg22.

Genome-wide association studies (GWAS): systematically screen a genome-wide array of markers against the phenotypes of interest to identify statistical associations between markers and phenotypes.

Pathogen-associated molecular pattern (PAMP): conserved pathogen molecules recognized by the plant; also known as Microbe-associated molecular pattern (MAMP)s.

PAMP-triggered immunity (PTI): plant defence responses activated following the recognition by the plant of PAMPs.

Quantitative trait locus (QTL): a genetic region that contributes to a phenotype displaying a continuous distribution.

Receptor-like kinase (RLK): a protein containing a receptor-recognition and a functional kinase domain.

MAMPs: microbe-associated molecular patterns.

Signal transduction: a process by which a chemical or physical signal is transmitted through a cell by means of series of molecular events such as protein phosphorylation that results in a cellular response.

Transcription activator-like effector nucleases (TALEN): a fusion protein between the plant gene DNA recognition repeats of the TAL effector protein and the DNA cleavage domains of FoKI, a bacterial type IIS restriction endonuclease.

Transcription activator-like effectors (TALE): TALEs bind to TALE-specific DNA sequences within the promoter regions of plant genes, activating gene transcription.

Effector: a virulence protein injected into a host cell by a pathogen to suppress host defence and cause disease.

(continued)

Box 18.1 (continued)

Effector-triggered immunity (ETI): a set of defence responses triggered by specific pathogen effectors upon recognition by their cognate host resistance proteins.

Hypersensitive response: the phenotypic response generated as a result of ETI, characterized by well-defined necrotic areas where infected cells have undergone programmed cell death.

***PR* (*pathogenesis related*) genes:** a group of genes induced after pathogen infection that encode small, secreted, or vacuole-targeted proteins with antimicrobial activities.

System Acquired Resistance (SAR): a broad-spectrum plant disease resistance induced after a local pathogen infection.

NPR1 (non-expresser of *PR* genes 1): a protein first identified in *Arabidopsis thaliana* that is required for *PR* gene expression, local defence, SA signalling and SAR.

Mobile signal: a signal transmitted from the local infection site to the systemic tissue to induce systemic resistance.

Salicylic acid (SA): plant hormone essential for the immune response against biotrophic pathogens.

Durability: a property that enables resistance to remain effective when deployed over a large area under substantial disease pressure over a long time.

R-genes: resistance genes of large effect that are inherited in a Mendelian fashion and typically, but not always, encode nucleotide-binding leucine-rich repeat proteins.

Pathosystems: ecological subsystems defined by a specific disease. A plant pathosystem includes one or more host plant species along with the pathogen(s) that cause(s) the disease.

Nucleotide-binding domain leucine-rich repeat containing (NLR) genes: a family of plant genes involved in pathogen recognition. Many resistance genes of large effect are NLR genes.

Races: variants within a pathogen species that elicit differential responses from resistance genes.

An array of morphological, genetic, biochemical and molecular processes are involved towards resistance to various pathogens and insect pests. Such mechanisms may be expressed continuously (constitutively) as preformed resistance, or they may be inducible (i.e. deployed only after attack). Recently, it is revealed that plant mechanisms of disease/insect resistance or susceptibility are related to mechanistic animal immunity. This has significantly thrown light on plant immunity. The identification of plant pattern recognition receptors (PRRs) that sense pathogen or insect pest conserved molecules termed pathogen-associated molecular patterns or microbe-associated molecular patterns or herbivore-associated molecular patterns

(PAMPs/MAMPs/HAMPs) – and the subsequent PAMP-triggered immunity (PTI) is a new paradigm for plant-pathogen interaction studies (see later).

The ability of pathogens/insect pests to suppress or evade PTI has augmented research on the so-called "gene-for-gene" effector-induced resistance in plants. It is now established that effectors with pathogen can successfully evade the plant's ability towards PAMPs/HAMPs. On the other hand, plants have effector-induced resistance or vertical resistance (otherwise known as effector-triggered immunity – ETI) that can be a successful means of controlling pathogens that are able to evade PTI. The defence against pathogens is boosted through selective transcription of genes. This is accomplished as ETI engages a compensatory mechanism. Through ETI, the resistance (R) genes undertake endogenous nucleotide-binding and leucine-rich repeat (NB-LRR) protein products. R gene-mediated resistance is generally not durable. However, the pyramiding of several resistance (R) genes is now effectively utilized in the same cultivar that increases durability of resistance.

18.1.1 Host Defence Responses to Pathogen Invasions

Plants have intricate and dynamic defence system to respond to various pathogens. Such defence can be classified as either innate or systemic plant response. The overview of plant defence response is presented in Fig. 18.1. An innate defence is exhibited by the plant in two ways, viz. specific (cultivar/pathogen race specific) and non-specific (non-host or general resistance). Though not well studied, the molecular basis of non-host resistance involves a large array of proteins and other organic molecules produced prior to infection or during pathogen attack. Constitutive defence includes morphological and structural barriers (cell walls, epidermis layer, trichomes, thorns, etc.), chemical compounds (metabolites, phenolics, nitrogen compounds, saponins, terpenoids, steroids and glucosinolates) and proteins and enzymes. Such compounds provide strength and rigidity that confer tolerance or resistance. The inducible defences (production of toxic chemicals or pathogen-degrading enzymes like chitinases, glucanases) and deliberate cell suicide are used by plants. Chitinases and glucanases demand high energy costs and higher nutrient requirements associated with their production and maintenance. In response to pathogen attack, such compounds become active which are inactive otherwise. Such compounds can fall in as either innate or systemic acquired resistance (SAR). Innate immunity is an efficient mechanism and a common form of plant resistance to microbes. Both these defence strategies depend on the ability of the plant to distinguish between self and non-self-molecules.

18.1.2 Vertical and Horizontal Resistance

Vertical resistance is also known as race-specific, pathotype-specific or simply specific resistance. Major genes govern vertical resistance. It is characterized by

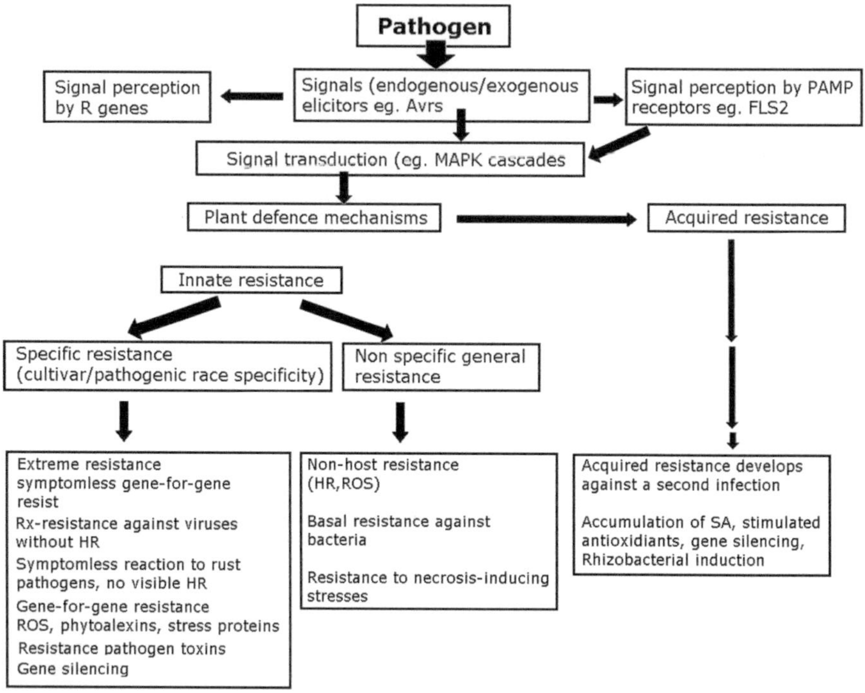

Fig. 18.1 Overview of cellular mechanisms of biotic stress response leading to innate immunity and systemic acquired resistance. Plant PRRs or R genes perceive PAMPS/DAMPs and effectors, respectively. Inside the cell, an overlapping set of downstream immune responses result from the PTI/ETI continuum. This includes the activation of multiple signalling pathways involving reactive oxygen species (ROS), defence hormones (such as salicylic acid, jasmonic acid and ethylene), mitogen-activated protein kinases (MAPK) and transcription factor families, e.g. AP2/ERF, WRKY, MYB, bZIP, etc. These signals activate either innate response or acquired immune response or both

pathotype specificity. The host becomes susceptible when attacked by a pathotype which is virulent towards that resistant gene lodged by the host. But to all other pathotypes, the host will be resistant. Generally, a single (monogenic) dominant gene or a few dominant genes govern vertical resistance. There is a chance that some of these genes may have multiple alleles as in leaf rust gene, Lr2, that accords resistance to *Puccinia recondite tritici*. Here, four genes designated as Lr2a, Lr2b, Lr2c and Lr2d are present and are tightly linked. Each of these genes accords resistance to a different spectrum of races and hence can be differentiated from one another. Such multiple alleles exist on Sr9 locus of wheat for *P. graminis tritici* and gene Pi-k in rice for resistance to *Pyriculariva grisea*. It is convenient that such tightly linked multiple alleles can be transferred in one attempt.

Horizontal resistance has many synonyms, e.g. race-non-specific, partial, general and field resistance. Horizontal resistance is generally controlled by polygenes and is pathotype non-specific. Thus, it is also known as general resistance. Horizontal resistance slows down the rate of spread of disease in the population. Horizontal resistance (HR) reduces the rate of disease spread and is evenly spread against all races of the pathogen. HR results from polygenes. Morphological features such as size of stomata, stomatal density per unit area, hairiness, waxiness and several others influence the degree of resistance expressed. Partial resistance, dilatory resistance, lasting resistance are some other terms coined for denoting horizontal resistance.

18.2 Biochemical and Molecular Mechanisms

Plant cells are generally protected by several layers of physical barriers, including the waxy cuticle on the leaf surface, the cell wall and the plasma membrane, which deny access to most microbes. Plants can also produce a wide range of chemicals as barriers against microbes and pests. Plant species produce saponins and glycosylated triterpenoids that can resist microbes. Their soap-like properties can disrupt the growth of fungal pathogens. The cell surface-localized pattern-recognition receptors (PRRs) through highly conserved pathogen-associated molecular patterns (PAMPs) can recognize different classes of pathogens (e.g. gram-positive as opposed to gram-negative bacteria). Plants independently evolve PAMP-triggered immunity (PTI) as the first layer of active defence at the cellular level. Such an immune mechanism can prevent potential pathogen infection.

18.2.1 Systemic Acquired Resistance (SAR)

In addition to triggering defence responses, the host also induces the production of signals such as salicylic acid (SA), methyl salicylic acid (MeSA), azelaic acid (AzA) and glycerol-3-phosphate (G3P). These signals induce expression of antimicrobial *PR* (*pathogenesis-related*) genes in the uninoculated distal tissue to protect the rest of the plant from secondary infection. This phenomenon is called systemic acquired resistance (SAR). SAR can also be induced by exogenous application of the defence hormone SA or its synthetic analogues 2,6-dichloroisonicotinic acid (INA) and benzothiadiazole S-methyl ester (BTH). SAR provides broad-spectrum resistance against pathogenic fungi, oomycetes, viruses and bacteria. SAR-conferred immunity can last for weeks to months and possibly even the whole growing season. Unlike ETI, SAR is not associated with programmed cell death (PCD). Instead, it promotes cell survival. A massive transcriptional reprogramming is responsible for SAR. This is dependent on the transcription cofactor NPR1 (non-expresser of *PR* gene 1) and its associated transcription factors (TFs). A battery of antimicrobial PR proteins that induce significant enhancement of endoplasmic reticulum (ER) function is responsible for this function (Fig. 18.2). However, SAR signalling pathway is not well understood despite intense research. How an avirulent pathogen induces the biosynthesis of the essential immune signal, SA, is not clear yet. The nature of the mobile signal for SAR is also unclear.

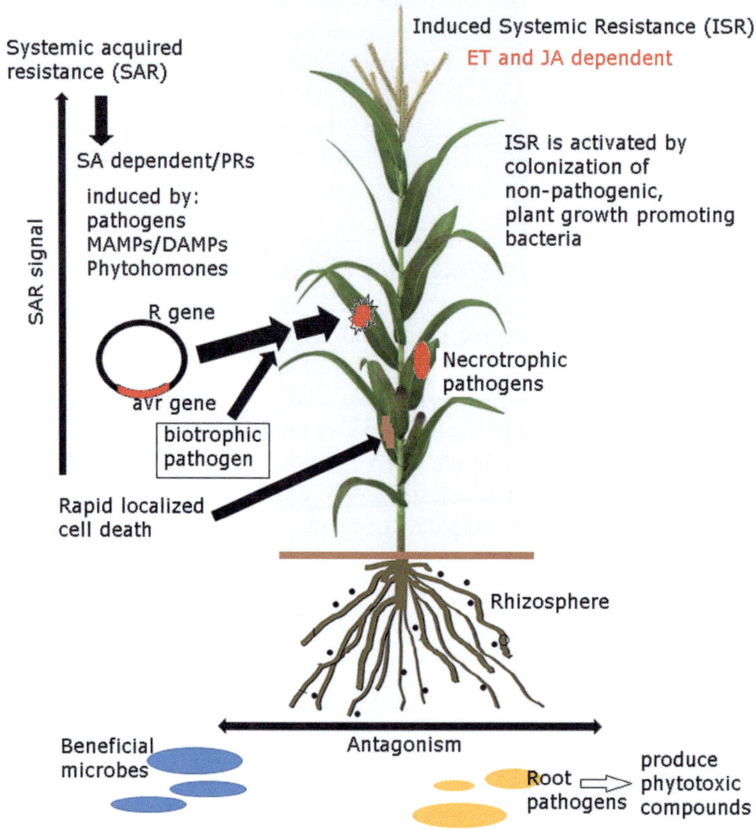

Fig. 18.2 Schematic representation of systemically induced immune responses. Systemic acquired resistance starts with a local infection and can induce resistance in yet not affected distant tissues. Transport of salicylic acid (SA) is essential for this response. Induced systemic resistance can result from root colonization by non-pathogenic microorganisms and, by long-distance signalling, induces resistance in the shoot. Ethylene (ET) and jasmonic acid (JA) are involved in the regulation of the respective pathways. Depending on the pathogen, JA/ET can also be involved in SAR. They induce pathogenesis-related genes different from those induced by SA (courtesy: Springer Verlag)

18.2.2 Induced Systemic Resistance (ISR)

Induced systemic resistance is the phenomenon by which biological or chemical inducers protect non-exposed plant parts against future attack by pathogenic microbes and herbivorous insects. Plants can develop induced resistance as a result of infection by a pathogen, upon colonization of the roots by specific beneficial microbes or after treatment with specific chemicals (Fig. 18.3). ISR can express not only at the site of induction but also systemically in other plant parts that are spatially separated from the inducer. ISR leads to an enhanced level of protection against a

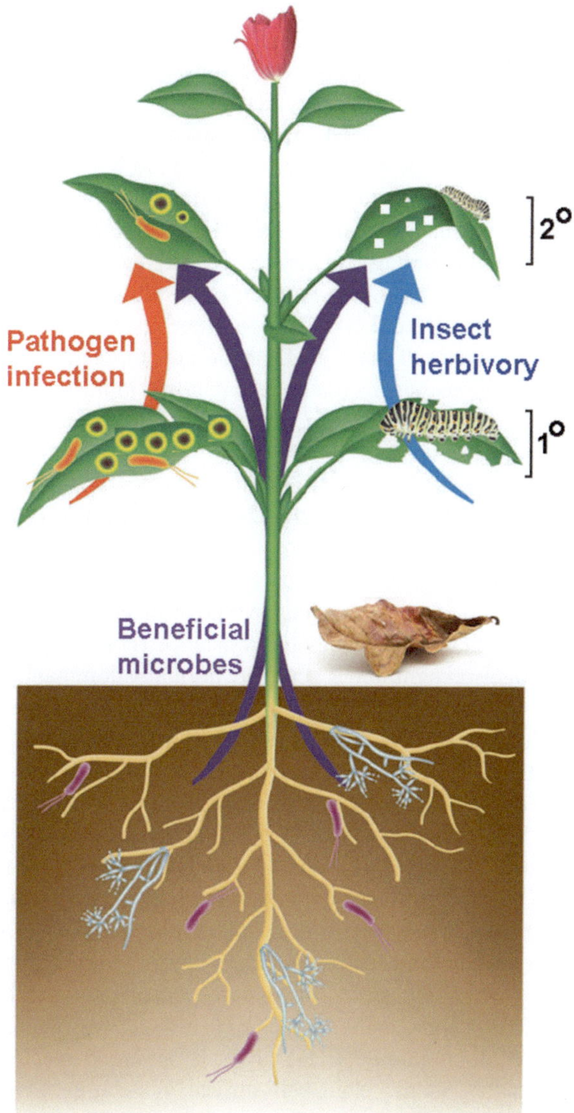

Fig. 18.3 Schematic representation of biologically induced resistance triggered by pathogen infection (*red arrow*), insect herbivory (*blue arrow*) and colonization of the roots by beneficial microbes (*purple arrows*). Induced resistance involves long-distance signals that are transported through the vasculature or as airborne signals and systemically propagate an enhanced defensive capacity against a broad spectrum of attackers in still healthy plant parts. Consequently, secondary (2°) pathogen infections or herbivore infestations of induced plant tissues cause significantly less damage than those in primary (1°) infected or infested tissues

broad spectrum of attackers. An array of interconnected signalling pathways regulate ISR. Here, the plant hormones play a major regulatory role.

In the plant immune system, pattern-recognition receptors (PRRs) have evolved to recognize common microbial compounds, such as bacterial flagellin or fungal chitin, called pathogen- or microbe-associated molecular patterns (PAMPs or

MAMPs). Plants also respond to endogenous plant-derived signals that arise from damage caused by invasion of enemy called damage-associated molecular patterns (DAMPs). Pattern recognition is translated into a first line of defence called PAMP-triggered immunity (PTI), which keeps most potential invaders on check. Successful pathogens have evolved a special mechanism to minimize host immune stimulation and utilize virulence effector molecules to bypass this first line of defence. This is achieved either by suppressing PTI signalling or preventing detection by the host. In turn, plants have acquired a second line of defence in which resistance (R) NB-LRR (nucleotide-binding-leucine-rich repeat) receptor proteins mediate recognition of attacker-specific effector molecules, resulting in effector-triggered immunity (ETI). ETI is a manifestation of gene-for-gene resistance, which is often accompanied by a programmed cell death (PCD) at the site of infection that prevents further progress of biotrophic pathogens (pathogens that live in host cells but do not kill the cells).

18.3 Qualitative and Quantitative Resistance

Resistance is either qualitative or quantitative. This is based on both phenotypic expression of resistance and the type of inheritance. Studies on qualitative resistance showed that major genes for resistance (not always) encode proteins involved in pathogen recognition. R genes are normally dominant, but recessive resistance genes can also occur. On the other hand, quantitative disease resistance (QDR) is with multiple genes of small effects. Genes governing QDR are known as minor genes. A continuum of phenotypic variation is expressed in a cross between a strong QDR and a weak QDR. The genetic dissection of QDR is challenging. The molecular mechanisms of QDR that govern a particular phenotype are not well understood as against qualitative resistance. Many QDR genes have roles in pathogen recognition like qualitative genes. Even though qualitative and quantitative resistance are dealt separately, the system can be continuous. Studies on *Arabidopsis*, tomato and rice have revealed mechanisms underlying immunity. There are two main mechanisms involved in the plant immune response: pathogen-associated molecular pattern (PAMP)-triggered immunity (PTI; also known as basal resistance) and effector-triggered immunity (ETI). PTI is a broad-spectrum resistance. PAMPs are recognized at the plant cell surface via conserved pattern recognition receptors (PRRs), which are typically membrane-localized receptor-like kinases (RLKs) or wall-associated kinases (WAKs) (Fig. 18.4a, b). PTI is a phenomenon by which most plants are resistant to most microbial pathogens. It can also contribute to quantitative resistance. By contrast, ETI forms the basis of qualitative resistance. Most commonly observed characteristics of qualitative and quantitative resistance are available in Table 18.4.

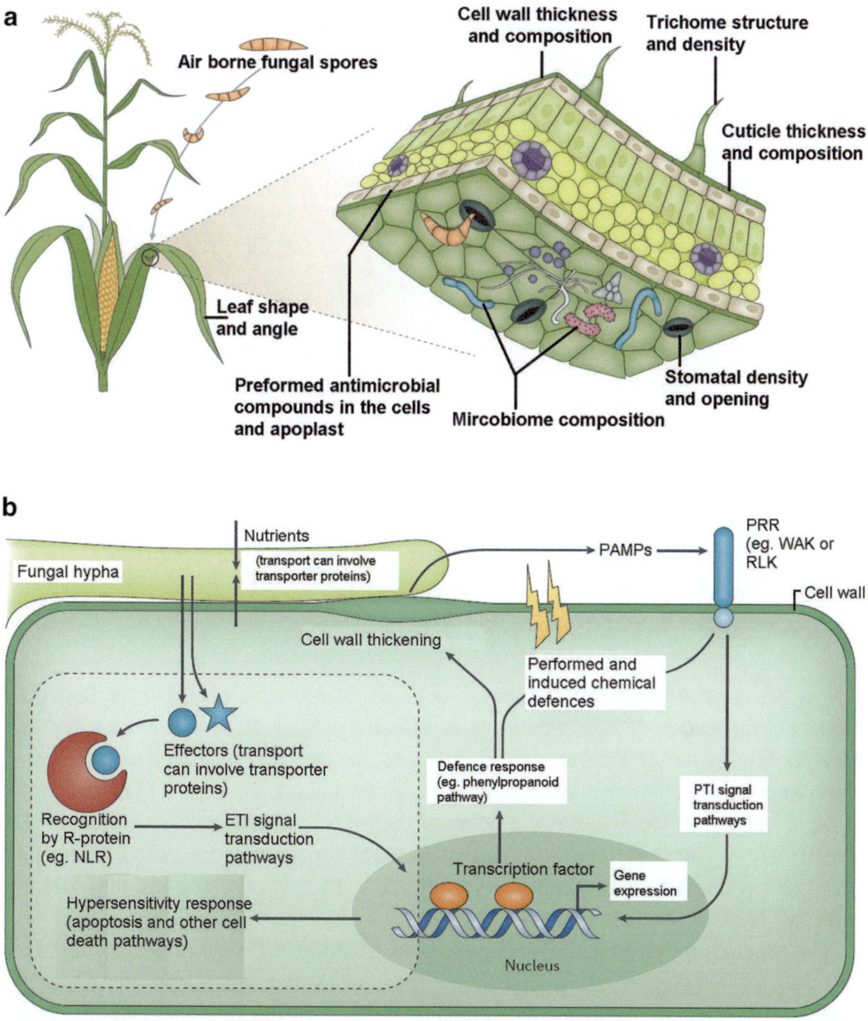

Fig. 18.4 Resistance mechanisms at the tissue and cellular levels. (**a**) At the organismal and tissue levels, the success of a pathogen can be influenced by a range of features of the morphology, biochemistry and microbiome of the plant. (**b**) At the cellular level, factors that affect the ability of a pathogen to infect its plant host include defence responses triggered by recognition events in the host via pattern recognition receptors (PRRs), such as wall-associated kinases (WAKs) or receptor-like kinases (RLKs), and resistance proteins (R-proteins), such as nucleotide-binding domain leucine-rich repeat containing (NLR) proteins; nutrient availability in the apoplast and cytoplasm; pre-existing chemical factors; and cell wall constitution. These factors are affected by host genotype and are potential causes of quantitative variation. Qualitative variation in resistance usually, though not always, occurs at the level of resistance gene-effector interactions. ETI, effector-triggered immunity; PAMPs, pathogen-associated molecular patterns; PTI, PAMP-triggered immunity (courtesy: Nature Reviews Genetics)

Table 18.4 Most commonly observed characteristics of qualitative and quantitative resistance

Category	Qualitative resistance	Quantitative resistance
Synonyms	Vertical, differential	Horizontal, uniform, general
Pathogen specify	Race-specific	Race-non-specific
Symptoms	No disease	Varying degree of disease
Degree of resistance	Complete, absolute	Incomplete, partial
Mechanism	Hypersensitivity	Diverse
Plant growth stage	All-stage resistance (seedling resistance)	Different in each stage (adult-plant resistance, APR)
Assessment	Infection type	Disease severity
Durability	Low	High
Inheritance	Mono-, digenic	Oligo-, polygenic
Gene effect	Major	Minor
Breeding strategy	Backcross breeding	Multi-stage/recurrent selection

Courtesy: Springer International

18.3.1 Genes for Qualitative Resistance

ETI is activated when plant resistance proteins (R proteins, encoded by R genes) recognize their corresponding pathogenic effector protein. Research has shown that they confer resistance by a range of different mechanisms. For example, some R genes encode detoxification enzymes, while others encode WAKs. ETI often results in rapid cell death localized at the point of pathogen penetration. While the hypersensitive response (HR) can be effective in blocking disease caused by biotrophic pathogens, cell death can benefit necrotrophic pathogens (pathogens that kill the cells and feeds on them).

The product of avirulence (*Avr1*) gene by the pathogen is recognized by the plant encoded by a corresponding resistance gene (*R1*), leading to an incompatible reaction that leads to resistance (Fig. 18.5a). If the plant has only susceptible alleles at this locus (*r1*), the reaction is always compatible (susceptible) that is independent of the genotype of the pathogen. Likewise, if the pathogen is virulent for *R1,* all reactions are compatible that leads to disease susceptibility. These patterns are described by the gene-for-gene hypothesis put forth by Flor in 1956 indicating that each resistance gene in the plant has a matching avirulence gene in the pathogen. Since then, this *hypothesis* has been verified in many plant-pathogen interactions with a qualitative inheritance of resistance. If the resistance gene is dominantly inherited, one resistance allele is enough to promote resistance (Fig. 18.5b). Most R genes that govern resistances to fungi and viruses belong to the largest class of R genes with a nucleotide-binding site plus leucine-rich repeat (NB-LRR). Fast production of oxidants is a typical indicator for HR. R and Avr genes are mostly dominant though in some cases resistance is recessive.

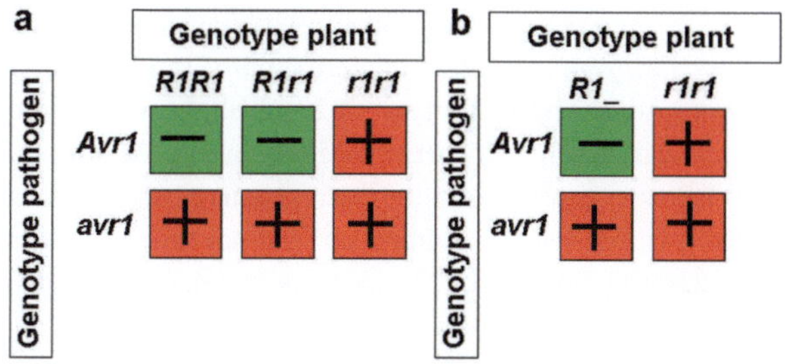

Fig. 18.5 Explanation of the gene-for-gene interaction for a diploid plant with one dominant resistance gene (*R1*) and a haploid pathogen with avirulence (*Avr1*) and virulence (*avr1*); + denotes a compatible reaction (susceptibility), – an incompatible reaction (resistance). (**a**) Full scheme with all possibilities; (**b**) quadratic check for dominantly inherited resistance genes (courtesy: Springer International)

Pathogen populations are capable of forming new virulent (*avr*) pathotypes by mutation of the *Avr* gene. Such a mechanism can evade recognition by hosts. Virulent races have the capacity to attack cultivars that are previously resistant. This is often called *breakdown of resistance.* Here, the pathogen is capable of making R gene ineffective through mutation of its gene to virulence. Gene-for-gene relationships have been identified in many plant-pathogen interactions, including bacteria, fungi, nematodes, viruses and insects. Mostly, biotrophs are included, like rusts (*Puccinia* spp.), powdery mildew (*Blumeria graminis*), smuts (*Ustilago* spp.), bunts (*Tilletia* spp.) and potato blight (*Phytophthora infestans*). Necrotrophs like rice blast (*Magnaporthe grisea*) or northern corn leaf blight (*Setosphaeria turcica*) are also evident.

Breeding for race specificity may lead to susceptibility in a few years that results in yield losses. Each pathosystem contains many R genes. For example, in wheat, there are about 70 formally and 11 temporarily designated genes for leaf rust (*Lr*) caused by *Puccinia triticina,* 58 genes for stem rust (*Sr*) caused by *P. graminis* and at least 53 formally and 39 temporarily designated genes for yellow rust (*Yr*) caused by *P. striiformis*. Most of them are race-specific. The high resistance level, simple inheritance and easy incorporation into commercial cultivars make them attractive to breeders. The best way to judge qualitative adult-plant resistances is to grow breeding populations in a spectrum of climatic conditions in order to rate which hosts are not infected or low infected. Monitoring differential sets in the same experiment will give indications on the pathogenic population at each environment and confirm which R genes are still effective.

18.3.2 Genes for Quantitative Resistance

Quantitative resistances offer higher durability. In some pathosystems, it can be expressed to the extent that it can offer complete resistance. Quantitative resistances are inherited by several genes that can interact with each other (epistasis) and with the environment. They are specific for plant growth stages and/or plant tissues. For example, *Fusarium culmorum* can infect all cereal parts, but ranking of genotypes in their resistances to seedling blight, foot rot or head blight is different. Quantitative resistances are selected in the field by artificial inoculation. Additionally, the time of rating is crucial. While a complete, qualitative resistance can just be rated at the end of the epidemic, for quantitative resistances, an optimal time for genotypic differentiation exists. The assessment can be done by area under disease progress curve (AUDPC).

To avoid confounding effects with effective major genes segregating in the breeding population, a seedling test should be applied first. Screening either with all effective avirulence/ irulence combinations present in the region or a highly virulent race would remove all major genes from the host population. Afterwards, progenies can be analysed in the field for adult-plant resistance. Quantitative resistances are usually characterized to be race-non-specific. However, some QTLs are effective only against a subset of pathogen isolates. In the rice/*Pyricularia grisea* pathosystem, only 2 out of 12 QTLs had an effect on all 3 tested isolates. There might be three types of quantitative resistances:

(a) Basal (overall) resistance governed by many QTLs in the classical sense, i.e. race-non-specific, and largely conserved across host species and even pathogens (broad-spectrum QTLs).
(b) Quantitative resistance mediated by QTLs that are specific for a pathosystem and might be effective only against a subset of isolates.
(c) Qualitative, hypersensitivity-based R genes. It can be speculated whether QTLs of the type (b) are just defeated race-specific resistance genes with some residual effect.

Linkage analysis and genome-wide association studies (GWAS) are used to identify the genomic loci influencing resistant phenotypes. A typical quantitative resistance locus (QRL) identified through linkage analysis encompasses hundreds of genes, and it is very difficult to identify the true causal gene. GWAS provide much higher-resolution mapping. Mapping studies reveal that resistance is often a polygenic trait (also known as a complex trait) that produces a continuous distribution of phenotypes. A synthesis of 16 mapping studies for diseases of rice found 94 QRLs that collectively covered more than half the rice genome. In maize, a similar synthesis identified 437 QRLs covering 89% of the maize genome. The underlying resistance mechanisms are unknown for most QRLs. Many of the genes identified to date are similar in sequence to NLR genes, PRR genes or defence genes that can be controlled by these recognition-related genes.

18.4 Pathogen Detection and Response

Pathogen resistance is because of a suite of cellular receptors that perform direct detection of pathogenic molecules. Pattern recognition receptors (PRRs) within the cell membrane detect pathogen-associated molecular patterns (PAMPs), and wall-associated kinases (WAKs) detect damage-associated molecular patterns (DAMPs) that result from cellular damage during infection (see Fig. 18.6a, b). Receptors with nucleotide-binding domains and leucine-rich repeats (NLRs) detect effectors that pathogens use to facilitate infection. PRRs, WAKs and NLRs initiate one of many

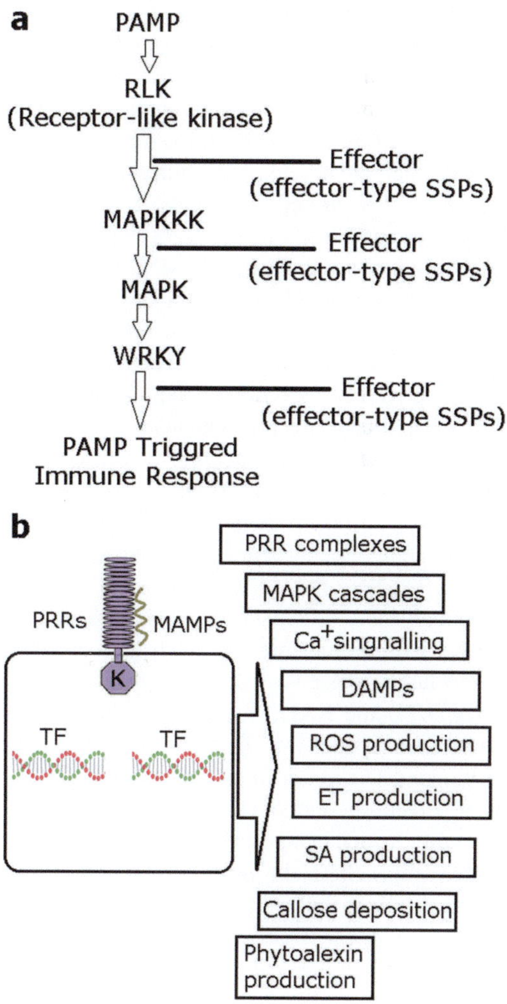

Fig. 18.6 (**a**) Pathogen-associated molecular patterns (PAMP)-triggered immunity in both plant defence and symbiosis. (**b**) Plant PTI signalling and outputs are regulated by transcription perception of different MAMPs by the cognate PRRs that controls various PTI responses *via* transcriptional regulation. TF = transcription factor; SSPs = small secreted proteins

signalling cascades that are yet to be explained. Mitogen-activated protein kinases (MAPKs), G-proteins, ubiquitin, calcium, hormones, transcription factors (TFs) and epigenetic modifications regulate the expression of pathogenesis-related (PR) genes. Hypersensitive response (HR), production of reactive oxygen species (ROS), cell wall modification, closure of stomata or the production of various anti-pest proteins and compounds (e.g. chitinases, protease inhibitors, defensins and phytoalexins) are the later reactions. Pathogen resistance in plants involves various organelles and classes of both proteins and nonprotein compounds. These organelles and proteins regulate defence response. Factors in each of these affect other signalling systems, such as growth and abiotic stress response.

PRRs can recognize a range of microbial components, including fungal carbohydrates, bacterial proteins and viral nucleic acids. These receptors often possess leucine-rich repeats (LRRs) that bind to extracellular ligands, transmembrane domains necessary for their localization in the plasma membrane, and cytoplasmic kinase domains for signal transduction through phosphorylation. LRRs are extremely divergent, with ability to bind to diverse elicitors. Many PRRs rely on the regulatory protein brassinosteroid insensitive 1-associated receptor kinase 1 (BAK1) and other somatic embryogenesis receptor-like kinases (SERKs). Some PRRs while activated can release kinase domains that enter the nucleus and can trigger transcriptional reprogramming. Molecules detected by PRRs are diverse: bacterial (flagellin, elongation factor EF-Tu and peptidoglycan), fungal (chitin, xylanase), oomycete (β-glucan and elicitins), viral (double stranded RNA) and insect (aphid-derived elicitors). Even though these studies were conducted in *Arabidopsis*, they are applicable in crops like wheat. Wheat PRRs are associated with resistance to rust (fungi of the genus *Puccinia*) via detection of fungal PAMPs.

WAKs like WAK1 andWAK2 perceive oligogalacturonic acid, resulting from plant cell wall pectin degradation by fungal enzymes. Plant lectins can recognize carbohydrates arising from pathogens or from damage incurred during infection. Many PAMPs and DAMPs contain carbohydrates (i.e. lipopolysaccharides, peptidoglycans, oligogalacturonides and cellulose) and are recognized by PRRs/ WAKs with lectin domains, such as lectin receptor kinases. Plants detect many extracellular molecules that indicate pathogen infection. These are extracellular DNA, ATP and NAD(P). Pathogens have evolved to interfere in the detection of PAMPs and reduce the efficacy of PTI. *Cladosporium fulvum* (causing tomato leaf mould) and *Magnaporthe oryzae* produce chitin-binding proteins in order to prevent plant perception. Pathogens also produce effectors to thwart many aspects of plant immunity, which plants have developed ways to overcome, as outlined in the zig-zag model (Fig. 18.7). In order to recognize these infection-facilitating pathogen effectors, plants utilize other, more varied class of proteins.

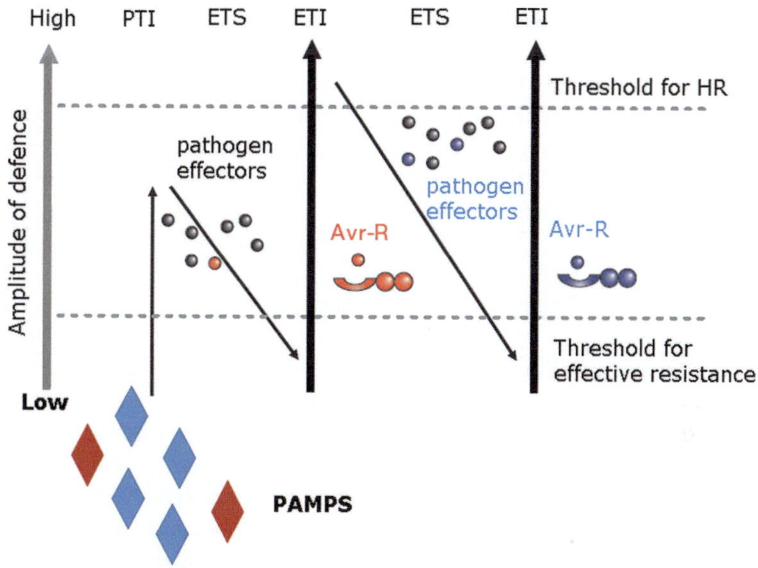

Fig.18.7 A zigzag model illustrates the quantitative output of the plant immune system. In this scheme, the ultimate amplitude of disease resistance or susceptibility is proportional to [PTI – ETS1ETI]. In phase 1, plants detect microbial/pathogen-associated molecular patterns (MAMPs/ PAMPs, red diamonds) *via* PRRs to trigger PAMP-triggered immunity (PTI). In phase 2, successful pathogens deliver effectors that interfere with PTI, or otherwise enable pathogen nutrition and dispersal, resulting in effector-triggered susceptibility (ETS). In phase 3, one effector (indicated in red) is recognized by an NB-LRR protein, activating effector-triggered immunity (ETI), an amplified version of PTI that often passes a threshold for induction of hypersensitive cell death (HR). In phase 4, pathogen isolates are selected that have lost the red effector and perhaps gained new effectors through horizontal gene flow (in blue) – these can help pathogens to suppress ETI. Selection favours new plant NB-LRR alleles that can recognize one of the newly acquired effectors, resulting again in ETI (courtesy: Nature publishing)

18.5 Signal Transduction

Signal transduction is a process by which a series molecular events ensure transmission of chemical or physical signal through a cell. Most common among these is protein phosphorylation catalysed by protein kinases that ultimately results in cellular response. Such proteins detecting stimuli are known as receptors. These stimuli lead to signalling cascade with chain of biochemical events. By interaction of more than one signalling pathway, they form a network. These networks ensure alteration in transcription or translation of genes and post-translational changes in proteins. Such molecular changes control cell growth and development. Initial stimuli are ligands or first messengers, and ligands can in turn activate receptors or signal transducers. Signal transducers can activate primary effectors. Primary effectors can activate secondary effectors and the chain of reactions continues. The new computational biology has the sophistication of analysing signalling pathways and networks to unravel the mechanism of disease spread and also the responses to drug/chemical being administered to control the disease. The initial contact of pathogen and plant would rapidly trigger the signal transduction process on the plasma membrane and cytoplasm of plant cells.

18.5.1 Resistance Through Multiple Signalling Mechanisms

Receptors activate signalling mechanisms that are common to many cellular processes, including MAPKs, G-proteins, ubiquitin and calcium fluctuations. In the general model of MAPK signalling, membrane-bound Ras proteins facilitate the conversion of GTP to GDP, phosphorylating MAPKKK (Raf) proteins, which then phosphorylate MAPKK (MEK) proteins, leading to the phosphorylation of MAPK (ERK) proteins. The involvement of MAPK in many cellular processes has led to the identification of MAPK genes in *Arabidopsis*, which contains 60 MAPKKKs, 10 MAPKKs and 20 MAPKs. Pathogen pectin degradation detected by WAK1 and WAK2 also initiates a MAPK cascade. Defence responses can also be downregulated by MAPK signalling, and pathogens develop effectors that interfere with MAPK signalling to suppress resistance responses. Similarly, the heterotrimeric G-protein (a membrane associated protein) and G-protein-coupled receptor (GPCR) system has been heavily studied due to its involvement in numerous cellular processes. Extracellular ligands bind to the transmembrane GPCR, causing the exchange of GDP for GTP in α-subunit of the G-protein complex, causing a dissociation of α-subunit from the β-γ subunit complex, initiating further signalling. Hydrolysis of GTP by α-subunit then causes the subunits to reassociate. Ubiquitination and subsequent protein degradation by the proteasome also have activity in many signalling systems, including defence. Pathogens have evolved effectors to interfere with the ubiquitin proteasome system in an attempt to disrupt this signalling and facilitate infection. Small ubiquitin-like modifiers (SUMOs) are also utilized by plants to regulate response, and pathogens disrupt this signalling as well. Receptors triggering fluctuations in calcium ions (Ca^{2+}) act as signalling mechanisms to trigger responses to symbiotic or pathogenic microbes. All these molecular signals can be transmitted through hormones that have roles in many different stress and developmental responses. Similar to calcium signalling, fluctuations in hormones drive differential expression of defence response genes.

Recent advances in genomic technology are contributing to the identification of both R genes and genes underlying QTLs. The increasing availability of effector-targeted strategy involves sequencing the existing pathogen population to characterize the relevant effectors and then deploying R genes that recognize those effectors. Effector genes in a pathogen genome are usually identified using a combination of bioinformatic and functional approaches. Once a set of putative or known effectors have been identified, they can be transiently expressed in the host to identify R genes that lead to a resistance (hypersensitive) response. Diverse germplasm (including wild relatives of crop species) can be screened for R genes that recognize the effectors that are most important for pathogenesis.

18.6 Classical Breeding Strategies

Breeding for disease resistance includes:

(a) Identification of resistant breeding sources (plants which carry a useful disease resistance trait). Ancient varieties and wild relatives are the resources of enhanced disease resistance.
(b) Crossing of a desirable but disease susceptible plant variety to another variety that is a source of resistance.
(c) Growth of the breeding populations in a disease-conducive setting. This may require artificial inoculation of pathogen onto the plant population.
(d) Selection of disease-resistant individuals. Breeders try to sustain or improve numerous other plant traits related to plant yield and quality, including other disease resistance traits, while they are bred for improved resistance to any particular pathogen.

Basically, three breeding strategies are possible that depend on the availability of resistance sources and the type of resistance. All methods can be used in self- and cross-pollinated crops. They are:

(a) Backcross breeding: Qualitative resistances from foreign, non-adapted material or wild species.
(b) Recurrent selection: Quantitative resistances from own breeding populations/ adapted cultivars with a low initial resistance level.
(c) Multi-stage selection: Qualitative or quantitative resistances from adapted sources that can directly be combined with agronomic and quality traits.

Breeders often use resistance sources from the adapted gene pool at first in order to avoid introgression of genome segments with negatively acting loci from foreign materials. There is every likelihood that the agronomic performance of progenies might drop drastically in the initial backcross generations when exotic resistance sources are used *via* backcross breeding. In fact, while breeding for quantitative resistances controlled by several genes, such drastic reduction in agronomic performance occurs.

18.6.1 Backcross Breeding

Backcross (BC) breeding is the introgression of target gene from a donor to a recipient genotype used as recurrent parent. This is the classical method for

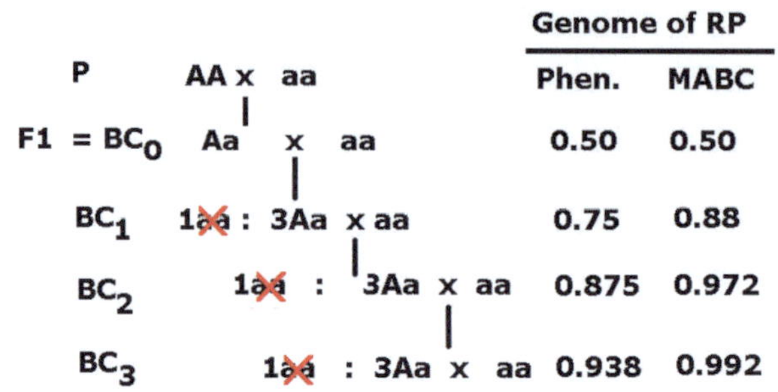

Fig. 18.8 Principle of backcrossing (BC) a single, dominant resistance gene (AA) with a recurrent parent (RP, aa); the average genome proportion of RP is given for phenotypic and marker-assisted backcrossing. After each BC susceptible genotype aa must be discarded by resistance tests or marker selection (see Chap. 10 for details)

introgressing individual R genes from foreign sources into elite breeding material (Fig. 18.8). With each backcrossing step, the recurrent parent genome enriches. Starting with BC_1, after each backcrossing, a selection for the desired resistant phenotype (Aa) is necessary. When deriving inbred lines, selfing must be done in the last BC to ensure homozygous progeny (AA) in the recurrent parent background. At the end, near-isogenic lines are produced that mainly differ in the resistance gene. In practical breeding, often the recurrent parent is changed from generation to generation to keep up with the general selection gain. Total backcross generations needed depend on the genetic difference between donor and recurrent parent. If the gap is more between donor and recurrent parent, more backcross generations are necessary to ensure agronomically reasonable near-isogenic line. Backcrossing of recessive genes takes more time, because after each BC generation, a selfing step has to be performed to produce resistant, homozygous (aa) progeny for selection (see Chap. 10 for details).

18.6.2 Recurrent Selection

Recurrent selection (RS) increases the frequency of desired alleles for quantitatively inherited traits by repeated cycles of selection and recombination. This also maintains genetic diversity. In cross-pollinated crops, test crosses are done to analyse and derive plants for dominant resistance genes. On the other hand, in self-pollinated crops, additional selfing steps are necessary to increase additively inherited genes. The main advantages of RS are:

(a) The possibility to test in several locations and/or years in early generations
(b) To simultaneously improve disease resistances and other agronomic and quality traits
(c) The direct use of selected progenies in breeding commercial cultivars

In barley (self-pollinated), exercise of selection cycles within one cycle could reduce disease severity to less than 10%. In wheat/FHB (*Fusarium* head blight) pathosystem, after two cycles of phenotypic selection, disease severity rates were 3.2% and 2.1% per year in spring and winter wheat, respectively. The task is challenging when agronomic traits are negatively associated with quantitative resistance. Progressive farmers prefer early and short genotypes. This is made possible by substantially increasing population size and a reduced selection intensity for resistance, earliness and shortness.

18.6.3 Multi-stage Selection

In breeding programmes, selection is a continuous process. In a single generation, several successive resistance screenings may be applied. Depending on heritability, degree of dominance and seed availability, different combinations of traits are selected in successive generations. Figure 18.9 gives selection steps for resistance traits in a modern breeding scheme for line cultivars using doubled haploids (DHs). DH lines have been adopted by barley and maize breeders worldwide and are under development in wheat breeding. They are produced either by in vivo parthenogenesis (maize, wheat) or by androgenesis (barley) and involve tissue-culture techniques (embryo rescue or plating of anthers/microspore, respectively). This procedure allows achieving fully homozygous lines after chromosome doubling in one step (see Box 18.2). The main advantages are saving time and higher selection intensity and accuracy, especially for quantitative traits. The main disadvantages are higher costs in some crops and only one round of recombination. Quantitative resistances with lower heritability are selected in DH_2 and DH_3 generations together with grain yield, when larger plots and more environments are available. In multi-stage selection, chances are higher in getting rare recombinants, uniting multiple resistances and superior agronomic traits.

Box 18.2: Doubled Haploids in Maize

Many maize breeding programmes adopted doubled haploid (DH) technology in recent years. It ensures development of completely homozygous lines in less than half of the time compared to conventional breeding. The technology involves the induction of haploidy and subsequent chromosome doubling of haploids. The induction of haploidy can be achieved by in vitro or in vivo methods. In vivo method is being widely applied since it does not require the species to be responsive to tissue culture. In both methods, heterozygous plants from crosses between two or multiple elite inbred parents within heterotic groups form the basis for developing new DH lines. Steps pertaining to in vivo haploid induction are:

(continued)

Box 18.2 (continued)

(a) Maternal haploidy is induced by pollinating with pollen from a haploid inducer. For production of paternal haploids, specific inducers are used as female parent.
(b) A suitable haploid identification system is employed for distinguishing putative haploid seeds (seeds with haploid embryo) from those with regular diploid embryo.
(c) Haploid seeds thus produced are treated with mitotic inhibitors to artificially double their chromosomes to produce doubled haploids.
(d) Putative DH plants are confirmed using a stalk colour marker and true DH plants are self-pollinated to produce DH line seed for further use in breeding and maintenance (Fig. 18.10).

Successful DH production depends on the availability of a haploid inducer genotype. Thus, a haploid identification system, an artificial chromosome-doubling procedure and suitable facilities to raise treated plants for maintenance and seed multiplication are required for DH production.

Commonly used genotypes for in vivo induction of maternal haploids in maize are the (inbreds) RWS, UH400, RWK-76 and UH402. The inducers have an average haploid induction rate of 8–12%. They carry the dominant marker gene *R1-nj* whose phenotypic expression is a purple colouration of the scutellum and the aleurone of seeds, which can be used as embryo and endosperm markers, respectively, to identify putative haploid seeds. In addition, both inducers carry a dominant purple stalk marker that enables the detection of "false positives" among putative haploid plants in the late-seedling stage. Seeds with a haploid embryos and diploid embryo can be visually separated using the *R1-nj* marker system. The haploid seeds have unpigmented (haploid) embryo and purple-coloured (triploid) endosperm, whereas normal F_1 seeds have a purple-coloured (diploid) embryo and a purple-coloured (triploid) endosperm. Further, completely unpigmented seeds will also be present at very low frequency.

18.7 Marker-Assisted Breeding Strategies

MAS has the advantage of compilation of several desired traits in one genotype through fewer breeding cycles. The main questions to be solved are the identification of genes/QTLs with high effects. Ideally, the marker is based on the sequence of the gene of interest (*perfect marker*). For single-marker assays, the competitive allele-specific PCR (KASPar) assay has quite recently emerged. KASPar is an SNP detection system, which is cost-effective for genotyping small subsets of SNP markers. For high-throughput screening, whole-genome array-based assays, like

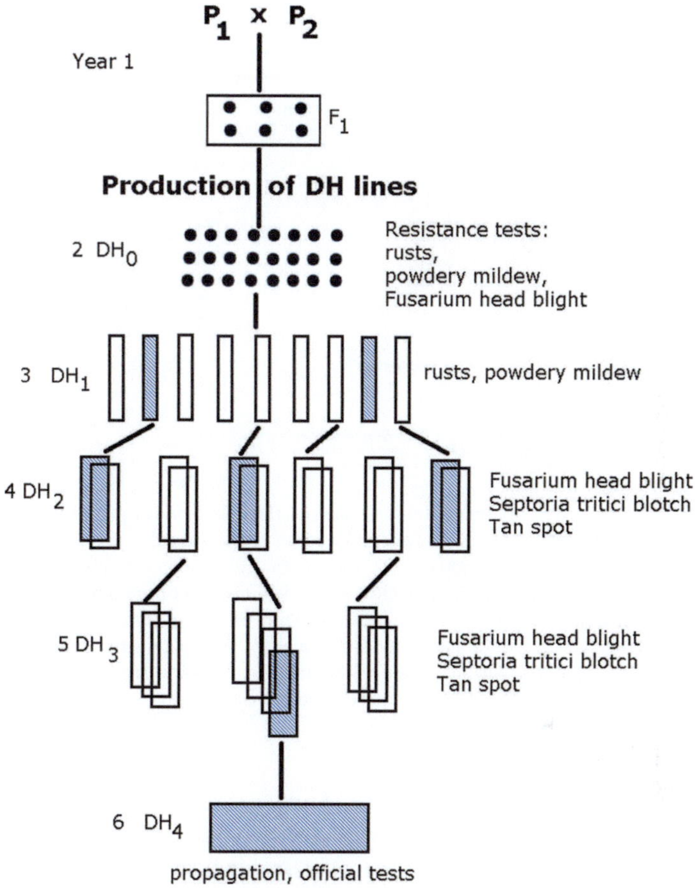

Fig. 18.9 Breeding scheme for self-pollinating crops using doubled haploid (DH) lines and possible selection steps for disease resistance in wheat

the diversity array technology (DArT) or the Infinium HD assays, have been developed. Since both techniques are based on the same marker technique, they can be combined when an SNP set has been established. Older marker techniques, like the single-sequence repeat marker, are still widely used but more expensive per data point and less versatile (see Chaps. 23 and 24 for details).

18.7.1 Monogenic vs. QTLs

For monogenic traits, modern marker detection is straightforward. Based on rather small segregating populations, (either F_2 derived, recombinant inbred lines (RIL) or DH populations), a low-density SNP assay will be sufficient to chromosomally

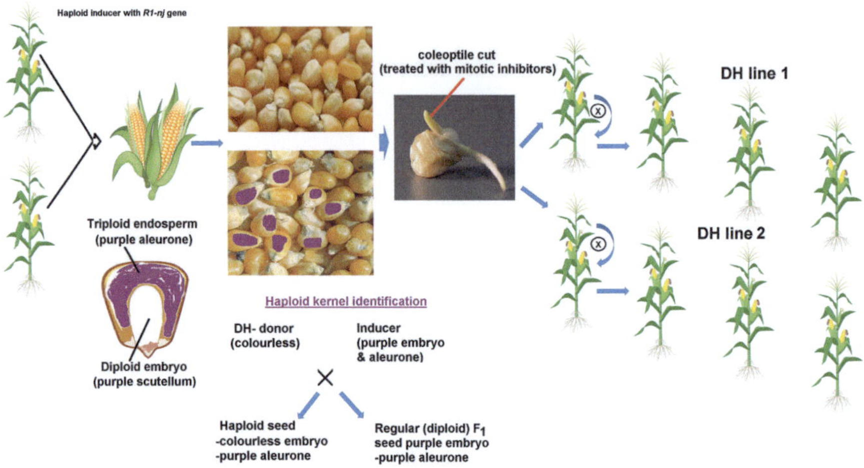

Fig. 18.10 Schematic description of doubled haploid (DH) line development with the in vivo haploid induction approach. (1) Haploidy is induced by pollinating the source germplasm with pollen from a haploid inducer genotype. (2) The pollinated ears of the source germplasm are harvested, and a seed marker system is employed for identification of the putative haploid seeds. (3) The haploid seeds are germinated and, after cutting 2 mm off the tip of the coleoptile with a razor blade, they are treated with mitotic inhibitors. Subsequently, the seedlings are transplanted to the field to produce DH plants. (4) DH plants are self-pollinated to produce seeds for maintenance and multiplication of the DH line (figure diagrammatic and representative)

localize the underlying resistance gene. Further, the genome segment can be enriched by additional SNP markers. Most closely linked SNPs should be analysed for their independence. They can be used afterwards in breeding populations. A QTL is a section of a chromosome that affects a phenotypic trait. For QTL detection, each individual of a segregating progeny is genotyped for DNA markers and phenotyped for quantitative resistance. The resulting data sets are analysed biometrically to identify significant associations between marker and traits. For QTL, mapping is more resource demanding than detection of monogenic traits, because population size should be bigger and several locations and/or years are necessary for phenotypic analyses. Markers across the whole genome are needed. The power of QTL detection does not considerably increase if the distance between adjacent polymorphic markers is smaller than 10 cM. This indicates that rather than marker density, population size is a limiting factor for QTL detection. Currently, two basic techniques are available: biparental mapping and association mapping. While biparental mapping employs structured segregating populations with only a few recombinations, association mapping uses a large array of genetically unrelated entries and historical recombination events.

18.7.2 Marker-Assisted Backcross Breeding (MABC)

Markers are an ideal tool for accelerating the timely backcross (BC) procedure. Backcrossing with monogenically inherited traits is simple and fast. The objectives are:

(a) Tagging the gene of interest (foreground selection).
(b) Selecting individuals that are homozygous for a maximum of recurrent parent alleles in a given BC generation (background selection).
(c) Reducing linkage drag. MABC is of special advantage when recessive alleles should be backcrossed and the target gene is expressed at a later stage in plant development (adult-plant resistance).

While backcross breeding with phenotypic selection is mainly restricted to monogenic resistances, MABC can also be used for introgression of several genes/QTLs. The aim is to introduce the target gene into the elite background and to recover a maximum percentage of recurrent parent genome as early as possible with minimum costs. A genome proportion of 99.2% can be reached by MABC in BC_3 generation. Conventional BC has to be prolonged till BC_6 to gain the same. The cost-effectiveness for gene introgression can be increased with two steps:

(a) In early backcross generations, when a high number of marker data points are needed, high-throughput assays are advantageous
(b) In advanced backcross generations, single-marker assays are more effective.

During BC, the donor chromosome segment around the target gene can remain long over subsequent backcross generations (linkage drag). For example, lengths up to 51 cM of the segment are attached to a resistance gene after six backcross generations in tomato. There are instances that undesirable traits are tightly linked to a gene of interest that was introgressed together with the gene of interest. This is an undesirable situation when the donor is fairly different from the elite recurrent parent in agronomic performance. In order to avoid linkage drag, the sequential analysis of several markers surrounding the target gene can be done. First, a fairly distant flanking marker should be analysed to search for a single or double recombinant. To find out the individual with the shortest intact chromosome segment, subsequent analysis of more tightly linked markers can be used. In summary, disease resistance must be introduced from foreign sources.

Pyramiding Resistance Genes Gene pyramiding is the accumulation of several R genes drawn from multiple parents into a single genotype. They are homozygous for all target loci. The objective is prosperous higher durability that can act

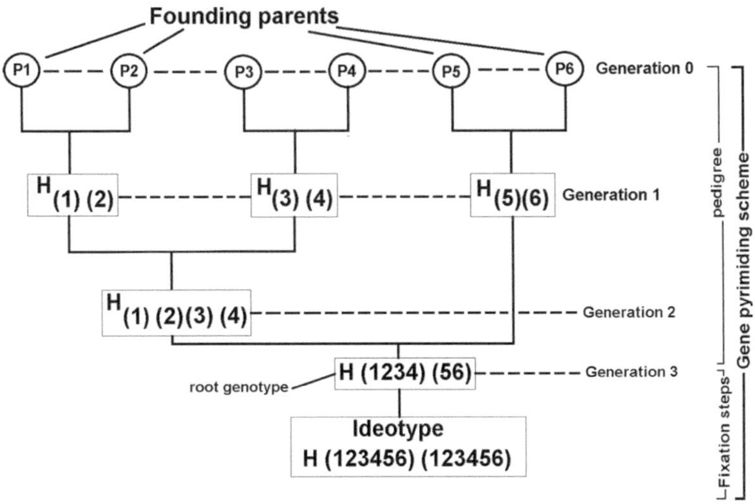

Fig. 18.11 Example of a gene-pyramiding scheme cumulating six target genes. Two parts for the gene-pyramiding scheme can be distinguished. The first part is called a *pedigree* and is aimed at cumulating one copy of all target genes in a single genotype (called *root genotype*). The second part is called the fixation steps and is aimed at fixing the target genes into a homozygous state, that is, to derive the ideotype from the root genotype

simultaneously in one variety with resistance against the same disease having many races. For this, fast progress is possible using molecular markers. Pyramiding genes/ QTLs involves two steps:

(a) Assembling all target genes in a single genotype by multiple crossings
(b) Fixation of the target genes in a single, homozygous genotype rr

The easiest way to combine multiple genes is by a symmetrical crossing scheme involving several single and double crosses and selection of the target genes in a heterozygous state (Fig. 18.11). For fixation of genes, a F_2 enrichment strategy is proposed to counter the demand for large population sizes due to the extreme low frequencies of the desired genotype. For example, the estimated frequency of individuals with eight genes in a homozygous state in one generation equals $(0.25)^8 = 0.00001526$ (=0.001526%). Using F_2 enrichment, in the first selfing generation genotypes with all target genes either in homozygous or heterozygous state are selected. In a second selfing generation, those genotypes with all genes in a homozygous state are selected. Then, probabilities for seldom occurring recombinants are much higher (Fig. 18.12). This procedure is also used for combining several *Bacillus thuringiensis* (*Bt*)-derived toxin genes through transgenesis. In all pyramiding projects, breeders ensure that the target genes are inherited independently and provide different resistance mechanisms or avirulence patterns. Pyramiding strategies are extremely useful in perennial crops due to their longevity. For *Fusarium* head blight resistance, for example, each of three different QTLs has

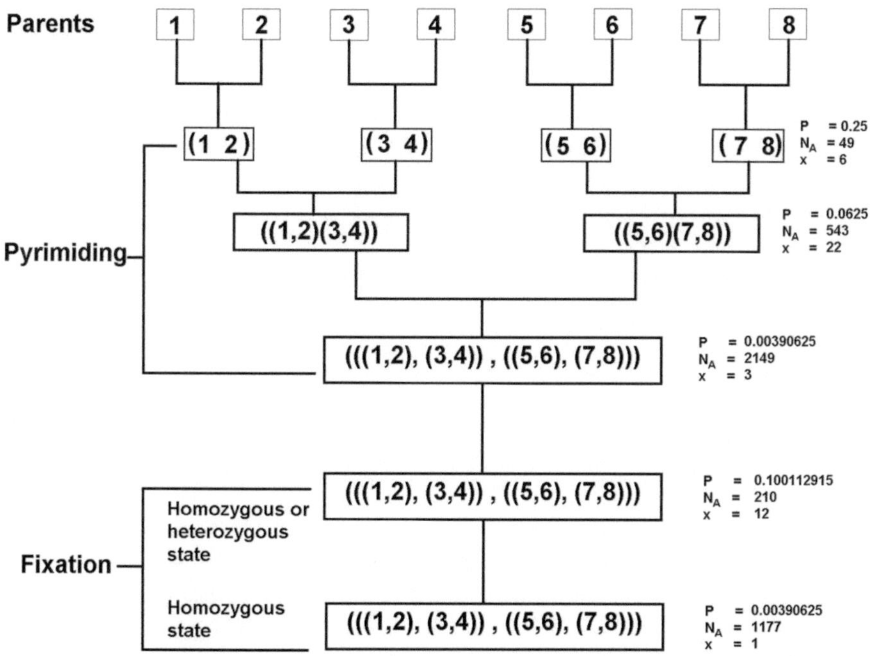

Fig. 18.12 Pyramiding eight genes (1–8) in a single genotype with the frequencies of the desired genotype (p), required population size is adjusted for seed needs in the next generation (N_A), number of selected individuals (x) assuming a 99% success rate and a complete linkage between marker and target gene. Using the seed chipping (SC) + self-pollination (SELF) breeding strategy as an example, the crossing schedule for event pyramiding and trait fixation is shown, featuring for each generation: the frequencies of the desired genotype (p), required population size (N) adjusted for seed needs in the next generation (N_A) and the number of selected individuals (x; also adjusted for seed needs in the next generation), assuming a 99% success rate. The generational goals for trait fixation are specified; for event pyramiding, the goal of each generation is to recover specified events in a heterozygous state. (Courtesy: Springer International)

been stacked in spring and winter wheat, respectively. Lines with different combinations of resistance alleles are created to analyse the effect of QTL individually and stacked in spring and winter wheat. Also in winter wheat, two QTLs on chromosomes 2B and 6A gave the greatest reduction in disease severity. Interestingly, disease reduction by stacked QTLs was lower than that expected from adding the individual QTL effects, revealing epistatic interactions.

Marker-Assisted Selection (MAS): In the past decade, massively parallel serial sequencing (MPSS: a procedure that is used to identify and quantify mRNA transcripts) platforms have become popular. These platforms are made for producing molecular markers cost-effective. Whole-genome re-sequencing, RNA sequencing, whole-genome exome capture sequencing and reduced representation sequencing (e.g. restriction site-associated DNA sequencing). Genotyping by sequencing and specific-locus amplified fragment sequencing with or without a reference genome are all advances in recent years which facilitate the discovery of SNPs and presence/absence of variation (PAV). Once SNPs and/or PAVs are identified, markers can be designed to detect the variation. Using the data from genetic mapping studies and the SNP resources identified, SNP assays can be developed for use in MAS. A customized genotyping system can be developed using customizable assays from several commercial biotechnology companies. Common assays include the Illumina GoldenGate, Kompetitive Allele Specific PCR (KASP™) (LGC, Middlesex, UK) and TaqMan®(Life Technologies, Carlsbad, CA, USA) (see chapter for details of MAS).

Benefits of MAS: MAS is more cost-efficient than expensive field or greenhouse trials. Also, MAS can be more reliable than phenotypic selection. Phenotypic selection for resistance is solely based on the presence of disease/insect where the environment plays a pivotal role in expressivity of disease symptoms. MAS relies on genetic markers that are independent of the environment and traits can be tracked outside of the target environment. MAS allow breeders to select for multiple independent resistance genes and stack them into a variety with more resilience.

Limitations of MAS: The main limitation is that the causal gene(s) (or a narrowly defined QTL) must be known. This can be identified by genetic mapping or can be taken from scientific literature. Marker must be close to the causal gene; otherwise, there is a chance of meiotic crossover occurring between the marker and the gene. In such a circumstance, MAS will fail to identify the causal gene, and the molecular marker will be said to be "broken". Application of multiple molecular markers is one remedy. There can be rare events of double crossover that can break both flanking markers from the causal gene. Additionally, for MAS to be effective, the causal genes need to account for a large effect of the phenotypic variance. The effect of causal genes can also be confounded by genotype x environment interactions. Causal genes can also perform differently in different genetic backgrounds. For these reasons, caution should be taken while employing MAS in a breeding programme. Breeder must periodically confirm that the selections carry the desired trait. Some of the disease-resistant varieties released worldwide are presented in Table 18.5.

18.8 Modern Approaches to Biotic Stress Tolerance

Though conventional breeding methods still play an important role in biotic stress, emerging tools in biotechnology are much needed to maximize the gains. Molecular marker-assisted breeding (MAB) has already gained momentum. There are major gaps in the improvement of traits controlled by a large number of small effects, epistatic QTLs displaying significant genotype × environment (G × E) interactions. Genome sequences for more than 55 plant species have been produced, and many more are being sequenced. This would enable the identification and development of genome-wide markers. Availability of markers covering the whole genomic regions has already shown promise in the development of special populations, such as recombinant inbred lines (RILs), near-isogenic lines (NILs), introgression lines (ILs) or chromosome segment substitution lines (CSSLs). Recently, heterogeneous

Table 18.5 Disease-resistant varieties released across globe (list neither exclusive nor exhaustive)

Variety	Origin	Disease/insect resistance
Novaspy	Canada	Apple scab
McShay	USA	Apple scab
Primevère	Canada	Apple scab
Golden Gopher	USA	Watermelon mosaic virus
Silver Slicer	USA	Cucumber mosaic virus
CaledoniaResel-L	USA	Wheat fusarium head blight
Atlantic	USA	Common bean mosaic virus
Honey Gold	USA	Common bean mosaic virus
Senator	USA	Summer squash powdery mildew
Black Pride	USA	Eggplant verticillium wilt
Pik-Red	USA	Tomato fusarium wilt
Pilgrim	USA	Tomato fusarium wilt
Kaseberg	USA	Wheat stripe rust
VSM (HD 2733)	India	Wheat rusts
Urja (HD 2864)	India	Wheat brown and black rust
HD 2967	India	Wheat leaf blight
HD 3043	India	Wheat stripe and leaf rust
Pusa Sugandh-5	India	Rice brown spot, leaf folder and blast
Pusa Composite 4	India	Maize stalk borer
Pusa 1088	India	Chickpea fusarium wilt
Pusa 5023	India	Chickpea fusarium wilt
PARC-298	Pakistan	Rice bacterial leaf blight
PARC-299	Pakistan	Rice bacterial leaf blight
PARC-301	Pakistan	Rice bacterial leaf blight
Pusa Vishal	India	Mungbean yellow mosaic virus
Pusa 9814	India	Mosaic virus, soybean mosaic virus
Eagle-10	Kenya	Wheat stem rust
Robin	Kenya	Wheat stem rust

inbred family (HIFs) and MAGIC (multi-parent advanced generation intercross) populations, which can serve the dual purpose of permanent mapping populations for precise QTL mapping, have shown promise. Also, genome-wide association (GWA) analysis has been successfully applied to rice, maize, barley and wheat. GWA has also been adapted to the "breeding by design" approach, often referred to as genome selection, which predicts the outcome of a set of crosses on the basis of molecular marker information. Development of "Green Super Rice", possessing resistance to multiple insects and diseases, high nutrient efficiency and drought resistance was achieved through this approach.

Gene expression studies also present a major area of interest for breeders. Through next-generation sequencing (NGS) technologies, direct sequencing of genomes and comparison with reference sequences are increasingly becoming more feasible. Re-sequencing was done in model species like *Arabidopsis*, to ultimately discover single-nucleotide polymorphisms (SNPs). Similar exercises

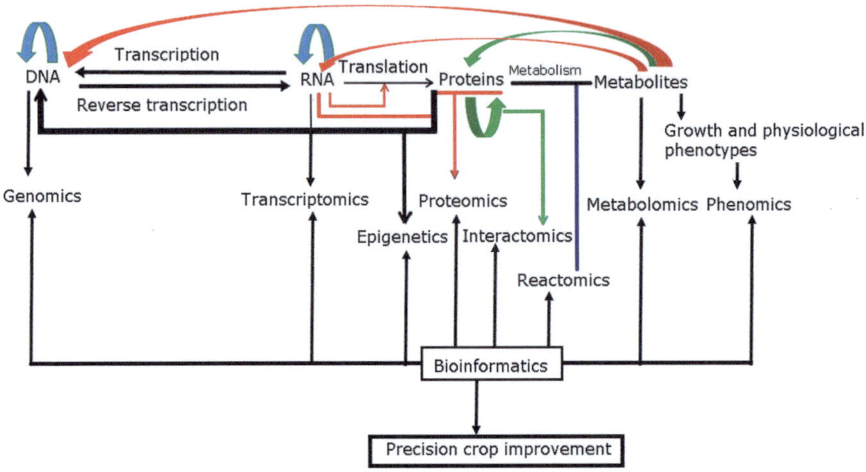

Fig. 18.13 Supportive omic tools for increasing plant breeding efficiency against biotic stresses. Green lines indicate interactions; largest bold black lines indicate epigenetic regulation; red lines indicate regulation; and blue line indicates metabolic reactions

have been carried out in rice, maize and soybean. Combining re-sequencing with the recent developments in omic biology, including transcriptomics, proteomics, metabolomics, epigenetics and physiological and biochemical methods, will remarkably provide novel possibilities to understand the biology of plants and consequently to precisely develop stress-tolerant crop varieties (Fig. 18.13). Recent invention of genotyping by sequencing (GBS) has enabled SNP marker detection, exposition of QTLs and the discovery of candidate genes controlling stress tolerance. So, in the coming future, genome/transcript profiling combined with genome variation analysis is to be a potential area of research.

Another newly developed approach, which combines genomics and bulk segregant analysis (BSA – technique to identify genetic markers associated with a mutant phenotype) to identify markers linked to genes, shows the possibility of coupling BSA to high-throughput sequencing methods. This method has been proved to be useful in identifying stress tolerance genomic regions in crop plants. A more recent modification that exploits SNP markers involving efficiency of BSA analysis is called target-enriched TEXQTL mapping. Here, by combining a large F_2 population and deeply sequenced markers, most QTLs can be identified within two generations. TEX-QTL method is a potentially useful development in plant breeding. Desirable alleles are also being identified by means of targeting induced local lesions in genomes (TILLING) or ecotype TILLING (EcoTILLING) methodologies (see Box 18.3 for RNAi and Chap. 16 for TILLING). These strategies predict gene functions and allow efficient prediction of the phenotype associated with a given gene – the so-called reverse genetics approach.

Box 18.3: RNAi-Mediated Plant Defence

RNA interference or silencing is the sequence-specific gene regulation by small non-coding RNAs. They are of two categories: small interfering RNA (siRNA) and microRNA (miRNA). They differ in their biogenesis, but regulate the target gene repression through ribonucleoprotein silencing complexes. There are four basic steps in plant RNA silencing:

(a) Introduction of double-stranded RNA (dsRNA) into the cell
(b) Processing of dsRNA into 18–25-nt small RNA (sRNA)
(c) sRNA methylation
(d) sRNA incorporation into effector complexes that interact with target RNA or DNA

Before cleavage of the target mRNA, formation of RNA-induced silencing complex (*RISC*) and its incorporation into the antisense strand of siRNAs happen. This complex then interacts with Argonaute and other effector proteins. For sRNA to meet the target mRNA, it has to move from the point of initiation to the target. Here, two main movement categories occur. These are cell-to-cell (short-range; symplastic movement through the plasmodesmata) and systemic (long-range; through the vascular phloem). These mobile silencing strategies use sRNAs to target mRNA in a nucleotide sequence-specific manner. Such systematic movements enhance systemic silencing of viruses. Resistance to *cassava mosaic virus* (CMV) was achieved in transgenic cassava plants through this method. A similar strategy was successful in transgenic tomato resistance against *potato spindle tuber viroid* (PSTVd). RNAi targeting of the virus coat protein has also been successfully engineered into plants to induce resistance against viruses. Virus-induced gene silencing (VIGS) has emerged as one of the most powerful RNA-mediated post-transcriptional gene silencing (PTGS) methods. It would be even better if interaction between sRNAs and their targets is validated in several backgrounds. However, mechanisms governing RNAi require further investigations.

Craig Cameron Mello and Andrew Z. Fire of the University of Massachusetts Medical School were awarded Nobel Prize for Physiology and Medicine in 2006 for the discovery of RNA interference.

The use of improved recombinant DNA techniques to introduce new traits in early phases of cultivar selection is also currently gaining momentum. Techniques such as oligonucleotide-directed mutagenesis (oDM) (see Chap. 16) as well as those based on zinc finger nuclease (ZFN), transcription activator-like effector nuclease (TALEN) and clustered regularly interspaced short palindromic repeat (CRISPR)/ CRISPR-associated protein 9 (Cas9) system (see Chap. 24) are all capable of specifically modifying a given target sequence leading to genotypes not substantially different from those obtained through traditional mutagenesis. The practical use of these techniques is yet to be fully demonstrated (Box 18.4).

Box 18.4: Systems Biology and Plant Defence

A successful pathogen has to conquer passive defence mechanisms. These include structural barriers such as the cuticle, the cell wall and constitutively produced antimicrobial compounds. In addition to these passive mechanisms, plants possess a two layered actively induced immune system. The first layer of the immune response is termed pathogen-associated molecular-pattern (PAMP)-triggered immunity (PTI). The second layer of plant defence, called effector-triggered immunity (ETI), is mediated by intracellular resistance (R) proteins that recognize molecules injected by pathogens into plant cell designated effectors. While PTI confers resistance against a broad group of microorganisms, ETI is specific to isolates of microorganisms producing a given effector and leads to a complete resistance response often accompanied by a rapid programmed cell death reaction called the hypersensitive response (HR). Systems biology aims at understanding the properties of living organisms emerging at the network (also called *emergent properties).* Emergent properties arise from the interaction between multiple components. As a methodology, systems biology aims at integrating observations on multiple components of the system (cell, organs or populations) by using mathematical models. Systems biology has emerged as a broadly used methodology with the development of the so-called omics techniques. This envisages progress in techniques like DNA and RNA sequencing for genes and mass spectrometry (MS) for proteins and metabolites.

Further Reading

Kushalappa AC et al (2016) Plant innate immune response: Qualitative and quantitative resistance. Crit Rev Plant Sci 35(1):38–55. https://doi.org/10.1080/07352689.2016.1148980

Fritsche-Neto R, Borém A (eds) (2012) Plant breeding for biotic stress resistance. Springer, Heidelberg

Shen Y et al (2018) The early response during the interaction of fungal phytopathogen and host plant. Open Biol 7:170057. https://doi.org/10.1098/rsob.170057

David J, Schneider DJ, Collmer A (2010) Studying plant-pathogen interactions in the genomics era: beyond molecular Koch's postulates to systems biology. Annu Rev Phytopathol 48:457–479

Collinge DB Transgenic crops and beyond: how can biotechnology contribute to the sustainable control of plant diseases? Eur J Plant Pathol 152:977–986. https://doi.org/10.1007/s10658-018-1439-2

Boyd LA (2013) Plant–pathogen interactions: disease resistance in modern agriculture. Trends Genet 29:233–240

Breeding for Abiotic Stress Adaptation 19

Keywords

Types of abiotic stresses · Drought tolerance · Salinity tolerance · Temperature tolerance · Macro- and microelements · Physiological and biochemical responses · Breeding for abiotic stresses · Breeding for drought tolerance/WUE · Photosynthesis under drought stress · Breeding for heat tolerance · Drought vs. heat tolerance · Salinity tolerance · Salinity tolerance mechanisms · Breeding strategies · Marker-assisted selection (MAS) · MABA for abiotic stress in major crops (rice, wheat, maize) · "omics" and stress adaptation · Comparative genomics tools · Transcript"omics" · Combining QTL mapping · GWAS and transcriptome profiling · Prote"omics" to unravel stress tolerance · Metabol "omics" · Phen"omics" for dissection of stress tolerance.

Abiotic stress is defined as the negative impact of non-living factors on the living organisms in a specific environment. The literal meaning of the word "stress" is coercion, that is, force in one direction. In Physics, stress is tension produced within a body by the action of an external force. Biologically, stress is a significant deviation from ideal conditions. Stress prevents plants from expressing their full genetic potential for growth, development and reproduction. Stress is a stimulus that surpasses the usual range of homeostatic regulation (homeostasis is stability or balance of the plant body – it is the body's attempt to maintain a constant internal environment) in any living being. Abiotic stresses (water deficit, high temperature, low temperature and high salinity) pose a serious threat to the food security world-wide. It poses a negative influence on the plant's survival and can reduce biomass and yield by up to 50–70%. Any stress above the threshold level can activate a cascade of responses at physiological, biochemical, morphological and molecular levels. This cascade of responses helps to withstand the stress. Stress tolerance is a quantitative trait with complex gene regulations. Molecular mechanisms and various

P. M. Priyadarshan, *PLANT BREEDING: Classical to Modern*,
https://doi.org/10.1007/978-981-13-7095-3_19

complex signalling pathways govern such gene regulations, and such a process involves activation and deactivation of stress responses.

19.1 Types of Abiotic Stresses

In broader sense, abiotic stress encompasses a spectrum of multiple stresses such as heat, cold, excessive light, drought, water logging, UV-B radiation, osmotic shock and salinity (Fig. 19.1). All these can dramatically affect the plants' growth leading to loss of yield. Such stresses initiate stress signals in plants to combat and to adapt to the stress situation through maintaining homeostasis. Based on their response to salinity stress, plants have been classified as glycophytes (stress-susceptible) and halophytes (stress-tolerant). Majority of the plants are glycophytes as they ensure their survival through tolerance, avoidance or resistance. Tolerance to any stress and avoidance prevents plant from getting exposed to stressful conditions. Both tolerance and avoidance spares the plant from any damage. Resistance is yet another complex phenomenon that is getting studied.

Water deficit (drought) that affects 64% of the global land area is the major stress. This is followed by flood (anoxia) affecting 13% of the land area, salinity 6%, mineral deficiency 9%, acidic soils 15% and cold 57%. Soil erosion, soil degradation and salinity affect 3.6 billion ha out of the world's 5.2 billion ha of dry land agriculture. Soil salinity has an impact upon 50% of total irrigated land in the world costing US$12 billion in terms of loss. Plants need light, water, carbon and mineral nutrients for their optimal growth, development and reproduction. Stress as extreme conditions (below or above the optimal levels) would limit plant growth and development. Plants can sense and react to stresses in many ways that favour their sustenance. Water deficit adversely affects photosynthetic capability; decreases leaf water potential and stomatal opening; reduces leaf size; suppresses root growth; reduces seed number, size and viability; delays flowering and fruiting; and limits plant growth and productivity (Fig. 19.2). Plants have the inherent capacity to minimize consumption of water and adjust their growth till they face adverse

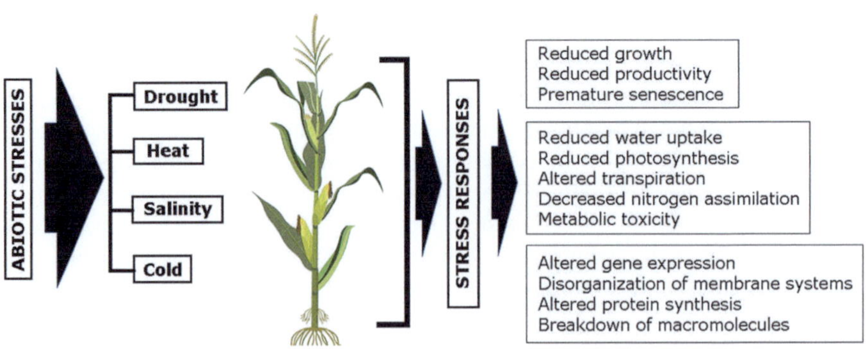

Fig. 19.1 Various stresses and stress responses

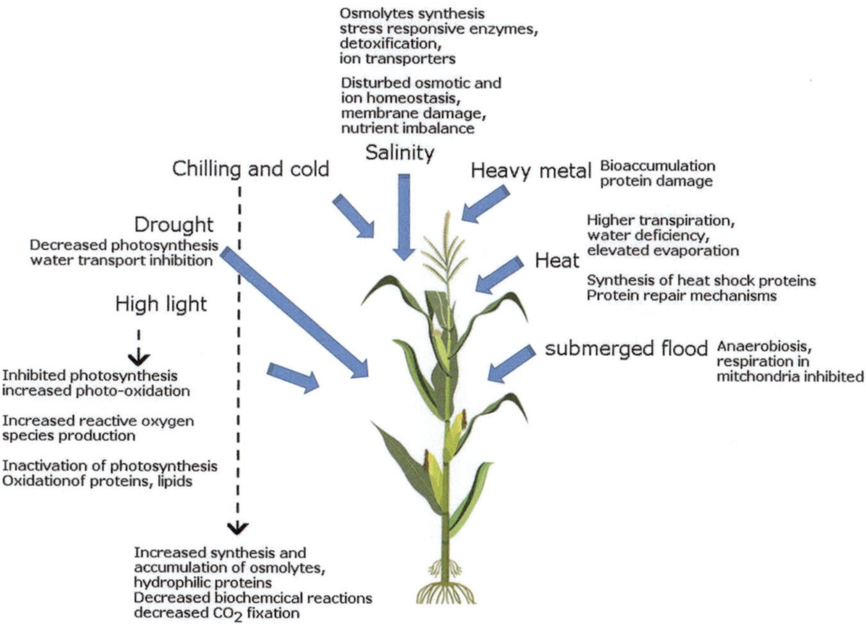

Fig. 19.2 Diverse abiotic stresses and the strategic defence mechanisms adopted by the plants. Though the consequences of heat, drought, salinity and chilling are different, the biochemical responses seem more or less similar. High light intensity and heavy metal toxicity also generate similar impact, but submergence/flood situation leads to degenerative responses in plants where aerenchyma is developed to cope with anaerobiosis. It is therefore clear that adaptive strategies of plants against variety of abiotic stresses are analogous in nature. It may provide an important key for mounting strategic tolerance to combined abiotic stresses in crop plants

conditions. Exposure to excess light induces photo-oxidation that increases the production of highly reactive oxygen intermediates to manipulate biomolecules and enzymes. Different levels of acidic conditions can negatively influence soil nutrients. Acidic conditions can also limit ease of availability of nutrients, and because of this, plants become nutrient deficient disrupting normal physiological pattern of growth and development. Tolerance to salinity stress calls for quick adjustment of both cellular osmotic and ionic homeostasis. One of the common strategies by plants to combat salinity is to avoid high saline environments. One way of accomplishing this is to keep sensitive plant tissues away from the zone of high salinity. Plants can also exude ions from the roots or compartmentalize ions away from the cytoplasm of physiologically active cells.

19.1.1 Drought Tolerance

Tolerance to drought stress in plants is indicated by leaf rolling, stomatal closure, photochemical quenching, photo inhibition resistance, water use efficiency (WUE), osmotic adjustment, membrane stability, epicuticular wax content, mobilization of water-soluble carbohydrates and increased root length. These traits are often used for phenotyping under drought stress. Leaf rolling can reduce transpiration rate and canopy temperature. Retention of higher relative water content (RWC) under water-deficit conditions is a strategy followed by drought-tolerant plants. The impact of drought on photosynthesis can be either direct or indirect. The direct effect is to reduce CO_2 diffusion via stomatal closure that limits CO_2 supply inside leaves thus reducing the availability of CO_2 to Rubisco. Indirect effects are to alter the biochemistry and metabolism of the photosynthetic apparatus, membrane permeability and the promotion of oxidative stress.

The aforesaid reactions can lead to poor grain development. Drought stress exerts osmotic pressure on plants. Proline plays an important role in the stabilization of cellular proteins and membranes under high osmotic concentrations. Secondary responses, such as oxidative stress, induce membrane damage during water stress. Roots are directly connected with soil and are the first potential organ to perceive water deficit. Most recently, next-generation phenotyping platforms with highly efficient software like PHENOPSIS and WIWAM are used to study drought tolerance.

19.1.2 Salinity Tolerance

Salinity induces both ion toxicity and osmotic stress in crop plants. Salinity alters ionic homeostasis of cells and delays germination. During vegetative stages, it reduces leaf area, total chlorophyll content, biomass and root length. Osmotic stress reduces the water absorption capacity of root systems and in addition increases water loss from the leaves. Other important physiological changes caused by the osmotic stress include membrane interruption, nutrient imbalance, impaired ability of ROS (reactive oxygen species are chemically reactive chemical species containing oxygen) detoxification, differences in antioxidant enzymes, decreased photosynthetic activity and reduced stomatal aperture. Ion toxicity occurs due to higher accumulation of Na^+ and Cl^- ions. ROS formation interrupts vital cellular processes through causing oxidative damage to various cellular components like proteins, lipids and DNA. Plants also develop various physiological and biochemical mechanisms to survive in high salt concentration (Fig. 19.3).

19.1.3 Temperature Tolerance

High or chilling/freezing temperatures induces poor germination, poor seedling emergence, abnormal seedling development, poor seedling vigour, reduced radicle and plumule growth, inhibition of photosystem II (*Psi* I) activity and ROS production. Cold stress influences the reproductive stage the most. Complete yield loss can be the result due to a rise in few degrees of temperature. Scorching and sunburns and abscission of leaves and inhibition of shoot and root growth are the permanent

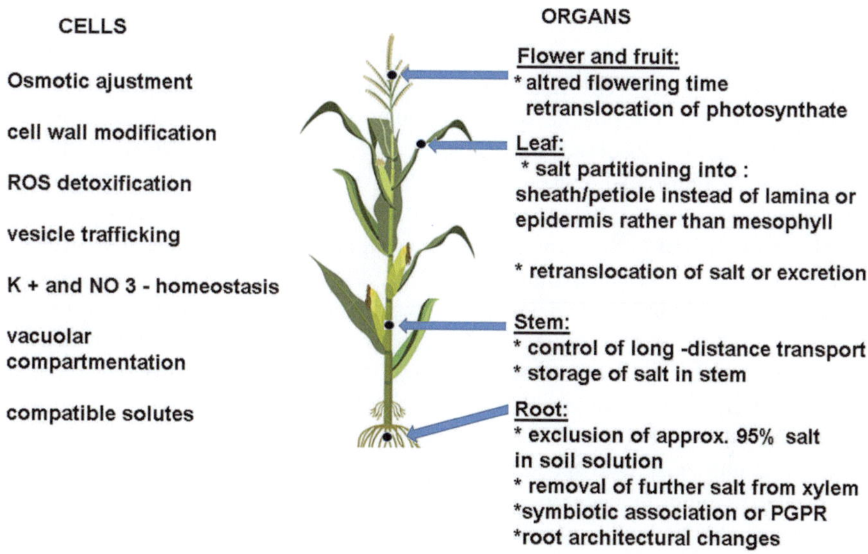

Fig. 19.3 Adaptive mechanisms of salt tolerance. On the left are listed the cellular functions that would apply to all cells within the plant. On the right are the functions of specific tissues or organs. Exclusion of at least 95% (19/20) of salt in the soil solution is needed as plants transpire 20 times more water than they retain. ROS = reactive oxygen species; PGPR = plant growth-promoting rhizobacteria

damages caused by heat stress. Biochemical changes due to high-temperature stress are irreversible damage to photosynthetic pigments and Rubisco-enhanced rate of photorespiration. It also exerts influence on ROS accumulation due to the inhibition of non-cyclic electron transport. Elevated temperature causes programmed cell death (PCD) in specific cells or tissues within minutes or even seconds.

Based on temperature range, cold stress is either chilling stress (<20 °C) and/or freezing stress (<0 °C). Chilling stress reduces the rate of enzymatic reactions and membrane transport activities. Freezing stress results in the formation of ice crystals and membrane damage. Indirectly, cold stress induces osmotic imbalance, oxidative stress and, in the case of chilling stress, the formation of water uptake barriers. Freezing stress causes cellular dehydration. Genotypes differ in their ability to tolerate chilling and freezing stresses. Cold acclimation results in altered gene expression, biomembrane lipid composition and accumulation of small molecules. Tropical and subtropical plants are more sensitive to chilling stress and lack cold acclimation mechanism. Low-temperature resistance is a complex mechanism.

19.1.4 Macro- and Microelements

There are elements essential for plant to complete its cycle. They are divided into macro- and micronutrients. The macronutrients are composed of nitrogen (N), phosphorus (P), potassium (K), calcium (Ca), magnesium (Mg) and sulphur (S). Large amount of these elements are required for plants to develop and meet their physiological activity. The macronutrients play a vital role in plant structure. Micronutrients are responsible for the regulatory activity of the cell organelles. These nutrients are absorbed and found in lower concentrations in plant tissues and supply the nutritional exigency of the plant. Some of them are zinc (Zn), boron (B), copper (Cu), iron (Fe), manganese (Mn), molybdenum (Mo) and chlorine (Cl).

Micronutrients at higher concentrations are toxic and provoke negative effects. This toxicity reduces photosynthetic pigments, affects permeability of membranes, increases the accumulation of reactive oxygen species (ROS) and increases the activities of antioxidant enzymes. Such a process leads to cell death. Stress caused by the excessive supply of nutrients induces overproduction of reactive oxygen species (ROS) as superoxide radical (O_2^-) and hydrogen peroxide (H_2O_2). There are mechanisms to explain the tolerance of plants to toxicity induced by heavy metals and nutrients. Two specific processes are metal ion homeostasis and compartmentalization of metals into the vacuole.

Once the stress stimulus is sensed, cells initiate a complex stress-specific signalling cascade. Following reactions will happen:

(a) Synthesis of phytohormones like abscisic acid, jasmonic acid, salicylic acid and ethylene.
(b) Accumulation of phenolic acids and flavonoids.
(c) Elaboration of various antioxidants and osmolytes and activation of transcription factors (TFs) along with the expression of stress-specific genes to mount appropriate defence system. Many mechanisms related to stress tolerance in plants are known. But, the "on-field response" to multiple stresses is still unclear.

Soil is a multiphasic system with different nutrient concentrations and plant is able to uptake a small fraction. Nutrient absorption ability of a plant is influenced by soil physical and chemical characteristics like structure, texture water content, pH, fertility level and nutrient content. A soil pH of 6–7.5 is ideal for nutrient content absorption. Higher or lower than this optimum pH can affect nutrient availability status under suboptimal conditions. For example, in sodic (alkaline) soils, phosphorus, iron and molybdenum deficiency are usually observed. In acid soils, plants suffer from phosphorus deficiency. Root exudate compounds such as sugars, organic acids, secondary metabolites and enzymatic compositions increase plant nutrient uptake ability under various stress conditions.

19.2 Physiological and Biochemical Responses

The creation of water deficiency within cells is the direct impact of drought, frost, salinity and heat stresses. This is followed by a parallel development of biochemical, molecular and phenotypic responses against stresses. Severe water deficits can result

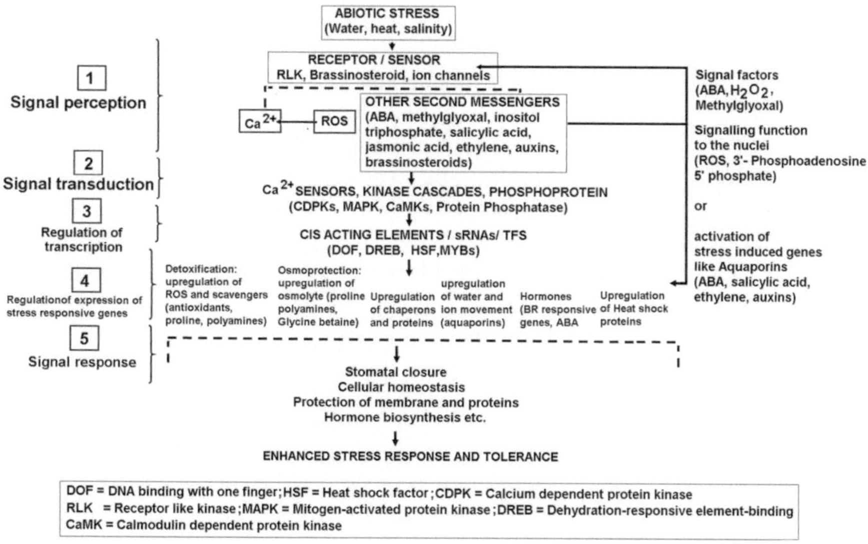

Fig. 19.4 Signalling pathways involved in plant abiotic stress responses. (Courtesy: Frontiers in Chemistry)

in peroxidation that negatively affects antioxidant metabolism. But, the level of peroxidation decreases upon rewatering and restores growth and development of new plant parts and stomatal opening. Both drought and rewatering lead to high accumulation of H_2O_2 in roots. Superoxide dismutase (SOD) plays a central role in antioxidant metabolism, and drought responses vary from plant to plant in terms of SOD enzyme.

The continued rise in global temperature can adversely impact morpho-anatomical, physiological, biochemical and genetic changes in plants. Heat reduces seed germination, leading to loss in photosynthesis and respiration and decrease in membrane permeability. Again, some prominent responses of plants against heat stress are alterations in the level of phytohormones, primary and secondary metabolites, enhancement in the expression of heat shock and related proteins and production of reactive oxygen species (ROS). (Fig. 19.4). Response of plants against heat stress involves maintenance of membrane stability and induction of mitogen-activated protein kinase (MAPK) and calcium-dependent protein kinase (CDPK) cascades.

19.2.1 Physiological Responses

Stress avoidance mechanisms are increased root system, reduced stomatal number and conductance, decreased leaf area, increased leaf thickness and leaf rolling or folding to lessen the evapotranspiration. Epicuticular wax biosynthesis, on the surfaces of the aerial plant parts, is also an adaptive response. The other tolerance mechanism is the maintenance of tissue hydrostatic pressure mainly through osmotic adjustments. Under drought, root hydraulic conductivity is reduced that prevents water losses from the plant to the dry soil. Water transport within a plant is determined by soil water availability and the atmospheric vapour pressure deficit, creating a turgor pressure within the cells. Water transport in roots is affected by various components such as root anatomy, water availability and salts in the soil. All of these factors are influenced by the activity of aquaporins, which are integral membrane proteins that function as channels to transfer select small solutes and water (see Box 19.1).

Box 19.1 Aquaporins

Plant aquaporins (AQP-water channels) are a large family of proteins that facilitate the transport of water and small neutral molecules across biological membranes. Depending on membrane-type localization and permeability to specific solutes, they are divided into several subfamilies. AQPs play a central role in acquiring abiotic stress tolerance. There are aquaporins belonging to PIP subfamily (plasma membrane intrinsic proteins) that are permeable to water and/or carbon dioxide. Isoforms of AQPs transporting water are involved in hydraulic conductance regulation in the leaves and roots, and those transporting carbon dioxide control stomatal and mesophyll conductance in the leaves. Changes in PIP aquaporins abundance/activity in stress conditions allow maintaining the water balance and photosynthesis adjustment. Studies have shown that tight control between water and carbon dioxide supplementation mediated by AQPs influences plant productivity, especially in stress conditions.

AQPs have remarkable features to transport water into and out of the cells along the water potential gradient. Plant AQPs are classified into five main subfamilies including the plasma membrane intrinsic proteins (PIPs), tonoplast intrinsic proteins (TIPs), nodulin 26-like intrinsic proteins (NIPs), small basic intrinsic proteins (SIPs) and X intrinsic proteins (XIPs). AQPs are localized in the cell membranes and are found in all living cells. However, most of the AQPs that have been described in plants are localized in tonoplast and plasma membranes. Regulation of AQP activity and gene expression are part of adaptation mechanisms to stress conditions. They rely on signalling pathways and complex transcriptional, translational and post-transcriptional factors. Regularizing AQPs through different mechanisms, such as phosphorylation, tetramerization, pH, cations, reactive oxygen species and phytohormones could play a key role in plant responses to environmental stresses.

Stress has a direct impact on photosynthesis. Photosynthesis comprises various components, including the photosystems and photosynthetic pigments, the electron transport system and CO_2 reduction pathways. An effect on any of these components can lead to reduction in photosynthesis. Both heat and water stresses are reported to decrease electron transport, degrade proteins and release magnesium and calcium ions from their protein-binding partners. High temperature can reduce chlorophyll content, increased amylolytic activity, thylakoid grana disintegration and disruption of assimilates' transport. Reduction in the net photosynthetic rate is associated with stomatal closure resulting in increased WUE (= net CO_2 assimilation rate/transpiration). Stomatal closure is having a more inhibitory effect on transpiration than on CO_2 diffusion into the leaf tissues. Carbon isotope discrimination that reflects both CO_2 exchange and water economy can assess phenotypic variation within a large breeding population.

In wheat, maize and sorghum, the reproductive phase is the most sensitive to high-temperature stress. The reproductive processes involving pollen and stigma viability, pollination, anthesis, pollen tube growth and early embryo development are especially vulnerable to heat stress. Male reproductive tissues are more sensitive to high-temperature stress than female reproductive tissues.

19.2.2 Biochemical Responses

As a first step of response to stress, signals from the environment activate signalling cascades in plants. There are receptors that perceive signals and stimuli from the environment. The first receptor kinase protein in plants, the receptor-like kinase (RLK), was described in the 1990s. A subfamily of RLKs known as WAKs (wall-associated kinases) receives signals from the environment and other adjacent cells to activate appropriate signalling cascades. Here, aquaporin proteins are key factors contributing to hydraulic conductivity. Aquaporin proteins are regulated by environmental stimuli with changes like phosphorylation, cytoplasmic pH and calcium. These are further re-localized into intracellular compartments. Abscisic acid (ABA) is the most critical hormone that regulates tolerance to abiotic stresses like drought, salinity, cold, heat and wounding. ABA is the root-to-shoot stress signal-inducing inhibition of leaf expansion and stomatal closure. Stomatal closure is a short-term response. ABA synthesis is closely related to osmotic stress. Osmotic stress induces synthesis of several other growth regulators, including auxin, cytokinins, ethylene, gibberellins, brassinosteroids and jasmonic acid. These growth regulators act as signal molecules in signalling networks. Increased intracellular Ca^{2+} levels are also induced by signal molecules like inositol trisphosphate, inositol hexaphosphate, diacylglycerol and reactive oxygen species (ROS). Calcium-binding proteins function as Ca^{2+} sensors that lead to the activation of calcium-dependent protein kinases. The activated kinases or phosphatases can phosphorylate or dephosphorylate specific transcription factors (TFs), thus regulating the expression levels of stress-responsive genes. The activated Ca^{2+} can interact with DNA-binding proteins resulting in their activation or suppression. This can lead to calcium-dependent protein kinase signalling cascades. This signalling cascade leads to production of antioxidants and compatible osmolytes (for osmotic adjustment) and the expression of heat shock proteins.

The main impacts of heat stress are protein denaturation, instabilities in nucleic acids, increased membrane fluidity, inactivation of the synthesis and degradation of proteins and loss of membrane integrity. At moderately high temperatures, cellular injuries can occur over a longer period. This reduces ion flux that leads to production of ROS and other toxic compounds that severely affect plant growth. Expression of heat shock proteins and other protective proteins is an effective adaptive strategy. Various abiotic stresses induce overproduction of ROS causing damage to proteins, lipids, carbohydrates and DNA and ultimately resulting in oxidative stress. The metabolite 3′-phosphoadenosine 5′-phosphate accumulates during high light and drought moving from chloroplast to nucleus to regulate ABA signalling and stomatal closure during the oxidative stress. This movement results in the activation of the high light transcriptome. Transcription factors (TFs) play important roles in stress tolerance. Many abiotic stress-related genes and TFs have been isolated from different plant species and overexpressed in transgenic plants to improve stress tolerance. The stress-inducible TFs include the members of the dehydration-responsive element-binding (DREB) protein, WRKY (pronounced "worky") and DNA-binding one zinc finger (DOF) protein families.

19.3 Breeding for Abiotic Stresses

Yield potential can be explained as the potential of a crop to yield maximum when all inputs are non-limiting. An assessment of yield stability can quantify the negative deviations away from the yield potential. Yield gap is the difference between the yield potential and the actual yield. Due to stress events, crops rarely reach their yield potential in most agricultural systems. Two basic genetical approaches currently being utilized to improve stress tolerance are (a) utilization of natural genetic variations either through direct selection under stressful environments or through the mapping of QTLs and MAS and (b) production of transgenic plants with novel genes or altered expression of existing genes to affect the degree of stress tolerance.

In principle, the change-induced responses at all functional levels of the organism are reversible (elastic deformation) but may become permanent (plastic deformation). Brief exposure to stress does not cause only temporary changes, and prolonged exposure only results in permanent changes (Fig. 19.5). Thus, after recovery, the dry matter returns to the original rate (angle of inclination α). However, in the case of chronic stress, the growth rate is reduced at a continuous angle ($\beta < \alpha$), and the loss in productivity is significantly higher. The use efficiency (UE) of water or nutrients is defined as the ratio between the yield per unit of resource available to the plant. As an example, water use efficiency (WUE) is the ratio between water used and the actual amount of water withdrawn. In the early stages of plant development, yield is usually replaced by the mass of shoot dry weight to estimate the UE. A genotype will be considered efficient if it produces well with minimum resource. In case of tolerance

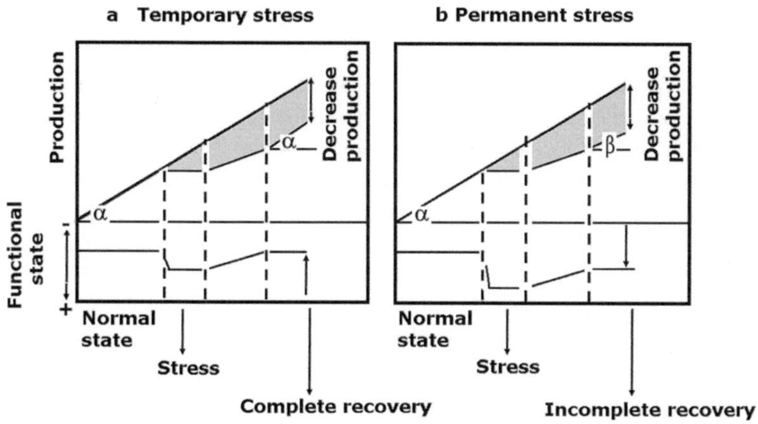

Fig. 19.5 Effect of environmental stress on productivity. (**a**) Temporary stress and (**b**) permanent stress. (Courtesy: Springer-Verlag)

and efficiency, plants use physiological and anatomic mechanisms to tide over the effect of stress. Plants use three main strategies to cope with stress:

(a) Specialization (when a genotype is adapted to a specific environment)
(b) Generalization (when a genotype has moderate suitability in most environments)
(c) Phenotypic plasticity (when signals from the environment interact with the genotype and stimulate the production of alternative phenotypes)

With the aforesaid general account on breeding for abiotic stress, we shall discuss breeding for drought tolerance/WUE, breeding for heat tolerance and breeding for salinity tolerance.

19.3.1 Breeding for Drought Tolerance/WUE

A drought-resistant ideotype is not always well defined. There is a widely accepted norm that a high-yielding variety will yield consistently in most environments. This norm is taken for granted. WUE is often equated with drought resistance. But this is not a generalized observation. Yield potential is defined as the maximum yield realized under non-stress conditions. Generally, drought resistance in physiological terms is "dehydration avoidance" and/or dehydration tolerance. WUE is mostly discussed in terms of plant production rather than gas exchange. Yield under water-limited conditions can be determined by the genetic factors controlling yield potential, drought resistance and/or WUE.

Under specific environmental stress, varieties with high yield potential produce lesser yield than varieties that have lower yield potential. Development of selection programs using programmed stress environments and other selection tools became necessary for deriving drought-tolerant varieties. While selection is exercised under low-yielding stress conditions, large differences among different years and locations are noticed. Heritability for yield under stress depends on (a) presence of genes for drought resistance under stressed environment and (b) the degree of control over the homogeneity and general stress conditions. Under stress, when selection for yield is exercised, a genetic shift occurs towards a dehydration-avoidant plant type. Dehydration avoidance is defined as the capacity to sustain high plant water status or cellular hydration under drought. It is not based on only one physiological factor or one gene that the design of dehydration-avoidant genotype is considered. Such a design will be successful through understanding the full spectrum of interactions among plant development like phenology, water use, penalty in yield potential and the specific dry land ecosystem. There is ample evidence that under water-limited conditions, there is association between high rate of osmotic adjustment (OA) and sustained yield. The plant will be meeting transpirational demand by reducing its LWP under stress situations. OA helps to maintain higher leaf relative water content (RWC) at low leaf water potential (LWP). Osmotic adjustment governs the turgor maintenance.

Water use efficiency (WUE) is the most important component of drought adaptation. Its relationship with yield is often confused with drought tolerance. For selection of tolerance and WUE, strategies are different. WUE can be evaluated from both the physiologic and agronomic point of view. Physiologically, WUE is the relationship between the CO_2 photosynthetic assimilation rate (A) and the plant's transpiration rate:

$$\mathrm{WUE} = \frac{(\mathrm{PA} - \mathrm{P1})}{1.6\,(VP1 - VPA)},$$

where:

PA is partial pressure of CO_2 in the air.
P1 is partial pressure of CO_2 inside the leaf.
VP1 is vapour pressure of the water inside the leaf.
VP_A is vapour pressure of the water in the air.
Agronomically, WUE is the relationship between the dry mass produced and the volume of water used in the cycle (precipitation plus irrigation water):

$$\mathrm{WUE} = \frac{\mathrm{GY}}{V},$$

where:

WUE: water use efficiency in agronomic terms.
GY: grain yield or dry mass yield
V: total water volume used in the cycle by the culture

19.3.2 Photosynthesis Under Drought Stress

Drought stress induces changes in photosynthetic pigments and components and also damages photosynthetic apparatus and activities of Calvin cycle enzymes. There will be loss of balance between the production of ROS and antioxidant defence. This causes the accumulation of ROS, which induces oxidative stress in proteins, membrane lipids and other cellular components. Components of photosynthesis affected by drought are shown in Fig. 19.6. Upon reduction in the available water, plants close their stomata (plausibly via ABA signalling), which decreases the CO_2 influx. Reduction in CO_2 is the reduction in carboxylation that leads to more electrons to form ROS. Severe drought conditions limit photosynthesis through a decrease in the activities of ribulose-1,5-bisphosphate carboxylase/oxygenase (Rubisco), phosphoenolpyruvate carboxylase (PEPCase), NADP-malic enzyme (NADP-ME), fructose-1,6-bisphosphatase (FBPase) and pyruvate, orthophosphate (Pi) dikinase (PPDK). Reduced tissue water increases the activity of Rubisco-binding inhibitors. Moreover, non-cyclic electron transport is down-regulated to match the reduced requirements of NADPH production and thus reduces ATP synthesis (see Box 19.2).

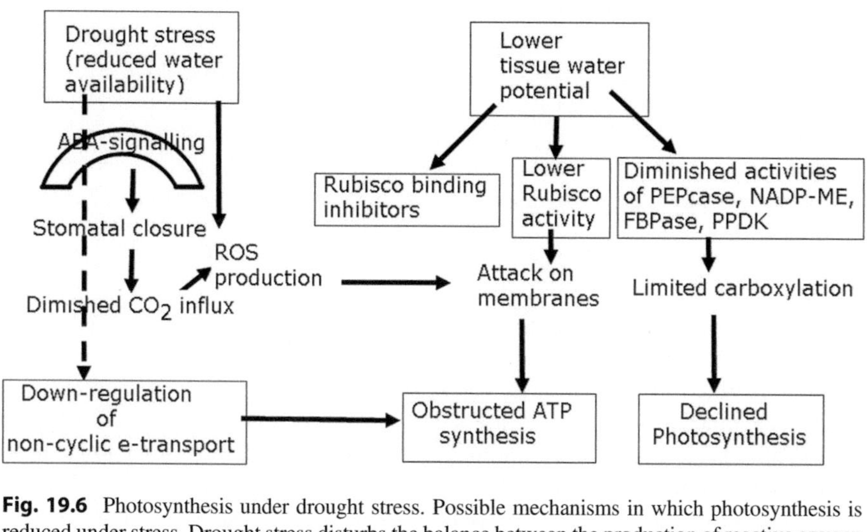

Fig. 19.6 Photosynthesis under drought stress. Possible mechanisms in which photosynthesis is reduced under stress. Drought stress disturbs the balance between the production of reactive oxygen species and the antioxidant defence, causing accumulation of reactive oxygen species, which induces oxidative stress. Upon reduction in the amount of available water, plants close their stomata (plausibly via ABA signalling), which decreases the CO2 influx. Reduction in CO_2 not only reduces the carboxylation directly but also directs more electrons to form reactive oxygen species. Severe drought conditions limit photosynthesis due to a decrease in the activities of ribulose-1, 5-bisphosphate carboxylase/oxygenase (Rubisco), phosphoenolpyruvate carboxylase (PEPCase), NADP-malic enzyme (NADP-ME), fructose-1, 6-bisphosphatase (FBPase) and pyruvate orthophosphate dikinase (PPDK). Reduced tissue water contents also increase the activity of Rubisco-binding inhibitors. Moreover, non-cyclic electron transport is down-regulated to match the reduced requirements of NADPH production and thus reduces the ATP synthesis. *ROS* reactive oxygen species

Box 19.2 Photosynthesis, Plant Productivity and Abiotic Stress Tolerance

World population will touch 9.1 billion by 2050. It was 7.4 billion in 2016. World has to produce 71% more food to feed this population. Resources are dwindling. Of the 31,000,000 km^2 arable land, 100,000 km^2 are lost every year. With climate change looming large, with increasing global population, there is a great need for more productive and stress-tolerant crops. Genetic engineering of plants is the answer to increase productivity as traditional methods of crop improvement have probably reached its limits. The advantages are multiple, when potential genes and metabolic pathways when genetically modified could result in improved photosynthesis and biomass production. Photosynthesis, as the sole source of carbon for the growth and development of plants, plays a central role. The most promising direction for increasing CO_2 assimilation is to implement carbon concentrating mechanisms found in cyanobacteria and algae into crop plants. This is because experiments on improving the CO_2 fixation versus oxygenation reaction catalysed by Rubisco are less encouraging. On the other hand, introducing the C4 pathway into C3 plants is a formidable challenge. Other attractions are increased biomass production through engineering of metabolic regulation, certain proteins, nucleic acids or phytohormones. Enhancing sucrose synthesis and assimilate translocation to sink organs are crucial. As abiotic stress tolerance limits crop productivity, efforts to produce transgenic plants with elevated stress resistance are prime. This can be accomplished due to elevated synthesis of antioxidants, osmoprotectants and protective proteins. Transcription factors that play a key role in abiotic stress responses of plants are also crucial.

Breeding Strategies: The traits conferring stress tolerance are governed by a variety of genes acting additively making the genetic manipulation for increased drought tolerance difficult. Drought can reduce crop yield by up to 50%. Efforts to enhance the efficiency for drought stress based on physiological traits have met with limited success since GE interaction for yield is always large. This is due to the lack of precise screening techniques that are not influenced by the environment. Several main traits show correlations with yield. Assessing these traits in a large set of germplasm is an uphill task, because major traits (such as plant height and days to heading or to maturity) can interact with traits like canopy temperature. Conventional breeding follows either visual assessment or traditional phenotyping process. They are slow, destructive, labour intensive and often inaccurate because of high chances of committing errors.

Table 19.1 Frequently used drought-tolerance indices in crops. Ys and Yp are stress and optimal (potential) yield of a given genotype, respectively. $\bar{Y}s$ and $\bar{Y}p$ are average yield of all genotypes under stress and optimal conditions, respectively

Indices	Formulae and description
Mean productivity (MP)	$(Yp + Ys^{\dagger})/2$ Selects for improved yields under stressed and non-stressed conditions
Geometric mean productivity (GMP)	$\sqrt{(Ys \times Yp^{\ddagger})}$ Selects for comparatively high yield under stressed and non-stressed
Stress tolerance (TOL)	$Y_p - Y_s$ Selects for low yield in favourable conditions with high yields under stress
Stress tolerance index (STI)	$(Y_p \times Y_s)/(\bar{Y}_p)^2$ Selects for high-yielding genotypes under stressed and non-stressed conditions
Stress susceptibility index (SSI)	$[1-(Y_s/Y_p)]/[1-(\bar{Y}_s^{\S}/\bar{Y}p)]$ Selects for low-yielding genotypes with high yield under stress
Yield stability index (YSI)	Y_s/Y_p Selects for yield stability under stressed and non-stressed conditions
Yield index (YP)	$Ys/\bar{Y}s$ Selects for yield stability under stress

Genes for Drought Tolerance: Drought is a complex trait and drought-tolerance response is carried out by various genes, transcription factors, miRNAs, hormones, proteins, cofactors, ions and metabolites. This complexity posed limitations to classical breeding. Adaptation to drought induces an active molecular response. For the last two decades, many stress-related genes have been characterized in many crop species. Recently, large transcriptome analysis revealed the molecular response. The optimal targets for engineering drought tolerance are transcription factors and components of the signal transduction pathways. In wheat, genes encoding the dehydration-responsive element-binding (DREB)/C-repeat binding factor (CBF) transcription factors were engineered into new wheat varieties. The transgenic plants showed increased stress tolerance and higher levels of proline. A higher expression of gene for mannitol-1-phosphate dehydrogenase (*mtlD*) gene that increases the level of mannitol improves drought tolerance.

Equations for Assessing Drought Tolerance: Several indices were proposed to describe yield performance of a given genotype under stress and non-stress conditions. Commonly used drought-tolerance indices in crops are available in Table 19.1. Relative yield performance of genotypes in drought-stressed and non-stressed environments can be used as an indicator of drought tolerance. In this way, genotypes can be categorized into four groups: (a) genotypes with high performance under both stress and non-stress conditions, (b) genotypes with high yield in non-stress conditions, (c) genotypes with high yield in stress conditions and (d) genotypes with low yield in both stress and non-stress conditions.

A stress susceptibility index (SSI) that measures yield stability in both potential and actual yields under varied environments is the acceptable technique. Stress tolerance index (STI) is a useful tool for predicting high yield and stress tolerance potential of genotypes. Since drought stress varies in severity over years, geometric mean is often used to assess relative performance. Yield index (YI) and yield stability index (YSI) help to evaluate genotypic stability in the both stress and non-stress conditions. However, multi-environment selection, covering a wide range of climatic variability, seemed more suitable to identify stress-tolerant and high-performing genotypes. Each index has to be interpreted according to its physiological meaning and optimal value. For example, good performance under both drought and irrigated conditions leads to high values of STI, mean productivity, geometric mean productivity, YSI and YI and to generally low values of SSI.

19.3.3 Breeding for Heat Tolerance

Heat stress is an increase in temperature (above the critical value) that is sufficient to cause irreversible damage to plants. A transitory rise of 10–15 °C in temperature is considered heat shock or heat stress. High daily temperatures leads to high evapo-transpiration rate. Its effect can be on high respiration rate during night and flower abortion and pollen sterility in some species. High temperature can totally inhibit germination, reducing the stand, and thus a reduced final crop yield. In other development phases, high temperature damages the photosynthesis apparatus and affects respiration, water ratios, cell membrane stability, hormone levels and the primary and secondary metabolites. For example, in wheat, high temperatures during grain swelling alter the protein composition and starch/protein ratio. This alteration has a direct bearing on the physical and chemical properties of the wheat flour. In cowpea, high daytime temperatures can present asymmetrical cotyledons and pig-mentation loss in the seed coat affecting the market value. High temperature and intense solar radiation can damage the surface and internal tissues of tomato and citrus fruit. In potato, a series of physiological disturbances can occur such as uneven growth, splits, internal cavities, alteration in the internal colouring and necrosis.

Tolerance Mechanisms and Traits Used: Physiological mechanisms that contribute to the heat-tolerant mechanisms are:

(a) Tolerance mechanisms: higher photosynthesis rate, stay-green, cell membrane thermal stability and heat shock proteins
(b) Escape mechanisms: canopy temperature depression (CTD) and earliness

Physiological methods are not always feasible while assessing many populations. Some of the traits used are cell membrane thermal stability, chlorophyll fluorescence, canopy temperature depression, triphenyl tetrazolium test, morphological characters, heat shock proteins and transcription factors.

Cell membrane thermal stability is an indicator of stress from high temperature. Membrane rupture enables the electrolytes to flee from the cells to the medium, and their concentration can be quantified by electric conductance. The greater the electrical conductivity, the greater will be the heat stress tolerance. It has a high genetic correlation with yield. However, cell membrane thermal stability per se is not advisable to be used as a criterion for heat stress.

The chlorophyll fluorescence technique could be used to quantify the effects of high temperatures on plants. This can assess the quantity of light absorbed and transduced between Photosystems I and II.

Canopy temperature depression (CTD) is the difference between air and leaf temperature. Infrared thermometry is one of the techniques used to measure CTD. Genotypes with greater heat stress tolerance can maintain their organ temperature, respiration and transpiration activities at normal levels, even under stress conditions. However, CTD can be limited by the influence of environmental factors like water availability in the soil, air temperature, relative humidity and radiation. CTD is suitable for the selection of superior lines in warm environments with low relative air humidity.

Triphenyl tetrazolium chloride (TTC) estimates mitochondrial electron transport chain. Reduced TTC indicates the level of mitochondrial respiration, determined by spectrophotometry, and it reflects the relative viability of the cell. TTC can predict heat tolerance using seedlings under controlled environment to save time and space.

Morphological characteristics indicating heat tolerance are plant vigour, leaf senescence, stay-green, number of emerged seedlings, tillering capacity, mean grain weight, number of grains per ear, harvest index and yield. In wheat, greater heat stress tolerance leads to high swelling rates indicating thereby swelling rate and grain weight could be used as selection criteria for tolerance. Due to high correlation values observed, number of grains per ear, biomass and harvest index could be used as selection criteria. Induced acceleration of reproductive development is another technique to improve heat tolerance. The reproductive period can be altered through modifying growth habit, progression of plant nodes, branches and reproductive nodes. In species like rice, sorghum and wheat, reproductive development is not easily manipulative. Time consumed for the accumulation of photoassimilates and their translocation for grain development is also minimal. But the translocation of photoassimilates stored before anthesis can be a mechanism of heat stress tolerance.

While many proteins are inhibited at high temperatures, heat shock proteins (HSPs) can have their synthesis increased. These proteins, together with their heat shock transcription factors (HSFs) are acting as molecular chaperones to maintain protein partitioning homoeostasis. The HSPs have been classified in five groups: HSP100, HSP90, HSP70, HSP60 and small HSP. Some HSFs have been identified and characterized associated with heat stress in cereals such as rice, wheat, corn, sorghum, rye, barley and oats.

19.3.4 Drought Versus Heat Tolerance

Drought and heat can reduce crop productivity and yields. When 40% water was reduced, in maize 40% yield was reduced and in wheat, it was 21%. Crop productivity will be further influenced by the impacts of climate change. The Intergovernmental Panel on Climate Change (IPCC – established under United Nations Environment Programme and World Meteorological Organization) report says that the air and ocean temperatures have warmed. Based on multiple independently produced data sets, it was inferred that land and ocean temperature data showed an average global warming of 0.85 °C has occurred between 1880 and 2012. The atmospheric concentrations of the greenhouse gases like carbon dioxide (CO_2), methane (CH_4) and nitrous oxide (N_2O) have also increased, with net emissions approaching 300 ppm in the recent years. When soil and atmospheric humidity is low with high ambient temperature, drought stress becomes imminent. This is due to an imbalance between evapotranspiration flux and water intake from the soil. Heat stress is a rise in soil and air temperature beyond a threshold level for a given time so as to hamper growth and development. Heat and drought stress are controlled by multiple genes. Water shortage trigger oxidative, osmotic and temperature stresses. As leaf temperature rises, reduced stomatal conductance and transpiration may induce heat stress. Higher temperature modifies leaf structure and leaf anatomy.

19.3.5 Salinity Tolerance

A standard definition says that saline soils are those which have an electrical conductivity (EC) of the saturation soil-paste extract of more than 4 dS/m at $25^{0\circ}$C, which corresponds to approximately 40 mM NaCl and generates an osmotic pressure of approximately 0.2 MPa. When grown on soils with an EC value above 4, crops significantly reduce their yield. Salts may include chlorides, sulphates, carbonates and bicarbonates of sodium, potassium, magnesium and calcium. The diverse ionic composition of salt-affected soils results in a wide range of physiochemical properties. In the case of saline-sodic soils, growth is hindered by a combination of high alkalinity, high Na^+, and high salt concentration. In this regard, it is important to distinguish between soil salinization and soil sodicity. Soil salinization is referred as the accumulation of soluble salts in the soils. This is particularly favoured by arid and semi-arid climates with evapotranspiration volumes being greater than the precipitation volumes along the year. Soil sodicity is a term given to the amount of Na^+ detained in the soil. High sodicity (more than 5% of Na^+ of the overall cation content) causes clay to swell excessively when wet, hence limiting severely air and water movements and resulting in poor drainage.

To tolerate salinity stress, plants utilize a variety of traits like cell function and development through signal perception, signal integration and processing. Several signalling molecules were discovered by the use of high-throughput sequencing technologies. ROS is a versatile signalling molecule. The mitogen-activated protein kinase (MAPK) can trigger plant response to biotic and abiotic stresses by activating the antioxidant enzymes. ROS accumulation activates many different MAPKs cascades. These include the ROS-responsive MAPKKK (MAPK kinase kinases), MEKK1, MPK4 and MPK6. Increased generation and accumulation of ROS such as superoxide (O_2), hydrogen peroxide (H_2O_2) and nitric oxide (NO) have an extensive impact on ion homeostasis by interfering with ion fluxes. Higher ROS levels lead to accumulation of salicylic acid contributing to plant defence, cell death and induced stomatal closure. Recent advances in considering the important role of ROS in plant salt responses was the discovery of a coupled function of plastid hemeoxygenases and ROS production in salt acclimation. These findings strongly suggest the involvement of the chloroplast to nucleus signalling pathway in salinity adaptation.

Potassium is needed for growth and development. Saline conditions induce increment in cytoplasmic Na^+ that results in reduction in K^+ that leads to changes of membrane potential, osmotic pressure, turgor pressure, calcium signalling, ROS signalling, etc. Maintenance of K^+ homeostasis is essential for enzyme activities, ionic and pH homeostasis, and cytosolic K^+ is an attribute of plant adaptive response to a broad range of environmental constraints. There is a correlation between the root's K^+ retention ability and plant salinity stress tolerance (e.g. wheat, barley and *Brassica* species). Electrolyte leakage, a hallmark of plant cell response to abiotic (including salinity) and biotic stresses, is based mainly on K^+ efflux. This stress-induced K^+ leakage is often accompanied by ROS generation and leads to cell death. Under stress, K^+ leakage, ROS and plant cell death (PCD) seem to be intimately connected. Plant responses to salinity stress are summarized in Fig. 19.7.

Breeding Strategies: Two main approaches used to impart salinity tolerance are (a) exploring natural genetic variation, either through selection under stress conditions or through quantitative trait loci (QTL) followed by marker-assisted selection (MAS), and (b) transgenic technology by modifying the expression of endogenous genes or introducing novel genes (of plant or non-plant origin). Conventional breeding needs diversified and well-characterized germplasm but met with limited success due to complexity of the trait. The primary step before proceeding to make transgenics is the identification of functional and regulator genes serving to control different metabolic pathways, including ion homeostasis, antioxidant defence system, osmolyte synthesis and other signalling pathways.

The candidate genes for salt tolerance are categorized into genes with functional and regulatory role. Functional genes are those involved in osmolyte biosynthesis, ion transporters, water channels, antioxidant systems, sugars, polyamines, heat shock proteins and late embryogenesis abundant proteins. Regulatory genes control transcriptional and post-transcriptional machinery. Some of these are transcription factors (TFs), protein kinases and phosphatases. Several state-of-the-art genomics-assisted approaches like transgenic overexpression, RNAi, microRNA, genome editing and genome-wide association studies are being used for improving salt tolerance. Overexpression of these genes has been shown as a successful strategy.

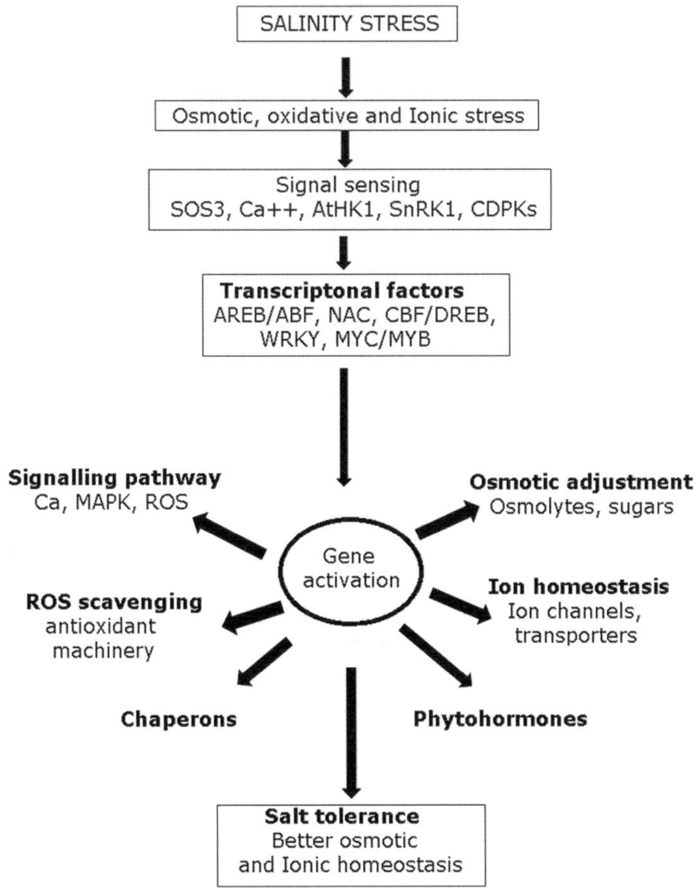

Fig. 19.7 Plant responses to salinity stress

19.4 MAB for Abiotic Stress in Major Crops

The complexity of abiotic stress tolerance has rendered slow progress through conventional breeding. Marker-assisted selection (MAS) is an indirect and accurate selection based on tightly linked molecular markers, viz. restriction fragment length polymorphisms (RFLP), amplified fragment length polymorphisms (AFLP), random amplified polymorphic DNA (RAPD), simple sequence repeats (SSR) or microsatellites, sequence-tagged microsatellite site (STMS), single nucleotide polymorphisms (SNPs), etc. It enables screening of traits which are difficult to score quantitative trait loci (QTL) analysis. MAS offers advantage over the other tools as having relaxed biosafety regulations and their wider public acceptance. QTLs identified through MAS in various crops are available in Table 19.2.

Table 19.2 QTLs identified for abiotic stress tolerance in various crop plants

Crop	QTLs/loci	Mapping population	Cross(s)	Genotyping markers	Environment	Chromosome	Stress
Rice	3 QTLs (physiological and yield traits)	RILs	IR20 × Nootripathu	SSRs	Field	1, 4, and 6	Drought
Rice	6 QTLs (ratio of deep rooting)	RILS	Zhenshan97B × IRAT109	SNP	Field	1, 2, 4, 7 and 10	Drought
Rice	4 QTLs (root length and root dry weight)	BC_2F_2	OM1490 × WAB880–1–38-18-20-P1-HB	SSRs	Greenhouse	2, 3, 4, 8, 9, 10,12	Drought
Rice	15 QTLs (1000 grain weight, leaf temperature, relative water content, grain, weight per plant, relative water content, productive tillers, grain number per plant, panicle weight, productive tillers, and spikelet fertility)	BIL	Swarna × WAB 450	SSR	Poly house	1, 2, 3, 7, 8 and 9	Drought
Rice	11 QTLs (spikelet fertility, daily flowering time, and pollen shedding level)	CSSLs	Sasanishiki × Habataki	SSR	Field	2, 4, 3, 8, 10, 11, 5 and 7	Heat
Rice	8 QTLs (spikelet fertility)	Three-way cross	(IR64 × Milyang23) × Giza178	SNP	Net house	4	Heat
Rice	5 QTLs (submergence tolerance beyond SUB1)	RILs	IR42 × FR13A	SSR	Greenhouse	1, 4, 8, 9 and 10	Water logging/ submergence
Rice	32 QTLs (shoot length, root length,	BILs	(Nipponbare×Kasalath) × Nipponbare	RFLPs	Glasshouse	1, 3, 4, 6 and 7	Water logging/ submergence

(continued)

Table 19.2 (continued)

Crop	QTLs/loci	Mapping population	Cross(s)	Genotyping markers	Environment	Chromosome	Stress
	and shoot fresh weight)						
Rice	4 QTLs (submergence)	F2:3	IR72 × Madabaru	SNP	Net house	1, 2, 9 and 12	Water logging/ submergence
Rice	85 QTLs (shoot potassium concentration, sodium–potassium ratio, salt injury score, plant height, and shoot dry weight.)	RILs	Bengal × Pokkali	SNP	Greenhouse	1, 2, 3, 4, 6, 7, 8, 10, 11 and 12	Salinity
Rice	16 QTLs (pollen fertility, Na+ concentration, and Na/K ratio in the flag leaf	F_2	Cheriviruppu × Pusa Basmati1	SSR	Net house	1,7,8 and 10	Salinity
Rice	28 QTLs (different morphological and physiological traits)	BC_3DH	(Caiapo × O. glaberrima) × Caiapo	SSR	Growth chamber	5 and 10	Iron (Fe) toxicity
Rice	7 QTLs (leaf bronzing)	RILs	IR 29 × Pokkali	SSR and SNP	Hydroponics	1, 2, 4, 7, 12	Iron (Fe) toxicity
Rice	3 QTLs (leaf bronzing)	BILs	Nipponbare × Kassalath	SSR and SNP	Hydroponics	1, 3,8	Iron (Fe) toxicity
Rice	9 QTLs (culm length, shoot dry weight, and root dry weight)	CSSLs	Koshihikari × Kasalath	SSR	Greenhouse	3,8	Iron (Fe) toxicity
Wheat	3 QTLs (yield and biomass)	NILs	Wild emmer wheat (*Triticum turgidum* ssp. *dicoccoides*) and	SNP	Net house	1BL, 2BS and 7AS	Drought

			durum (*T. turgidum* ssp. *durum*) and bread wheat (*T. aestivum*)				
Wheat	4 QTLs (net photosynthesis, water content, and cell membrane stability)	F_2	Chakwal-86 × 6544–6	SSR	Hydroponics	2A	Drought
Wheat	13 main QTLs (ABA content)	F_4	Yecora Rojo × Pavon 76	TRAP, SRAP, and SSR	Field	3B, 4A and 5A	Drought
	22 QTLs (coleoptile length, seedling height, longest root length, root number, seedling fresh weight, stem and leaf fresh weight, root fresh weight, seedling dry weight, stem and leaf dry weight, root dry weight, root-to-shoot fresh weight ratio, and root-to-shoot dry weight ratio)	RILs	Weimai 8 × Luohan 2 Weimai 8 × Yannong 19	SSR, ISSR, STS, SRAP, and RAPD	Laboratory	1B, 2A, 2B, 3B, 4A, 5D, 6A, 6D, 7B and 7D	Drought
Wheat	6 QTLs (seminal root angle and seminal root number)	DHs	SeriM82 × Hartog	DArT and SSR	Gel chambers	2A, 3D, 6A, 5D, 4A, 1B, 3A, 3B and 6B	Drought
Wheat	20 major and minor QTLs (1000 grain weight, grain weight per spike, number of grains per spike, spike	F_3 and F_4	Oste-Gata × Massara-1	SSR	Field	3B, 7B,1B, 2B, 1B and 3B	Drought

(continued)

Table 19.2 (continued)

Crop	QTLs/loci	Mapping population	Cross(s)	Genotyping markers	Environment	Chromosome	Stress
	number per m^2, spike weight, spike harvest index, and harvest index)						
Wheat	37 QTLs (parameters of chlorophyll fluorescence kinetics)	DH	Hanxuan 10 × Lumai 14	AFLP and SSR	Growth chamber	1A, 1B, 2B, 4A and 7D	Heat
Wheat	5 QTLs (thylakoid membrane damage, plasma membrane damage, and chlorophyll content)	RILs	Ventnor × Karl 92	SNP	Greenhouse	6A, 7A, 1B, 2B and 1D	Heat
Wheat	7 stable QTLs (grain yield, thousand grain weight, grain filling duration, and canopy temperature)	DH	Berkut × Krichauff	SSR	Field	1D, 6B, 2D and 7A	Heat
Wheat	3 QTLs (grain yield, thousand grain weight, grain filling duration, and canopy temperature)	RILs	NW1014 HUW468	SSR	Field	2B, 7B and 7D	Heat
Wheat	14 QTLs (three main spike yield components; kernel number, total kernel weight, and single kernel weight)	RILs	Halberd Karl 92	SSR	Greenhouse	1B, 2D, 3B, 4A, 5A, 5B, 6D, 7A and 7B	Heat
Wheat	1 QTL (proportion of dead leaves)	Varieties	Durum wheat	SSR	Glasshouse	4B	Salinity

Wheat	18 additive and 16 epistatic QTLs (the root, shoot and total dry weight, K^+, Na^+ concentration, and K^+/Na^+ ratio)	RILs	Chuan 35,050 × Shannong 483	SSR	Glasshouse	1A, 2A, 4B, 5D, 1B, 3A, 6D, 7B, 1D, 2B, 5A, 5B, 7A, 4A, 6A and 6B	Salinity
Maize	169 QTLs (grain yield per plant, ear length, kernel number per row, ear weight, and hundred kernel weight)	NAM	11 biparental families (2000 RILs)	SNPs	Field	1, 3 and 10	Drought
Maize	203 QTLs (ASI, ears per plant, stay-green and plant-to-ear height ratio) (ASI = anthesis silking interval)	RILs F2:3 F2:3	CML444 × MALAWI CML440 × CML504 CML444 × CML441	SNPs and SSRs	Field	1, 3, 4, 5, 7 and 10	Drought
Maize	45 QTLs (grain yield per plant and yield components)	F2:3	B73 × DTP79	SSRs	Field	1, 2, 3, 4, 5, 6, 7, 8 and 10	Drought
Maize	145 QTLs (grain yield, ASI), 7 mQTL for grain yield, and 1 mQTL for ASI	RILs F2:3 F2:3	CML444 × MALAWI CML440 × CML504 CML444 × CML441	SNPs	Field	1, 2, 3, 4, 5, 7, 8 and 10	Drought
Maize	64 QTLs (grain yield, number of kernels per row, number of rows per ear, ear length, ASI, visually scored drought score, relative water content, osmotic	F2:3	B73 × DTP79	RFLPs, SSRs, and AFLPs	Field	1, 2, 3, 4, 5, 7, 8 and 10	Drought

(continued)

Table 19.2 (continued)

Crop	QTLs/loci	Mapping population	Cross(s)	Genotyping markers	Environment	Chromosome	Stress
	potential, and relative sugar content)						
Maize	43 QTLs (QTLs associated with grain yield, leaf width, plant height, ear height, leaf number, tassel branch number, and tassel length)	F_2	B73 × DTP79	RFLPs, SSRs, and AFLPs	Field	1, 2, 3, 4, 5, 6, 7, 8 and 10	Drought
Maize	17 QTLs (leaf chlorophyll, plant senescence, electric root capacitance)	RILs	CML444 × SC-Malawi	SSRs	Field	1, 2, 4, 5, 6 and 10	Drought
Maize	25 QTLs (ASI, plant height, grain yield, ear height, and ear setting)	F2:3	D5 × 7924	SSRs	Rain shelter	1, 2, 3, 4, 6, 8, 9 and 10	Drought
Maize	22 QTLs (sugar concentration, root density, root dry weight, total biomass, relative water content, and leaf abscisic acid content)	F2:3	DTP79 × B73	RFLP	Greenhouse	1, 3, 5, 6, 7 and 9	Drought
Maize	6 QTLs (ph 6–1, rl1–2, sdw4–1, sdw7–1, tdw4–1, and tdw7–1)	F2:3	HZ32 × K12	SSRs	Glasshouse	1, 4, 6, 7 and 9	Waterlogging
Maize	18 QTLs (yield, brace roots, chlorophyll	RILs	CML311-2-1-3 × CAWL-46-3-1	SNP markers	Field	1, 2, 3, 4, 5, 7, 8 and 10	Waterlogging

	content, % stem, and root lodging)			using KASP platform			
Maize	15 QTLs (seedling height, root length, shoot fresh weight, root fresh weight, shoot dry weight, and root dry weight)	BC_2F_2	K12 × HZ32	SSRs and SNPs	Greenhouse	5, 6 and 9	Waterlogging
Maize	2 QTLs for aerenchyma formation (Qaer1.06–1.07 and Qaer7.01)	S1 and S2	*Z. nicaraguensis*	SSRs	Greenhouse	1 and 7	Waterlogging
Maize	25 QTLs (total brace root tier number and effective brace root tier number)	RILs and immortalized F_2	Huangzao 4 × CML288	SSRs	Field	1, 2, 3, 5, 6, 7, 8, 9 and 10	Waterlogging
Maize	27 QTLs (germination and early growth)	RILs	B73 × P39 and B73 × IL14 h	SNP	Field	1, 2, 3, 4, 5, 6, 7, 8 and 9	Cold
Maize	6 QTLs (days to emergence)	Inbred populations	Two large panels of flint inbred lines	SNP	Growth chamber	3,4,5,7,10	Cold
Maize	15 QTLs (shoot length, root length, ratio of root length and shoot length shoot fresh weight, root fresh weight, plant fresh weight, plant dry weight, shoot dry weight, root dry weight, ratio root dry weight, and shoot dry weight)	F2:3	B73 and CZ-7	SSRs	Greenhouse	1, 2, 4, 5, 6, 7, 8, 9 and 10	Salinity

For more details of transgenic and MAS methods of breeding, please see Chaps. 22 and 23 respectively. Genetic information has been applied for salt and drought tolerance in different crops such as *Arabidopsis*, rice, wheat, maize and *Brassica*. MAS has also developed waterlogged-tolerant lines in different crop plants. An account of progress achieved in MAS for abiotic stress tolerances in some of the crops are presented here.

19.4.1 Rice

Drought stress is a major constraint in rice production under rainfed conditions. Identification and introgression of consistent QTLs can be an effective strategy to induce drought tolerance. Although a number of QTLS have been identified in rice for drought resistance, the progress on marker-assisted backcrossing (MAB)-based introgression of the identified QTLS is limited (Table 19.2). Three QTLs mapped on chromosome 1 (*RM8085*), chromosome 4 (*I12S*), and chromosome 6 (*RM6836*) for physiological and yield traits can be effectively utilized for introgression into elite rice lines for stable yield production under drought stress-prone ecologies. QTL for deep rooting is an important trait for imparting drought tolerance. SNP-based genotyping resulted into mapping of six QTLS for RDR (ratio of deep rooting) on chromosomes 1, 2, 4, 7 and 10. Ten SSR genotyping-based QTLs for physiological and productivity-related traits under drought using backcross inbred lines (BILs) were derived from the cross of Swarna and WAB 450. Four QTLs related to root length and root dry weight were identified in a BC_2F_2 population derived from a cross of OM1490 and WAB880–1-38-18-20-P1- HB.

Heat stress threatens global rice production in this era of climate change. Two different populations (biparental F_2 population and three-way F_2 population) derived from the cross of heat-tolerant variety Giza178 × IR64 and IR64 × Milyang23 × Giza178, respectively, resulted in four QTLs, namely, qHTSF1.2, qHTSF2.1, qHTSF3.1 and qHTSF4.1. In a population of chromosome segment substitution lines derived from a cross of Sasanishiki (*japonica* ssp. heat susceptible) and Habataki (*indica* spp. heat tolerant), 11 QTLs were mapped through SSR markers on chromosomes 1, 2, 3, 4, 5, 7,8, 10 and 11 for spikelet fertility, daily flowering time and pollen shedding under heat stress.

Submergence is a problem of serious concern in rice-growing ecologies particularly in South and Southeast Asia. *SUB1* gene (from *O. sativa* ssp. *indica* cultivar FR13A) has been utilized enabling rice to survive under complete submergence for 15 days. More novel QTLs are required for longer-term submergence. A cross between IR72 and Madabaru was made to develop $F_{2:3}$ population, and using SNP markers, four QTLs were identified on chromosomes 1, 2, 9 and 12. Recombinant inbred lines (RILs) derived from a cross of IR42 and FR13A led to the detection of five QTLs on chromosomes 1, 4, 8, 9 and 10. These novel QTLs have tremendous potential to augment *SUB1* for better rice production under submerged conditions.

For salinity resistance, QTL mapping in F_2 population derived from a cross of salinity-tolerant Cheriviruppu with sensitive cultivar Pusa Basmati 1 (PB1) using

131 SSR markers were mapped for 16 QTLs for different traits such as pollen fertility, Na^+ concentration and Na/K ratio on chromosomes 1, 7, 8 and 10. Such QTLs could be used for improving salinity tolerance. Lowland rice facing the problem of iron (Fe) toxicity can be improved with African rice (*Oryza glaberrima*) genes for resistance to iron toxicity. Therefore, SSR-based QTL mapping carried out in BC_3DH lines derived from the backcross of *O. sativa* (Caiapo)/*O.glaberrima* and (MG12)//*O. sativa* (Caiapo) under Fe^{2+} condition in hydroponics resulted in the identification of 28 QTLs for 11 morphological and physiological traits on chromosome 5 and 10.

19.4.2 Wheat

Moisture stress tolerance in wheat can be tackled through introgression of drought-tolerant QTLs. Three QTLs from RILs were raised from a cross of wild emmer wheat (*Triticum turgidum* ssp. *dicoccoides*) and durum (*T. turgidum* ssp. *durum*) and bread wheat (*T. aestivum*) on chromosomes1BL, 2BS and 7AS. Wild emmer wheat is a source of drought resistance. Thirteen QTLs for abscisic acid content in F_4 population were derived from a cross of drought-sensitive (Yecora Rojo) and drought-tolerant (Pavon 76) using different markers (sequence-related amplification polymorphism (SRAP), target region amplification polymorphism (TRAP) and SSR). The QTLs mapped on chromosomes 3B, 4A and 5A through linked markers (Barc164, Wmc96 and Trap9) can be used for breeding for drought tolerance. Similarly, QTL mapping conducted in F_2 population derived from cross of tolerant cultivar, Chakwal-86, with sensitive cultivar, 6544–6, using SSR markers mapped four QTLs for photosynthesis, cell membrane stability and relative water content on chromosome 2A. Twenty-two QTLs on chromosomes 1B, 2A, 2B, 3B, 4A, 5D, 6A, 6D, 7B and 7D for different traits like coleoptile length, seedling height, longest root length, root number, seedling fresh weight, stem and leaf fresh weight, root fresh weight, seedling dry weight, stem and leaf dry weight, root dry weight, root-to-shoot fresh weight ratio, and root-to-shoot dry weight ratio were identified in two RIL populations derived from Weimai8 × Luohan 2 and Weimai 8 × Yannong 19, respectively. Six QTLs found to be major source for drought tolerance.

Root architectural traits can play an important role in imparting resistance to drought in wheat. Four QTLs and two QTLs for seminal root angle and seminal root number, respectively, were mapped through DArT (diversity arrays technology) in a doubled haploid population derived from a cross of SeriM82 and Hartog. Four QTLs for seminal root angle were located on chromosomes 2A, 3D, 6A and 6B, while two QTLs for seminal root number on 4A and 6A. Wheat is affected due to high temperature during grain filling and is a major production constrain globally. Parameters of chlorophyll fluorescence kinetics (PCFKs) can be utilized for the identification of heat stress-tolerant cultivars. QTL mapping was done in a DH population derived from a cross of Chinese cultivars, Hanxuan 10 and Lumai 14, using SSR and AFLP markers under controlled conditions. Seven QTLs were mapped on chromosomes 1A, 1B, 2B, 4A and 7D for traits related to PCFKs such as initial fluorescence, maximum fluorescence, variable fluorescence and maximum quantum efficiency of photosystem II. Similarly, in a population of Ventnor/Karl 92 cross, mapping of QTLs for thylakoid membrane damage (TMD), plasma membrane damage (PMD) and SPAD chlorophyll content (SCC) in RIL population were developed. In DH population derived from across of Berkutwith and Krichauff, seven stable QTLs were identified on chromosomes 1D, 6B, 2D and 7A. Three, two and one QTLs were identified for grain filling duration, thousand grain weight, grain yield and canopy temperature.

19.4.3 Maize

In sub-Saharan Africa (SSA) and Asia, maize yields remain variable due to climate shocks. In 2016, over 70,000 metric tons of drought-tolerant maize seeds were commercialized in 13 countries in SSA, benefiting an estimated 53 million people. More than 230 drought-tolerant maize varieties have been released by CIMMYT (*Centro Internacional de Mejoramiento de Maíz y Trigo*; International Maize and Wheat Improvement Center) and its allied partners. The overall estimated economic value of increased maize production due to climate-resilient maize in Ethiopia was almost 30 million USD. During 2015–2017, more than 50 elite heat stress-tolerant, CIMMYT-derived maize hybrids have been licenced to public and private sector partners for varietal release, seed scale-up and deployment in the region.

Evaluation of three tropical biparental populations under water stress (WS) and well-watered (WW) regimes to identify genomic regions responsible for grain yield (GY) and anthesis-silking interval (ASI) identified a total of 83 and 62 QTLs, respectively, through individual environment analyses. Six constitutively expressed meta-QTLs were mapped on chromosomes 1, 4, 5 and 10 for GY. One meta-QTL on chromosome 7 for GY and one on chromosome 3 for ASI were found to be "adaptive" to WS conditions. Another evaluation of 5000 inbred lines using 365 SNPs for genome-wide association-derived SNPs associated with drought-related traits were seen located in 354 candidate genes. Fifty-two of these genes showed significant differential expression in the inbred line B73 under the well-watered and water-stressed conditions.

Waterlogging is an important abiotic stress in maize. MAS-based incorporation of QTLs for waterlogging tolerance in cultivars is the most sustainable and viable approach to tackle this issue. Recombinant inbred lines (RILs) were derived from a cross of waterlogging-tolerant line (CAWL-46-3-1) and a sensitive line (CML311-2-1-3). Significant range of variation for grain yield under waterlogging along with a number of other secondary traits such as brace roots (BR), chlorophyll content and root lodging were isolated from among the RILs. Genotyping with 331 polymorphic single SNP markers using KASP (Kompetitive Allele Specific PCR) platform revealed a total of 18 QTLs on chromosomes1, 2, 3, 4, 5, 7, 8 and 10.

Low temperature or cold is yet another type of abiotic stress in maize. Analysis of two independent RIL populations from the crosses of B73 × P39 and B73 × IL14h identified a total of 27 QTLs for germination and early growth under field condition.

SNP genotyping mapped the QTLs on chromosomes 1, 2, 3, 4, 5, 6, 7, 8 and 9. A genome-wide association analysis in temperate maize inbred lines for pyramiding of cold tolerance genes was also made successful. Salinity also affects the maize production. Different traits related to salt tolerance such as shoot length, root length, ratio of root length and shoot length, shoot fresh weight, root fresh weight, plant fresh weight and plant dry weight are the traits that can be targeted for mapping the QTLs. SSR genotyping mapped 15 QTLs for target traits on chromosomes 1, 2, 4, 5, 6, 7, 8, 9 and10.

19.5 "Omics" and Stress Adaptation

An array of "omics" approaches emerged in due course of time since the need for developing improved genotypes with abiotic stress tolerance was recognized. These approaches, viz. genomics, proteomics, transcriptomics and metabolomics, are four axes of plant system biology that can decipher the complexity of stress response. Genomics is the study of the genome; transcriptomics explains functions of both sense and the nonsense RNA or transcriptome; proteomics addresses structural and functional analysis of proteins and regulatory pathways of post-translational protein modification; and metabolomics analyses various metabolites. A unified approach shall be competent enough to explain the intricate networks underlying abiotic stress tolerance.

19.5.1 Comparative Genomics Tools

Genomics is of two types: structural and functional. Structural genomics deals with genome sequencing, mapping and cloning of the traits. Functional genomics addresses gene functions (see chapter on Genomics for further details). Only comparative genomics (when genomic features of different organisms are compared) tools will be briefly discussed here.

The availability of sequenced plant genomes, expression data and stress-related cDNA libraries has made the discovery of stress-related genes easy. Genes of interest from model crop species can be now transferred to the newly sequenced crops. The basic requirement for comparative genomic studies is the availability of the orthologous data sets having a common ancestor. The stress-associated transcription factors (TFs) from orthologs of different plant species have similar sequences and expression patterns. This makes the possibility to identify the orthologous genes having the same functions in the crop species whose functional analysis is at a preliminary stage. Comparative genomics has been successfully applied to predict the stress-responsive TFs in soybean, maize, sorghum, barley and wheat using the known stress-responsive TFs in *Arabidopsis* and rice. So, it has been concluded that the comparative genomic studies will widen the potential of development of stress-tolerant crop species by incorporating the necessary information from model plants.

19.5.1.1 Transcript"omics"

Identification of candidate genes involved in various stress regulatory networks via genome-wide expression profiling is one novel method to study the stress response in plants. This is done through transcriptome profiling. Earlier, this was being done by northern blotting but was inefficient to analyse the entire set of genes. Several high-throughput techniques like expressed sequence tags (ESTs) sequencing, serial analysis of gene expression (SAGE) and massively parallel signature sequences (MPSS) could utilize the nucleotide sequence information to understand the level of transcription. Microarray technology allows indirect assessment of gene expression using the principle of nucleic acid hybridization of mRNA or cDNA fragments. The next-generation sequencing (NGS) strategies like RNA-seq for sRNAs have revolutionized the field of transcriptomics and thus paving the way for the improvement of plant genomic resources.

Expressed sequence tags (ESTs) make use of the cDNA libraries having about 10,000 clones of the genes involved. EST technology has enabled to generate a huge amount of data that can be further used for studying the plant stress tolerance mechanisms. Approximately, 449,101 ESTs have been reported for drought stress. ESTs associated low temperature, high temperature, nutrient deficiency and light stresses, respectively, have been made available on the National Center for Biotechnology Information browser (http://www.ncbi.nlm.nih.gov/).

SAGE is a high-throughput and cost-effective technique used for differential analysis of the expressed genes. The technique involves mRNA extraction, cloning and sequencing. Specific tags are used to identify the relevant genes within the database, and the pattern of expression of differential genes is determined by the relative amount of the individual tags.

Massively parallel signature sequencing (MPSS) is a genome-wide transcriptional profiling approach which makes use of the cloning technique, a technology developed by Lynx Therapeutics Inc., California. The cDNA molecules are cloned onto microbeads which are then sequenced for the generation of short cDNA tags. The ability of MPSS to generate ample amount of good-quality data with effective data management makes it superior than SAGE in terms of speed and information.

DNA microarray is a technique based on northern hybridization. Two types of DNA microarrays are available: cDNA arrays and oligoarrays. The difference between them is that in cDNA arrays, robotics is used to immobilize the spotted cDNA fragments onto the slides, whereas in the case of oligoarrays, photolithographic mask is used to directly synthesize the oligonucleotides on a solid matrix. Oligoarrays are preferred as they can be effectively used for SNP detection and do not require large-scale maintenance, PCR reactions as well as clone validation as in cDNA microarrays. Microarray technology is powerful, but limitations like time, labour intensiveness, contamination of DNA, etc. limit its use. Since a huge amount of data is generated, the statistical analysis becomes a challenging task.

RNA-seq is an advanced approach used for transcriptome profiling. RNA-seq is a cost-effective and high-throughput technology. RNA-seq technique is independent of gene information. It uses available genomic information for designing probes which identifies novel transcripts to study non-coding RNAs. The RNA-seq approach is being used for mapping the start site of transcription for developing an idea of tissue specific gene expression.

19.5.1.2 Combining QTL Mapping, GWAS and Transcriptome Profiling

Massive amount of genes are expressed differentially in plants; transcriptome profiling is also difficult. Combining GWAS, QTL mapping and transcriptome profiling becomes good exercise to study candidate QTLs. In soybean, near-isogenic lines (NILs), using the Affymetrix Soy GeneChip and high-throughput Illumina whole transcriptome sequencing, 13 candidate genes have been identified in the QTL segment of ~8.4 Mb (8400 nucleotides). The transcriptome technologies provide better insight of a gene for which the genome sequence is unavailable. There could be a discrepancy on the amount of protein and the levels of gene transcripts. This calls for analysis of proteome for further validation. Different ways by which the transcriptomics approaches are applied for studying abiotic stress tolerance in crops are enlisted in Table 19.3.

19.5.2 Prote"omics" to Unravel Stress Tolerance

Proteome is the link between its transcriptome and metabolome. There is a disparity between mRNA abundance and level of protein accumulation. So, it is logic to use proteomics for evaluation of plant stress responses. Proteins governing stress response are translated from the functional portion of the genome. This research started with the introduction of two-dimensional (2D) gel electrophoresis to separate crude protein mixtures. Several new technologies like mass spectrometry, fluorescent 2D differential gel electrophoresis, gel-free approaches such as multidimensional protein identification technology (MudPIT) isotope-coded affinity tags (ICAT), stable isotope labelling by amino acids in cell culture (SILAC), isobaric tags for relative and absolute quantitation (iTRAQ) have augmented this research. These are introduced to reduce the errors and to perform large-scale protein analysis in a single gel for the identification of post-translationally modified proteins.

19.5.3 Metabol"omics"

Metabolomics determines and quantifies metabolites in a biological system. Since metabolism varies with the type of abiotic stress, metabolomics is a comprehensive approach for unravelling the metabolic pathways and metabolites that regulate the response of crop plants towards various abiotic stresses.

Several approaches like metabolic fingerprinting, metabolite profiling and targeted analysis are used in metabolomics. Metabolic fingerprinting approach has been extensively used for generating specific metabolic signatures associated with a specific stress response from a mass of samples. A number of techniques like nuclear magnetic resonance (NMR), mass spectrometry, Fourier transform ion cyclotron

Table 19.3 Applications of transcriptomics approaches for understanding abiotic stress tolerance mechanisms

Crop	Technology used	Outcome
Rice	SAGE	24 differentially expressed genes were identified of which 18 genes were an aerobically induced and 6 genes were repressed
Salt-tolerant (FL478) and salt-sensitive (IR29) rice varieties	Rice oligoarray	Response of IR 29 was strikingly different from FL478 with induction of a large number genes induced in the former. Salt stress activated a number of genes in flavonoid pathway in IR 29 but not in FL 478 during vegetative growth stage
Soybean	Custom array containing 9728 cDNAs	Genes involved in DNA repair and RNA stability were induced; 48 differentially expressed genes were identified
Chickpea (*Cicer arietinum* L.)	High-resolution power of super SAGE coupled to the Roche 454 life/APG GS FLX titanium NGS technology	Characterized the complete transcriptome of chickpea plant's roots and nodules under drought stress and control conditions
Soybean	HiCEP (29,388) high-coverage expression profiling	97 genes and 34 proteins differentially expressed genes during flood stress were identified
Soybean seven tissues and seven stages during seed development	RNA-seq (RNA-Seq, also called whole transcriptome shotgun sequencing (WTSS), uses next generation sequencing (NGS) to reveal the presence and quantity of RNA in a biological sample at a given moment)	Expression atlas for soybean genes has been generated
Chickpea (*Cicer arietinum* L.)	Combined high-throughput next-generation sequencing and transcript profiling for GWAS	363 and 106 transcripts showed increased and decreased expression (over threefold) in roots and nodules, respectively, during salt stress
Sweet potato	Illumina paired-end RNA-seq	Temperature stress-responsive genes were identified from transcriptome sequence, such as abscisic acid-responsive element-binding factors (AREB) and CBF TFs*
Switchgrass cultivar Alamo	Affymetrix gene chips	5365 differentially expressed probe sets during heat stress
Cotton seedlings	Comparative microarray analysis	The functional genes and abiotic stress-related pathways were identified
Transgenic rice plants	RNA sequencing-mediated expression profiling	Provided valuable information about the ER stress response in

(continued)

Table 19.3 (continued)

Crop	Technology used	Outcome
		rice plants and led to the discovery of new genes related to ER stress (ER = endoplasmic reticulum)
Chenopodium quinoa	RNA-seq analysis	Drought stress-tolerant genes were identified
EST collections of chickpea	NGS (next-generation sequencing) platforms (Illumina and FLX/454)	A more extensive chickpea transcriptome assembly (CaTA v2) was developed

Courtesy: Springer Nature; *CBF = C-repeat/DRE-Binding Factor; DRE = dehydration-responsive element; TF = transcription factor

resonance mass spectrometry or Fourier transform infrared (FT-IR) spectroscopy can be used for generating fingerprints. Metabolite profiling quantifies total metabolome, and it helps in generating a snapshot of all the metabolites in a sample. Analytical techniques like NMR, liquid chromatography-mass spectrometry (LC-MS), capillary electrophoresis-mass spectrometry (CE-MS) and gas chromatography-mass spectrometry (GC-MS) are used for metabolite profiling. GC-MS is the highly sought after technique for metabolite profiling. The last approach, i.e. the targeted analysis, is aimed at precise identification of a specific metabolite or a target using a particular analytical technique for best results.

19.5.4 Phen"omics": For Dissection of Stress Tolerance

Phenomics addresses measurement of all the physical and biochemical parameters that often change with the changing environment. It is an integrated technology involving several technologies like photonics, biology, computers and robotics. Recent advancements and developments in the fields of image processing and automation technology have encouraged the researchers towards the real-time analysis of plant growth and developmental stages. Some of the technologies are discussed here.

Infrared and hyperspectral imaging is a technique to study phenomics. This technique is based on the principle that the movement of molecules within an object leads to the emission of characteristic infrared radiations. Two most popular devices that screen infrared radiations are a near-infrared (NIR, wavelength of approximately 0.9–1.7 mm) imaging device and a far-infrared (far-IR, wavelength of approximately 7.5–13.5 mm) imaging device. Another device is a crop phenology recording system (CPRS) which makes use of both visible light and infrared imaging to establish the relationship between camera-derived indices and agronomic traits. Apart from these two, one more imaging technique, ie. hyperspectral imaging technique, is widely used for studying plant architecture, health conditions and growth characteristics.

Imaging techniques like 3D structural tomography and functional imaging have been introduced for better visualization of living plants. The X-ray computed tomography (CT) scanners equipped with an acceleration algorithm using the adaptive minimum enclosing rectangle (AMER) and graphics processing unit (GPU) is effectively used for estimating the tiller number in rice. Optical coherence tomography (OCT) is a new technology based on photonics, has an approximately 1 mm spatial resolution and is used for in vivo 3D imaging of plant structures. Another tomography, optical projection tomography (OPT) with greater penetration and capability of detecting non-fluorescent signals can be applied to visualize plant developmental stages and gene expression. Magnetic resonance imaging (MRI) provides information regarding the structural organization and the internal processes occurring in vivo by imaging the water protons. The structural imaging and functional imaging technologies (such as fluorescence imaging and positron emission tomography, PET) reveal the alterations occurring in plants at physiological levels. Among all the imaging techniques, chlorophyll fluorescence imaging is widely used in plant phenomics. Chlorophyll fluorescence imaging is one such technique that determines the photosynthetic efficiency and stress encountered by plants.

Imaging technologies like Scanalyzer 3D which efficiently analyse all the tolerance mechanisms in plants during salinity stress like Na^+ exclusion, osmotic tolerance and tissue tolerance have been used in cereals. Another modification of the Scanalyzer 3D enables the researchers to more accurately estimate the cereal biomass under salt stress conditions. Martrack Leaf, a marker tracking approach, has made it possible to perform the high-resolution and accurate 2D analysis of leaf expansion in soybean. Infrared thermal imaging technique has been successfully used for quantification of osmotic stress response in cereal crops. The visible and the near-infrared (NIR) digital imaging techniques enable the high-throughput screening of crops during nitrogen stress. Furthermore, it has been concluded that the combination of precise phenomics/phenotyping approaches and high-resolution genetic dissection can explain the functional gene polymorphisms and abiotic stress tolerance mechanisms to a greater extent.

"Omics" technologies are being tremendously used in research activities which give rise to huge amount of data. Information handling is a tedious job. For storage, organization and easy accessibility of the available data, several computational resources act as the repositories. These databases are a storehouse of information on molecular markers, genes, microRNAs, siRNAs, proteins, metabolites and phenomics. Such databases on genomics-, transcriptomics-, proteomics- and metabolomics-related databases are enlisted in Table 19.4.

An overall assessment of genetic attributes of abiotic stresses, their constraints and effective survival strategies are available in Table 19.5.

A list of stress-tolerant rice varieties released in Asia and Africa are available in Table 19.6.

Table 19.4 Online databases associated with various omics research in crop plants

Genomics databases	Transcriptomics databases	Proteomics databases	Metabolomics databases
National Center for Biotechnology Information	Soybean knowledge base, University of Missouri	Proteome analysis at EBI	The soybean metabolome database
Gramene	Soybean transcription factors database, Missouri	Soybean transcription factors database, Missouri	BRENDA
The Arabidopsis Information Resource (TAIR)	TIGR Arabidopsis arrays	Soybean proteins database	Platform plant metabolomics
The Oryza Tag Line mutant database	Gene expression omnibus	ExPASy *A. thaliana* 2D-proteome Database	Metabolic modelling
TIGR rice genome Annotation	NSF rice oligonucleotide array project	Swissprot	Iowa gene expression toolkit
Maize genome Resources	Zeamage	PlantsP: functional genomics of plant	Solcyc Solanaceae metabolic pathway annotations
Geneontology	Tomato expression database	Functional genomics of plant	Plant metabolome database
Maize genetics and genomics database	Soybean transposable elements database	ExPASy:SIB Bioinformatics Resource portal	AraCyc Arabidopsis metabolic pathway annotations
An integrated soybean genome database including BAC-based physical maps	Virtual centre for cellular expression profiling in rice	Database of *A. thaliana* annotation	MetAlign tool for GC- or LC-MS data analysis
SoyBase and the soybean breeder's toolbox	PLEXdb	PlantPReS	Plant metabolic network

Courtesy: Springer Nature

Table 19.5 Abiotic stresses, their constraints and effective survival strategies

Stress	Constraint	Tolerance and survival strategy[a]	Transient solution	Chronic solution
Flooding	Reduced energy owing to lower photosynthesis rate and/or low O2 levels	Energy conservation or expenditure	Growth quiescence	Rapid growth for avoidance Aerenchyma for aeration
Drought	Low water potential	Limited water loss Improved water uptake	Hydrotropism Reduced transpiration	Deeper roots Reduction of leaf area Adjusted osmotic status
Salinity	Elevated salt levels (e.g. NaCl) cause ion cytotoxicity and reduce osmotic potential	Reduced root ion uptake Vacuolar ion compartmentalization osmotic adjustment	Limited ion movement to transpiration stream Reduced shoot growth	Limited root ion flux to shoots Vacuolar ion compartmentalization in shoot cells
Ion toxicity	Cytotoxicity	Limited uptake Vacuolar ion compartmentalization	Efflux of organic acids to apoplast and immobilization of ions by chelation Intracellular chelation	Compartmentalization of ions (e.g. vacuole and apoplast) Efflux of organic ions to chelate toxic ions in soil
Ion deficiency	Inadequate nutrient acquisition	Enhanced uptake by transporters and developmental adaptations	Transport protein induction and activation Reduced growth	Transport protein function Root sensing and architecture remodelling for acquisition Partitioning for storage

Low and sub-freezing temperatures	Membrane damage Low water potential	Low-temperature acclimation Induction of stress protection genes	Acclimation Dormancy Osmoprotection	Acclimation and dormancy Altered membrane composition Increased compatible solutes
High temperature	Reduced photosynthesis Reduced transpiration Impaired cellular function	Maintenance of membrane function and reproductive viability	Leaf cooling Molecular chaperones	Altered membrane composition Molecular chaperones
Ozone (>120 nL)	Reduced photosynthesis ROS	Increased capacity to control ROS	Stomatal closure Elevated antioxidant capacity	Elevated antioxidant capacity

ROS reactive oxygen species

[a]Depending on the species and developmental state, the effective survival strategy may be tolerance (e.g. metabolic acclimation for survival) or avoidance (e.g. escape of drought by deeper rooting). Avoidance strategies may be constitutive or stress-induced evolutionary adaptations

Courtesy: Springer Nature

Table 19.6 Stress-tolerant rice varieties that have been released in South Asia and Africa

Name of the variety	Designation	Code/ GID	IRRI parent(s)/ background variety	Country/ states or provinces	Month/ year released	Stress	Ecosystem
WITA-9				Uganda	2014	High yield	Irrigated
WAC18-WAT15-3-1				Guinea Conakry	2014	High yield	Lowland
WAB 95-B-B-40-HB				Guinea Conakry	2014	Drought	Upland
Varsha Dhan	CLRC 899		IR 31342-8-3-2/ IR31406-3-3-3-1// IR 26940-3-3-3-1	India	2005	Submergence	Shallow deep water (stagnant flood)
Tripura Khara Dhan 2	IET 22835		IR87707-182-B-B-B	Tripura, India	18-Oct-14	Drought	
Tripura Khara Dhan 1	IET 22837		IR87707-446-B-B-B	Tripura, India	18-Oct-14	Drought	
Tripura Hakuchuk 2	TRC 2013-5		IR 82589-B-B-138-2	Tripura, India	18-Oct-14	Drought	
Tripura Hakuchuk 1	TRC 2013-4		IR 83928-B-B-56-4	Tripura, India	18-Oct-14	Drought	
Tripura Aus Dhan	TRC 2013-12		IR 83928-B-B-42-4	Tripura, India	18-Oct-14	Drought	
Tai	IR03A262	1111689	IR 71606-1-2-1-3-2-3-1/ IRRI 118	Tanzania	2013		Rainfed/Irrigated
Swarnali	IET23148			West Bengal, India	2017	submergence	
Swarna-Sub1	IR 05F102 (IR82810-407)	1847271	IR49830-7-1-2-2, Swarna	Nepal	2012	Submergence	
Swarna-Sub1 (improved Swarna)	IR 05F102 (IR82810-407)	1847271	IR49830-7-1-2-2, Swarna	India	2009	Submergence	

Swarna Shreya				India	2016	Drought	
Sukkha Dhan 6	IR 83383-B-B-129-4			Nepal, rainfed lowland	2014	Drought	
Sukkha Dhan 5	IR 83388-B-B-108-3			Nepal, rainfed lowland	2014	Drought	
Sukkha Dhan 4	IR 87707-446-B-B-B			Nepal, rainfed lowland	2014	Drought	
Sukkha Dhan 3				Nepal	2012	Drought	
Name of the variety	Designation	Code/ GID	IRRI parent(s)/ background variety	Country/ States or Provinces	Month/ Year released	Stress	Ecosystem
NERICA 16				Sierra Leone	2014	Drought	Upland
NERICA 15				Sierra Leone	2014	Drought	Upland
NDRK 5088 (Narendra Usar Dhan 2008)	TCCP 266-249-B-B-3/IR 262-43-8-1		Introduction of line from IRRI	UP, India	2009	Saline	Sodic
NDR 8011				Uttar Pradesh, India	2016	Submergence	
M'ziva	IR 77080-B-34-3	1192189	IR 70179-1-1-1-1/IRRI 134	Mozambique	2013		Rainfed
Mugwiza	IR91028-115-2-2-2-1			Burundi	2016		Irrigated
Makassane	IR 80482-64-3-3-3	2595051	MEM BERANO/PADI ABANG GOGO	Mozambique	2011		Irrigated
MPATSA	IR 82077-B-B-71-1			Malawi	2015		Irrigated

(continued)

Table 19.6 (continued)

Name of the variety	Designation	Code/ GID	IRRI parent(s)/ background variety	Country/ states or provinces	Month/ year released	Stress	Ecosystem
Komboka	IR05N221	1265595	IR 74052-297-2-1/IR 71700-247-1-1-2	Tanzania	2013		Rainfed lowland/ Irrigated
Komboka	IR05N221	1265595	IR 74052-297-2-1/IR 71700-247-1-1-2	Kenya, Uganda	2014		Rainfed lowland/ Irrigated
Kolondieba 2				Mali	2015	Submergence	Deep flooded lowland
Kadia 24				Mali	2015	Submergence	Deep flooded lowland
KATETE	IR 80411-B-49-1			Malawi	2015		Irrigated

Further Reading

Ali J et al (2017) Harnessing the hidden genetic diversity for improving multiple abiotic stress tolerance in rice (*Oryza sativa* L.). PLOS One. https://doi.org/10.1371/journal.pone.0172515

Dresselhaus T, Hückelhoven R (2018) Biotic and abiotic stress responses in crop plants. Agronomy 8:267. https://doi.org/10.3390/agronomy8110267

Frascaroli (2018) Breeding cold-tolerant crops: physiological, molecular and genetic perspectives. In: Wani SH, Herath V (eds) Cold tolerance in plants. Springer, Cham, pp 159–177. https://doi.org/10.1007/978-3-030-01415-5_9

He M et al (2018) Abiotic stresses: general defenses of land plants and chances for engineering multistress tolerance. Front Plant Sci 9:1771. https://doi.org/10.3389/fpls.2018.01771

Munns R, Gilliham M (2015) Salinity tolerance of crops – what is the cost? New Phytologist 208:668–673

Negrão S et al (2017) Evaluating physiological responses of plants to salinity stress. Ann Bot 119 (1):1–11. https://doi.org/10.1093/aob/mcw191

Rahman AMNRB, Zhang J (2018) Preferential geographic distribution pattern of abiotic stress tolerant rice. Rice 11:10. https://doi.org/10.1186/s12284-018-0202-9

Raza A et al (2019) Impact of climate change on crops adaptation and strategies to tackle its outcome: a review. Plants 8:34. https://doi.org/10.3390/plants8020034

Wani SH (2018) Biochemical physiological and molecular avenues for combating abiotic stress in plants. Academic, London

Genotype-by-Environment Interactions 20

Keywords
Statistical models for assessing $G \times E$ interactions · Genotypes and environments · Basic ANOVA and regression models · Multiplicative models · AMMI analysis · Pattern analysis · GGE biplot · Measures of yield stability · Software

Abbreviations

AMMI	Additive main effect and multiplicative model
BLUP	Best linear unbiased prediction
COMM	Completely multiplicative model
FAMM	Factor analytic multiplicative mixed model
FR	Factorial regression
G × E	Genotype × environment interaction
GREG	Genotype regression model
LR	Linear regression
M × E	Marker × environment interaction
MET	Multi-environment trial
NCOI	Non-crossover interaction
PCA	Principal component analysis
PLSR	Partial least square regression
QTL	Quantitative trait locus
Q × E	QTL × environmental interaction
SHMM	Shifted multiplicative model
SVD	Singular value decomposition
SREG	Sites regression model
TPE	Target population of environments

P. M. Priyadarshan, *PLANT BREEDING: Classical to Modern*,
https://doi.org/10.1007/978-981-13-7095-3_20

A phenotype is the function of a genotype, the environment and the differential response of genotypes to different environments. This is known as genotype-by-environment ($G \times E$) interaction. $G \times E$ is a statistical decomposition of variance and provides a measure of the relative performance of genotypes grown under different environments. These interactions were managed and analysed by the plant breeders during the history of crop domestication, crop improvement and dispersal.

A conceptual $G \times E$ interaction is commonly depicted as the slope of the line when genotype performance is plotted against an environmental gradient (Fig. 20.1). When cultivar performs the same across environments, non-parallel and non-intersecting lines are made available. When lines intersect, the indication is that the rank of cultivars changes across environments where the optimum cultivar will be location specific.

The first step to investigate GEI is to obtain phenotypic observations on a set of genotypes exposed to a range of environmental conditions. Genotypes can include advanced lines of a breeding programme, cultivars and segregating offspring from a specific cross such as F_2, a backcross or recombinant inbred lines (RILs). Genotypes can be subjected to different agri-management regimes that include levels of a particular stress or a combination of stresses. In multi-environment trials (METs), genotypes are evaluated over a number of geographical locations for several years.

Data from METs are collected in the form of two-way tables of means, with genotypes in rows and environments in columns. Each cell of such a table will have an estimate of the performance (adjusted mean) of a particular genotype in a specific environment. To identify genotypes and environments beyond doubt, indices are used, the letter i for genotypes and the letter j for environments.

20.1 Statistical Models for Assessing $G \times E$ Interactions

Genotype-by-environment interaction (GEI) is an important phenomenon in plant breeding. Statistical models for describing, exploring, understanding and predicting GEI are available. All models depart from a two-way table of genotype by environment means. Finlay-Wilkinson model, AMMI model and GGE biplot models use only means of two-way table. However, factorial regression model is an approach to explicitly introduce genotypic and environmental covariates for describing and explaining GEI. In QTL modelling, as a natural extension of factorial regression, the marker information is transformed into genetic predictors. Tests for regression coefficients corresponding to these genetic predictors are tests for main effect QTL expression and QTL-by-environment interaction (QEI). QEI is based on environmental covariables for predicting GEI for genotypes and environments. When multiple environments are considered, the necessity of sophisticated mixed models

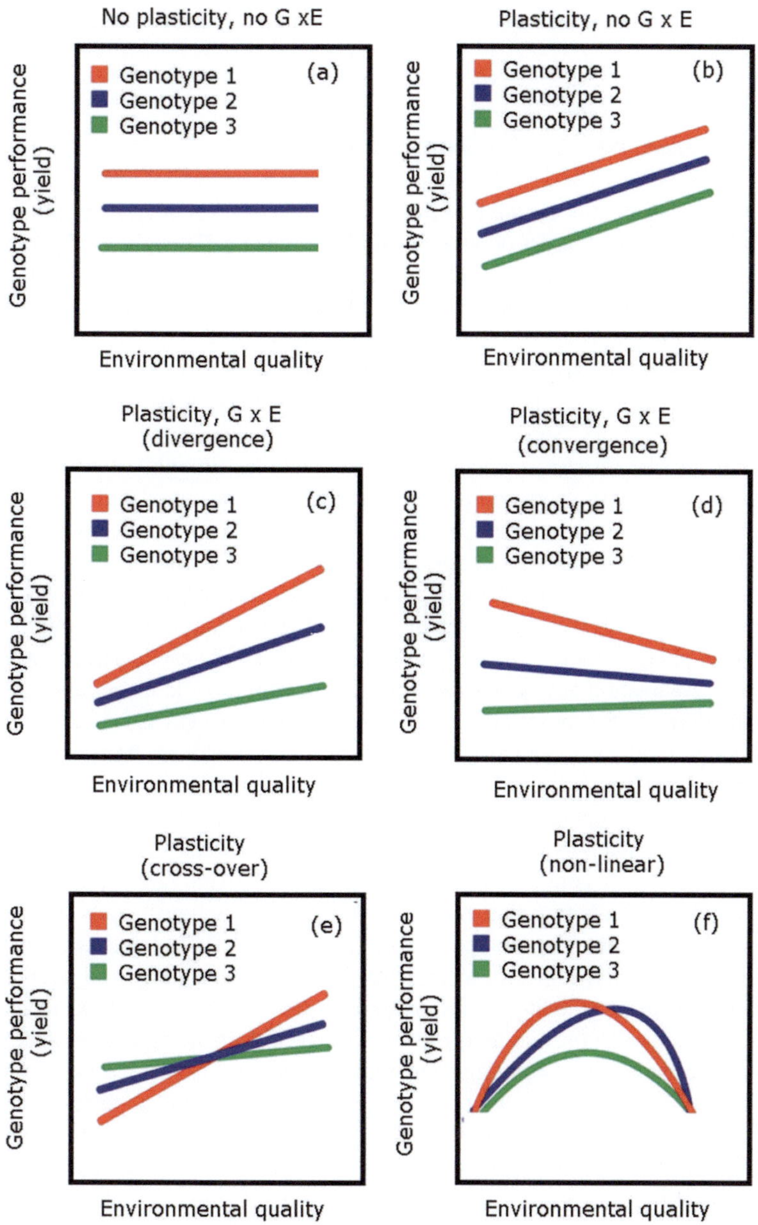

Fig. 20.1 Reaction norms for three genotypes that illustrate various forms of plasticity and genotype × environment interaction ($G \times E$). No plasticity in (**a**) versus plasticity in (**b**) to (**f**), no $G \times E$ in (**a**) and (**b**) versus various forms of $G \times E$ in (**c**) till (**f**)

is needed to allow heterogeneity of genetic variances and correlations across environments.

20.1.1 Genotypes and Environments

In $G \times E$ breeding, it is essential to understand the concepts of target population of genotypes (TPG) and target population of environments (TPE). The TPG contains all possible genotypes and the TPE delineates the future growing conditions. The TPE can be defined by geography, soil and meteorological conditions, management choices and the incidence of biotic and abiotic stress. TPE must be reflected in the environmental design space of prediction models.

Phenotype is the result of an outcome of interactions between genetic and environmental factors. So, TPG and TPE cannot be chosen as independent. For example, for abiotic stress breeding, if TPE includes drought and well-watered conditions, genotypes performing better under drought stress and well-watered conditions are to be developed. In short, TPG consists of genotypes with wider adaptation. Also, for biotic stress, the same logic will be applicable.

The reaction norm for yield depends on the reaction norms for the yield components. The joint reaction norm of yield and yield components is a multivariate function of phenotypes that mutually affect each other and genetic and environmental inputs. Adaptation and adaptedness usually pertain to yield or biomass. A good understanding of the processes leading to adaptedness and $G \times E$ interaction requires observations on yield together with its main component traits as a function of time.

A reaction norm defines a genotype-specific function that translates environmental inputs into a phenotype. $G \times E$ interaction occurs when these reaction norms intersect, diverge or converge (compare Fig. 20.1a, b with Fig. 20.1 c to f). The presence of $G \times E$ interaction makes phenotypic prediction models to be more elaborate and to contain genotype-specific parameters. The intercepts, slopes and curvatures as genotype-specific parameters are called sensitivity and adaptability parameters in plant breeding. Good analytical procedures are required to unravel $G \times E$ interaction and reveal their causes. This facilitates the breeder in making an informed decision while selecting a superior genotype for a TPE.

The basic parametric (a test which has information about the population parameter) analytical approaches used to study $G \times E$ interactions can be classified into two types: ANOVA-based and regression-based statistical models. The ANOVA-based concept was first developed by R.A. Fisher in 1924 and was adopted in plant breeding to distinguish genetic from environmental sources of variance. This basic model can be improved by using bilinear terms for $G \times E$ analysis. Regression models for $G \times E$ analysis were introduced by Finlay and Wilkinson in 1963.

ANOVA and linear regression (LR) models are used to partition and analyse $G \times E$. A term accounting for the deviation from genotypic and environmental main effects versus the slope of the genotypic regression line on environmental means is used to explain $G \times E$ in ANOVA and LR models, respectively. When LR models are used, genotypes with moderate slope and above-average performance can be matched to those environments as the TPE. However, it is every likely to miss a lot of information when using these $G \times E$ methods.

When interaction from more than one dimension occurs, multiplicative models such as the additive main effect and multiplicative interactive (AMMI) model put forth by Gauch in 1992, the site regression model (SREG, which is also called genotype + genotype × environment) proposed by Cornelius and co-workers in 1996, shifted multiplicative (SHMM) model by Seyedsadr and Cornelius in 1992, the genotype regression (GREG) model by Cornelius and others in 1996 and the completely multiplicative model (COMM) by Cornelius and other in 1996 were introduced. These models can be considered as modifications of the ANOVA model, where $G \times E$ are decomposed into multiple linear orthogonal components that explain the interaction in more than one dimension. Using these models, TPE can be identified by using biplot visualization. But these models cannot identify the causes of $G \times E$

To detect and measure causes of $G \times E$, factorial regression (FR) model is used. This is done by estimating genotypic sensitivity to explicit environmental covariates that can statistically test the influence of those environmental variables on $G \times E$. Factorial regression models are sensitive to multicollinearity when a large number of correlated external variables are used. This sensitivity can be mitigated by using partial least square regression (PLSR). One or few PLSR factors can explain the variance of the **X** matrix (containing predictor variables) as well as the covariance between matrices **X** and **Y** (containing a response variable or variables). PLSR is a parsimonious model for analysing METs with a large number of external variables.

The aforementioned models are normally used for modelling fixed effects. In fixed effect models, it is assumed that the estimate is the same in all trials as well as estimated in the trial under study. Since this is not realistic, the estimates from fixed effect models are normally only used in the trial under study. On the other hand, estimation of random effects assumes that the effects obtained from the trial under study are a representative of similar trials. Therefore, $G \times E$ analysis can be appropriately performed using a mixed effect methodology where fixed and random effects are present. Mixed effect models allow the modelling of independent random effects with a variance parameter, and they also consider heterogeneity in variance across environments, correlations between environments and relationships among genotypes.

Random effects in mixed effect models can be computed by using best linear unbiased prediction (BLUP) put forth by Robinson in 1991. Here, the correlations between estimates of the realized values and the true values of the random effects are maximized. This can increase the accuracy of estimation and thus identification of the TPE. Heterogeneity of variance across and covariance between pairs of environments can be modelled using different variance–covariance structures such as compound symmetry, where variance and covariance are assumed to be constant among environments. Or, it can be done with unstructured covariance matrix, where heterogeneous variance and covariance are assumed for each environment and pair of environments. While dealing with using unbalanced data, a parsimonious factor analytic can be used.

20.1.2 Basic ANOVA and Regression Models

The ANOVA model first developed by R.A. Fisher in 1924 can be used for analysing $G \times E$.

$$Y_{ijk} = \mu + \alpha_i + \beta_j + (\alpha\beta)_{ij}\epsilon_{ijk} \tag{20.1}$$

Y_{ijk} is yield response variable; μ is the overall mean; α_i is the genotypic effect for the ith genotype; $I = 1, 2K\ldots I$; $j = 1, 2K\ldots J$; $(\alpha\beta)_{ij}$ is the interaction of the ith genotype with the jth environment; ϵ_{ijk} is the residual error; $\epsilon_{ijk} \sim N(0, \sigma 2)$.

The ANOVA model though quantifies the magnitude of classifiable main effects and interactions, it fails to describe the characteristics of the $G \times E$ term. Thus, the model can be considered as a base model for identifying the presence of $G \times E$ and quantifying it in a single dimension. This can be used in identifying environments as TPE when (a) no significant $G \times E$ is present and (b) the magnitude of $G \times E$ is found in the presence of significant $G \times E$. Moreover, the model requires replications within environments, which is a challenge especially when a large number of genotypes are need to be tested and land is limited.

A LR model with regression of individual genotype performance over environmental means put forth by Yates and Cochran in 1938 and Finlay and Wilkinson in 1963 can also be used to analyse $G \times E$. The model can be represented as:

$$Y_{ijk} = \mu + oi + \beta_j(1 + bi) + \delta_{ij} + \epsilon_{ijk} \tag{20.2}$$

where:

Y_{ijk} is the yield response variable; μ is the overall mean; β_j is the environmental effect for the jth environment, bi is the genotypic slope on environmental means such that each genotype has an intercept oi; and the slope bi and δ_{ij} is the residual deviation of interaction so that the total interaction is $\beta_j bi + \delta_{ij}$. The slopes can be related to the ANOVA model's interaction term, and the heterogeneity of the lines illustrates interactions. In the case of genotypes with a moderate slope (moderate sensitivity or stable genotypes) and above-average performance across environments, the environments can be grouped as a representative TPE for those genotypes. However, it often fails to explain a large proportion of variation caused by $G \times E$. The model also assumes a linear relationship between $G \times E$ and environmental means. Unless a high proportion of $G \times E$ can be attributed to the model, the linear relationship assumption is violated, and the results do not explain enough. In fact, when a few extreme environments are involved in the analysis, the fit of the model will be influenced by the performance of genotypes in the extreme environments. The model can be used to identify the TPE where genotypes react similarly but cannot identify the reasons for $G \times E$. When experimentation is being carried out in geographically distant regions where $G \times E$ is too complex, this model is not useful.

20.1.3 Multiplicative Models

Multiplicative models are modifications of ANOVA. Principal component analysis (PCA), developed by Pearson in 1901, is a widely used technique in MET analysis. Singular value decomposition (SVD) is the basis of PCA.

$$Y_{ij} = \mu + \sum \lambda_l \zeta_{il} \eta_{jl} + \epsilon_{ij} \tag{20.3}$$

Y_{ij} is the yield response variable from a balanced data set (a dot is used to replace the subscript, indicating that the data have been summed over that subscript; in this case, the replications); μ is the overall mean subtracted from the $G \times E$ matrix of means; λ_l indicates singular values (the square roots of the eigenvalues); ζ_{il} and η_{jl} are the left singular vectors (genotype scores, which summarize the relationships among genotypes) and right singular vectors (environmental loadings, which summarize the relationships among environments), respectively; ϵ_{ij} is calculated using Eq. (20.4):

$$\epsilon_{ij} = \tilde{\ } N\left(O, \frac{\sigma^2}{k}\right) \tag{20.4}$$

where k is the number of replicates.

The five multiplicative models that are most commonly used for evaluating $G \times E$ are the AMMI models (Eq. 20.5), the SREG model (Eq. 20.6), the SHMM (Eq. 20.7), the GREG model (Eq. 20.8) and the COMM (Eq. 20.9):

$$Y_{ij} = \mu + \alpha_i + \beta_j + \Sigma \lambda_l \zeta_{il} \eta_{jl} + \epsilon_{ij}. \tag{20.5}$$

$$Y_{ij} = \imath j + \Sigma \lambda_l \zeta_{il} \eta_{jl} + \epsilon_{ij}. \tag{20.6}$$

$$Y_{ij} = \theta + \Sigma \lambda_l \zeta_{il} \eta_{jl} + \epsilon_{ij}. \tag{20.7}$$

$$Y_{ij} = pi + \Sigma \lambda_l \zeta_{il} \eta_{jl} + \epsilon_{ij}. \tag{20.8}$$

$$Y_{ij} = \sum \lambda_l \zeta_{il} \eta_{jl} + \epsilon_{ij}. \tag{20.9}$$

where Y_{ij} is the yield response variable; μ is the overall mean; α_i is the genotypic effect for the ith genotype; β_j is the environmental effect for the jth environment; $\imath j$ is the environment mean; θ is the shift parameter; pi is the genotypic mean; λ_l is the singular value; ζ_{il} and η_{jl} are the genotype and environment singular vectors, respectively; and $\in_{ij}$ is the residual error (Eq. 20.4). The results of the multiplicative models can be expressed in the form of biplots. Environments and genotypes that are similar cluster together in the biplot. Genotypes that are clustered in the centre of the plot have average responses from all environments (broad adaptation). Genotypes that are clustered with specific environments are having specific adaptation. Only AMMI and GGE biplot will be discussed here.

20.1.4 AMMI Analysis

The two main purposes of AMMI analysis of a yield trial's treatment design are (a) understanding complex $G \times E$ interactions, including delineating mega-environments and selecting genotypes to exploit narrow adaptations, and (b) increasing accuracy to improve recommendations, repeatability, selections and genetic gains. The main purposes of an experimental design are assigning experimental units to treatments, quantifying errors and gaining accuracy.

Analysis of variance (ANOVA) of a yield trial's treatment design partitions its variance into three sources: genotype main effects (G), environment main effects (E) and genotype x environment interaction effects (GE). For breeders, manipulating genotypes, G and GE are relevant because only they affect genotype rankings. AMMI first applies ANOVA to partition the variation into G, E and GE, and then it applies principal components analysis (PCA) to GE (Fig. 20.2). Accordingly, both G and GE are analysed, but separately and without confounding. Broad adaptations are associated with G and are beneficial everywhere, whereas narrow adaptations are associated with GE and require subdividing the environments into two or more mega-environments. A mega-environment is defined as a subset of the environments having the same or at least similar genotypes. There are four steps in AMMI analysis, they are ANOVA, model diagnosis, mega-environment delineation and selection and recommendation. These steps will be briefly dealt here.

ANOVA: Three attributes from the ANOVA provide preliminary indications on whether AMMI analysis will be worthwhile: the sum of squares (SS) for genotypes (G), GE signal (GES) and GE noise (GEN). The SS values for G and GE are direct outputs from ANOVA. To estimate the SS for GEN, simply multiply the error mean square (from replication) by the number of degrees of freedom (df) for GE. Then obtain GES by subtracting GEN from GE. AMMI analysis is appropriate for data sets having substantial G and substantial GES. When the SS for GES is at least as large as that for G, AMMI analysis will be acceptable. On the other hand, occasionally GE is buried in noise, with the SS for GEN approximately equal to that for GE. In that case, GE should be ignored, so AMMI analysis is inappropriate.

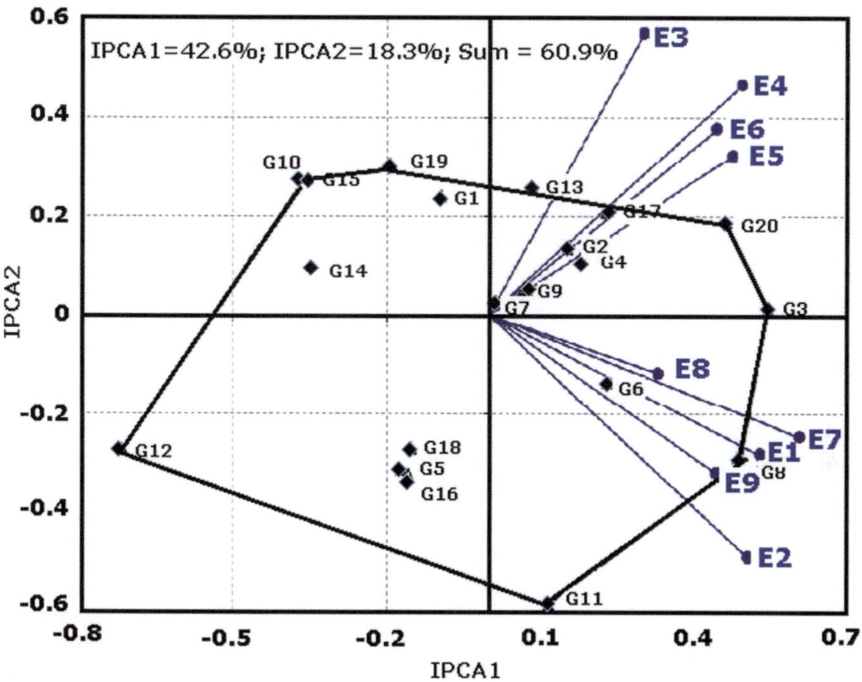

Fig. 20.2 Based on genotype and environment scores, AMMI biplot for 20 bread wheat cultivars using the mean grain yield obtained from 9 environments

Model Diagnosis: The AMMI model equation given by Gauch in 2013 is:

$$Y_{ge} = \mu + \alpha_g + \beta_e + \Sigma_n \lambda_n \gamma_{gn} \delta_{en} + p_{ge} \tag{20.10}$$

where Y_{ge} is the yield of genotype g in environment e, μ is the grand mean, α_g is the genotype deviation from the grand mean, β_e is the environment deviation, λ_n is the singular value for interaction principal component (IPC) n and correspondingly $\lambda^2{}_n$ is its eigenvalue, γ_{gn} is the eigenvector value for genotype g and component n, ũFF;$_{en}$ is the eigenvector value for environment e and component n, with both eigenvectors scaled as unit vectors, and p_{ge} is the residual.

Successive IPCs are denoted by IPC1, IPC2, IPC3 and so on, and the number of these components is 1 less than the minimum of the number of genotypes and number of environments. The member of the AMMI model family retaining 0 components is denoted by AMMI 0, and the following members retaining 1 or more components are denoted by AMMI1, AMMI2.......and so on, up to the full model retaining all components denoted by AMMIF. The fitted values of the full model automatically equal the raw data Yge exactly, so the residual term disappears. But reduced models leave a residual p_{ge}. A yield trial with an experimental design has additional terms in its model equation. For instance, the equation for the AMMI model applied to a yield trial with the popular RCB experimental design is:

$$Y_{\text{ger}} = \mu + \alpha_g + \beta_e + \Sigma_n \lambda_n \gamma_{\text{gn}} \delta_{\text{en}} + p_{\text{ge}} + K_{r(e)} + \epsilon_{\text{ger}} \quad (20.11)$$

where Y_{ger} is the yield of genotype g in environment e for replicate r, and the two additional terms beyond those in Eq. (20.10) are $\kappa_{r(e)}$, which is the block effect for replication r within environment e, and ϵ_{ger}, which is the error. For the RCB design, the yields Y_{ge} of the raw data AMMIF are simply the averages over the R replicates, $(\Sigma_r Y_{\text{ger}})/\text{R}$, although some other experimental designs make adjustments to the raw data.

Mega-environment Delineation: As the selected member of the AMMI model family changes, the mega-environments also change tending to define a larger number of mega-environments. For instance, AMMI1 delineates 2 mega-environments, AMMI2 delineates 3, AMMI3 delineates 4 and AMMI4 to the full model; AMMI8 delineates 5 or 6. Consequently, mega-environments cannot be delineated meaningfully or reliably without first performing a model diagnosis to select the best member of the AMMI model family for a given data set.

It is also important for mega-environments to have predictive potential for locations and years. Predictable environmental factors associated with locations or management practices increase the number of usable mega-environments, whereas unpredictable environmental factors associated with years decrease the number and usefulness of mega-environments.

Subdividing the environment into several mega-environments is costly for breeding programmes, and only a practical portion GE_P of the interaction signal GE_S may be available for exploiting narrow adaptations to increase yields. This limitation necessitates selecting a low-order model such as AMMI1 for delineating a small and manageable number of mega-environments. But fortunately, merely 2 or 3 mega-environments often suffice to allow GEP to capture a sizable portion of GES.

Mega-environments can be displayed by both tables and graphs. A ranking table shows the ranks for the best several genotypes in each environment. Listing the environments in order by their IPC1 scores makes these tables more structured and informative. The advantages of the tables are (a) ranking tables can show any member of the AMMI model family readily, whereas graphs can accommodate only AMMI1 and AMMI2; (b) ranking tables can identify the best genotypes, whereas mega-environment graphs identify only the top-ranked genotype for each environment; and (c) a single ranking table can list several AMMI models side by side to facilitate comparisons and to serve multiple research purposes. But biplot and mega-environment graphs using AMMI1 and AMMI2 can also be helpful for visualizing complex patterns in yield trial data, so graphs and tables are complementary (Fig. 20.2).

Selection and Recommendation: The ultimate aim of yield trial is selection of best genotypes for a breeding programme or recommendation of the same for a growing region. Normally, selection pursues both high yield and stability. But this approach has five problems:

(a) there are dozens of stability parameters, making a choice difficult. However, a specific stability concept is stability across years within a given location or mega-environment because it reduces susceptibility to unpredictable GE interactions;
(b) there are manifold ways to integrate high yield and stability, but many fail to optimize outcome;
(c) stability is a meaningful objective only within an individual mega-environment, not across multiple mega-environments and selecting for stability across mega-environments may lead to sacrificing potential yield gains from narrow adaptations;
(d) at least eight trials within each mega-environment are necessary for reasonably reliable estimate stability and
(e) instability (GE) presents plant breeders with both problems and opportunities.

So, an alternative is to determine which genotypes win in which environments according to a parsimonious AMMI model.

20.1.5 Pattern Analysis

The availability of largely unbalanced data sets, each one relating to a specific year and having several test locations is relatively frequent in multi-environment trials. The combined analysis of this information for location classification may be realized using a procedure that requires different steps:

(a) Estimation of the phenotypic correlations among test locations for genotype original yields in each individual data set
(b) Averaging across data sets of the correlations for each pair of locations
(c) Transformation of the similarity matrix (as provided by correlation coefficients) into a dissimilarity matrix of squared Euclidean distances, inputting it (rather than the genotype by location matrix of standardized yields) into the cluster analysis

A weighted average should be used for correlations based on a variable number of genotypes using the following formula (expressed for z values):

$$z = \sum (n_i - 3) z_i / \sum (n_i - 3) \qquad (20.12)$$

where z is the weighted average and z_i and n_i are the z value and the associated number of genotypes for the correlation, respectively, in the data set i. For example, the weighted average of the three phenotypic correlations $r_1 = 0.50$, $r_2 = 0.80$ and $r_3 = 0.90$, with the respective number of genotypes $n_1 = 16$, $n_2 = 10$ and $n_3 = 15$, can be obtained through the z transformation as:

$$z = [(13 \times 0.55) + (7 \times 1.10) + (12 \times 1.47)]/(13 + 7 + 12) = 1.01 \quad (20.13)$$

which, once back transformed, provides the average phenotypic correlation $r_{\mathrm{p}} = 0.77$ for insertion in the similarity matrix. Of course, pairs of locations may differ for number of correlations contributing to the average r_{p} value since some sites may be absent in some data sets. At least one individual correlation is needed for each pair of locations to allow both sites in the analysis.

20.1.6 GGE Biplot

Biplot technique was originally developed by Gabriel in 1971. Through singular value decomposition (SVD), a $g \times e$ matrix of mean yield of g cultivars in e environments can be approximated as the product of a genotype matrix and an environment matrix so that yield of genotype i at environment (location) j, Y_{ij}, is approximated as:

$$\hat{Y}_{ij} = + \sum\nolimits_{n=1}^{r} \lambda_n \xi_{\mathrm{in}} \eta_{\mathrm{in}} \qquad (\lambda_1 \geq \lambda_2 \geq {}_3\lambda^3 \geq \lambda_r) \quad (20.14)$$

where r is the number of PCs required to approximate the original data, with $r \leq \min(g, e)$, and λ_n is the singular value of PCn, the square of which is the sum of squares explained by PCn. ξ_{in} and ξ_{jn} are the ith genotype score and the jth environment score, respectively, for PCn. The SVD allows the $g \times e$ table of means to be displayed in a plot having g points for the genotypes plus e points for the environments. Each genotype is represented by a point, called a marker, defined by the genotype's scores on all PCs, and each environment is represented by a marker defined by the environment's scores on all PCs. Such a plot is called a biplot because both the genotypes and the environments are plotted in a single plot. Biplots can be multidimensional, but two-dimensional biplots, using only the first and the second PCs, are most common, both for biological reasons and for easy comprehension. To achieve symmetric scaling between the genotype scores and the environment scores, Eq. (20.14) is usually written in the form:

$$\hat{Y}_{ij} = \sum_{n=1} \xi^*_{\mathrm{in}} \eta^*_{\mathrm{in}} \quad (20.15)$$

where, $\xi^*_{\mathrm{in}} = \lambda^{0.5}{}_n \xi_{\mathrm{in}}$ and $\eta^*_{\mathrm{in}} = \lambda^{0.5}{}_n \eta_{\mathrm{jn}}$

The mean yield of genotype i in environment j is commonly described by a general linear model:

$$\hat{Y}_{ij} = \mu + \alpha_i + \beta_j + \Phi_{ij} \tag{20.16}$$

where μ is the grand mean, α_i is the main effect of ith genotype, β_j is the main effect of jth environment and Φ_{ij} is the interaction between genotype i and environment j. Deletion of α_i and/or β_j or all of $\mu + \alpha_i + \beta_j$ allows variation explainable by the deleted term(s) to be absorbed into the Φ_{ij} term. It is the matrix of Φ_{ij} values that is subjected to SVD. Subjecting the Φ_{ij} in Eq. (20.16) to SVD results in the additive main effects and multiplicative interaction (AMMI) model.

20.2 Measures of Yield Stability

High yield stability refers to a genotype's ability to perform consistently across a wide range of environments. Stability measures may be either "static" (Type 1) or "dynamic" (Type 2). Static stability is analogous to the biological concept of homeostasis. A stable genotype tends to maintain a constant yield across environments. Dynamic stability implies a stable genotype with a yield response in each environment that is always parallel to the mean response of the tested genotypes, i.e. zero GE interaction. Type 4 stability relates to consistency of yield exclusively in time, i.e. across years (or crop cycles) within locations, whereas Type 1 stability relates to consistency in both time and space, i.e. across environments belonging to the same or different sites (see Chap. 14 also).

There are two major stability measures that can be ascribed to the static, Type 1 stability concept:

(a) The environmental variance S^2, i.e. the variance of genotype yields recorded across test or selection environments (i.e. individual trials). For the genotype i:

$$S_i^2 = \sum \left(R_{ij} - m_i\right)^2 / (e - 1) \tag{20.17}$$

where R_{ij} = observed genotype yield response in the environment j (the m_{ij} notation may also be appropriate since values are averaged across experiment replicates), m_i = genotype mean yield across environments, e = number of environments. Greatest stability is $S^2 = 0$. Derived stability measures include the square root value (S) and its coefficient of variation.

(b) The regression coefficient of genotype yield in individual environments as a function of the environment mean yield (m_j), adopting Finlay and Wilkinson's b coefficient. The modelled genotype response:

$$R_{ij} = a_i + b_i m_j \tag{20.18}$$

where a_i = intercept value, is analogous to equation:

$$R_{ij} = m + G_i + Lj + (b_i - 1)Lj = a_i + b_i mj \tag{20.19}$$

reported for joint regression analysis of adaptation, but genotype responses to environments (rather than to locations) are of concern here. Greatest stability is $b = 0$.

The following measures are probably the most popular in the context of the dynamic, Type 2 stability concept:

(a) Shukla's stability variance made available during 1972 and Wricke's ecovalence published during 1962, which give the same results for ranking genotypes. Wricke's ecovalence is simpler to calculate and is for the genotype *i*:

$$W_i^2 = \sum (R_{ij} - m_i - m_j + m)^2 \tag{20.20}$$

where R_{ij} is the observed yield response (averaged across experiment replicates), m_i and m_j correspond to previous notations and m is the grand mean. Greatest stability is $W^2 = 0$.

(b) Finlay and Wilkinson's regression coefficient across environments (as above), assuming greatest stability for $b = 1$. Therefore, instability can be evaluated as the distance in absolute value from the unity coefficient, $|b_i - 1|$.

Eberhart and Russell in 1966 proposed the estimated variance of genotype deviations from regressions (s_d^2) as a further stability measure for consideration in conjunction with the *b* parameter. This is a Type 3 stability concept and an indicator of the goodness of fit of the regression model for describing the stability response. It is argued that poor fit (i.e. large s_d^2 values) simply points towards the adoption of other Type 2 measures (such as Wricke's or Shukla's) rather than bothering with two stability parameters (b and s_d^2), whereas good fit implies no practical usefulness of s_d^2 estimates.

Type 4 stability concept relates to stability only in time (i.e. across test years or crop cycles), averaged across test locations, rather than stability also in space (as implied by stability analysis across environments) as proposed by Lin and Binns in 1988. The stability measure can be derived from an ANOVA that is limited to data of the genotype under assessment. The ANOVA can be performed on yield values averaged across experiment replicates, including just two factors, i.e. location and year within locations. The stability measure is represented by the ANOVA MS for the latter factor ($M_{y(l)}$). High stability is indicated by low $M_{y(l)}$ value, i.e. low temporal variation of genotype yield values (hence, the similarity with the Type 1, homeostatic concept of stability). In fact, the estimate of this variation as provided by $M_{y(l)}$ is inflated by the experimental error variance. The actual variance of this effect ($S_{y(l)}^2$) could be estimated as:

$$S_{y(l)}{}^2 = M_{y(l)} - (M_{\text{err}}/r) \tag{20.21}$$

where M_{err} = pooled error (i.e. average experimental error for the genotypes) in the combined ANOVA and r = number of experiment replicates. While $S_{y(l)}{}^2$ and $M_{y(l)}$ values are equivalent for ranking genotypes, the former are more appropriate for adoption in yield reliability indices. $S_{y(l)}{}^2$ values could also be estimated through a hierarchical ANOVA performed on plot values of each genotype. This includes the MS for the replicate within years source of variation ($M_{r(y)}$). In this case:

$$S_{y(l)}{}^2 = \left(M_{y(l)} - \left(M_{r(y)}\right)/r\right. \tag{20.22}$$

The current estimate of $S_{y(l)}{}^2$ values may differ slightly from the estimate obtained with formula (20.21).

20.2.1 Software

The values of environmental variance (for original or relative yields) and the derived reliability indices can easily be calculated through a worksheet (as available in IRRISTAT). The comparison of environmental variance values, requiring also correlation analysis, and the calculation of Type 4 stability measures, requiring the execution of simple one-way ANOVAs, can be performed by IRRISTAT or any ordinary statistical software. In particular, the ANOVA for each genotype performed on plot yields for estimation of $S_{y(l)}{}^2$ values (as per formula [20.18]) can easily be carried out through IRRISTAT. All these estimations can be done by SAS (Statistical Analysis System by SAS Institute).

Further Reading

Annicchiarico P (1992) Cultivar adaptation and recommendation from alfalfa trials in northern Italy. J Genet Breed 46:269–278

Annicchiarico P (1997) STABSAS: a SAS computer programme for stability analysis. Ital J Agron 1:7–9

Annicchiarico P (1997a) Joint regression vs AMMI analysis of genotype-environment interactions for cereals in Italy. Euphytica 94:53–62

Annicchiarico P (1997b) Additive main effects and multiplicative interaction (AMMI) of genotype-location interaction in variety trials repeated over years. Theor Appl Genet 94:1072–1077

Annicchiarico P (2002) Defining adaptation strategies and yield stability targets in breeding programmes. In: Kang MS (ed) Quantitative genetics, genomics, and plant breeding. CABI, Wallingford, pp 365–383

Cooper M, DeLacy IH, Basford KE (1996) Relationships among analytical methods used to study genotypic adaptation in multi-environment trials. In: Cooper M, Hammer GL (eds) Plant adaptation and crop improvement. CABI, Wallingford, pp 193–224

Cornelius PL, Crossa J, Seyedsadr MS (1996) Statistical tests and estimators of multiplicative models for genotype-by-environment interaction. In: Kang MS, Gauch HG (eds) Genotype-by-environment interaction. CRC Press, Boca Raton, pp 199–234

Des Marais DL, Hernandez KM, Juenger TE (2013) Genotype-by-environment interaction and plasticity: exploring genomic responses of plants to the abiotic environment. Annu Rev Ecol Evol Syst 44:5–29

Gauch HG, Zobel RW (1996) AMMI analysis of yield trials. In: Kang MS, Gauch HG (eds) Genotype-by-environment interaction. CRC Press, Boca Raton, pp 85–122

Grishkevich V, Yanai I (2013) The genomic determinants of genotype × environment interactions in gene expression. Trends Genet 29:479–487

Gauch HG Jr (1992) Statistical analysis of regional yield trials: AMMI analysis of factorial designs. Elsevier, Amsterdam

Malosetti M, Ribaut J-M, van Eeuwijk FA (2013) The statistical analysis of multi-environment data: modeling genotype-by-environment interaction and its genetic basis. Front Physiol. https://doi.org/10.3389/fphys.2013.00044

Piepho HP, Möhring J, Melchinger AE, Büchse A (2008) BLUP for phenotypic selection in plant breeding and variety testing. Euphytica 161:209–228

Saïdou A-A, Thuillet A-C, Couderc M, Mariac C, Vigouroux Y (2014) Association studies including genotype by environment interactions—prospects and limits. BMC Genet 15:3

Yan W, Hunt LA, Sheng Q, Szlavnics Z (2000a) Cultivar evaluation and mega-environment investigation based on GGE biplot. Crop Sci 40:596–605

Yan W (2014) Crop variety trials: data management and analysis. Wiley/Blackwell, Hoboken

Yan W, Kang MS (2003) GGE Biplot analysis: a graphical tool for breeders, geneticists and agronomists. CRC Press, Boca Raton

Yan W, Hunt LA, Sheng Q, Szlavnics Z (2000b) Cultivar evaluation and mega-environment investigation based on the GGE biplot. Crop Sci 40:597–605

Part V

Breeding for New Millennium

Tissue Culture

21

Keywords
History · Components of Tissue Culture Media · Preparing the Plant Tissue Culture Medium · Transfer of Plant Material to Tissue Culture Medium · Micropropagation · Protoplast Culture · Somatic Embryogenesis and Synthetic Seeds · Plant Tissue Culture Terminology

Tissue culture is the in vitro aseptic (sterile) culture of cells, tissues and organs under controlled nutritional and environmental conditions. Two concepts, plasticity and totipotency (ability of a cell to give rise to new organism or part), are central to understanding plant tissue culture. It involves the use of small pieces of plant tissue (*explants*) which are cultured in a nutrient medium under sterile conditions. Using the appropriate growing conditions for each explant type, tissues can be induced to rapidly produce new shoots and roots. These plantlets can also be divided, usually at the shoot stage, to produce large numbers of new plantlets. The new plants can then be placed in soil and grown in the normal way.

21.1 History

The science of plant tissue culture began with the discovery of cell, when in 1838, Schleiden and Schwann proposed that cell is the basic structural unit of all living organisms. Cell is also capable of autonomy so as to regenerate into a whole plant. Based on this, in 1902, a German physiologist, Gottlieb Haberlandt, for the first time attempted to culture isolated single palisade cells from leaves in Knop's salt solution added with sucrose. The cells were alive for 1 month but failed to divide. Though unsuccessful, he was instrumental in laying the foundation of tissue culture technology. He is regarded as the father of plant tissue culture. After that, some of the landmark discoveries that took place in tissue culture are:

P. M. Priyadarshan, *PLANT BREEDING: Classical to Modern*,
https://doi.org/10.1007/978-981-13-7095-3_21

1926 – Went discovered the first plant growth hormone, indole acetic acid.
1934 – White introduced vitamin B as a growth supplement in tissue culture media for tomato root tip.
1939 – Gautheret, White and Nobecourt established endless proliferation of callus cultures.
1941 – Overbeek was the first to add coconut milk for cell division in *Datura*.
1946 – Ball raised whole plants of *Lupinus* by shoot tip culture.
1955 – Skoog and Miller discovered kinetin as cell division hormone.
1957 – Skoog and Miller gave concept of hormonal control (auxin/cytokinin) of organ formation.
1959 – Reinert and Steward regenerated embryos from callus clumps and cell suspension of carrot (*Daucus carota*).
1960 – Cocking was first to isolate protoplast by enzymatic degradation of cell wall.
1962 – Murashige and Skoog developed MS medium with higher salt concentration.
1964 – Guha and Maheshwari produced first haploid plants from pollen grains of *Datura* (anther culture).
1966 – Steward demonstrated totipotency by regenerating carrot plants from single cells of tomato.
1970 – Power et al. successfully achieved protoplast fusion.
1971 – Takebe et al. regenerated first plants from protoplasts.
1972 – Carlson produced the first interspecific hybrid of *Nicotiana tabacum* by protoplast fusion.
1974 – Reinhard introduced biotransformation in plant tissue cultures.
1977 – Chilton et al. successfully integrated Ti plasmid DNA from *Agrobacterium tumefaciens* in plants.
1978 – Melchers et al. carried out somatic hybridization of tomato and potato resulting in pomato.
1981 – Larkin and Scowcroft introduced the term somaclonal variation.
1983 – Pelletier et al. conducted intergeneric cytoplasmic hybridization in radish and grape.
1984 – Horsh et al. developed transgenic tobacco by transformation with *Agrobacterium*.
1987 – Klien et al. developed biolistic gene transfer method for plant transformation.
2005 – Rice genome was sequenced under the International Rice Genome Sequencing Project.

A summary of applications of tissue culture in crop improvement is available in Fig. 21.1.

The culture medium is composed of macronutrients, micronutrients, vitamins, other organic components, plant growth regulators, carbon source and some gelling agents in case of solid medium. Murashige and Skoog medium (MS medium) is most extensively used for the vegetative propagation of many plant species in vitro. The pH of the media is vital that regulates both growth of plants and activity of plant growth regulators. pH is adjusted between 5.4 and 5.8. Both the solid and liquid media can be used for culturing. The composition of the medium, particularly plant

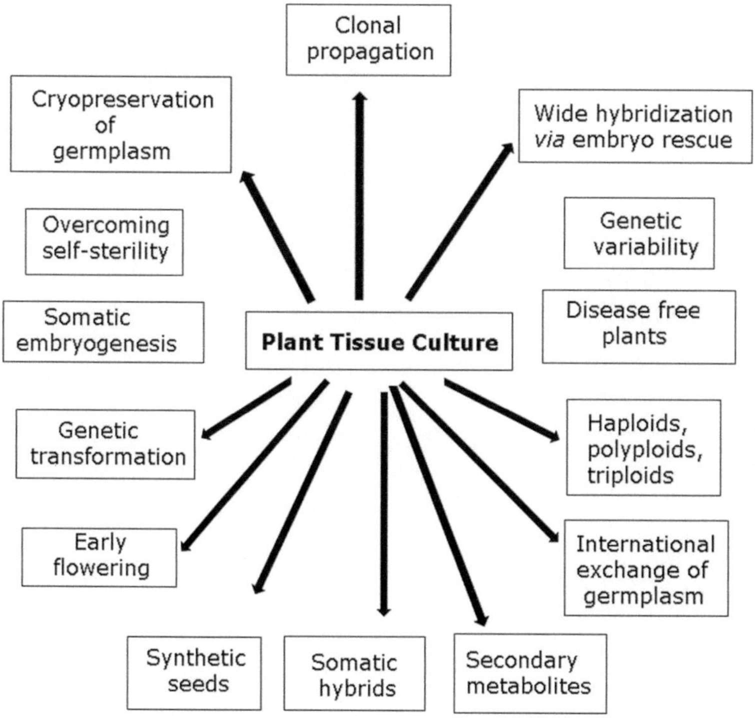

Fig. 21.1 Various facets of tissue culture

hormones and the nitrogen source, has profound effects on the response of the initial explant. Plant growth regulators (PGRs) play an essential role in determining the growth of cells and tissues in culture medium. Auxins, cytokinins and gibberellins are most commonly used plant growth regulators. The type and the concentration of hormones vary with the tissues and species. While high concentration of auxins favours root formation, high concentration of cytokinins promotes shoot regeneration. Development of mass of undifferentiated cells known as *callus* can be achieved with a balance of auxin and cytokinin.

21.2 Components of Tissue Culture Media

The composition of culture medium governs growth and morphogenesis of plant tissues. Several media formulations are commonly used for the majority of all cell and tissue culture work. These media formulations include those described by Murashige and Skoog, Gamborg's B5, Schenk and Hildebrandt, Nitsch and Nitsch and Lloyd and McCown (woody plant media). Murashige and Skoog's MS medium, Schenk and Hildebrand's SH medium and Gamborg's B5 medium are

Table 21.1 Components of various tissue culture media

Component	MS	B5	SH	N&N	WPM
Ammonium phosphate monobasic	–	–	300.00	–	–
Ammonium nitrate	1650	–	–	720.0	400.0
Ammonium sulphate	–	134.0	–		
Boric acid	6.2	3.0	5.00	10.0	6.2
Calcium nitrate	–	–	–	–	386.0
Calcium chloride.$2H_2O$	–	150.0	151.00	–	–
Calcium chloride, anhydrous	332.2	–	–	–	72.5
Cobalt chloride•$6H_2O$	0.025	0.025	0.10	–	–
Cupric sulphate•$5H_2O$	0.025	0.025	0.20	0.025	0.25
Na_2-EDTA	37.26	37.3	19.80	37.25	37.3
Sodium phosphate monobasic	–	130.42	–	–	–
Ferrous sulphate•$7H_2O$	27.8	27.8	–	27.85	27.85
Magnesium sulphate	180.7	122.09	195.05	90.34	180.7
Manganese sulphate•H_2O	16.9	10.0	10.0	18.94	22.3
Molybdic acid, sodium salt, $2H_2O$	0.25	0.25	0.10	0.25	0.25
Potassium iodide	0.83	0.75	1.00	–	–
Potassium nitrate	1900	2500.0	2500.0	950.0	–
Potassium sulphate	–	–	–	–	990.0
Potassium phosphate monobasic	170	–	–	68.0	170.0
Zinc sulphate•$7H_2O$	8.6	2.0	1.00	10.0	8.6
Myo-inositol	100.0	100.0	1000.0	100.0	100.0
Nicotinic acid	1.0	1.0	5.0	5.0	0.5
Pyridoxine HCl	1.0	1.0	0.5	0.50	0.5
Folic acid	–	–	–	0.50	–
Thiamine HCl	10.0	10.0	5.0	0.50	1.0
Glycine	–	–	–	2.0	2.0
Biotin	–	–	–	0.05	–

All ingredients mg/l; MS, Murashige and Skoog; B5, Gamborg's B5 medium; SH, Schenk and Hildebrandt; N&N, Nitsch and Nitsch; WPM, Lloyd and McCown (woody plant medium)

all high in macronutrients, while the other media formulations contain considerably less of the macronutrients (Table 21.1).

Macronutrients: Macronutrients provide six major elements: nitrogen (N), phosphorus (P), potassium (K), calcium (Ca), magnesium (Mg) and sulphur (S). The optimum concentration of each nutrient required varies with species to species. Culture media should contain at least 25–60 mM of inorganic nitrogen for adequate plant cell growth (see end of the chapter for calculation of molar, millimolar and micromolar solutions). Though cells can grow on nitrates alone, considerably better results are achieved when the medium is fortified with both a nitrate and ammonium nitrogen source. Certain species require ammonium or another source of reduced nitrogen for cell growth to occur. Nitrates are usually supplied in the range of

25–20 mM and ammonium between 2 and 20 mM. Potassium is required for cell growth of most plant species. Most media contain K, in the nitrate or chloride form, at concentrations of 20–30 mM. The optimum concentrations of P, Mg, S and Ca range from 1 to 3 mM when all other requirements for cell growth are satisfied. Calcium and magnesium slats are added at last to avoid precipitation of the media.

Micronutrients: The essential micronutrients for plant cell and tissue growth include iron (Fe), manganese (Mn), zinc (Zn), boron (B), copper (Cu) and molybdenum (Mo). Chelated forms of iron and zinc are commonly used in culture media. Iron may be the most critical of all but is difficult to dissolve and frequently precipitate after media are prepared. Murashige and Skoog used an ethylenediaminetetraacetic acid (EDTA)-iron chelate to bypass this problem. Cobalt (Co) and iodine (I) may also be added to certain media, but growth requirements for these elements have not been well understood. Sodium (Na) and chlorine (Cl) are also used in some media but are not essential for cell growth.

Carbon and Energy Source: The source of carbohydrate is sucrose, often substituted by glucose and fructose. Glucose is effective as sucrose and fructose are somewhat less effective. Other carbohydrates that have been tested include lactose, galactose, raffinose, maltose and starch. Sucrose ranges between 2% and 3%. Use of autoclaved fructose can be detrimental to cell growth. Carbohydrates must be supplied to the culture medium because cell lines are not fully autotropic, that is, capable of supplying their own carbohydrate needs by CO_2 assimilation during photosynthesis.

Vitamins: Vitamins are required by plants as catalysts in various metabolic processes. Some vitamins may become limiting factors for cell growth. Frequently used vitamins are thiamine (B_1), nicotinic acid, pyridoxine (B_6) and myo-inositol. Thiamine that is basically required by all cells is normally used between 0.1 and 10.0 mg/l. Nicotinic acid and pyridoxine are often added but are not essential. Nicotinic acid is normally used at concentrations of 0.1–5.0 mg/l; pyridoxine is used at 0.1–10.0 mg/l. Although myo-inositol is a carbohydrate not a vitamin, it stimulates growth in certain cell cultures. Though not essential, its presence in small quantities stimulates cell growth in most species and is used at a range of 50 to 5000 mg/l. Other vitamins such as biotin, folic acid, ascorbic acid, pantothenic acid, vitamin E, riboflavin and *p*-aminobenzoic acid have been included in some cell culture media. Effect of vitamins is generally negligible and is not considered growth-limiting factors.

Amino Acids or Other Nitrogen Supplements: Though cultured cells are capable of synthesizing amino acids, addition of certain amino acids or amino acid mixtures can be used to stimulate cell growth. Amino acids provide source of nitrogen that can be taken up by the cells more rapidly than inorganic nitrogen.

The most common sources of organic nitrogen used in culture media are amino acid mixtures (e.g. casein hydrolysate), L-glutamine, L-asparagine and adenine.

Casein hydrolysate is generally used at concentrations between 0.05% and 0.1%. Examples of amino acids included in culture media to enhance cell growth are glycine at 2 mg/l, glutamine up to 8 mM, asparagine at 100 mg/l, L-arginine and cysteine at 10 mg/l and L-tyrosine at 100 mg/l. Tyrosine has been used to stimulate morphogenesis in cell cultures but should only be used in an agar medium. Addition of adenine sulphate can greatly enhance shoot formation.

Undefined Organic Supplements: Addition of a wide variety of undefined organic extracts in the media often stimulates favourable tissue responses. Such supplements are protein hydrolysates, coconut milk, yeast extracts, malt extracts, ground banana, orange juice and tomato juice. However, supplements should only be used as a last resort. Only coconut milk and protein hydrolysates are used to an extent now. Protein (casein) hydrolysates are generally added to culture media at a concentration of 0.05–0.1%, while coconut milk is commonly used at 5–20% (v/v).

The addition of activated charcoal (AC) can absorb inhibitory compounds, absorption of growth regulators from the culture medium or darkening of the medium. The inhibition of growth in the presence of AC is generally attributed to the absorption of phytohormones to AC. 1-Naphthaleneacetic acid (NAA), kinetin, 6-benzylaminopurine (BA), indole-3-acetic acid (IAA) and 6-γ-γ-dimethylallylaminopurine (2iP) all bind to AC, with the latter two growth regulators binding quite rapidly. AC stimulates cell growth because of its ability to bind to toxic phenolic compounds in culture. Activated charcoal is generally acid-washed prior to addition to the culture medium at a concentration of 0.5–3.0%.

Solidifying Agents or Support Systems: Agar is the widely used gelling agent for semisolid and solid media. Agar is mixed with water which forms a gel that melts at approx. 60–100°C and solidifies at approximately 45°C. Agar gels are stable at all feasible incubation temperatures. Also, agar gels have no reaction with media constituents and are not digested by plant enzymes. The firmness of an agar gel is governed by the concentration and brand and also pH of the medium. Agar concentrations usually range between 0.5% and 1.0%.

Another gelling agent is Gelrite. This product is synthetic and is used at 1.25–2.5 g/l that gives a clear gel to detect contamination. Alternative supporting systems are perforated cellophane, filter paper bridges, filter paper wicks and polyurethane foam. The suitability of agar gel or other systems depends on the species.

Growth Regulators: Four broad classes of growth regulators are important for the culture media: the auxins, cytokinins, gibberellins and abscisic acid. Skoog and Miller were the first to report that the ratio of auxin to cytokinin determined the type and extent of organogenesis in plant cell cultures. Both an auxin and a cytokinin are usually added to culture media in order to obtain morphogenesis, although the ratio of hormones required for root and shoot induction is not universally the same.

Considerable variability exists among genera, species and even cultivars in the type and amount of auxin and cytokinin required for induction of morphogenesis.

The auxins commonly used in plant tissue culture media are 1H-indole-3-acetic acid (IAA), 1H-indole-3-butyric acid (IBA), 2,4-dichlorophenoxyacetic acid (2,4-D) and 1-naphthaleneacetic acid (NAA). The only naturally occurring auxin found in plant tissues is IAA. Other synthetic auxins that have been used in plant cell culture include 4-chlorophenoxyacetic acid or *p*-chlorophenoxyacetic acid (4-CPA, PCPA), 2,4,5-trichlorophenoxyacetic acid (2,4,5-T), 3,6-dichloro-2-methoxybenzoic acid (dicamba) and 4-amino-3,5,6-trichloropicolinic acid (picloram).

Various auxins differ in their physiological activity and in the extent to which they move through tissue and are bound to the cells or metabolized. Naturally occurring IAA has been shown to have less physiological activity than synthetic auxins. Based on stem curvature assays, 2,4-D has 8 to 12 times the activity, 2,4,5-T has 4 times the activity, PCPA and picloram have 2 to 4 times the activity, and NAA has 2 times the activity of IAA. Although 2,4-D, 2,4,5-T, *p*-chlorophenoxyacetic acid (PCPA) and picloram are often used to induce rapid cell proliferation, exposure to high levels or prolonged exposure to these auxins, particularly 2,4-D, results in suppressed morphogenic activity. Auxins are generally included in a culture medium to stimulate callus production and cell growth, to induce roots and to initiate somatic embryogenesis.

The cytokinins commonly used in the media include 6-benzylaminopurine or 6-benzyladenine (BAP, BA), 6-γ-γ-dimethylaminopurine (2iP), N-(2-furanylmethyl)-1H-puring-6-amine (kinetin) (kinetin is also known as 6-furfurylaminopurine) and 6-(4-hydroxy-3-methyl-trans-2-butenyl) aminopurine (zeatin). While zeatin and 2iP are considered naturally occurring, BAP and kinetin are synthetically derived. Adenine, another naturally occurring compound, has a base structure similar to that of cytokinins and has shown cytokinin-like activity in some cases. Many plant tissues demand absolute requirement for a specific cytokinin for morphogenesis. Some tissues are considered to be cytokinin independent.

Cytokinins are required to shoot formation and axillary shoot proliferation and to inhibit root formation. The type of morphogenesis depends upon the ratio and concentrations of auxins and cytokinins. Root initiation of plantlets, embryogenesis and callus initiation generally occur when the ratio of auxin to cytokinin is high, whereas adventitious and axillary shoot proliferation occur when the ratio is low.

Gibberellins (GA_3) and abscisic acid (ABA) are two other growth regulators occasionally used, and certain species require these hormones for enhanced growth. Generally, GA_3 is added to promote the growth of low-density cell cultures, to enhance callus growth and to elongate dwarfed or stunted plantlets. Depending on the species, abscisic acid is either to inhibit or stimulate callus or to manipulate callus growth.

Table 21.2 Material requirement for preparing one litre tissue culture medium

Two litre Erlenmeyer flask for one litre preparation	(Stirrer and stirbar, optional) balance for weighing out sucrose and agar
Distilled water (and squirt bottle or water dropper)	Droppers
One litre packet of pre-mixed medium (MS salts) (generally stored in refrigerator). Bring to room temperature before opening. Shake down well and cut opening cleanly and all the way across very close to the sealed edge with a scissors. The powder is very fine and somewhat hygroscopic so it sticks all over the inside of the foil-lined package	NaOH, HCl at 1 M each for adjusting pH
Sucrose (25 or 30 g)	pH paper (range 5–7) or pH meter calibrated to pH 4 and pH 7
Agar (7–8 g)	Large baggies for storing tube racks or sleeves of plates to keep them moist free

21.3 Preparing the Plant Tissue Culture Medium

For preparing one litre tissue culture medium, the materials needed are given in Table 21.2.

The procedure for preparing one litre medium is as follows:

1. Add about 800 ml of distilled water to the two litre flask. You need a two litre flask for one litre of medium to contain boil-overs that will occur during the sterilization process.
2. Add first the macroelements, microelements, etc. one by one (if one prefers to make media himself). Add calcium and magnesium at last only to avoid precipitation.
3. Add sucrose and swirl or stir to dissolve the sucrose completely.
4. Check the pH (*do not* add agar until you adjust the pH). The pH will be around 5–5.5. Though plants generally like acid soil, this pH is too low for the agar to gel.
5. Adjust the pH to 5.7. *Note*: Add the base or acid in small portions (about 1 ml per dose).
6. Add distilled water to one litre line on the Erlenmeyer flask.
7. Add the agar (7–8 g/l). The agar will *not* dissolve.
8. Cover with two layers of aluminium foil and put a piece of autoclave tape on the label area of the flask. Autoclave for 15 min at 15 psi (standard autoclave conditions). If available, use slow exhaust.
9. If you are using glass tubes with Magenta-brand or Kimax brand plastic closures, rinse the tubes out with distilled water (OK to have a tiny bit of water residue in the tubes) and autoclave with the culture medium (with caps ON of course). Avoid using disposable 50 ml centrifuge tubes or plastic petri plates. *Do not* autoclave these for they will melt and smell terrible inside the autoclave.

10. Cool to about 60 °C.
11. Aseptically, pipette or pour the warm liquid medium into the sterile plastic or sterilized glass culture vessels in a hood. The gel will set in about 1 h.

Auxins, kinetins and gibberellins are the main types of plant hormones; one stimulates roots, another stimulates shoots, and gibberellins stimulate internode growth. Generally these hormones can be added *before* the medium is sterilized. Usually stock solutions that are stored frozen are used. Generally 1 mg/ml or 10 mg/ml stocks work fine, since most of the hormones are needed in very low concentrations, like 1 mg/l. Making up the stock solution varies for each hormone. Some have to be dissolved in a very concentrated way using acid or base and then brought to volume with distilled water.

21.4 Transfer of Plant Material to Tissue Culture Medium

Use the sterile gloves and equipment for all of the following steps:

1. Place the plant material in the Clorox bleach in a sterilized container (period of sterilization varies with plant material). The containers of sterile water, sterilized forceps and blades, some sterile paper towel to use as a cutting surface and enough tubes containing sterile medium are to be kept into the laminar air flow that gives sterile air flow. The outside surfaces of the containers, the capped tubes and the aluminium wrapped supplies should be briefly sprayed with 70% alcohol before moving them into the chamber.
2. The gloves can be sprayed with a 70% alcohol solution for sterilization. Once this is done, one may not touch anything that is outside of the sterile chamber.
3. Carefully open the container containing the plant material and pour in enough sterile water to half fill the container. Replace the lid and gently shake the container to wash tissue pieces (explants) thoroughly for 2–3 min to remove the bleach. Pour off the water and repeat the washing process three more times.
4. Remove the sterilized plant material from the sterile water; place it on the paper towel or a sterile petri dish. Cut the plant tissue into smaller pieces to about 2 to 3 mm. If using rose, cut a piece of stem about 10 mm in length with an attached bud. Any pale-coloured tissue damaged by the bleach shall be avoided.
5. Take a prepared section of plant material with sterile forceps and place onto the medium in the polycarbonate/glass tube.
6. Replace the cap tightly on the tube and preferably seal it.

21.5 Micropropagation

Micropropagation has become an important part of commercial multiplication of many species. Several techniques for in vitro plant propagation have been devised, including the induction of axillary and adventitious shoots, the culture of isolated meristems and plant regeneration by organogenesis and/or somatic embryogenesis. Using axillary and apical meristems, plants can be regenerated. Adventitious buds and shoots are formed de novo; meristems are initiated from explants, such as those of leaves, petioles, hypocotyls, floral organs and roots. The following are the stages of micropropagation:

Stage I: Establishment of axenic cultures – introduction of the surface-disinfected explants into culture, followed by initiation of shoot growth. For this, apical and axillary buds, adventitious meristems, leaves, bulb scales, flower stems or cotyledons shall be used. Usually 4–6 weeks are required to complete this stage or even 12 months in some woody species. A culture is stabilized when explants produce a constant number of normal shoots after subculture.

*Stage II: S*hoot proliferation and multiple shoot production. Each explant has expanded into a cluster of small shoots. Multiple shoots are transplanted to new culture medium. Shoots are subcultured every 2–8 weeks. To maximize the quantity of shoots, subcultures may be done.

Stage III: Root formation – special root induction media are used (auxin enriched) for root induction. This stage may involve not only rooting of shoots but also conditioning of plantlets to increase their potential for acclimatization.

Stage IV: Acclimatization – transfer of regenerated plants to soil under natural environmental conditions. Plants transferred from in vitro to ex vitro conditions undergo gradual modification of leaf anatomy and morphology, and their stomata begin to function (the stomata are usually open when the plants are in culture). Plants also form a protective epicuticular wax layer over leaf surface. Only gradually the regenerated plants become adapted to new environment (Fig. 21.2).

21.6 Protoplast Culture

Protoplast is the entire cell minus cellulosic cell wall. Though the culture of protoplasts started during the 1970s, only by the 1990s, protoplast-based technologies were used for *Agrobacterium* and biolistics-mediated gene delivery to plants. Use of hypertonic solutions makes plasma membranes of cells contract from their walls. Subsequent removal of the cell wall releases large populations of spherical, osmotically fragile protoplasts (naked cells). Viable protoplasts are potentially totipotent (totipotency is the ability of a single cell to divide and produce all of the differentiated cells in an organism). Cellulase enzymes digest the cellulose in plant cell walls, while pectinase enzymes break down the pectin holding cells together. In 1960, E.C. Cocking demonstrated the feasibility of enzymatic degradation of plant cell walls to obtain large quantities of protoplasts.

Digestion of cell wall is usually carried out after incubation in an osmoticum (a solution of higher concentration than the cell contents which causes the cells to plasmolyse). This makes the cell walls easier to digest. Debris is filtered and/or centrifuged out of the suspension and the protoplasts are then centrifuged to form a pellet. On re-suspension, the protoplasts can be cultured on media which induce cell

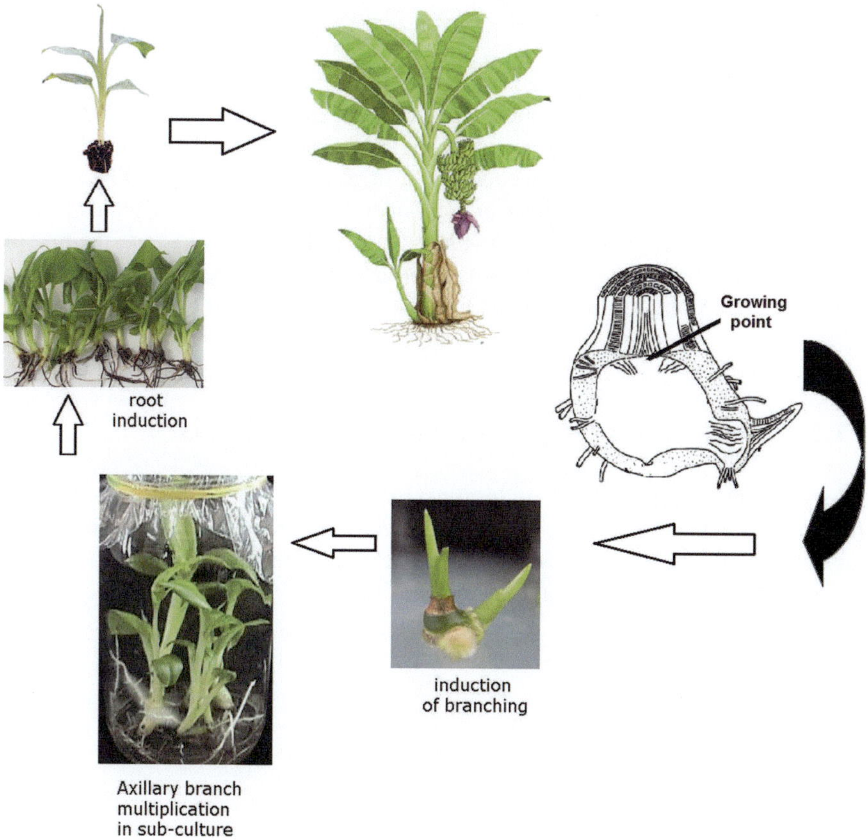

Fig. 21.2 A scheme for micropropagation of banana (diagrammatic)

division and differentiation. A large number of plants can be regenerated from a single experiment. For example, a gram of potato leaf tissue can produce more than a million protoplasts.

Protoplasts can be isolated from a range of plant tissues: leaves, stems, roots, flowers, anthers and even pollen. Protoplasts are used in a variety of ways like electroporation, incubation with bacteria, heat shock and high pH treatment to induce them to take up DNA. The protoplasts can then be cultured and plants regenerated. In this way, genetically engineered plants can be produced more easily than is possible using intact cells/plants.

Plants from distantly related or unrelated species are unable to reproduce sexually because of incompatibility. Protoplasts of unrelated species can be fused to produce cybrids combining desirable characteristics like disease resistance, good flavour and cold tolerance. Fusion is carried out through application of electric current or by treatment with chemicals like polyethylene glycol (PEG). Fusion products can be selected media containing antibiotics or herbicides. These can then be induced to form whole plant that can be tested for desirable traits.

21.7 Anther Culture

Haploid plants are with gametic or *n* number of chromosomes. Doubled haploids, or dihaploids, are chromosome doubled haploids or 2n plants. Androgenesis is the process by which haploid plants can develop from male gametophyte. The ability to produce haploid plants is a tremendous asset in genetic and plant breeding studies. Doubling the chromosome number of haploids to produce doubled haploids results in completely homozygous plants. In 1964, Guha and Maheshwari were the first to produce haploid plants by placing immature anthers of *Datura innoxia* Mill. into culture. To date, androgenic haploids have been produced in over 170 species.

21.8 Somatic Embryogenesis and Synthetic Seeds

This is an artificial process (done in vitro) by which a plant or embryo is derived from a single somatic cell or group of somatic cells that are not normally destined for the development of embryos. No endosperm or seed coat is formed around a somatic embryo. Applications of this process include:

Clonal propagation of genetically uniform plant material
Elimination of viruses
Provision of source tissue for genetic transformation
Generation of whole plants from single cells (protoplasts)
Development of synthetic seed technology

Cells from the source tissue are cultured to form an undifferentiated mass of cells called a callus (Fig. 21.3). The main PGRs used are auxins but can contain small amount of cytokinins. Shoots and roots are monopolar, while somatic embryos are bipolar, allowing them to form a whole plant without culturing on multiple media types. The first documentation of somatic embryogenesis was by Steward and colleagues in 1958 and Reinert in 1959 with carrot cell suspension cultures.

Somatic embryogenesis can occur directly or indirectly. Direct embryogenesis occurs when embryos originate directly from the explant creating an identical clone. Indirect embryogenesis occurs when explants produce undifferentiated, or partially differentiated, callus cells from where somatic embryos originate. Factors and mechanisms controlling cell differentiation in somatic embryos are unclear. Various polysaccharides, amino acids, growth regulators, vitamins, low molecular weight compounds and polypeptides are responsible for somatic embryogenesis. Several signalling molecules known to influence or control the formation of somatic embryos have been found and include extracellular proteins, arabinogalactan proteins (AGPs = family of extensively glycosylated hydroxyproline-rich glycoproteins that influence plant growth and development) and Lipochitin oligosaccharides (LCOs = signaling molecules required by ecologically and agronomically important bacteria and fungi to establish symbioses with diverse

Fig. 21.3 Somatic embryogenesis. (**a**) Callus culture with somatic embryos, (**b**) induction of somatic embryogenesis, (**c**) bilobed somatic embryos developing and (**d**) growing somatic embryo (figure representative)

Fig. 21.4 Synthetic seeds

land plants). Temperature and lighting can also affect the maturation of the somatic embryo (Fig. 21.4).

Artificial seeds, otherwise known as "synseeds" and "synthetic seeds", were described by Murashige in 1977. He defined artificial seeds as "an encapsulated single somatic embryo". Redenbaugh and colleagues in 1986 were the first to produce synthetic seeds encapsulating somatic embryos. Artificial seeds are confined to those species in which somatic embryos could be produced. In addition to somatic embryos, other vegetative parts like shoot buds, cell aggregates, axillary buds or any other micropropagules could also be encapsulated. This is only possible if they own the capacity to be sown as a seed and converted into a plant under in vitro or ex vitro conditions. Artificial seeds offer the exclusion of acclimatization step needed in micropropagation that gives breeders greater flexibility. Tissues used for artificial seed production are somatic embryos, shoot tips, axillary buds, nodal segments, protocorm-like bodies (PLBs), microshoots and embryogenic calluses.

Two types of artificial seeds (encapsulated somatic embryos) are commonly produced: desiccated and hydrated. Desiccated artificial seeds are derived through encapsulation in polyoxyethylene glycol followed by desiccation. Desiccation can be done by leaving artificial seeds in unsealed petri dishes overnight to dry, or they can be passed through slowly over a more controlled period of reducing relative humidity. This is possible where somatic embryos are desiccation-tolerant. Induction of desiccation tolerance can be done using a high osmotic potential of the maturation medium. The osmotic potential could be increased with mannitol, sucrose, etc. Hydrated artificial seeds are made by encapsulating somatic embryos in hydrogel capsules. Encapsulation provides protection and also assists in converting the in vitro micropropagules into "artificial seeds" or "synseeds".

Alginate matrix was discovered to be the optimal encapsulation for artificial seed production because of its sensible thickness, weak spinnability of solution, low toxicity of microorganism, low expense, bio-suitability characteristics and fast gelation. The major principle for alginate encapsulation formation depends on the exchange of ions between Na^+ in sodium alginate and Ca^+ in $CaCl_2$ $2H_2O$, which happens when sodium alginate droplets involving the artificial embryos or any other plant propagule are dropped into the $CaCl_2$ $2H_2O$ solution, producing stable explant beads. The solidity and rigidity of the capsule (explant beads) depends upon the two gelling agents' (sodium alginate and $CaCl_2$ $2H_2O$) concentrations and mixing duration. Nutrients and growth regulators are required to be added to the artificial endosperm that are essential for embryo survival.

21.9 Plant Tissue Culture Terminology

Adventitious – Developing from unusual points of origin, such as shoot or root tissues, from callus or embryos, from sources other than zygotes.

Agar – A polysaccharide powder derived from algae used to gel a medium. Agar is generally used at a concentration of 6–12 g/l.

Aseptic – Free of microorganisms.

Aseptic technique – Procedures used to prevent the introduction of fungi, bacteria, viruses, mycoplasma or other microorganisms into cultures.

Autoclave – A machine capable of sterilizing wet or dry items with steam under pressure. Pressure cooker is a type of autoclave.

Auxin – A group of plant growth regulators that promotes callus growth, cell division, cell enlargement, adventitious buds and lateral rooting. Endogenous auxins are auxins that occur naturally. Indole-3-acetic (IAA) is a naturally occurring auxin. Exogenous auxins are auxins that are man-made or synthetic. Examples of exogenous auxins include 2,4-dichlorophenoxyacetic acid (2,4-D), indole-3-butyric acid (IBA), α-naphthaleneacetic acid (NAA) and 4-chlorophenoxyacetic acid (CPA).

Callus – An unorganized, proliferate mass of differentiated plant cells, a wound response.

Chemically defined medium – A nutritive solution for culturing cells in which each component is specifiable and ideally of known chemical structure.

Clone – Plants produced asexually from a single source plant.

Clonal propagation – Asexual reproduction of plants that are considered to be genetically uniform and originated from a single individual or explant.

Contamination – Being infested with unwanted microorganisms such as bacteria or fungi.

Cytokinin – A group of plant growth regulators that regulate growth and morphogenesis and stimulate cell division. Endogenous cytokinins, cytokinins that occur naturally, include zeatin and 6-γ,γ-dimethylallylaminopurine (2iP). Exogenous cytokinins, cytokinins that are man-made or synthetic, include 6-furfurylaminopurine (kinetin) and 6-benzylaminopurine (BA or BAP).

Explant – Tissue taken from its original site and transferred to an artificial medium for growth or maintenance.

Gibberellins – A plant growth regulator that influences cell enlargement. Endogenous growth forms of gibberellin include gibberellic acid (GA_3).

Horizontal laminar flow unit – An enclosed work area that has sterile air moving across it. The air moves with uniform velocity along parallel flow lines. Room air is pulled into the unit and forced through a HEPA (high-energy particulate air) filter, which removes particles 0.3 μm and larger.

Hormones – Growth regulators, generally synthetic in occurrence, that strongly affect growth (i.e. cytokinins, auxins and gibberellins).

Internode – The space between two nodes on a stem.

In vitro – To be grown in glass (Latin); propagation of plants in a controlled, artificial environment using plastic or glass culture vessels, in a defined growing medium.

In vivo – To be grown naturally (Latin).

Medium – A nutritive solution, solid or liquid, for culturing cells.

Micropropagation – In vitro clonal propagation of plants from shoot tips or nodal explants, usually with an accelerated proliferation of shoots during subcultures.

Node – A part of the plant stem from which a leaf, shoot or flower originates.

Passage – The transfer or transplantation of cells or tissues with or without dilution or division, from one culture vessel to another.

Pathogen – A disease-causing organism.

Pathogenic – Capable of causing a disease.

Petiole – A leaf stalk; the portion of the plant that attaches the leaf blade to the node of the stem.

Plant tissue culture – The growth or maintenance of plant cells, tissues, organs or whole plants in vitro.

Regeneration – In plant cultures, a morphogenetic response to a stimulus that results in the products of organs, embryos or whole plants.

Shoot apical meristem – Undifferentiated tissue, located within the shoot tip, generally appearing as a shiny dome-like structure, distal to the youngest leaf primordium and measuring less than 0.1 mm in length when excised.

Somaclonal variation – Phenotypic variation, either genetic or epigenetic in origin, displayed among somaclones.

Somaclones – Plants derived from any form of cell culture involving the use of somatic plant cells.

Sterile – (a) Without life. (b) Inability of an organism to produce functional gametes. (c) A culture that is free of viable microorganisms.

Sterile techniques – The practice of working with cultures in an environment free from microorganisms.

Subculture – With plant cultures. This is the process by which the tissue or explant is first subdivided then transferred into fresh culture medium.

Tissue culture – The maintenance or growth of tissue, in vitro, in a way that may allow differentiation and preservation of their function.

Totipotency – A cell characteristic in which the potential for forming all the cell types in the adult organism are retained.

Undifferentiated – With plant cells, existing in a state of cell development characterized by isodiametric cell shape, very little or no vacuole, a large nucleus and exemplified by cells comprising an apical meristem or embryo.

Molar Solutions

One **molar** (1 M) solution contains one **mole** of solute per litre of solution.

One **millimolar** (1 mM) solution contains one **millimole** of solute per litre of solution.

One **micromolar** (1 μM) solution contains one **micromole** of solute per litre of solution.

How to Prepare a Molar Solution?

A 1 molar solution (1 M) contains 1 mole of solute dissolved in a solution totalling 1 L. If you use water as the solvent, it must be distilled and deionized. Do not use tap water. A mole is the molecular weight (MW) expressed in grams (sometimes referred to as the "gram molecular weight" (gMW) of a chemical). Thus, 1 M = 1 gMW of solute per litre of solution.

To prepare 1 molar sodium chloride, we calculate the molecular weight (MW) of sodium chloride. Checking the Periodic Table of Elements, we find that the atomic weight of sodium (Na) is 23 and the atomic weight of chlorine (Cl) is 35.5. Therefore, the molecular weight of sodium chloride (NaCl) is: Na (23) + Cl (35.5) = 58.5 g/mole. To make a MS aqueous solution of NaCl, dissolve 58.5 g of NaCl in some distilled deionized water (the exact amount of water is unimportant; just add enough water to the flask so that the NaCl dissolves). Then add more water to the flask until it totals 1 L to have 1 molar solution.

How to Prepare a 70-mM (Millimolar) Sucrose Solution?
The molecular weight of sucrose can be determined from its chemical formula, namely, $C_{12}H_{22}O_{11}$, and the atomic weights of carbon, hydrogen and oxygen. The formula weight for sucrose is identical to its molecular weight, namely, 342.3 grams per mole. A 1-M solution would consist of 342.3 g sucrose in 1-L final volume. A concentration of 70 mM is the same as 0.07 moles per litre. Take 0.07 moles/l times 342.3 grams per mole and you have 23.96 grams needed per litre (i.e. 342.3 $\times$ 0.07) to make 70-mM sucrose solution.

Further Reading

Anis M, Ahmad N (2016) Plant tissue culture: propagation, conservation and crop improvement. Springer, Singapore

Bhojwani SS, Grover A (1996) Tissue culture a novel source of genetic variations. Botanica 46:1–6

Davey MR, Anthony P (2010) Plant cell culture: essential methods. Wiley-Blackwell, Hoboken

Dodds J (2004) Experiments in plant tissue culture. Cambridge University Press, Cambridge

George EF (1993) Plant propagation by tissue culture. In: Part 1, The Technology. Edington, Exegetics Ltd

Gray DJ, Purohit A, Triglano RN (1991) Somatic embryogenesis and development of synthetic seed technology. Crit Rev Plant Sci 10:33–61

Iliev et al. (2010) Plant micropropagation. In: Davey and Anthony P (eds.) Plant cell culture. Wiley

Kyte et al (2013) Plants from test tubes: an introduction to micropropagation. Timber press, Portland

Murashige T (1974) Plant propagation through tissue culture. Ann Rev Plant Physiol 25:135–166

Onishi N, Sakamoto Y, Hirosawa T (1994) Synthetic seeds as an application of mass-production of somatic embryos. Plant Cell Tissue Organ Cult 39:137–145

Redenbaugh K, Fujii JA, Slade D (1988) Encapsulated plant embryos. In: Mizrahi A (ed) Advances in biotechnological processes. Alan R. Liss Inc., New York

Rihan HZ et al (2017) Artificial seeds (Principle, aspects and applications). Agronomy 7:71. https://doi.org/10.3390/agronomy7040071

Sathyanarayana BN (2007) Plant tissue culture: practices and new experimental protocols. New Delhi, I. K. International

Shahzad A et al (2017) Historical perspective and basic principles of plant tissue culture. In: Plant biotechnology: principles and applications. Springer, pp 1–36

Smith R (2012) Plant tissue culture 3rd Edn. Techniques and experiments. Elsevier, Amsterdam

Trigiano RN, Gray DJ (2010) Plant tissue culture, development, and biotechnology. Taylor and Francis. https://doi.org/10.1201/9781439896143

Genetic Engineering

22

Keywords

Restriction Endonucleases · Techniques for Producing Transgenic Plants · Engineering Insect Resistance · Engineering Herbicide Tolerance · Site-Directed Nucleases · What and Why CRISPR?

Manipulating the genetic material of an organism as per the will of man is genetic engineering. Such manipulated organisms are genetically modified organisms (GMOs). One definition of GMO is *an organism whose genetic material has been modified in a way that is not made possible by nature*. Another acceptable definition is *artificial modification of an organism's genetic composition.* Such modifications are carried out through transfer of a gene taken from cells of another donor organism. Genes transferred are known as *transgenes*. Creation of genetically modified organisms requires *recombinant DNA*. Recombinant DNA is a combination of DNA from different organisms or different locations in a given genome that would not normally be found in nature. Recombinant DNA technology was first achieved in 1973 by Herbert Boyer of the University of California at San Francisco and Stanley Cohan of Stanford University who used *E. coli* restriction enzymes to insert foreign DNA into plasmids. Paul Berg of Stanford University invented assembling of recombinant molecule containing DNA from different organisms during 1971.

Genetic engineering offers the facility of introducing new traits like increased crop yields, secondary traits and nutritional quality. For example, herbicide-tolerant crops achieved through genetic engineering are capable surviving herbicides that allow farmers to spray herbicides without affecting yield. Similarly, GMOs producing insecticidal toxins resist attacks from insects. In this way, the process becomes cost-effective, reducing the use of synthetic insecticides. In the nutritional front, "golden rice" is engineered to produce beta-carotene.

The new traits expressed in such transgenic plants are derived from a variety of other organisms. Scientists have given a gene from the bacterium *Salmonella* to

P. M. Priyadarshan, *PLANT BREEDING: Classical to Modern*,
https://doi.org/10.1007/978-981-13-7095-3_22

cultivars of soybeans, corn, canola and cotton to degrade the herbicide glyphosate (Roundup™). Similarly, gene for insecticidal toxin from *Bacillus thuringiensis* (Bt) is introduced into cotton, potato and corn.

The derivation of golden rice was achieved through introduction of several genes for multi-step biochemical pathways. Rice is staple food for much of the world and lacks vitamin A. An estimated 100 million to 200 million children worldwide have vitamin A deficiency, a condition that causes blindness and increases susceptibility to diarrhoea, respiratory infection and childhood diseases like measles. Beta-carotene and other carotenes (the red, yellow and orange pigments found in carrots and other vegetables) are the precursor of vitamin A. Rice synthesizes beta-carotene in its chloroplasts but not in the edible seed tissue. Ingo Potrykus and his colleagues of ETH (Swiss Federal Institute of Technology), Zürich, found that geranyl geranyl diphosphate (GGPP), a precursor to carotenoid production, is present in rice seed. They genetically engineered golden rice to express the enzymes necessary for the conversion of GGPP to beta-carotene. Beta-carotene synthesis from geranyl geranyl diphosphate needs four biochemical reactions, and each reaction is catalysed by a different enzyme. The bacterium, *Agrobacterium tumefaciens*, containing three plasmids, was used to introduce all the genes necessary for the complete biochemical pathway for beta-carotene production. USFDA (US Food and Drug Administration) approved golden rice in 2018.

Early activities in genetic engineering were dominated by start-ups in the USA like Cetus Madison (Agracetus), Agrigenetics, Calgene, Advanced Genetic Systems, Molecular Genetics and others, as well as Plant Genetic Systems in Belgium and a number of larger, more-established agrochemical companies such as Monsanto, DuPont, Lilly, Zeneca, Sandoz, Pioneer, Bayer, etc. Genetic engineering is now dominated by a handful of big companies.

22.1 Restriction Endonucleases

It is reasonable to believe that genetic engineering was born in the early 1970s, with the popular discovery of restriction endonucleases – molecular scissors to cut DNA. Paul Berg in 1972 presented first studies on cloning, with which he used first restriction enzymes extracted from the bacterium *E. coli* known as *Eco* RI. Paul Berg and his colleagues combined the *E. coli* genome with the genes of a bacteriophage and the SV40 virus that gave way to new science - genetic engineering. Bacteria use such enzymes to neutralize parasitic bacteriophages. They cleave the sugar-phosphate backbone of DNA strands. In most practical settings, a given enzyme cuts both strands of duplex DNA within a stretch of just a few bases. These enzymes have specific recognition sites. Depending on their molecular structure, these enzymes fall in one of the three classes. Class I endonucleases have a molecular weight of around 300,000 Daltons, are composed of non-identical subunits and require Mg^{2+}, ATP (adenosine triphosphate) and SAM (S-adenosyl methionine) as cofactors for activity. Class II enzymes are much smaller, with molecular weights in the range of 20,000 to 100,000 Daltons. They have identical subunits and require only Mg^{2+} as a cofactor.

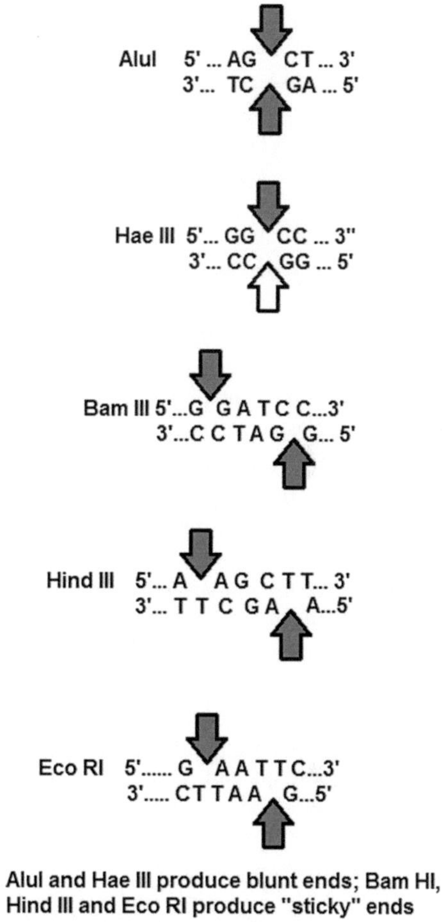

Fig. 22.1 Restriction enzymes and their cleavage sites

The Class III enzyme is a large molecule, with a molecular weight of around 200,000 Daltons, composed of non-identical subunits. These enzymes differ from enzymes of the other two classes. They require both Mg^{2+} and ATP but not SAM as cofactors. Class III endonucleases are the rarest of the three.

As an example, *Bam*HI searches for the sequence GGATCC in double-stranded DNA. When the sequence is located, the enzyme Bam*HI* digests the phosphodiester backbone in two specific places – between the pair of G nucleotides on each strand. That leaves us with a four-nucleotide single-stranded 5′ end on each side after separation as follows:

5′-ACAGGATAGGAGTCA**G** **GATCC**AGAGGACCTAGGATACCTC-3′
3′-GTCCTATCCTCAGT**CCTAG** **G**TCTCCTGGATCCTATGGAG-5′.

The specificity of other endonucleases is available in Fig. 22.1. Restriction recognition sites can be unambiguous or ambiguous. *Bam*HI recognizes the sequence G GATCC and no other endonuclease recognizes this sequence. This is what is meant by unambiguous. In contrast, *HinfI* recognizes a 5-bp sequence starting with GA, ending in TC and having any base in between. *HinfI* has an ambiguous recognition site. Similarly, *Xho*II will recognize and cut sequences of AGATCT, AGATCC, GGATCT and GGATCC. These enzymes are ambiguous.

22.2 Techniques for Producing Transgenic Plants

How a plant can take up a gene? Researchers working with rice often use the soil bacterium *Agrobacterium tumefaciens*. This bacterium, which causes crown gall disease in many fruit plants, is well known for its ability to infect plants with a tumour-inducing (Ti) plasmid. A section of the Ti plasmid, called T-DNA, integrates into chromosomes of the plant. *Recombinant DNA* can be added to the T-DNA through restriction endonuclease "cutting" of DNA and ligation of DNA with DNA ligase, and the T-DNA gets introduced into the chromosomes of a plant, thus leading to transfer of novel genes (Fig. 22.2).

All species are not susceptible to *Agrobacterium tumefaciens*. Researchers interested in modifying wheat and corn have practised other methods for delivering genes to plant cells. One approach is to use a "gene gun", or "microprojectile bombardment" or "biolistic gun", which fires plastic bullets filled with DNA-coated metallic pellets. An explosive blast or burst of gas propels the bullet towards a stop plate. The DNA-coated pellets are directed through an aperture in the stop plate and then penetrate the walls and membranes of their cellular targets. If projectiles penetrate

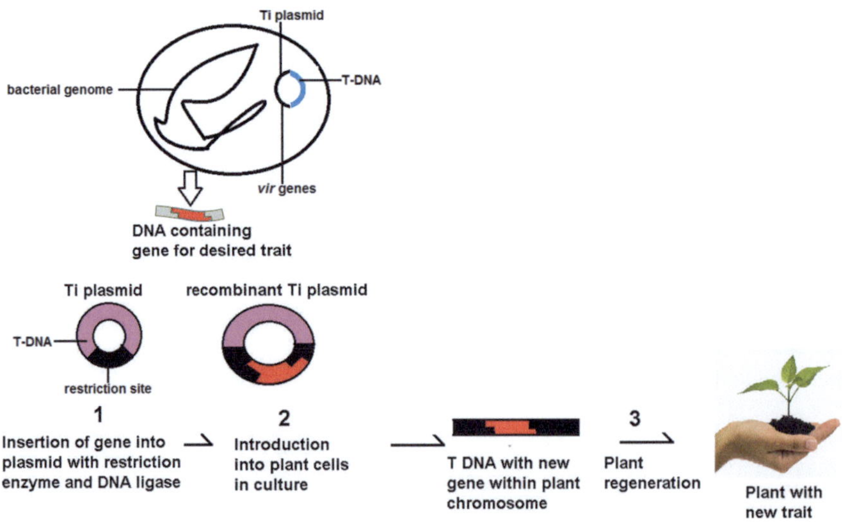

Fig. 22.2 *Agrobacterium*-mediated genetic transformation

the nuclei of cells, the introduced DNA integrates into the DNA of the plant genome. Transformed cells can then be cultured in vitro to raise whole plants.

Marker genes are included in DNA constructs so that the insertion of novel DNA can be identified and selected. When marker genes for herbicide resistance are included, plants that grow in the presence of the herbicide are assumed to possess the transgene of interest. All genes need not express in every tissue. As an example, derivation of golden rice ensured that the novel genes are expressed in the endosperm. It is necessary to introduce regulatory DNA sequences of the novel genes into the recombinant Ti plasmid in order to ensure expression of the introduced gene.

22.2.1 Engineering Insect Resistance

Insects damage agricultural crops that incur significant losses every year. Over 35% of the current global cotton production would be lost in the absence of insect control measures. However, insecticides used every year lead to production of resistant races of insects over time. Obviously, this situation forces farmers to use higher doses of insecticides which increases the costs and poses an environmental threat (see Box 22.1). Genetic engineering that can produce insecticides in plants can reduce use of insecticides. Genes for the production of insecticides derived from *Bacillus thuringiensis* (Bt for short), another common soil bacterium, have been used to introduce insect resistance in plants.

Box 22.1 Bt Cotton

Cotton, like any other monoculture crop demands intensive use of pesticides as pests incur extensive damage. Many pests have developed resistant races to pesticides over the past 40 years. Only successful approach to engineering crops for insect tolerance has been the addition of Bt toxin. Bt crops causes much less damage to the environment (no hazard to mammals and fish). Bt crops are now commercially available in corn, cotton and potato.

The Bt gene was isolated from a bacterium *Bacillus thuringiensis* and transferred to American cotton. The American cotton was subsequently crossed with Indian cotton to introduce the gene into native varieties. The Bt gene introduced genetically into the cotton seeds protects the plants from bollworm (*Helicoverpa armigera* of Lepidoptera), a major pest of cotton. The worm feeding on the leaves of a Bt cotton plant becomes lethargic and sleepy, thereby causing less damage to the plant.

Field trials demonstrated that Bt variety yielded 25–75% more cotton than normal variety. Also, Bt cotton demands only two sprays of chemical pesticide against eight sprays for normal variety. Data from the Indian Council of Agricultural Research India show that India uses about half of its pesticides on cotton to fight the bollworm menace. Bt cotton was created through the

(continued)

Box 22.1 (continued)

addition of genes encoding toxin crystals in the Cry group of endotoxin that can cause death of insect cells. Bt cotton was first approved for field trials in the USA in 1993, and approval for commercial use came in 1995. Bt cotton was approved by the Chinese government in 1997. In 2002, a joint venture between Monsanto and Mahyco introduced Bt cotton to India.

In 2011, India grew the largest GM cotton crop in over 10.6 million hectares. The US GM cotton crop was 4.0 million hectares, the second largest area in the world, followed by China with 3.9 million hectares and Pakistan with 2.6 million hectares. By 2014, 96% of cotton grown in the USA was genetically modified and 95% of cotton grown in India was GM. India is the largest producer of cotton, and GM cotton, as of 2014. The Punjab Agricultural University has developed the first genetically modified Bt cotton seeds that can be reused, resulting in saving of input cost to farmers. The new cotton variety is among few others identified by the Indian Council of Agricultural Research (ICAR) for cultivation in north region. The three Bt cotton varieties include PAU Bt 1, F1861 and RS 2013.

Bacillus thuringiensis subspecies *kurstaki* produces a toxin that kills the larvae of Lepidoptera (i.e. moths and butterflies) and a toxin from the subspecies *israelensis* is effective against Diptera such as mosquitoes and blackflies. Spore preparations derived from *Bacillus thuringiensis* have been used by organic farmers as an insecticide for several decades. When the target insect ingests the *Bt* spore, the protein crystal dissociates into several identical subunits. These subunits are a *protoxin*, i.e. a precursor of the active toxin. Under the alkaline conditions of the insect's gut, digestive enzymes (proteases) unique to the insect break down the protoxin to release the active toxin. The toxin molecules insert themselves into the membrane of the gut epithelial cells, setting in motion a series of processes that eventually stop the entire cell's metabolic activity. The insect stops feeding, becomes dehydrated and eventually dies. Several crops like tobacco, tomato, potato, cotton and maize are modified with Bt genes.

22.2.2 Engineering Herbicide Tolerance

A crop can be made tolerant to herbicide by inserting a gene that causes plants to become unresponsive to the toxic chemical. The herbicide glyphosate (also known as Roundup™) is the world's largest-selling herbicide. It is a broad-spectrum herbicide that kills a wide variety of monocot and dicot weeds. Roundup is transported downwards in plants and so has the advantage of killing the roots of perennial weeds.

Glyphosate inhibits EPSP synthase, an enzyme that is involved in the shikimic acid pathway. The enzyme catalyses the conversion of 3-phosphoshikimate to the compound EPSP (5-enolpyruvylshikimate-3-phosphate). EPSP is converted, via a series of biochemical reactions, into essential aromatic amino acids like phenylalanine, tyrosine and tryptophan. Glyphosate acts by binding with EPSP synthase and, in doing so, prevents the enzyme from catalysing the reaction. If the shikimic acid pathway is blocked in this way, the plant is deprived of these essential amino acids and cannot make the proteins it requires. The plant weakens and eventually dies (see Box 22.2 for a comprehensive list of GM crops).

Box 22.2 GM Crops

Maize

DK404SR is a cyclohexanedione herbicide-tolerant maize (under licence from BASF Inc.).

Star link is a Cry9c Bt corn produced by Plant Genetic Systems (now Aventis CropScience). Cry9c is a *Bacillus thuringiensis* protein.

MON 802 is an insect-resistant maize (under licence from Monsanto Co.). This was developed as tolerant to glyphosate herbicide and protects the plant from the European corn borer (*Ostrinia nubilalis*).

Rice

The first two GM rice varieties (with herbicide resistance), called **LLRice60** and **LLRice62**, that were produced by Bayer Crop science were approved in the USA in 2000. These were approved in Canada, Australia, Mexico and Colombia. However, none of these approvals triggered commercialization.

Golden rice with higher concentrations of vitamin A was originally created by Ingo Potrykus and his team (Professor Emeritus, Institute of Plant Sciences, Swiss Federal Institute of Technology, Zürich, Switzerland). This genetically modified rice is capable of producing beta-carotene, a precursor for vitamin A. Bt rice is modified to express the cryIA (b) gene of the *Bacillus thuringiensis.* This gene confers resistance to a variety of pests including the rice borer through the production of endotoxins. The benefit of Bt rice is that farmers do not need to spray their crops with pesticides to control fungal, viral or bacterial pathogens, which otherwise needs three to four times of spray per growing season to control pests. The Chinese government is doing field trials on such insect-resistant strains. Other benefits include increased yield and revenue from crop cultivation. China approved this rice for large-scale cultivation in 2009.

(continued)

Box 22.2 (continued)

Potato

The genetically modified **Innate potato** was approved by the USDA in 2014 and the FDA (Federal Drug Administration) in 2015. Developed by J.R. Simplot Co., it is designed to resist black spot bruising and contains less of the amino acid asparagine that turns into acrylamide during the frying of potatoes. Acrylamide is a probable human carcinogen. This is known as "innate" because it does not contain any genetic material from other species. "Innate" is a group of potato varieties that have had the same genetic alterations applied using the same process.

Agrobacterium mediated gene transfer and electroporation/particle bombardments randomly choose the sites of insertion and thus are problematic due to position effects. Remedy for this drawback is to modify genes in situ at their natural positions in the genome or to deliver foreign DNA into a predicted genomic location. For this, two major approaches for gene targeting in plants are being followed: (i) gene targeting through site-specific recombination (SSR) and (ii) gene targeting through homologous recombination (HR).

22.3 Site-Directed Nucleases

Site-directed nucleases (SDN) are suitable for cutting or otherwise modifying predetermined DNA sequences in the genome that are defined as "genome editing". Examples of SDNs are zinc finger nuclease (ZFN) and transcription activator-like effector nuclease (TALEN).

a

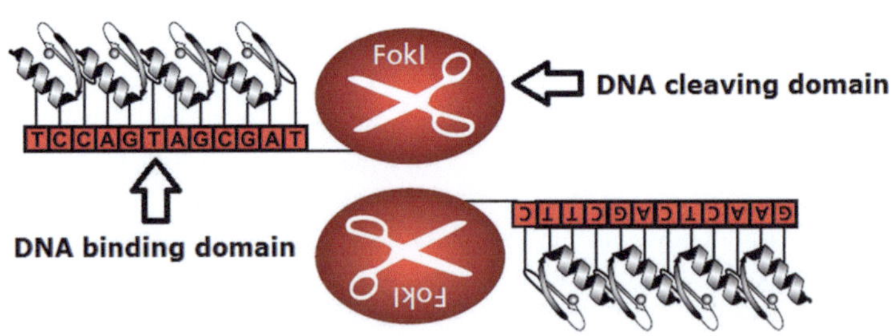

Fig. 22.3a Zinc finger nuclease. ZFN consists of two functional domains – One is a DNA-binding domain comprised of a chain of two-finger modules, each recognizing a unique hexamer (6 bp) sequence of DNA. Two-finger modules are stitched together to form a zinc finger protein, each with more than 12 bp. The other domain is of DNA-cleaving and is comprised of the nuclease domain of FokI. When a pair of ZFNs binds to adjacent sites on DNA with the correct orientation and spacing, a highly specific pair of *genomic scissors* is created

Zinc finger nucleases (ZFNs) are custom-designed proteins that cut at specific DNA sequences. Zinc finger (ZF) arrays have been the technology for targeting a specific DNA sequence since 2001 (Fig. 22.3a). A large number of zinc fingers that recognize various nucleotide triplets have been identified. ZF are capable of recognizing their specific targets with precision. However, ZFNs do have some drawbacks like every nucleotide triplet is not having corresponding zinc finger. ZFNs are of ~30 amino acid modules that interact with nucleotide triplets. ZFNs have been designed that recognize all of the 64 possible trinucleotide combinations, and by stringing different zinc finger moieties, one can create ZFNs that specifically recognize any specific sequence of DNA triplets. Each ZFN typically recognizes 3–6 nucleotide triplets. Since the nucleases to which they are attached only function as dimers, pairs of ZFNs are required to target any specific locus: one that recognizes the sequence upstream and the other that recognizes the sequence downstream of the site.

During 2009, Jens Boch of the Martin Luther University and Halle-Wittenberg and Adam Bogdanove of Iowa State University found out the nucleotide recognition code of the TAL (transcription activator-like) effectors, which were isolated from the plant bacterial pathogen *Xanthomonas*. *Xanthomonas* bacteria are pathogens of rice, pepper and tomato. They cause significant economic damage. The central TAL targeting domain is composed of 33–35 amino acid repeats. The bacteria were found to secrete effector proteins (transcription activator-like effectors, TALEs) to the cytoplasm of plant cells, which affect processes in the plant cell and increase its susceptibility to the pathogen. Effector proteins are capable of DNA binding and activating the expression of their target genes *via* mimicking the eukaryotic transcription factors.

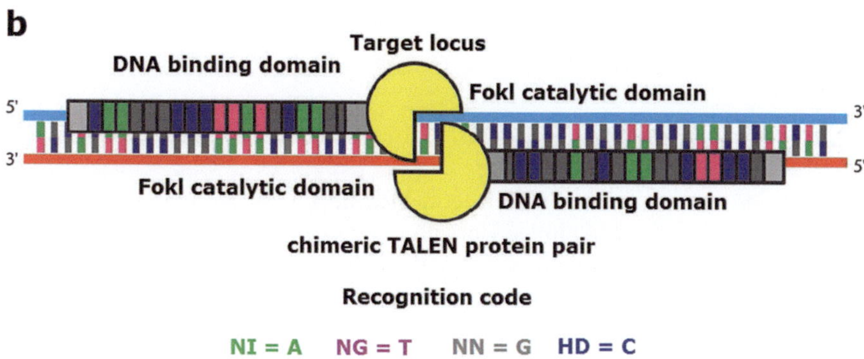

Fig. 22.3b Typical TALEN design. A scheme for introducing a double-strand break using chimeric TALEN proteins. One monomer of the DNA-binding protein domain recognizes one nucleotide of a target DNA sequence. Two amino acid residues in the monomer are responsible for binding. The recognition code (single-letter notation is used to designate amino acid residues) is provided. Recognition sites are located on the opposite DNA strands at a distance sufficient for dimerization of the FokI catalytic domains. Dimerized FokI introduces a double-strand break into DNA

TALE proteins are composed of a central domain responsible for DNA binding, a nuclear localization signal and a domain that activates the target gene transcription (Fig. 22.3b). Their capability to bind to DNA was first described in 2007, and a year later the code for recognition of the target DNA was deciphered. The DNA-binding domain consists of monomers, and these monomers bind one nucleotide in the target nucleotide sequence. Monomers are tandem repeats of 34 amino acid residues, 2 of which are located at positions 12 and 13 and are highly variable (repeat variable di-residue, RVD), and RVDs are responsible for the recognition of a specific nucleotide. This code is degenerate, i.e. some RVDs can bind to several nucleotides with different efficiencies.

Most studies use monomers containing RVDs such as Asn and Ile (NI), Asn and Gly (NG), two Asn (NN) and His and Asp (HD) for binding the nucleotides A, T, G and C, respectively. Since the NN RVD can bind both G and A, a number of studies were performed to find monomers that will be more specific. The first amino acid residue in the RVD (H and N) was found not to be directly involved in the binding of a nucleotide, but to be responsible for stabilizing the spatial conformation. The second amino acid residue interacts with a nucleotide, with the nature of this interaction being different: D and N form hydrogen bonds with nitrogenous bases, and I and G bind target nucleotides through van der Waals forces.

In principle, a double-strand break with known recognition sites can be introduced in any region of the genome artificial TALE nucleases. The need to have T before the 5′ end of the target sequence is the only limitation to TALE nucleases. However, site selection may be made in most cases by varying the spacer sequence length. The W232 residue in the N-terminal region of the DNA-binding domain was demonstrated to interact with 5′ T, affecting the efficiency of TALEN binding to the target site. This limitation could be overcome through selecting mutants of TALEN N-terminal domain that are capable of binding to A, G or C. ZFNs and TALENs are replaced by CRISPR technology in the recent past.

22.3.1 What and Why CRISPR?

Yoghurt and cheese are made from fermented milk with *Streptococcus* strains. Rodolphe Barrangou and Philippe Horvath, food scientists at Danisco USA, Inc., during 2007 observed chromosomes of these bacteria contain oddly repetitive sequences called "clustered regularly interspaced short palindromic repeats" or CRISPR. Between these repeats are the sequences from viruses that infect bacteria

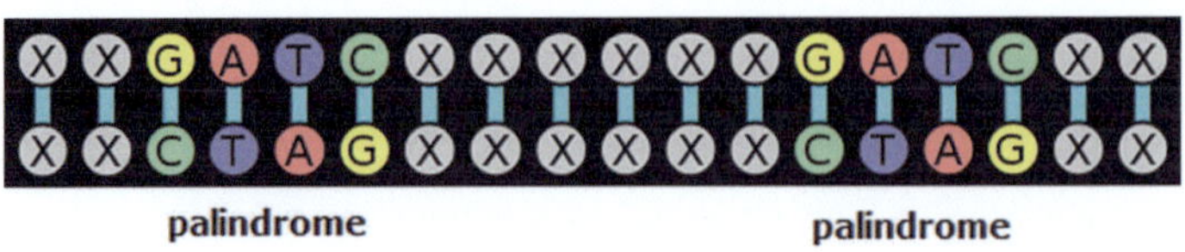

Fig. 22.4 Palindromic sequences

(Fig. 22.4). Such sequences are used as *mnemonic* (something like memory letters) to remember past invaders. If the same virus tries to infect again, the bacteria are ready with an immune response that includes a copy of the remembered sequences, called a crRNA, and a second RNA, dubbed tracrRNA, encoded near the CRISPR repeats. Together, these RNAs recruit the Cas9 protein to viral DNA, and the enzyme cuts the foreign DNA. DuPont acquired Danisco in 2011 and began using the insights to create bacteriophage-resistant *S. thermophilus* for yoghurt and cheese production. Today, yoghurt from Tel Aviv or California is a CRISPR-enhanced dairy product. That means people are consuming the yoghurt or cheese produced by a GMO.

During December 2008, Erik Sontheimer and his postdoc colleague Luciano Marraffini of the Northwestern University in Evanston, Illinois, were the first to show how CRISPR protected bacteria. It was during 2012 that Emmanuelle Charpentier of Max Planck Institute for Infection Biology in Berlin (she was with Umeå University, Sweden, then) and Jennifer Doudna of the University of California, Berkeley, could demonstrate a CRISPR/Cas9 system that could cut DNA in a test tube. During 2013, Feng Zhang of the Broad Institute published papers in *Science* showing that the CRISPR system could guide its bacterial enzyme, Cas9, to precisely target and cut DNA in human cells. In parallel, George Church, a Harvard geneticist, also demonstrated the same. Suddenly, it was possible to find and edit genes in the genome almost as simply as text in a word document. Now, Emmanuelle Charpentier, Jennifer Doudna, George Church and Feng Zhang are together known as heroes of CRISPR. This was a revolutionary achievement.

Thirty-five years have transformed plant molecular biology from *Agrobacterium*-mediated gene transfer and electroporation to site-directed genome editing with CRISPR. CRISPR could offer an easier path to genetically modified crops and livestock than other genetic engineering techniques do. Since foreign DNA is not involved, it is expected that the ethics relating to GMO may not stand as a road block for further commercializing the crop species thus modified through CRISPR.

The CRISPR/Cas9 system supersedes previous genome editing techniques such as ZFNs and TALENs, both of which rely on the nuclease domain of FokI endonucleases to break the double-strand DNA. Compared with ZFN and TALENs, CRISPR/Cas9 is much easier to manipulate and hence has broader application. ZFN, for example, consists of an array of Cys2–His2 ZF domains, with each finger binding to specific PAMs (protospacer adjacent motif), which make it difficult to select proper target sequences. When at work, two ZFNs form a dimer to locate a unique 18–24-bp DNA sequence. Owing to off-target risks, difficulty in engineering modular DNA-binding proteins and context-dependent binding requirements, the application of ZFN and TALEN technologies remains very limited.

As said earlier, the invading foreign DNA are cleaved by the Cas nucleases, then captured and integrated into the CRISPR locus in the form of spacer sequences interspaced by conserved repeated sequences. The acquired spacers serve as templates to create short CRISPR RNAs (crRNAs) which form a complex with the trans-activating crRNA (tracrRNA); together they function as guiding strands to direct the Cas9 nuclease to the complementary invading DNA (Fig. 22.5). Once

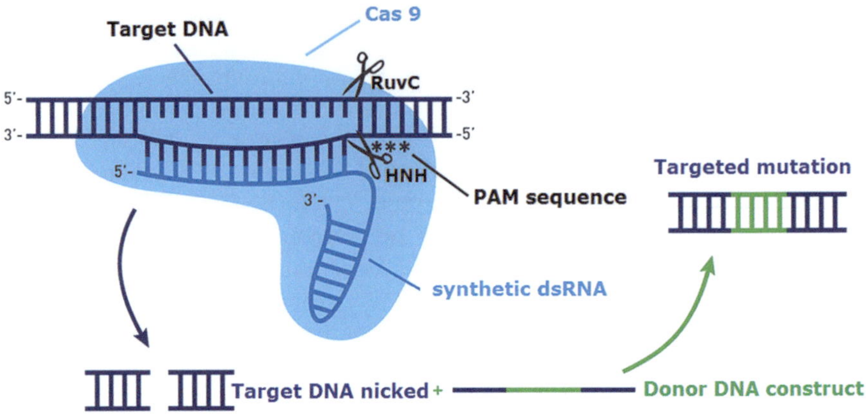

Fig. 22.5 CRISPR/Cas9 target recognition. Single chimeric sgRNA to introduce double-stranded breaks into the target loci. A complex of sgRNA and Cas9 is capable of introducing double-strand breaks into selected DNA sites. SgRNA is an artificial construct consisting of elements of the CRISPR/Cas9 system (crRNA and tracrRNA) combined into a single RNA molecule. A protospacer is a site that is recognized by the CRISPR/Cas9 system. A spacer is a sequence in sgRNA that is responsible for complementary binding to the target site. RuvC and NHN are catalytic domains causing breaks at the target site of the DNA chain. PAM is a short motif (NGG in the case of CRISPR/Cas9) whose presence at the 3′ end of the protospacer is required for introducing a break

Fig. 22.6 *Streptococcus pyogenes*

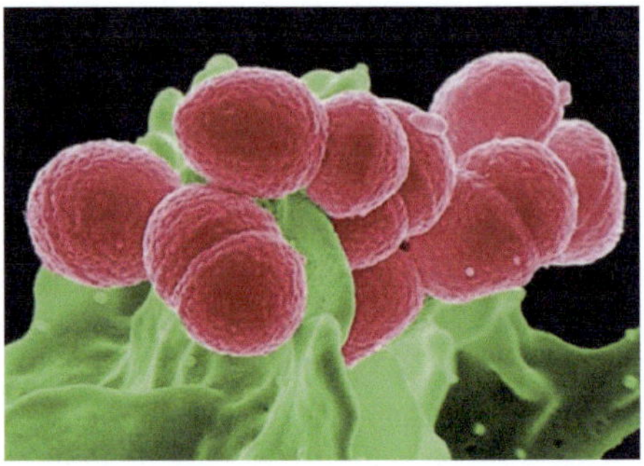

bound, the Cas9 protein cleaves the "crRNA complementary" and opposite strand through its NHN and RuvC1-like nuclease domains, respectively. The CRISPR/Cas9 system that is commonly used today for genome editing is a type II CRISPR/Cas system adapted from *Streptococcus pyogenes* (Fig. 22.6).

In the modern system, targeted genome editing using CRISPR Cas9 technology has two components: an endonuclease and a short guide RNA (Fig. 22.7). The endonuclease is the bacterial Cas9 nuclease protein from *Streptococcus pyogenes*. The Cas9 nuclease possesses two DNA-cleaving domains (the RuvC1 and HNH-like nuclease domains) that cleave double-stranded DNA, making double-strand breaks (DSB). The gRNA is an engineered single-stranded chimeric RNA, combining the scaffolding function of the bacterial tracrRNA with the specificity of the bacterial

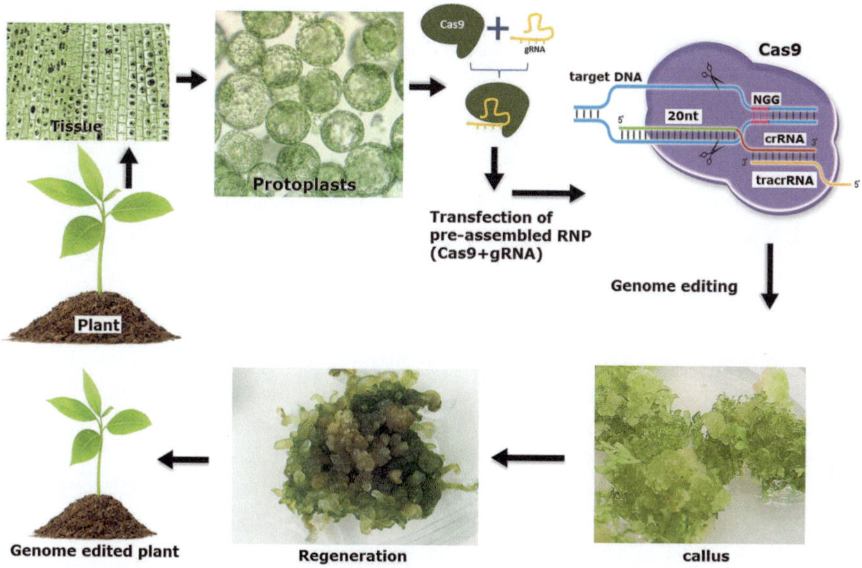

Fig. 22.7 Schematic representation of Cas9 protein-based genome editing in plant cells. Protoplasts are prepared by treatment with cell wall-digesting enzymes. Cas9 protein and gRNA were independently prepared and assembled in vitro before being introduced into the protoplasts. The protoplasts divided after recovering their cell wall. Dividing cells formed callus (a mass of undifferentiated plant cells). Independent calli derived from a single protoplast were tested for successful genome editing by polymerase chain reaction (PCR), restriction fragment length polymorphism (RFLP) and sequencing (see Chap. 23 on Molecular Breeding). Whole plants were regenerated from the mutation-bearing calli

crRNA. The last 20 bp at the 5′ end of the gRNA acts as a homing device, which recruits the Cas9/gRNA complex to a specific DNA target site, directly upstream of a protospacer adjacent motif (PAM), through RNA-DNA base pairing. The PAM sequence differs between different strains and types of CRISPR/Cas proteins, and the sequence for the *S. pyogenes* Cas9 is 5'-NGG. The adapted CRISPR/Cas9 system available today can, therefore, be directed towards any 5'-N20-NGG DNA sequence and create a precise double-strand break. The DSB is then repaired by one of two universal repair mechanisms found in nearly all cell types and organisms: the non-homologous end-joining (NHEJ) or the homology-directed repair (HDR).

CRISPR system of course is not involving a foreign DNA and probably is not coming under ethical scan. However, certain questions, such as the precise molecular mechanism, the influence on local chromatin context, the perfect length of sgRNA for best efficiency, the off-target probability of a given sgRNA and methods for efficient delivery in plants, remain to be addressed (see Box 22.3 for a comparison of ZFN, TALEN and CRISPR). CRISPR technology is being used in improving tomato, soybean, wheat, sunflower and banana by several firms in the private sector like Syngenta and Tropic Biosciences.

Box 22.3 Zinc Finger (ZFN), TALEN and CRISPR/Cas9

ZFN

Zinc finger nucleases (ZFNs) are a class of engineered DNA-binding proteins that facilitate targeted editing of the genome by creating double-strand breaks in DNA at user-specified locations. Each zinc finger nuclease (ZFN) consists of two functional domains:

(1) A DNA-binding domain comprised of a chain of two-finger modules, each recognizing a unique hexamer (6 bp) sequence of DNA. Two-finger modules are stitched together to form a zinc finger protein, each with specificity of ≥24 bp.
(2) A DNA-cleaving domain comprised of the nuclease domain of FokI. When the DNA-binding and DNA-cleaving domains are fused together, a highly specific pair of "genomic scissors" is created.

TALENs (Transcription Activator-Like Effector Nucleases)

TALENs are restriction enzymes that can be engineered to cut specific sequences of DNA. They are made by fusing a TAL effector DNA-binding domain to a DNA-cleaving domain (a nuclease which cuts DNA strands). Transcription activator-like effectors (TALEs) can be engineered to bind to practically any desired DNA sequence. When combined with a nuclease, DNA can be cut at specific locations.

CRIPSR/Cas9

CRISPR (clustered regularly interspaced short palindromic repeats) is a family of DNA sequences in bacteria (*Streptococcus pyogenes*). The sequences contain snippets of DNA from viruses that have attacked the bacterium. Such sequences are in turn used by the bacterium to detect and destroy DNA from similar viruses during subsequent attacks. These sequences play a key role in a bacterial defence system and form the basis of a technology known as **CRISPR/Cas9** that effectively and specifically changes genes within organisms. In a simple version of the CRISPR/Cas system, by delivering the Cas9 nuclease complexed with a synthetic guide RNA (gRNA) into a cell, the cell's genome can be cut at a desired location, allowing existing genes to be removed and/or new ones added.

Further Reading

Ara K, Peter BK (2009). Recent advances in plant biotechnology. Springer

Arencibia AD (2000) Plant genetic engineering: towards the third millennium. Elsevier, Amsterdam, New York

Barrangou R et al (2007) CRISPR provides acquired resistance against viruses in prokaryotes. Science 315:1709–1712

Daniel HH (2005) A guild of 45 CRISPR-associated (Cas) protein families and multiple CRISPR/Cas subtypes exist in prokaryotic genomes. PLoS Comput Biol 1:474–483

Frank K, Christian J (Eds.) (2010) Genetic modification of plants agriculture, horticulture and forestry. Series: Biotechnology in Agriculture and Forestry, Vol. 64. 675 p. Springer

Jackson JF, Linskens HF (2010) Genetic transformation of plants. Springer, New York

Scott NW, Fowler MR, Slater A (2008) Plant biotechnology: the genetic manipulation of plants. Oxford University Press, Oxford

Setlow JK (2004) Genetic engineering: principles and methods. Springer, New York

Songstad DD, Petolino JF, Voytas DF, Reichert NA (2017) Genome editing of plants. Crit Rev. Plant Sci

Molecular Breeding

23

Keywords
What Are Molecular Markers? · Genetic Markers · Classical Markers · DNA Markers · Summary of Major Classes of Genetic Markers · Prerequisites for Molecular Breeding · Activities of Marker-Assisted Breeding · What is Mapping? · MAS for Qualitative Traits · MAS for Quantitative Traits · QTL Detection (Statistical) · Next-Generation Molecular Breeding · Next-Generation Sequencing (NGS) · Genotyping-by-Sequencing (GBS) · RFLP, and AFLP as Tools to Map Genomes · RAPD Technique · Genetic Maps · Physical Maps

Application of molecular biology in plant breeding is molecular breeding. The process of developing new crop varieties through conventional means can take almost 25 years, but the application of biotechnology has considerably shortened the time to 7–10 years for deriving new crop varieties for commercial exploitation. One of the tools for easier and faster selection of plant traits is marker-assisted selection (MAS). The areas of molecular breeding include QTL mapping or gene discovery, marker-assisted selection and genomic selection, genetic engineering and genetic transformation.

Molecular breeding is used to describe several modern breeding strategies, including marker-assisted selection (MAS), marker-assisted backcrossing (MABC), marker-assisted recurrent selection (MARS) and genome-wide selection (GWS) or genomic selection (GS). In this chapter, we shall discuss fundamentals of marker-assisted breeding in plants and some issues related to the procedures in practical breeding. First, some of the fundamental concepts in molecular breeding are narrated:

What Are Molecular Markers?

Molecular markers (DNA markers) reveal neutral sites of variation at the DNA sequence level. While morphological markers show variation in the phenotype,

P. M. Priyadarshan, *PLANT BREEDING: Classical to Modern*,
https://doi.org/10.1007/978-981-13-7095-3_23

molecular markers do not show variation in the phenotype. A marker could be a single-nucleotide difference in a gene or a piece of repetitive DNA. Molecular markers are much more than morphological markers and do not disturb the physiology of the organism. Restriction enzymes, gel electrophoretic separation of DNA fragments, Southern hybridization, polymerase chain reaction (PCR) and labelled probes are the tools that allow us to access and use these markers (see Chap. 22 on "Genetic Engineering" for a detailed account on restriction enzymes).

Electrophoretic Separation of DNA Fragments: Gel electrophoresis is a process for separation and analysis of macromolecules (DNA, RNA and proteins) and their fragments, based on their size and electric charge. Nucleic acid molecules are separated by applying an electric field to move the negatively charged molecules through a matrix of agarose or other substances. Shorter molecules can migrate through the pores of the gel faster than longer molecules. This phenomenon is called sieving. Proteins are separated by charge in agarose because the pores of the gel are too large to sieve proteins. Nanoparticles can also be separated in gel electrophoresis. Gel electrophoresis is utilized after amplification of DNA via PCR. This technique can also be used as a preparative technique prior to use of mass spectrometry, RFLP, PCR, cloning, DNA sequencing or Southern blotting for further characterization.

Southern Hybridization (Southern Blotting): Southern blotting was named after Edwin Southern who developed this procedure at Edinburgh University in 1975. Here, the separated DNA molecules are transferred from agarose gel onto a membrane. Southern blotting locates a particular DNA sequence within a mixture. For example, it can be used to locate a specific gene within an entire genome. The amount of DNA needed for this technique is dependent on the size and specific activity of the probe. Short probes tend to be more specific. Under optimal conditions, one can expect to detect 0.1 pg of the DNA for which the probe is being done. The following are the steps:

1. DNA (genomic or other source) is digested with a restriction enzyme and separated by gel electrophoresis, usually an agarose gel. Because there are so many different restriction fragments on the gel, it usually appears as a smear rather than discrete bands. The DNA is denatured into single strands by incubation with NaOH.
2. The DNA is transferred to a membrane which is a sheet of special blotting paper. The DNA fragments retain the same pattern of separation they had on the gel.
3. The blot is incubated with many copies of a probe which is single-stranded DNA. This probe will form base pairs with its complementary DNA sequence and bind to form a double-stranded DNA molecule. The probe cannot be seen since it either is radioactive or has an enzyme bound to it (e.g. *alkaline phosphatase or horseradish peroxidase*).

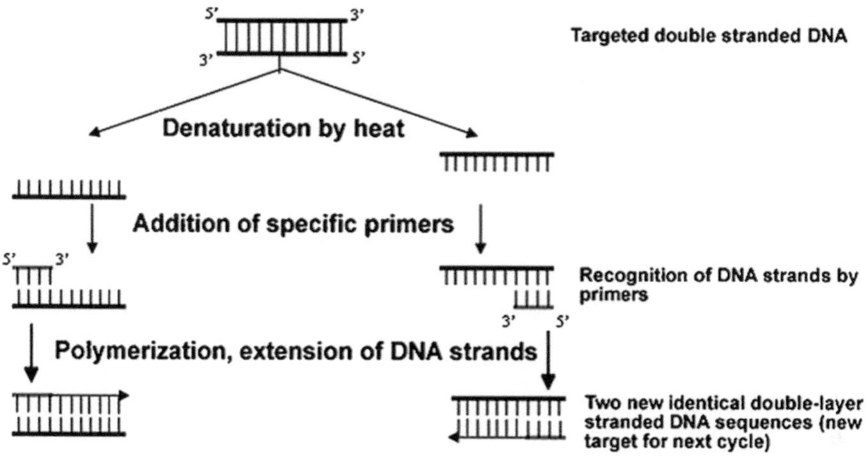

Fig. 23.1 Polymerase chain reaction technique

4. The location of the probe is revealed by incubating it with a colourless substrate that the attached enzyme converts to a coloured product that can be seen or gives off light which will expose X-ray film. If the probe was labelled with radioactivity, it can expose X-ray film directly.

Polymerase Chain Reaction (PCR): PCR is a revolutionary method developed by Kary Mullis (then with Cetus Corporation, California) in 1983. (Mullis received Nobel Prize for Chemistry in 1993 for this invention; he died of pneumonia on August 7, 2019). PCR is based on the principle that *DNA polymerase* has the ability to synthesize a new strand of DNA complementary to the template DNA. DNA polymerase can add a nucleotide only onto a pre-existing 3' OH group. For this, it needs a *primer* to which it can add the first nucleotide. Such a situation makes it possible to delineate a specific region of template sequence which the scientist wants to amplify. Billions of copies (amplicon) of the specific sequence will be made at the end of the PCR (Fig. 23.1).

At the beginning of the process, high temperature is applied to the double-stranded DNA to separate the strands. DNA polymerase synthesizes new complimentary DNA strand. Popularly used enzyme is *Taq* DNA polymerase (from *Thermus aquaticus*). *Pfu* DNA polymerase (from *Pyrococcus furiosus*) is also used because of its high fidelity in copying DNA. These polymerases are heat resistant. Primers that are short pieces of single-stranded DNA, complementary to the target sequence are also added. The polymerase begins synthesizing new DNA from the end of the primer. Here, nucleotides (dNTPs or deoxynucleotide triphosphates), single units of the bases A, T, G and C, are the essential "building blocks" for new DNA strands.

Reverse Transcription PCR: Reverse transcription PCR is PCR proceeded with conversion of sample RNA into cDNA with enzyme reverse transcriptase. This PCR starts to generate copies of the target sequence exponentially (more and more).

Brief Steps of Traditional PCR:

1. The DNA strands are denatured at high temperature, breaking the weak hydrogen bonds that bind one side of the helix to the other.
2. The temperature is lowered and primers (short bits of DNA) are added. The primers bond to their specific sites.
3. The temperature is brought back up to body temperature and Taq polymerase is added.
4. Repeat step 1 for *n* cycles, amplifying the DNA.

Real-Time PCR or Quantitative PCR: This instrumentation is used to monitor the progress of a PCR in real time. A relatively small amount of PCR product (DNA, cDNA or RNA) can also be quantified. This is based on the principle that the detection of the fluorescence produced by a reporter molecule increases as the reaction progresses. Fluorescence increases due to accumulation of the PCR product with each cycle of amplification. These fluorescent reporter molecules include dyes that bind to the double-stranded DNA (i.e. SYBR Green) or sequence-specific probes (i.e. molecular beacons or TaqMan probes). The process can begin with minimal amounts of nucleic acid and the end product can be quantified accurately. There is no post-PCR processing which saves resources and time. This technique has revolutionized PCR-based quantification of DNA and RNA. Real-time RT-PCR refers to additional cycle of reverse transcription that leads to formation of DNA from RNA. Based on the molecule used for the detection of fluorescence, the real-time PCR techniques can be categorically placed under two heads:

Non-specific Detection Using DNA-Binding Dyes: The fluorescence of the reporter dye increases as the product accumulates with each successive cycle of amplification. Through recording the amount of fluorescence emission at each cycle, it is possible to monitor the PCR during exponential phase. If a graph is drawn between the log of the starting amount of template and the corresponding increase in the fluorescence of the reporter dye, a linear relationship is observed.

SYBR® Green is the most widely used dye for real-time PCR. SYBR® Green binds to the minor groove of the DNA double helix. Unbound dye exhibits very little fluorescence. This fluorescence substantially increases when the dye is bound to double-stranded DNA. SYBR® Green remains stable under PCR conditions and the optical filter of the thermocycler can be affixed to harmonize the excitation and emission wavelengths. Ethidium bromide can also be used as dye but its carcinogenic property restricts its use. Though these dyes are simplest and cheapest, both specific and non-specific products generate signal. This is a drawback with these dyes.

Specific Detection Using Target Specific Probes: Specific detection of real-time PCR is done with some oligonucleotide probes labelled with both a reporter fluorescent dye and a quencher dye. Probes based on different chemistries are available for real-time detection, these include:

1. Molecular beacons
2. TaqMan probes
3. Scorpion primers
4. SYBR® Green

Molecular beacons are oligonucleotide probes that detect the presence of specific nucleic acids. Molecular beacons are hairpin-shaped molecules with an internally quenched fluorophore whose fluorescence is restored when they bind to a target nucleic acid (Fig. 23.2a). The loop portion of the molecule is a probe sequence complementary to a target nucleic acid molecule. The stem is formed by annealing of complementary arm sequences on the ends of the probe sequence. The end of one arm has a fluorescence moiety, and the end of the other arm has a quenching moiety. The stem keeps these two moieties in close proximity to each other, causing the fluorescence of the fluorophore to be quenched by energy transfer. Since the quencher moiety is a non-fluorescent chromophore and emits the energy that it receives from the fluorophore as heat, the probe is unable to fluoresce. When the probe encounters a target molecule, it forms a hybrid that is longer and more stable than the stem, and its rigidity and length preclude the simultaneous existence of the stem hybrid. Thus, the molecular beacon undergoes a spontaneous conformational reorganization that forces the stem apart and causes the fluorophore and the quencher to move away from each other, leading to the restoration of fluorescence.

Well-designed TaqMan probes require very little optimization. In addition, they can be used for multiplex assays by designing each probe with a spectrally unique quench pair. However, TaqMan probes can be expensive to synthesize, with a separate probe needed for each mRNA target being analysed (Fig. 23.2b).

With Scorpion probes, PCR product detection is achieved with a single oligonucleotide. The Scorpion probe maintains a stem-loop configuration in the non-hybridized state. The fluorophore is attached to the 5′ end and is quenched by a moiety coupled to the 3′ end. The 3′ portion of the stem also contains sequence that is complementary to the extension product of the primer. This sequence is linked to the 5′ end of a specific primer via a non-amplifiable monomer. After extension of the Scorpion primer, the specific probe sequence is able to bind to its complement within the extended amplicon, thus opening up the hairpin loop. This prevents the fluorescence from being quenched and a signal is observed (Fig. 23.2c).

SYBR® Green is the simplest and most economical for quantitating PCR products. SYBR® Green binds double-stranded DNA and upon excitation emits light. SYBR® Green is inexpensive, easy to use and sensitive. SYBR® Green will bind to any double-stranded DNA, and since the dye binds to double-stranded DNA, there is no requirement of a probe. However, detection by SYBR® Green requires extensive optimization. Since the dye cannot distinguish between specific and

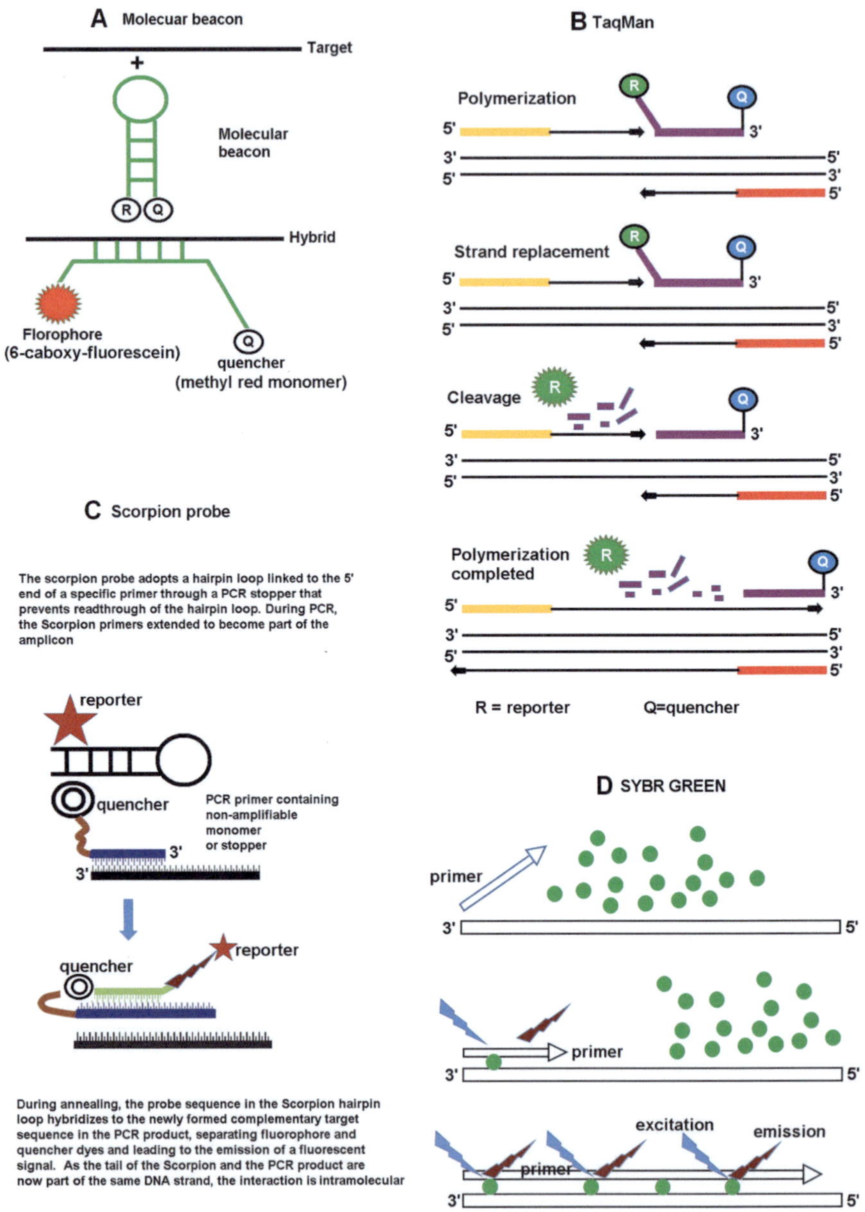

Fig. 23.2 Target specific probes. (**a**) Molecular beacons, (**b**) TaqMan probe, (**c**) Scorpion probe, (**d**) SYBR® Green probe

non-specific product accumulated during PCR, follow-up assays are needed to validate results (Fig. 23.2d).

23.1 Genetic Markers

Genetic markers are determined by allelic forms of genes or genetic loci. They are transmitted from one generation to another and can be used as experimental probes or tags to keep track of an individual, a tissue, a cell, a nucleus, a chromosome or a gene. Genetic markers are of two categories: classical markers and DNA markers. Classical markers include morphological markers, cytological markers and biochemical/protein markers. DNA markers, on the other hand, can be studied with polymorphism-detecting techniques or methods like Southern blotting (nucleic acid hybridization), PCR (polymerase chain reaction) and DNA sequencing such as RFLP, AFLP, RAPD, SSR, SNP, etc.

23.1.1 Classical Markers

Morphological Markers: During days of early plant breeding, the markers used were visible traits like leaf shape, flower colour, pubescence colour, pod colour, seed colour, seed shape, hilum colour, awn type and length, fruit shape, rind (exocarp) colour and stripe, flesh colour, stem length, etc. These morphological markers generally represent genetic polymorphisms that could be identified and manipulated with relative ease. Therefore, they are usually used in construction of linkage maps by classical two and/or three point tests. Since a few such markers are linked with other agronomic traits, they could be for indirect selection. Semi-dwarfism in rice and wheat led to the success of high-yielding cultivars. In wheat breeding, the dwarfism governed by gene *Rht10* was introgressed into *Taigu* nuclear male sterile wheat by backcrossing, and a tight linkage was generated between *Rht10* and the male sterile gene *Ta1*. Then the dwarfism was used as the marker to identify male sterile plants. Morphological markers are limited in number and are not linked with yield and quality.

Cytological Markers: In cytology, the structural features of chromosomes can be shown by chromosome karyotype and bands. The distributions of euchromatin and heterochromatin are displayed by colour, banding patterns, width, order and position. For example, Q bands are produced by quinacrine hydrochloride and G bands by Giemsa stain, and R bands are reversed G bands. Apart from characterization and detection of mutation, these processes are also used for physical mapping and linkage group identification. However, direct uses are very limited.

Biochemical/Protein Markers: Protein markers may also be categorized into molecular markers though molecular markers are mostly DNA markers. Isozymes are alternative forms of an enzyme. Isozymes differ in their molecular weights and electrophoretic mobility but are with same catalytic activity. Isozymes are products of different alleles. Their difference in electrophoretic mobility is caused by point mutation due to amino acid substitution. Hence, such markers can be genetically mapped onto chromosomes and then used to map genes. A number of isozymes are very limited so also their usage as markers.

An example of biochemical marker used in wheat is high molecular weight glutenin subunit (HMW-GS). A correlation between the presence of certain HMW-GS and gluten strength measured by the SDS-sedimentation volume test was achieved. On this basis, a numeric scale to evaluate bread-making quality as a function of the described subunits (*Glu-1* quality score) was designed. Assuming the effect of the alleles to be additive, the bread-making quality was predicted by adding the scores of the alleles present in the particular line. It was established that the allelic variation at the *Glu-D1* locus has a greater influence on bread-making quality than the variation at the *Glu-1* loci. Subunit combination 5+10 for locus *Glu-D1* (*Glu-D1* 5+10) renders stronger dough than *Glu-D1* 2+12. Therefore, breeders may enhance the bread-making quality in wheat by selecting subunit combination *Glu-D1* 5+10 instead of *Glu-D1* 2+12.

23.1.2 DNA Markers

DNA markers are fragments of DNA revealing mutations/variations. DNA marker is a small region of DNA sequence showing polymorphism (base deletion, insertion and substitution) between different individuals. There are two basic methods to detect the polymorphism: Southern blotting, a nuclear acid hybridization technique (by Southern in 1975), and PCR, a polymerase chain reaction technique. Using PCR and/or molecular hybridization followed by electrophoresis (e.g. PAGE, polyacrylamide gel electrophoresis; AGE, agarose gel electrophoresis; CE, capillary electrophoresis), polymorphism for a specific region of DNA can be identified based on band size and mobility. In addition to Southern blotting and PCR, refined detection systems have also been developed. For instance, several new array chip techniques use DNA hybridization combined with labelled nucleotides, and new sequencing techniques detect polymorphism by sequencing. Ideal DNA markers for marker-assisted breeding should meet the following criteria:

(a) High level of polymorphism
(b) Even distribution across the whole genome
(c) Co-dominance in expression (so that heterozygotes can be distinguished from homozygotes)
(d) Clear distinct allelic features
(e) Single copy and no pleiotropic effect
(f) Low cost to use
(g) Easy assay/detection and automation
(h) High availability and suitability to be duplicated genome-specific in nature
(i) No detrimental effect on phenotype

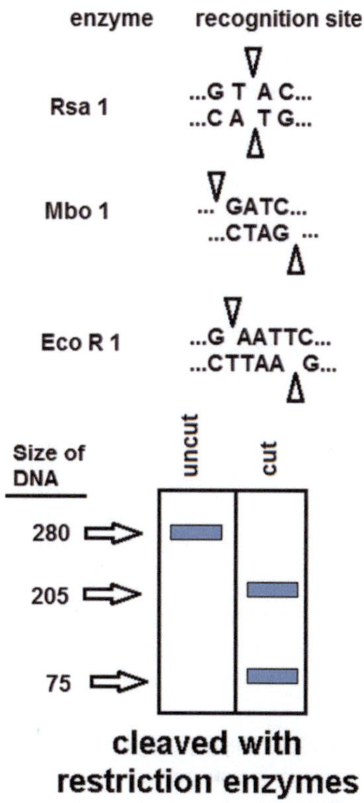

Fig. 23.3 RFLP technique. Uncut and cut samples of DNA. Note the sizes of the DNA fragments add up to size of uncut DNA

Extensively used polymorphisms are restriction fragment length polymorphism (RFLP), amplified fragment length polymorphism (AFLP), random amplified polymorphic DNA (RAPD), microsatellites or simple sequence repeats (SSR) and single-nucleotide polymorphism (SNP). These marker techniques assist in selecting multiple desired traits using F_2 and backcross populations, near-isogenic lines, doubled haploids and recombinant inbred lines.

RFLP Markers: RFLP markers are the first-generation DNA markers and one of the important tools for plant genome mapping (Fig. 23.3). They are a type of Southern blotting-based markers. RFLP was invented in 1984 by the English scientist Alec Jeffreys. Mutation (deletion and insertion) occurs at restriction sites or between adjacent restriction sites in the genome (see Chap. 22 on "Genetic Engineering" for restriction sites). The changes in base pair (insertions or deletions) within the restriction fragments could derive restriction fragments of different sizes. As a result of this, when homologous chromosomes are digested by restriction enzymes, the varied restriction products are detected by electrophoresis and DNA-probing techniques. RFLP markers are powerful tools for comparative and synteny mapping (mapping a set of genes on a specific chromosome). Most RFLP markers are co-dominant and locus-specific. By using an improved RFLP technique, i.e. cleaved amplified polymorphism sequence (CAPS), also known as PCR-RFLP, high-throughput markers can be developed from RFLP probe sequences. PCR (polymerase chain reaction) is a technology in molecular biology used to amplify a single copy or a few copies of a piece of DNA to generate thousands to millions of copies of a particular DNA sequence. CAPS technique consists of digesting a PCR-amplified fragment and detecting the polymorphism by the presence/absence of restriction sites. An advantage with RFLP is that the sequence of the probe need not be known. Only a genomic clone is needed to detect polymorphism. RFLP markers were predominant in the 1980s and 1990s, but a fewer direct uses of RFLP markers are reported these days.

RAPD Markers: RAPD is a PCR-based marker system. This was developed independently in 1990 by two different laboratories (Williams and co-workers of E.I. du Pont de Nemours & Co. USA and Welsh and McClelland of California Institute of Biological Research) and was called RAPD and AP-PCR (arbitrary primed PCR), respectively. In this technique, the total genomic DNA is amplified by PCR using a short, single primer (usually about ten nucleotides/bases), the primer which binds to different sites to amplify random sequences. Amplification can take place during the PCR, if two hybridization sites are similar to one another (at least 3000 bp) and in opposite directions. The amplified fragments generated by PCR depend on the length and size of both the primer and the target genome (Fig. 23.4).

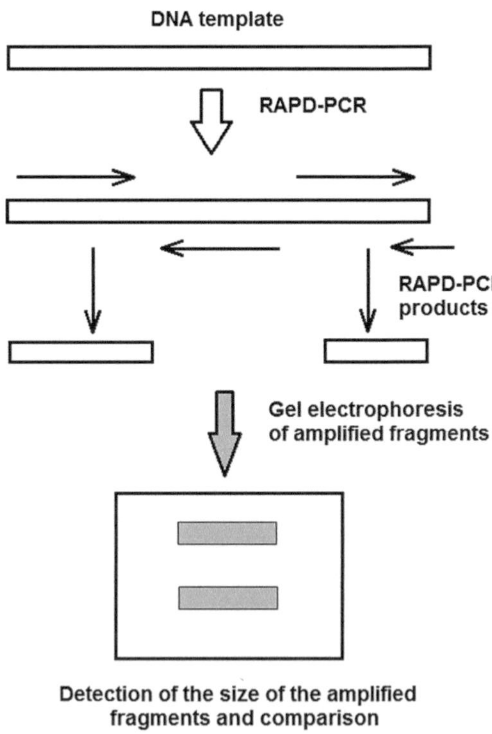

Fig. 23.4 The principle of RAPD-PCR technique. Arrows indicate primer annealing sites

The PCR products (up to 3 kb) are separated by agarose gel electrophoresis and imaged by ethidium bromide (EB) staining. Polymorphisms at the primer-binding sites are made visible in the electrophoresis as RAPD bands. RAPD predominantly provides dominant markers. RAPD gives high levels of polymorphism and is simple and easy as follows:

(a) No DNA sequence information is needed for the design of specific primers.
(b) No blotting or hybridization steps; hence it is quick, simple and efficient.
(c) Small amounts of DNA (about 10 ng per reaction) are needed and the process can be automated. Higher levels of polymorphism can be detected compared to RFLP.
(d) Primers are non-species specific and can be universal.
(e) The RAPD products of interest can be cloned, sequenced and then used to derive other types of PCR-based markers, such as sequence-characterized amplified region (SCAR), single-nucleotide polymorphism (SNP), etc.

However, RAPD also has some limitations/disadvantages, such as low reproducibility and incapability to detect allelic differences in heterozygotes.

AFLP Markers: AFLPs are PCR-based markers (Fig. 23.5a). It was developed by Keygene in the 1990s. An AFLP primer (17–21 nucleotides in length) consists of a synthetic adaptor sequence, the restriction endonuclease recognition sequence and an arbitrary, non-degenerate "selective" sequence (1–3 nucleotides). The primers are capable of annealing perfectly to their target sequences (the adapter and restriction sites) as well as a small number of nucleotides adjacent to the restriction sites. The first step in AFLP involves restriction digestion of genomic DNA (about 500 ng)

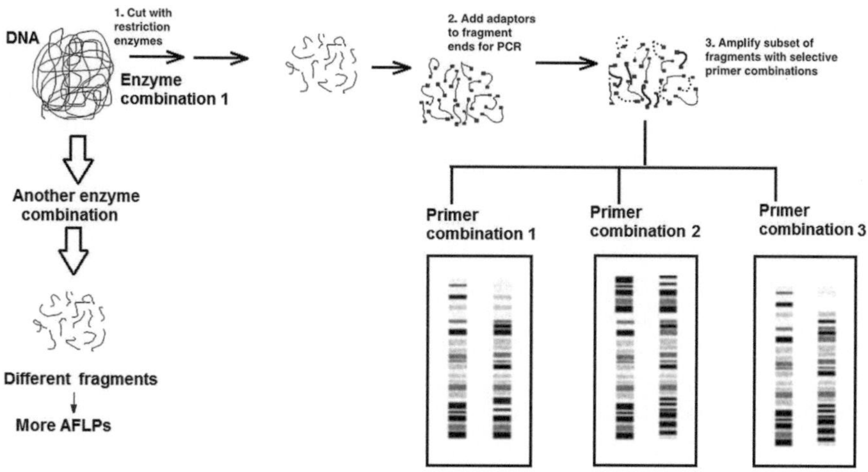

Fig. 23.5a Amplified fragment length polymorphism (AFLP)

Table 23.1 Order of magnitude (weight)

1 picogram	$= 10^{-12}$ g
1000 picograms	= 1 nanogram (ng) or 10^{-9} g
1000 nanograms	= 1 microgram (μg) or 10^{-6} g
1000 micrograms	= 1 milligram (mg) or 10^{-3} g
1000 milligrams	= 1 gram (g)
1000 grams	= 1 kilogram (kg)

(see Table 23.1) with two restriction enzymes, a rare cutter (6-bp recognition site, *Eco*RI, *Pts*I or *Hin*dIII) and a frequent cutter (4-bp recognition site, *Mse*I or *Taq*I). The adaptors are then ligated to both ends of the fragments to provide known sequences for PCR amplification (Fig. 23.5b). Only those fragments that are cut by the frequent cutter and rare cutter will be amplified. AFLP markers are reliable, robust and reproducible with high marker density separated by high-resolution electrophoresis systems. The fragments can be detected by dye-labelling primers radioactively or fluorescently.

A typical AFLP fingerprint (restriction fragment patterns) contains 50–100 amplified fragments, and up to 80% of these could serve as genetic markers. AFLP assays can be conducted using relatively small DNA samples (1–100 ng). AFLP has a high genotyping throughput and is relatively reproducible. Sequence information of probe is not required and a set of primers can be used for different species. The applications of AFLP markers include biodiversity studies, analysis of germplasm collections, genotyping of individuals, identification of closely linked DNA markers, construction of genetic DNA marker maps, construction of physical maps, gene mapping and transcript profiling.

SSR Markers (Microsatellites): SSRs (simple sequence repeats), are also called microsatellites, short tandem repeats (STRs) or sequence-tagged microsatellite sites (STMS) (Fig. 23.6). It was first characterized in 1984 at the University of Leicester by Weller Jeffreys and colleagues. They are PCR-based markers. They are random tandem repeats of short nucleotide motifs (2–6 bp/nucleotides long), di-, tri- and tetra-nucleotide repeats (e.g. $(GT)_n$, $(AAT)_n$ and $(GATA)_n$), that are widely distributed throughout the genomes of plants. The copy number is the source of polymorphism in plants. High level of allelic variation is the attribute of SSRs that makes them valuable genetic markers. The PCR-amplified products can be separated in high-resolution electrophoresis systems (e.g. AGE and PAGE), and the bands can be realized through fluorescent labelling or silver staining.

SSR markers have the attributes like hyper-variability, reproducibility, co-dominant nature, locus specificity and random genome-wide distribution. SSR markers can be easily analysed by PCR and detected by PAGE or AGE. SSR assays require small DNA samples (~100 ng) with low start-up costs for manual assays. On the other hand, SSRs require nucleotide information for primer design. Marker development process is labour intensive and higher start-up costs for automated

Fig. 23.5b AFLP flow chart. Adaptor DNA = short double-strand DNA molecules of 18–20 bp length representing two types of molecules. Each type is comparable with one restriction enzyme generated DNA end. Pre-amplifications use selective primers, which contain an adaptor DNA sequence plus one or two random bases at the 3′ end for reading into the genomic fragments. Primers for re-amplification primer sequence plus one or two additional bases at the 3′ end. A tag is attached at the 5′ end of one of the re-amplification primers for detecting amplified molecules

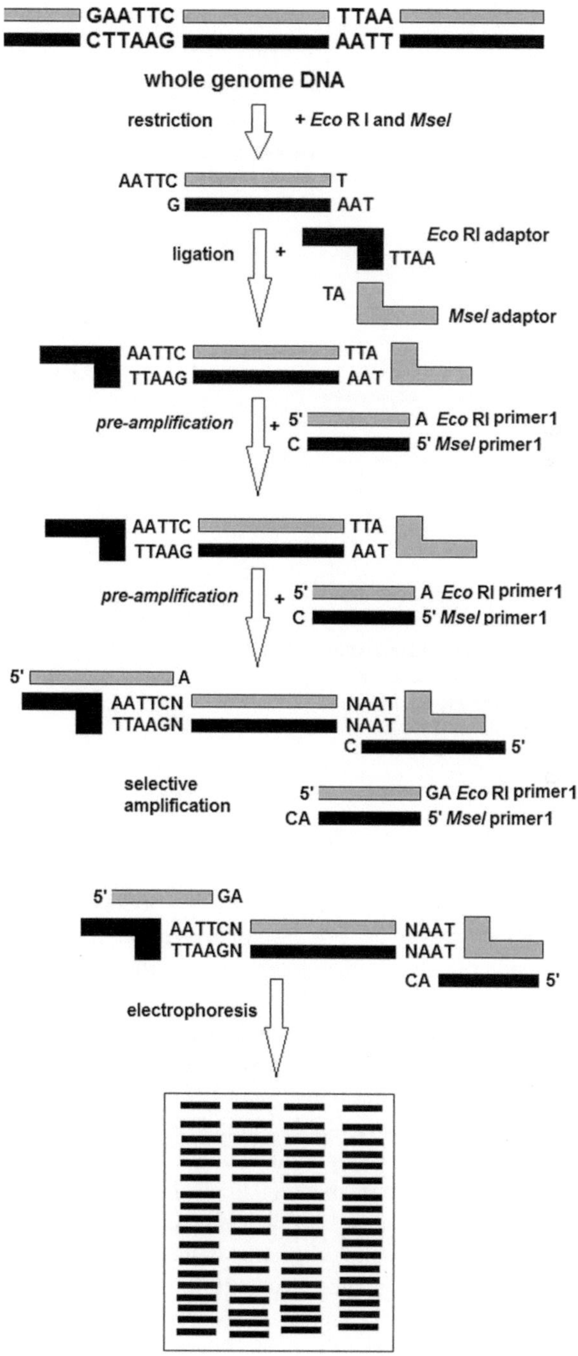

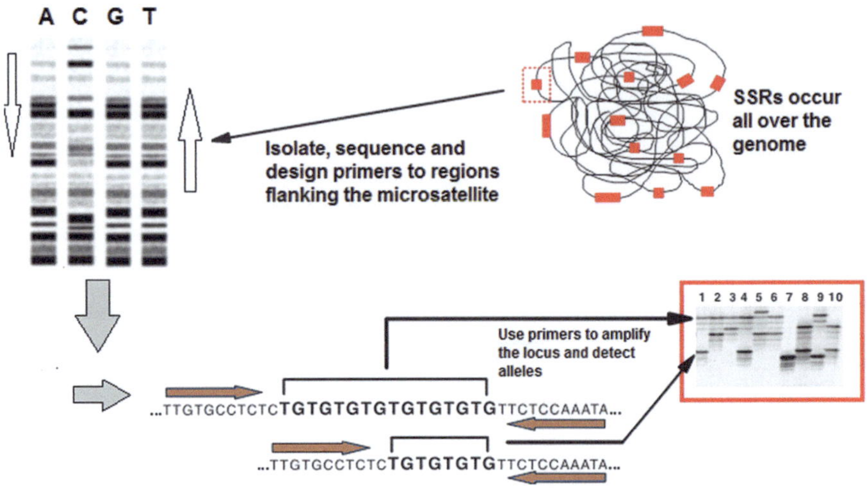

Fig. 23.6 How primers are designed and used to generate simple sequence repeats (SSRs)

Fig. 23.7 A pair of homologous chromosomes each with a single chromatid to illustrate the molecular basis of a single-nucleotide polymorphism (SNP)

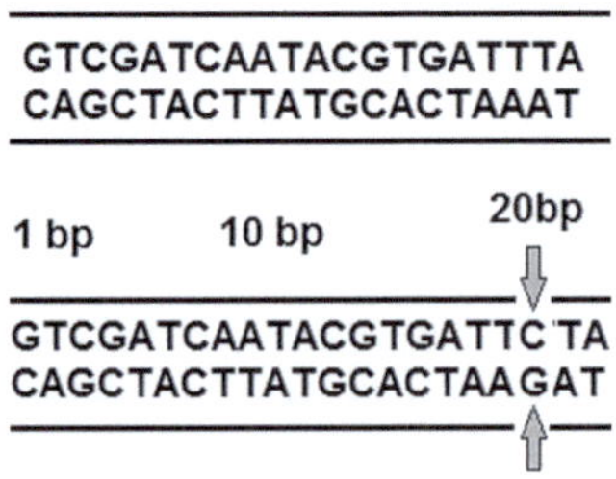

process are the disadvantages. Plenty of SSR markers have been developed in various crop species. For example, over 35,000 SSR markers are developed and mapped onto all 20 linkage groups in soybean.

SNP Markers: SNP is a single-nucleotide base difference between two DNA sequences or individuals. SNPs can be categorized according to nucleotide substitutions as either transitions (C/T or G/A) or transversions (C/G, A/T, C/A or T/G) (Fig. 23.7). In principle, single-base variants in cDNA (mRNA) are considered to be SNPs. Since a single-nucleotide base is the smallest unit of inheritance, SNPs can provide maximum markers. In plants, SNP frequencies are in a range of one SNP in every 100–300 bp. If one allele contains a recognition site for a restriction enzyme while the other does not, digestion of the two alleles will produce different fragments in length. The sequence available in a crop species with SNP markers can be compared with the sequence data stored in the major databases and identify SNPs. Four alleles can be identified when the complete base sequence of a segment of DNA is considered and these are represented by A, T, G and C at each SNP locus in that segment. SNPs are co-dominant markers. As the simplest/ultimate form for polymorphism, SNPs have emerged as potential genetic markers. High start-up cost of SNPs is the limitation. The choice of DNA markers is still a challenge for plant breeders.

23.1.3 Summary of Major Classes of Genetic Markers

Morphological Traits: Morphological markers like seed or flower colour are limited in number. The presence of dominance, late expression, deleterious effects, pleiotropy and epistasis restrict their usage.

Proteins: Isozyme markers are low in number. Newer techniques that can assay more than 50 seed storage proteins could provide a very cost-effective means.

Restriction Fragment Length Polymorphism (RFLP): It requires probe DNA and its hybridization with plant DNA. Provides high-quality data but has limited throughput potential.

Random Amplified Polymorphic DNA (RAPD): First new generation of markers based on the polymerase chain reaction (PCR). Using arbitrary primers to amplify random pieces of DNA, it requires no knowledge of the genome; but inconsistent among populations and laboratories.

Simple Sequence Repeat Length Polymorphism (SSRLP): Also known as microsatellite, variable number of tandem repeats (VNTR) or sequence-tagged microsatellite site (STMS) markers. It is a high-quality, highly consistent and a preferred assay for marker-assisted selection. They are expensive as they require extensive sequence data.

Amplified Fragment Length Polymorphism (AFLP): The sample DNA is enzymatically cut into small fragments, but due to selective PCR amplification only a fraction of fragments are studied. This assay provides much marker information, but not suited to high-throughput marker-assisted selection.

Expressed Sequence Tag (EST): This requires extensive sequence data of regions of DNA that are expressed. Once developed, they provide high-quality, highly consistent results since they are limited to expressed regions, thus providing information on functional genes.

Single-Nucleotide Polymorphism (SNP): The majority of differences between genotypes are point mutations raising from single-nucleotide polymorphisms. Extensive sequence data are needed to develop SNP markers. Their great advantage is that they do not require electrophoresis but managed with microarrays.

Table 23.2 Comparison of widely used molecular markers for plant genome analysis

Attribute	RFLP	RAPD	AFLP	SSR	SNP
Abundance	Medium	Very high	Very high	High	Very high
Types of polymorphism	Single-base change, insertion, deletion, inversion	Single-base change, insertion, deletion, inversion	Single-base change, insertion, deletion, inversion	Repeat length single base	Single-base change
No. of polymorphic loci analysed	1.0–3.0	1.5–5.0	20–100	1.0–3.0	1.0
PCR-based	No	Yes	Yes	Yes	Yes
DNA required (μg)	10	0.02	0.5–1.0	0.05	0.05
DNA quality	High	Medium	High	Medium	Medium
DNA sequence information	Not required	Not required	Not required	Required	Required
Level of polymorphism inheritance	Medium	High	High	High	High
Reproducibility	High	Low	Medium	High	High
Technical complexity	High	Low	Medium	Low	Medium
Developmental cost	High	Low	Moderate	High in start	High
Cost/analysis	High	Low	Moderate	Low	Low
Species transferability	Medium	High	High	Medium	Low
Automation	Low	Medium	Medium	High	High

A comparison of widely used molecular marker for genome analysis is available in Table 23.2.

In the recent past, there have been numerous developments in marker science with many new systems becoming available like cleavage amplification polymorphism (CAP), sequence-specific amplification polymorphism (S-SAP), inter-simple sequence repeat (ISSR), sequence-tagged site (STS), sequence-characterized amplification region (SCAR), selective amplification of microsatellite polymorphic loci (SAMPL), single-nucleotide polymorphism (SNP), sequence-related amplified polymorphism (SRAP), target region amplification polymorphism (TRAP), microarrays, diversity arrays technology (DArT), single-strand conformation polymorphism (SSCP), denaturing gradient gel electrophoresis (DGGE), temperature gradient gel electrophoresis (TGGE) and methylation-sensitive PCR.

23.1.4 Prerequisites for Molecular Breeding

Molecular breeding is the DNA marker-assisted breeding that calls for sophisticated instrumentation and facilities. The prerequisites are:

(a) Appropriate marker system and reliable markers: The success in the selection of the gene depends on the position of markers that are located in close proximity to the target gene or present within the gene. SSRs are the current markers of choice for many crop species. SNPs require more sequence data.
(b) Quick DNA extraction and high-throughput marker detection: Hundreds to thousands of genotypes are screened for desired marker patterns. Hence, a faster DNA extraction technique and a high-throughput marker detection system are essential to handle for a large-scale screening of multiple markers.
(c) Genetic maps: A high-density genetic linkage map is vital for MAS. When a trait is seen associated with markers, a dense molecular marker map will assist to identify makers that are close to (or flank) the target gene. A desirable map should have an adequate number of evenly spaced polymorphic markers to accurately locate desired QTLs/genes.
(d) Knowledge of marker-trait association: This is the most crucial factor for MAS. Markers that are closely linked to target traits can ensure success of MAS. Such information is retrieved through gene mapping, QTL analysis, association mapping, classical mutant analysis, linkage or recombination analysis, bulked sergeant analysis, etc.
(e) Quick and efficient data processing and management: Quick and efficient data process will ensure timely reports to breeders. In MAS in addition to a large number of samples, multiple markers are to be handled simultaneously. This situation requires an efficient and quick system for labelling, storing, retrieving, processing and analysing large data sets. The development of bioinformatics and statistical software packages provide useful tools for this purpose.

23.2 Activities of Marker-Assisted Breeding

(a) Planting the breeding populations with potential segregation for traits
(b) Sampling plant tissues (at early stages of growth), e.g. emergence to young seedling stage
(c) Preparing DNA samples of each genotype for PCR and marker screening
(d) Running PCR or other amplifying systems for the molecular markers linked to the trait of interest
(e) Scoring amplified products through PAGE, AGE, etc.
(f) Identifying individuals/families carrying the desired marker alleles
(g) Selection of best individuals/families with desired marker alleles
(h) Repetition of above process for several generations to ensure association of markers with traits

Selection of all QTLs or genes simultaneously is a difficult process due to limitation of resources and facilities. MAS will not be much effective for complex traits regulated by many genes compared to traits controlled by one or a few genes. More than three QTLs are not appreciable choice. In tomato, even five QTLs were used to improve fruit quality marker-assisted introgression. With SNP markers (especially rapid automated detection and genotyping technologies), selection of more QTLs at the same time might be preferred and practicable. Priority must be attached to major QTLs that can explain proportion of phenotypic variation and/or can be consistently detected and evaluated across a range of environments and vivid populations. When more markers associated with a particular QTL will ensure success in selecting the QTL of interest.

The favourable situations while adopting MAS in breeding are:

(a) The selected trait is expressed late in plant development, like fruit and flower features.
(b) The target gene is recessive (so that individuals which are heterozygous positive for the recessive allele can be selected and/or crossed to produce some homozygous offspring with the desired trait).
(c) Special conditions are required in order to ensure expression of the target gene (s), as in the case of breeding for disease and pest resistance, where inoculation is required.
(d) The phenotype of a trait is governed by two or more unlinked genes. For example, selection for multiple genes or gene pyramiding may be required to develop enhanced or durable resistance against diseases or insect pests.

23.2.1 What Is Mapping?

Arranging markers in definite order is mapping. The genetic concepts of segregation and recombination, as done with classical Mendelian markers showing full dominance, have to be refreshed while doing mapping. Dominant and recessive alleles are given as upper- and lowercase letters, respectively.

As a result of meiosis, two alleles of a locus will segregate (separate from one another) with equal frequencies into the gametes. If A and a are two such alleles, then a diploid individual heterozygous at this locus (genotype Aa) will give gametes, half of which are A and half of which are a. B and b at a separate locus will segregate 50:50 into the gametes. If the A/a locus and the B/b locus are unlinked (i.e. are on different chromosomes), then the alleles will undergo independent segregation, giving four possible combinations in the gametes: AaBb to AB, Ab, aB and ab. The simplest way to follow such events, and to introduce recombination, is first to make a cross between two homozygous parents (P_1 and P_2). The offspring of this cross are referred to as the first filial (F_1) generation:

P_1 AABB x aabb P_2

⇩

AaBb (F_1)

Next, carry out a testcross between F_1 and the double-recessive parent P_2. The F_1 segregates to give four kinds of gametes (AB, Ab, aB, ab). The phenotypes of the testcross progeny tell us the genotypes of the gametes:

Testcross progeny

AB ab Parental type
Ab ab Recombinant
aB ab Recombinant
ab ab Parental type

The four classes of testcross progeny will occur in equal numbers. The two phenotypes that differ from P_1 and P_2 are those phenotypically Ab and aB and are the recombinants. With independent segregation, these will comprise 50% of the testcross progeny. On the other hand, if the genes are linked (i.e. on the same chromosome), the recombinants will only arise when crossing over occurs between them, and then their frequency will be <50%, as a rule. It is 50% because crossing over happens at the four-stranded stage of meiosis and only involves two of the four chromatids. Therefore, the maximum crossover value we can get for linked genes is 50%, and this will only occur when the loci are far apart, like at opposite ends of the chromosomes, so that there is always at least one crossover point (*chiasma*) between them (Fig. 23.8).

Recombination is the process by which new combinations of parental genes or traits arise and, as seen in Fig. 23.8, occurs through independent segregation of unlinked loci or by crossover between linked loci. The percentage of recombinants is the recombination frequency or crossover value. This is an estimation of the distance

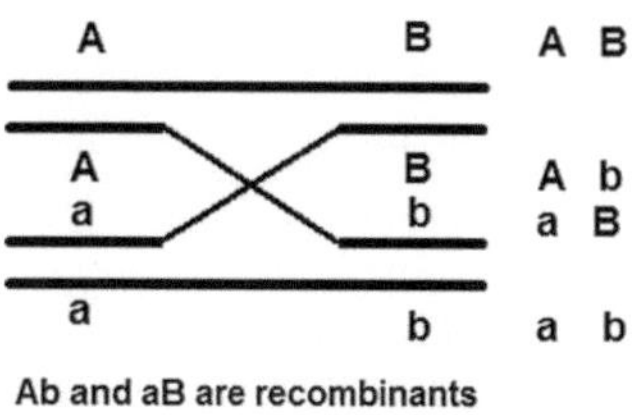

Fig. 23.8 Diagram of a bivalent at the four-strand (diplotene) stage of meiosis, showing how a chiasma involves only two of the four chromatids and can lead to a maximum of 50% recombination for genes at opposite ends of the chromosomes. When the two loci are closer together, chiasma formation will not always occur and recombination will be <50%

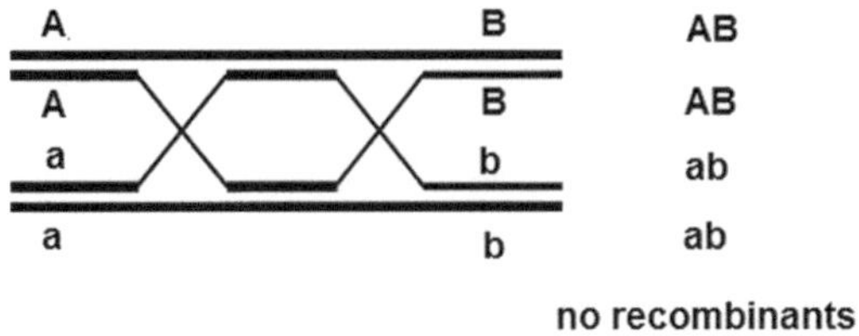

Fig. 23.9 Diagram of a bivalent at the four-strand (diplotene) stage of meiosis, showing how double crossovers involving the same pair of chromatids go undetected as recombinants and thus underestimate genetic distance

between two loci, on the assumption that the probability of crossing over is proportional to the distance between the loci.

23.2.1.1 Recombination and Linkage Maps

The recombination value for a pair of loci from a segregating backcross population is:

$$\frac{\text{No. of recombinants} \times 100}{\text{Total number of progeny}}$$

Supposing that the recombination between loci 1 and 2 = 6%, that between loci 2 and 3 = 20% and that between 1 and 3 = 24%, then we can order the loci along the chromosome:

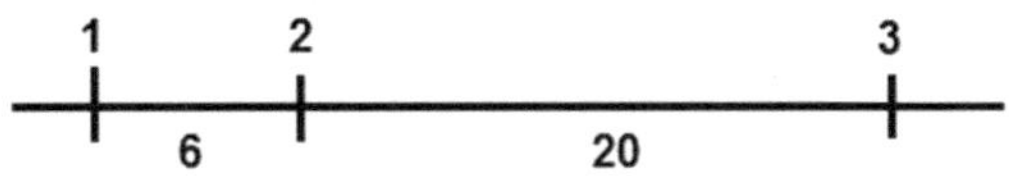

One percent recombination = one arbitrary map unit (centimorgan, or cM), and notice that in our map the genetic distances are not additive: 6 + 20 = 26 is the true distance between markers 1 and 3 (not 24). The underestimate based on the recombination between 1 and 3 is due to double (or multiple) crossovers, which go undetected as recombinants (Fig. 23.9). It is because of this reason that maps are made up by adding small intervals. Markers in one linkage group map together as they are all located in a single chromosome. The total number of linkage groups will correspond to the basic chromosome number of the species.

23.3 MAS for Qualitative Traits

Major genes/QTLs control several traits. They include resistance to diseases/pests, male sterility, self-incompatibility and others related to shape, colour and architecture of whole plants and/or plant parts. They inherit in a mono- or oligogenic way. Transfer of such genes to a specific line can lead to tremendous improvement. The tight linkage between markers and major genes can be selected which are sometimes more efficient than direct selection.

Soybean cyst nematode (SCN) (*Heterodera glycines* Ichinohe), the most economically significant soybean pest, may be taken as an example of MAS for major genes. Resistant cultivars are identified, but identifying resistant segregants in breeding populations is a difficult and expensive process. However, the SSR marker Satt309 has been identified to be located only 1–2 cM away from the resistance gene *rhg1*, which forms the basis of many public and commercial breeding efforts. Genotypic selection with Satt309 was 99% accurate in predicting lines that were susceptible. In yet another study, by using molecular markers, in a cross J05 × V94-5152, they developed five lines that were homozygous for all eight marker alleles linked to the genes/loci resistant to soybean mosaic virus (SMV). These lines exhibited resistance to SMV strains G1 and G7 and presumably carried all three resistance genes (*Rsv1*, *Rsv3* and *Rsv4*) that would potentially provide broad and durable resistance to SMV.

23.4 MAS for Quantitative Traits

Most of the important agronomic traits are polygenic or controlled by multiple QTLs. MAS for such traits is based on QTLs involved, QTL × environment interaction and epistasis. Therefore, repeated field tests are to be conducted in order to ensure exact characterization of the effects of QTLs and to estimate their stability across environments. However, there are factors that act as constrain to application of QTL mapping:

(a) Strong QTL-environment interaction makes phenotyping difficult since gene expression varies from location to another.
(b) Deficiencies in QTL statistical analysis.
(c) Sometimes there are no QTLs with major effects on the trait. This means a large number of QTLs have to be identified that becomes a tough goal to achieve.

QTLs can be of three types, viz. (a) major QTLs, (b) major + minor QTLs and (c) minor QTLs. Usually major QTLs control qualitative traits and have the Mendelian inheritance, whereas the other two types deviate the Mendelian nature of inheritance and make the situation difficult to trace them.

Linkage between a genetic marker and QTL was first demonstrated by Sax in 1923 by associating the seed size (a quantitative trait) with seed colour (a morphological marker) in *Phaseolus vulgaris*. Lack of more genetic markers was a major practical limitation. Later, the construction of saturated molecular marker maps that permits searching an entire genome for QTLs was made available. Prerequisites for QTL analysis are (a) an appropriate mapping population with segregation for the trait(s) of interest, (b) a saturated linkage map of molecular markers, (c) an acceptable phenotypic screening process to quantify the trait's manifestation and (d) powerful statistical packages to identify the QTLs.

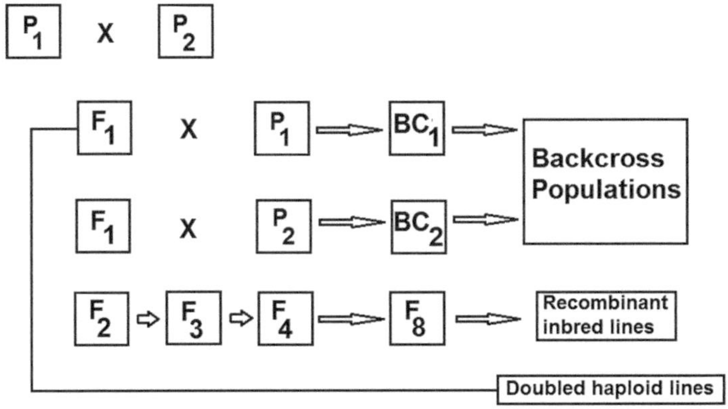

Fig. 23.10 Scheme to create various populations for mapping QTLs

Creation of Mapping Populations: Ways to create various mapping populations are given in Fig. 23.10. It would be always advantageous if accurate predictions are made by using early generations (e.g. F_2, F_3, BC population, etc.). It may be noted that predictions made during early generations may be misleading due to masking of minor genes. This masking can be avoided through continuous inbreeding to derive recombinant inbred lines (RILs). Thus, RILs can remain as the best choice of population for QTL analysis. As an alternative step, doubled haploid lines (DHL) can also be used.

Saturated Linkage Map: In tomato, entire genome for QTLs influencing a particular trait could be analysed with DNA markers. Subsequently, linkage maps were constructed with DNA markers in maize, lettuce, rice, potato, wheat and common bean. Such maps were based on RFLP markers. They were supplemented with RAPD, inter-simple sequence repeats, AFLP and SSRs. Currently, SSR markers are most popular for linkage map construction.

Reliable Phenotypic Screening Procedures: Phenotype of a trait is always dynamic. To adequately explore the QTL during the mapping phase, the phenotype must be evaluated in replicated trials in different environments. Such data will provide information about the magnitude of the effect of different QTL and whether there is interaction between QTL and environment.

Mapping Methods and Software: The first step in identification of a QTL is genotyping the individuals of a population by molecular marker survey. One can get three possible genotypes for each marker, i.e. A/A, A/a and a/a. The second step is phenotyping for the trait of interest. The third step is grouping the individuals based on the genotype of each marker, and finding out the group mean is the fourth step. Working out ANOVA to determine to test the significance of differences between the individual groups of each marker is the fifth step. The absence of significance indicates the absence of QTL near the marker. The presence of significance shows presence of QTL associated with the marker. There are several assumptions for QTL mapping: (1) genes for quantitative traits are available in the genome, just like simple genetic markers; (2) if the molecular markers occupy large portion of the genome, the genes for quantitative traits are linked with some of the genetic markers; and (3) if the genes and markers are segregating in a genetically defined population, then the linkage relationship among them may be resolved by studying the association between trait variation and marker segregation pattern. Single-marker analysis (SMA) and interval analysis can assist to study the association between trait variation and marker segregation pattern.

23.4.1 QTL Detection (Statistical)

QTL mapping detects QTL while minimizing the occurrence of false positive (type I error, i.e. declaring an association between a marker and QTL when in fact it does not exists). The tests for QTL or trait association are often performed by the following approaches:

Single-Marker Analysis (SMA): SMA is also referred as single-point analysis. It is the simplest method for detecting QTL associate with single markers. Tools like *t*-test, analyses of variance (ANOVA) and linear regression shall assist in undertaking single-point analysis. SMA is to be done for each marker locus separately. The drawbacks with the single-marker analysis are as follows: (a) The putative QTL genotypic means and QTL positions are confounded. This confounding causes the estimated QTL effects to be biased, and (b) QTL positions cannot be precisely determined, due to the non-dependence among the hypothesis tests for linked markers that confound QTL effect and position. The SMA is a well-acclaimed starting point for learning QTL mapping and practical data analysis. In single-marker analysis, only one marker is involved at a time to find the QTL-marker association (Fig. 23.11).

Interval Analysis or Interval Mapping: This is second level of QTL mapping but requires prior construction of a marker-based linkage map. This type of mapping is based on the joint frequencies of a pair of adjacent markers and a putative QTL in the middle (Fig. 23.12). Three types of interval mapping are (a) simple interval mapping (SIM), (b) composite interval mapping (CIM) and (c) multiple interval mapping (MIM).

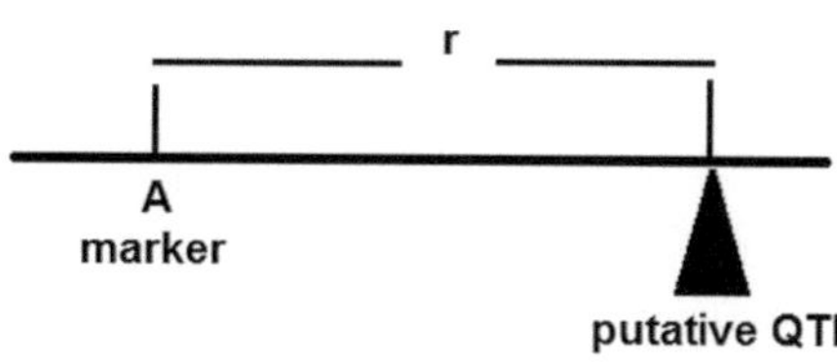

Fig. 23.11 Single-marker analysis. Association of a marker with a putative QTL

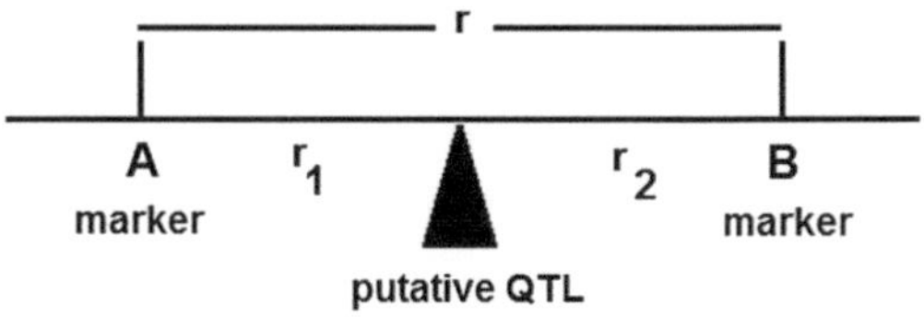

Fig. 23.12 Interval mapping. Association of a putative marker to tow flanking markers

Simple Interval Mapping (SIM): Simple interval mapping was first proposed by Lander and Botstein in 1989. SIM method makes use of linkage maps and analysis intervals between adjacent pairs of linked markers. Presence of a putative QTL is estimated if the logarithm of odds ratios (LOD) exceeds a critical threshold which is more often fixed as > or = 3. The use of linked markers for analysis compensates for recombination between the marker and the QTL and is considered statistically more powerful than SMA. Simple interval mapping (SIM) considers one QTL at a time. So, when multiple QTLs are located in the same linkage group, SIM can bias identification and estimation. SIM evaluates the association between the trait values and the expected contribution of hypothetical QTL (target QTL) at multiple analysis points between each pair of adjacent marker loci (the target interval). The flanking marker loci and their distance from the QTL direct the detection of QTL.

Composite Interval Mapping (CIM): Developed by Zeng in 1993, this combines internal mapping with linear regression. It considers a marker interval plus a few other well-chosen single markers in each analysis. It is more precise and effective than SMA and SIM, especially when linked QTLs are considered. Both single-marker analysis (SMA) and IM are biased when multiple QTL are linked to the marker/interval being considered. To deal with multiple QTL problems, a combination of SIM with multiple regression analysis in mapping is done. Multiple regression methods were integrated with IM to increase the probability of including all significant QTLs in the model. This method is named as composite interval mapping (CIM).

(c) Multiple Interval Mapping (MIM): MIM is the extension of interval mapping to multiple QTLs, just as multiple regression extends analysis of variance. MIM allows one to infer the location of QTLs to position between markers. MIM gives allowance for missing genotype data and can allow interaction between QTLs. Although CIM produces more accurate and precise estimates than IM, the inclusion of too many cofactors reduces its usefulness. But, MIM deals with the mapping of multiple QTLs more powerfully. MIM has the provision to use multiple marker intervals simultaneously to fit multiple putative QTLs for mapping QTLs. The MIM method is based on Cockerham's model for interpreting genetic parameters and the method of maximum likelihood for estimating genetic parameters. MIM improves precision and power of QTL mapping. Attributes like epistasis between QTLs, genotypic values of individuals and heritability of quantitative traits can also be analysed.

Linkage Analysis of Markers: Linkage map is prepared by using computer programs through coding data for each molecular marker on each individual. Numerous computer packages like Join Map, MAPMAKER/EXP, GMENDEL, LINKAGE and Map Manager QTX are available. Among them, Join Map is the most widely used.

Markers are assigned to linkage groups using the odds ratios (i.e. the ratio of linkage versus no linkage). This ratio is more conveniently expressed as the logarithm of the ratio and is called a logarithm of odds (LOD) value or LOD score. LOD values of >3 are typically used to construct linkage maps. A LOD value of 3 between 2 markers indicates that linkage is 1000 times more likely (i.e. 1000:1) than no linkage (null hypothesis). While higher critical LOD values will result in more number of fragmented linkage groups, the small LOD values will tend to have few linkage groups. Two markers if they are not linked are placed in distinct linkage groups. Linkage groups represent chromosomal segments or entire chromosomes. Polymorphic markers are clustered in some regions and absent in others. In addition to this, the frequency of recombination is not equal along chromosomes. The total individuals in the mapping population govern the accuracy of measuring the genetic distance and determining marker order.

Determination of Genetic (MAP) Distance: Mapping is building up of a map by adding loci one by one, starting from a pair of loci that is most informative. A marker is added further on the basis of total linkage information with markers that are already added. For each added locus, the best position is searched and a goodness-of-fit measure is calculated. Distance along a linkage map is measured in terms of the frequency of recombination between genetic markers. If the distance between genetic markers is greater, then the chance of recombination during the meiosis will be greater. Since recombination frequency and the frequency of crossing over are not linearly related, mapping functions are required to convert recombination fractions into centimorgans (cM). Two commonly used mapping functions are the Kosambi mapping function (that assumes one recombination event can influence the occurrence of adjacent recombination events) and the Haldane mapping function (assuming no interference between crossover events).

23.5 Next-Gen Molecular Breeding

The utility of molecular markers and mapping have been discussed in some detail in the previous sections. Markers are prerequisite for gene mapping and tagging, segregation analysis, genetic diagnosis, forensic examination, phylogenetic analysis and numerous biological applications. The use of most of the marker systems is restricted because of limited availability and high cost. SNPs are the most preferred markers. But, the development of high-throughput genotyping platforms for large numbers (thousands to millions) of SNPs is relatively time-consuming and costly. The greater demand for low-cost sequencing led to the development of high-throughput sequencing (or next-generation sequencing) that produces thousands or millions of sequences simultaneously. Such systems will be dealt here.

23.5.1 Next-Generation Sequencing (NGS)

Next-generation sequencing (NGS) relies on massively parallel sequencing and imaging techniques to yield several hundreds of millions to several hundreds of billions of DNA bases per run. Several NGS platforms, such as Roche 454 FLX Titanium, Illumina MiSeq and HiSeq2500, Ion Torrent PGM, have been developed and used during the last decade. Nanopore sequencing is the latest in this set of technologies. In ultra-high-throughput sequencing, as many as 500,000 sequencing-by-synthesis operations may be run concurrently.

All NGS strategies follow a similar protocol like (a) preparation of DNA templates (randomly sheared DNA fragments) with universal adapters ligated at both ends of the DNA template; (b) immobilization of clonally amplified DNA molecules on a synthetic surface to generate up to several billions of sequences in a massively parallel fashion; and (c) sequencing is done through incorporation of one or more nucleotides which is followed by the emission of a signal. This signal is detected by a sequencer.

NGS technologies commercialized by Illumina generate shorter reads, ranging from 50 to 300 bp, and the sequencing throughput ranges from 1.5 to 600 Gbp depending on the platform. The DNA strands are amplified by PCR to generate clusters of 1000 copies each (Fig. 23.13). Nucleotides added to the system will get paired at appropriate points in the single-stranded DNA. Each such attachment of nucleotide will emit a fluorescent signal. The amount of fluorescence emitted will be detected and measured by a sequencer (Fig. 23.13). The nature of the signal determines the identity of the base being incorporated. NGS is used for whole genome sequencing and re-sequencing to detect large numbers of SNPs for exploring diversity. Constructing haplotype maps (genetic variants are often inherited together in segments of DNA called haplotypes) and performing genome-wide association studies (GWAS – is an observational study of a genome-wide set of genetic variants in different individuals to see if any variant is associated with a trait) are other uses of NGS.

23.5.2 Genotyping-by-Sequencing (GBS)

While NGS has become cost-effective, GBS generates a large number of SNPs. Key components of this system include low cost, reduced sample handling, fewer PCR and purification steps, no size fractionation, no reference sequence limits, efficient bar coding and easiness to scale up. Figure 23.14 provides simplified GBS technology.

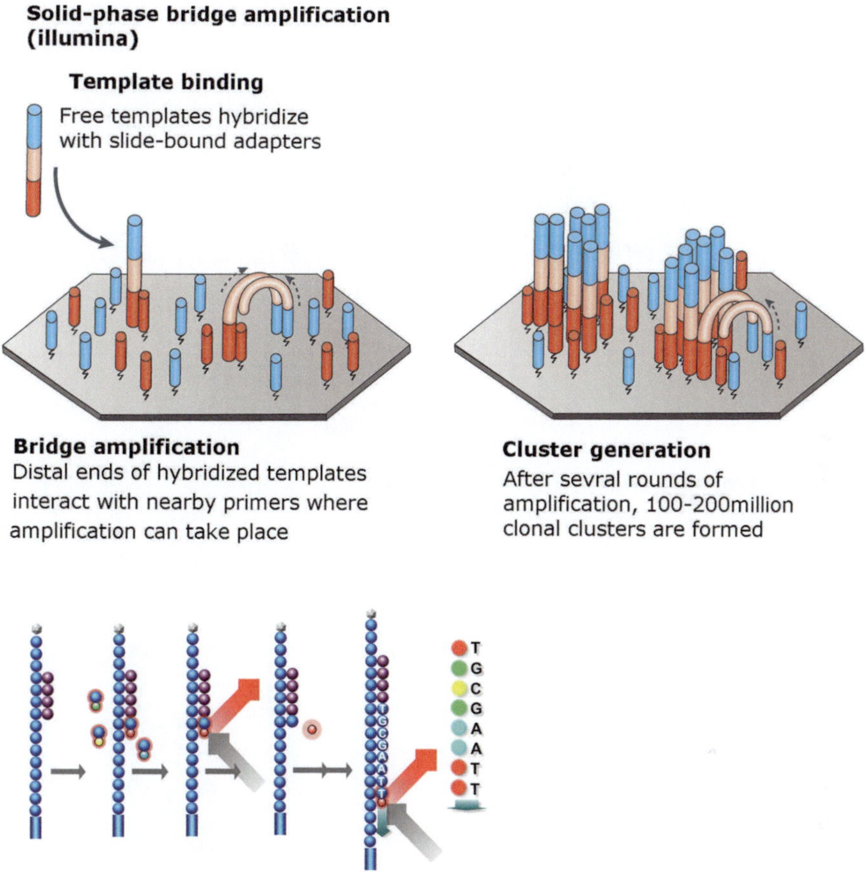

Fig. 23.13 Next-generation sequencing technology by Illumina. Tagged nucleotides are added in order to the DNA strand. Each of the four nucleotides has an identifying label that can be excited to emit a characteristic wavelength. A computer records all of the emissions, and from this data, base calls are made

Two different GBS strategies are as follows:

(a) Restriction enzyme digestion, in which no specific SNPs have been identified and ideal for discovering new markers for MAS programs. DNA is digested with one or two selected restriction enzymes prior to the ligation of adapters.
(b) Multiplex enrichment PCR, in which a set of SNPs has been defined for a section of the genome. Here, PCR primers amplify specific areas of interest.

GBS through the NGS approach has been used to re-sequence recombinant inbred lines (RILs). GBS is applied successfully in maize, wheat, barley, rice, potato and cassava. In maize, a collection of 5000 RILs have been re-sequenced using a restriction endonuclease-based approach and the Illumina sequencing technology. This generated 1.4 million SNPs and 200,000 indels (an insertion or deletion of

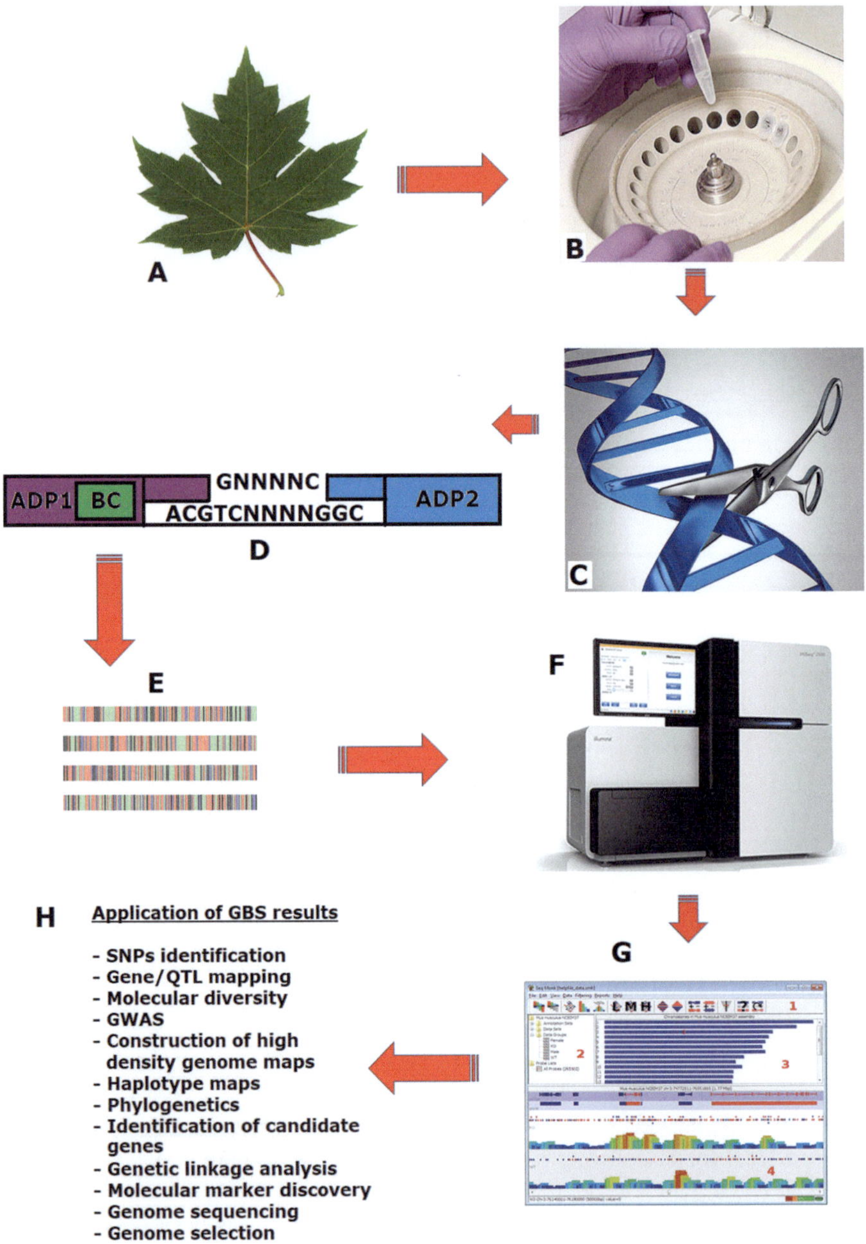

Fig. 23.14 Schematic steps of the genotype-by-sequencing (GBS) protocol. (**a**) Tissue is obtained from any plant species. (**b**) DNA extraction. (**c**) DNA digestion with restriction enzymes. (**d**) Ligations of adaptors (ADP) including a bar coding [BC] region in adapter 1 in random *PstI-MseI* restricted DNA fragments. (**e**) Representation of different amplified DNA fragments with different bar codes from different biological samples/lines. These fragments represent GBS library. (**f**) Analysis of sequences from library on a NGS sequencer. (**g**) Bioinformatic analysis of NGS sequencing data. (**h**) Possible application of GBS results

bases). A comprehensive genotyping of 2815 maize inbred accessions showed that 681,257 SNP markers are distributed across the entire genome, in which some SNPs are linked to the known candidate genes for kernel colour, sweetness and flowering time. In potato, 12.4 gigabases of high-quality sequence data and 129,156 sequence variants have been identified, which are mapped to 2.1 Mb of the potato reference genome with a median average read depth of 636 per cultivar.

23.5.3 Genetic Maps

A genetic map represents the ordering of molecular markers along chromosomes as well as the genetic distances, generally expressed as centimorgans (cM = a centimorgan is a unit used to measure genetic linkage; one centimorgan equals a one percent chance that a marker on a chromosome will become separated from a second marker on the same chromosome due to crossing over in a single generation), existing between adjacent molecular markers. Most frequently, genetic maps have been created from F_2, backcrosses and recombinant inbred lines. Although longer to develop, RILs offer a higher genetic resolution. Once a mapping population has been created, it takes only few months to produce a genetic map with a 10-cM resolution (Fig. 23.15). Genetic maps facilitate identification of quantitative trait loci and marker-assisted selection.

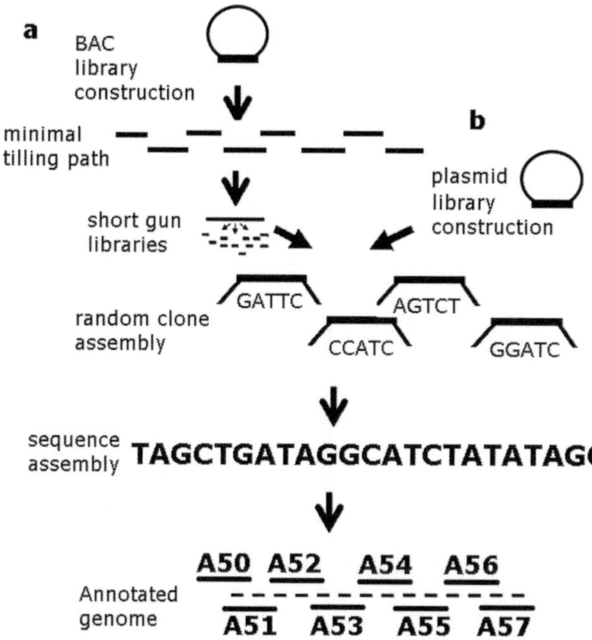

Fig. 23.15 Approaches of large-scale sequencing. (**a**) Clone-by-clone strategy and (**b**) short gun strategy

23.5.4 Physical Maps

Genetic maps provide gene location, but the kilobases per centimorgan (kb/cM) ratio is large, from 120 to 250 kb/cM in *Arabidopsis* and between 500 and 1.500 kb/cM in corn. Therefore, a 1-cM interval may harbour ~30 to 100 or even more genes. Physical maps bridge such gaps, representing the entire DNA fragment spanning the genetic location of adjacent molecular markers.

Physical maps can be defined as a set of large insert clones with minimum overlap encompassing a given chromosome. First-generation physical maps in plants were based on YACs (yeast artificial chromosomes). Chimaeras and stability issues, however, dictated the development of low-copy, *E. coli*-maintained vectors such as bacterial artificial chromosomes (BACs) and P1-derived artificial chromosomes. Although BAC vectors are relatively small (molecular weight of BAC vector pBeloBAC11 is 7.4 kb, for instance), they carry inserts between 80 and 200 kb on average and possess traditional plasmid selection features such as an antibiotic resistance gene and a polycloning site within a reporter gene allowing insertional inactivation. BAC clones are easier to manipulate than yeast-based clones. Once a BAC library is prepared, clones are assembled into contigs using fluorescent DNA fingerprint technologies and matching probabilities. Physical and genetic maps can be aligned, bringing along continuity from phenotype to genotype. Furthermore, they provide the platform clone-by-clone sequencing approaches rely upon. Figure 23.16 shows the relationship between genetic and physical maps and their

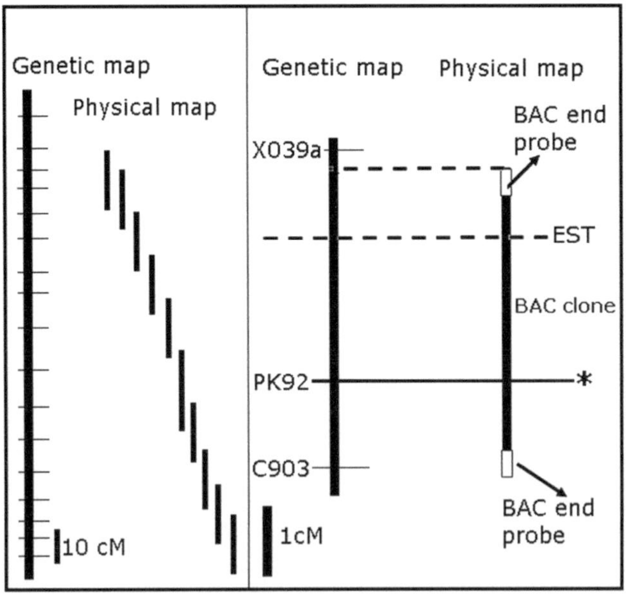

Fig. 23.16 Maps used in plant genetics. (**a**) Genetic and physical maps of a hypothetical chromosome. Horizontal lines on the genetic map represent loci targeted by a molecular marker; vertical lines represent overlapping BAC clones. (**b**) Alignment of genetic and physical maps using BAC ends sequence (dashed lines), ESTs (dotted line) and molecular markers (∗)

alignment. Physical maps provide the bridge needed between the resolution achieved by genetic maps and that needed to isolate genes through positional cloning.

Further Reading

Arif IA (2010) A brief review of molecular techniques to assess plant diversity. Int J Mol Sci 11:2079–2096. https://doi.org/10.3390/ijms11052079

Birchler JA, Han F (2018) Barbara McClintock's Unsolved Chromosomal Mysteries: Parallels to Common Rearrangements and Karyotype Evolution. Plant Cell 30:771–779

Collard BCY, Mackill DJ (2008) Marker-assisted selection: an approach for precision plant breeding in the twenty-first century. Phil Trans R Soc B 363:557–572. https://doi.org/10.1098/rstb.2007.2170

Dunwell JM (2011) Crop biotechnology: prospects and opportunities. J Agric Sci 149(S1):17–29. ISSN 1469-5146. https://doi.org/10.1017/S0021859610000833

Nybom et al (2014) DNA fingerprinting in botany: past, present, future. Investig Genet 5:1–35

Welsh J, McClelland M (1990) Fingerprinting genomes using PCR with arbitrary primers. Nucl Acids Res 18:7213–7218

Xu Y (2010) Molecular plant breeding. CABI

Williams JGK, Kubelik AR, Livak KJ, Rafalski JA, Tingey SV (1990) DNA polymorphisms amplified by arbitrary primers are useful as genetic markers. Nucl Acids Res 18:6531–6535

Genomics

24

Keywords

Genetic structure of plant genomes · Nuclear genomes and their size · Chemical and physical composition of plant DNA · The packaging of the genome · The genomic DNA sequence · Model plant species · Genome co-linearity/genome evolution · Whole genome sequencing · Transposable elements · DNA microarrays · Genomics-assisted breeding · Genome sequencing and sequence-based markers · High-throughput phenotyping · Marker-trait association for genomics-assisted breeding · From genotype to phenotype · Post-transcriptional gene silencing (PTGS) · The new systems biology

Abbreviations

bp, kbp	Base pairs, kilobase pairs
ddNTPs	Dideoxynucleotide triphosphates
DH	Doubled haploid
DiGE	Difference gel electrophoresis
DNA	Deoxyribonucleic acid
DSB	Double-strand break
dsRNA	Double-stranded RNAs
ELISA	Enzyme-linked immunosorbent assay
FT-MS	Fourier transform mass spectrometry
GBSS	Granule-bound starch synthase
GC	Gas chromatography
GFP	Green fluorescent protein
GM	Genetically modified
GMM	Genetically modified microorganism
GMO	Genetically modified organism
GUS	Beta-glucuronidase gene

P. M. Priyadarshan, *PLANT BREEDING: Classical to Modern*,
https://doi.org/10.1007/978-981-13-7095-3_24

GVA	Grapevine virus A
HILIC	Hydrophilic interaction chromatography
HPLC	High-performance liquid chromatography
hpRNA	Hairpin RNA
HR	Homologous recombination
HRM	High-resolution melting
LC	Liquid chromatography
LFD	Lateral flow devices
LNA	Locked nucleic acids
LOD	Limit of detection
LOQ	Limit of quantification
MALDI	Matrix-assisted laser-desorption ionization
MAS	Marker-assisted selection
miRNA	MicroRNA
mRNA	Messenger RNA
MS	Mass spectrometry
MS-HRM	Methylation-sensitive high-resolution melting
ncRNA	Non-coding RNA
NHEJ	Non-homologous end-joining
NMR	Nuclear magnetic resonance
NOS	Nopaline synthase
NPTII	Neomycin phosphotransferase gene
nt	Nucleotides
NTTF	New Techniques Task Force
NTWG	New Techniques Working Group
ODM	Oligonucleotide-directed mutagenesis
OECD	Organisation for Economic Co-operation and Development
ORF	Open reading frames
PAGE	Polyacrylamide gel electrophoresis
PAT	Phosphinothricin phosphotransferase
PCR	Polymerase chain reaction
PCT	Patent Cooperation Treaty
PEG	Polyethylene glycol
PTA	Plate-trapped antigen
PTGS	Post-transcriptional gene silencing
RdDM	RNA-dependent DNA methylation
RNAi	RNA interference
RP	Reversed-phase
rRNA	Ribosomal RNA
RT qPCR	Real-time quantitative PCR
siRNA	Small interfering RNA
SNPs	Single-nucleotide polymorphisms
TALEN	Transcription activator-like effector nucleases
TAS	Triple antibody sandwich
T-DNA	Transfer DNA

TFO	Triple helix-forming oligonucleotide
TGS	Transcriptional gene silencing
TOF	Time of flight
tRNA	Transfer RNA
UHPLC	Ultra-high-performance liquid chromatography
UV	Ultra-violet
ZFN	Zinc finger nuclease

Genomics is the study on how the complex sets of genes are expressed in cells (the term *genomics* was coined by Tom Roderick, a geneticist at the Jackson Laboratory, Bar Harbor, USA, in 1986). It's a discipline in genetics that applies recombinant DNA, DNA sequencing methods and bioinformatics to sequence, assemble and analyse the structure and function of genomes. Though the term genetic engineering is modification of plants and animals through recombinant DNA technology, human beings have been actually practising genetic engineering for thousands of years. The rate of crop improvement was increased because of an in-depth understanding of genetics during the beginning of the twentieth century. Introduction of hybrid corn was the most dramatic agricultural development. But highly inbred lines gave decreased yield because of homozygous deleterious recessive alleles. As per the observation of George Harrison Shull, crossing of two different inbred lines gave progeny with "hybrid vigour", with fourfold yield. Hybrid rice of the International Rice Research Institute in the Philippines gave 20% extra yield. Currently, breeders are looking for genes to optimize nutritional quality like golden rice. Rice is staple food for almost half the world's population, but it lacks vitamin A. Vitamin A deficiency causes reduced vision and immunity. Genetically engineered golden rice is with vitamin A. It has been named golden rice because of the gold-coloured beta-carotene, a precursor to vitamin A. The intensity of golden colour increases with the presence of pro-vitamin A. The commencement of the twenty-first century made new ways to understand genomes. The complexity of plant genomes is multi-fold compared to eukaryotic genomes with evolutionary flips and turns of DNA sequences. Chromosome numbers and ploidy levels are also widely different. The size of plant genomes (both number of chromosomes and total nucleotide base pairs) shows the greatest variation in the biological world. As an example, wheat contains over 110 times more DNA compared to *Arabidopsis thaliana* (Table 24.1). Plant DNA contains sequence repeats, sequence inversions or transposable element insertions that modify the genetic content further.

24.1 Genetic Structure of Plant Genomes

Nuclear genome consists of DNA and the nucleus is encased by a double membrane in each cell (Fig. 24.1). During mitosis, the genome condenses into chromosomes, the nuclear membranes break down, and the chromosomes divide, moving into the two daughter cells. Towards the end of the twentieth century, a small number of

Table 24.1 Nuclear genome size in different species

Common name	Scientific name	Nuclear genome size (in megabases)
Wheat	*Triticum aestivum*	15,966
Onion	*Allium cepa*	15,290
Garden pea	*Pisum sativum*	3947
Corn	*Zea mays*	2292
Asparagus	*Asparagus officinalis*	1308
Tomato	*Lycopersicon esculentum*	907
Sugar beet	*Beta vulgaris*	758
Apple	*Malus X domestica*	743
Common bean	*Phaseolus vulgaris*	637
Cantaloupe	*Cucumis melo*	454
Grape	*Vitis vinifera*	483
Arabidopsis	*Arabidopsis thaliana*	145
Man	*Homo sapiens*	2910

1 Mb = 1,000,000 bases

plant genomes were sequenced. Rice and *Arabidopsis* were the fully sequenced genomes. Well-characterized genomes include maize (corn), soybean, alfalfa, grape, citrus, sugar beet, sorghum, barley, potato, tomato, poplar tree and the pigeon pea. Plant cell also contains several mitochondria and plastids, both with their own genome (the cytoplasmic genomes) that constantly interact with the nuclear genome.

24.1.1 Nuclear Genomes and Their Size

Rice nuclear genome consists of 450 million base pairs (Mbp) of DNA distributed among 12 chromosome pairs that include genes encoding nearly 38,000 proteins. However, these genes represent less than 10% of the total amount of DNA, and the rest of the DNA consists of repetitive sequences in thousands. *Arabidopsis thaliana* has 157 Mbp with about 31,000 genes on 5 chromosome pairs.

All higher plants, at the diploid level, require approximately the same number of genes and regulatory DNA sequences for physiological processes like seed germination, growth, flowering and reproduction. However, nuclear genome sizes vary enormously between species. The amount of nuclear DNA can be given as an absolute weight of the DNA (in pg, picograms) or converted into the number of base pairs represented by that weight. The number of base pairs for 1C genome size ranges from 70 Mbp in the carnivorous plant *Genlisea* to more than 1,30,000 Mbp in the lily species *Fritillaria assyriaca.* This is a remarkable difference of 2000 times. One reason for size variation is polyploidy with multiple copies of chromosomes, and the popular belief is that 50% or more of angiosperms are polyploid in their origin. Another reason for genome size variation is the amount of repetitive DNA in the genome.

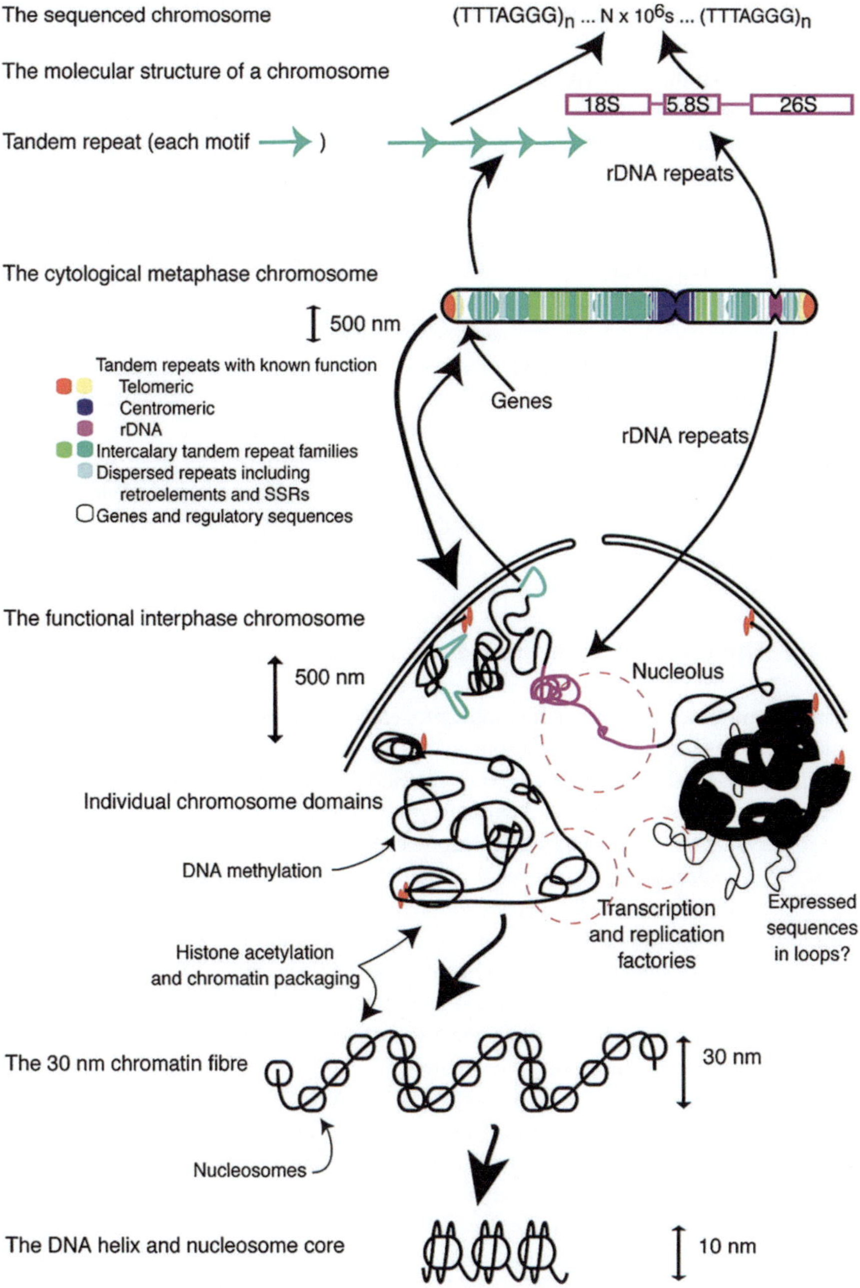

Fig. 24.1 Genome organization

24.1.2 Chemical and Physical Composition of Plant DNA

Each double-stranded DNA molecule is made up of four deoxynucleotides, viz. adenine, thymine, guanine and cytosine (A, T, G and C). The number of A residues equals the number of T residues (so also G with C) because of the pairing of bases, but the ratio of (G + C)/(A + T), or GC content, is a characteristic of the genome. Plant genes usually have higher GC content in exons (DNA that translates to protein) and lower contents in introns (regions that flank an exon). Spectrophotometry that works based on the absorbance of UV light at 260 nm by DNA is used to measure the concentration and purity of DNA. Enzymes of bacterial origin, called *restriction endonucleases*, are used for the site-specific cleavage (hydrolysis) of very large DNA molecules. These endonucleases recognize short sequence motifs of 4–8 bp and cleave the long DNA into defined fragments. Such fragments are separated in gel electrophoresis. Analysis of DNA involves denaturation of the double-stranded DNA into single-stranded molecules and labelling them with probes. Fluorescent in situ hybridization (FISH) is carried out on chromosome preparations using probes detected with fluorescence. For a detailed account of DNA, one may refer a book on molecular biology.

24.1.3 The Packaging of the Genome

Plant chromosomes are in pairs of homologues each chromosome originating from either male or female. The diploid chromosome number is referred to as $2n$, and the number in a gamete, the haploid number, would be n. Chromosome number is characteristic of each species, known to vary from $n = 2$ (*Haplopappus gracilis*) to $n = 630$ in Adder's tongue fern (*Ophioglossum reticulatum*). Each chromosome includes one or two double-stranded linear DNA molecules (after replication). The length of a DNA shall be from less than 20 Mbp to more than 900 Mbp depending on the species. When stretched to full length, the DNA molecule would be between 7 and 300 mm long. DNA is wrapped around nucleosomes made out of octamer core of histones. Around 50 bp of DNA wrap twice around each nucleosome. There is a spacer (typically 10–20 bp long) before the next nucleosome (Fig. 24.2). Since 2000, the significance of the histone proteins to gene expression has become increasingly recognized.

Little is understood about the packaging of DNA because of difficulty of imaging a complex structure where DNA together with salts, nuclear proteins and interaction of charges gives rise to the structure. The telomere protects the chromosome through a sequence TTTAGGG. This sequence is added to the end of the DNA molecule by *telomerase*, with *reverse transcriptase* activity (ability to produce DNA from RNA). Each chromosome is with a regional *centromere* consisting of hundreds of kilobase-long DNA. Centromere functions to hold the two DNA molecules that are condensed into *chromatids*. Centromere is where the *kinetochore* assembles and *spindle microtubules* attach to move the chromatids apart during division. The replication and transcription enzymes open the DNA to permit *DNA polymerase* to transcribe mRNA.

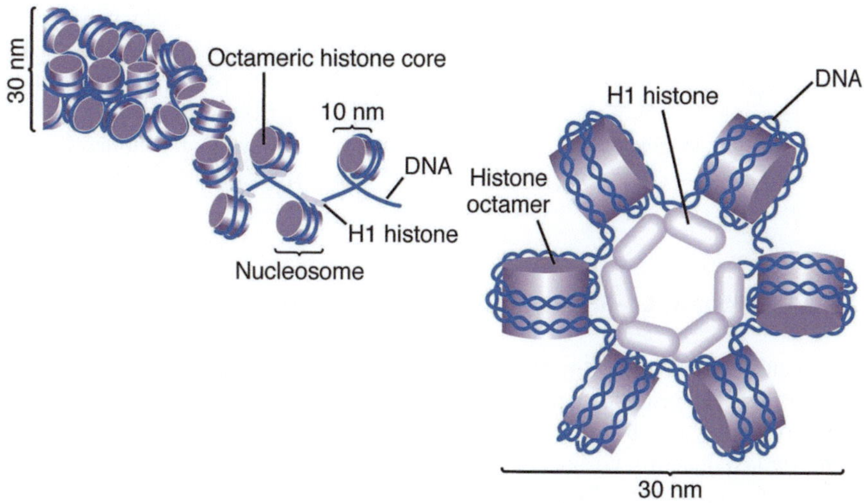

Fig. 24.2 Organization of chromatin

24.1.4 The Genomic DNA Sequence

The sequence of DNA includes exons, introns, regulatory sequences and repetitive DNA motifs. Repetitive DNA consists of sequence motifs from dinucleotides (such as the monotonic repetition GAGAGA) to motifs longer than 10,000 bp. These motifs are repeated in many hundreds to thousands. Such repetitive sequences are dispersed throughout the genome that make up around 50–75% of the entire DNA of a nucleus. Often referred to as junk DNA, repetitive DNA is vital for genome function and evolution. Repetitive DNA may change in sequence and abundance that becomes responsible for divergence of genomes and speciation. Satellite DNA is yet another set of DNA. Satellite DNA makes up large proportion of heterochromatin, the condensed form of chromatin during cell cycle that has some evolutionary significance.

24.1.5 Model Plant Species

Knowledge of plant genomes has been growing with the advent of new techniques to study DNA sequences, such as gene mapping and chromosome synteny (synteny is the condition of two or more genes being located on the same chromosome whether or not there is demonstrable linkage between them). Manipulation of genetic traits like crop yield, disease resistance, growth abilities, nutritive qualities or drought tolerance can be undertaken with increased understating of genome. Multiple genes are responsible for coding these traits. Genome mapping model plants could lead to better understanding of evolution at genetic level. Rice and *Arabidopsis* are such model systems (see Table 24.1). *Arabidopsis* has a small genome of 120 megabases

(Mb) and has only five haploid chromosomes. Rice has two main subspecies: *japonica* is mostly grown in Japan, while *indica* is grown in China and other Asia-Pacific regions. Rice also has very saturated genetic maps, physical maps, whole genome sequences as well as EST collections pooled from different tissues and developmental stages. It has 12 haploid chromosomes, with a genome size of 420 Mb. Both *Arabidopsis* and rice can be transformed through biolistics and *A. tumefaciens*.

24.1.6 Genome Co-linearity/Genome Evolution

Plant genomics has its ability to bring together more than one species for analysis. The comparative genome mapping of related plant species demonstrated that during evolution, the organization of genes gets conserved. This unequivocally demonstrated genome co-linearity between model crops (*Arabidopsis* for dicots and rice for monocots). Co-linearity can be defined as the conservation of gene order within a chromosomal segment between different species. A concept related to this is synteny. Synteny is the presence of two or more loci on the same chromosome irrespective of the fact that they are genetically linked or not.

Co-linearity is observed among cereals (corn, wheat, rice, barley), legumes (beans, peas and soybeans), pines and Cruciferae species (canola, broccoli, cabbage, *Arabidopsis thaliana*). Recently, the first studies at the gene level have demonstrated that micro co-linearity of genes is less conserved; small-scale rearrangements and deletions complicate micro co-linearity between closely related species. A 78-kb genomic sequence of sorghum around the locus *adh1* has shown micro co-linearity with homologous genomic fragment from maize. They share nine genes in common and also another five unshared genes reside in this genomic region.

24.1.7 Whole Genome Sequencing

The prevailing method of determining the sequence of a long DNA segment is the *shotgun sequencing* approach, in which a random sampling of short-fragment sequences is acquired and then assembled by a computer program to infer the sampled segment's sequence. In the early 1980s, segments of 5000–10,000 base pairs (5–10 kbp) were sequenced. By 1990, this became 40 kbp, and by 1995, the entire 1800-kbp *Haemophilus influenzae* bacterium was sequenced (see Chap. 23 for DNA sequencing).

24.1.8 Transposable Elements

Transposable elements (TEs), (jumping genes) or transposons, are sequences of DNA that move from one location to the other in the genome. Maize geneticist Barbara McClintock discovered TEs in the 1940s, and for several decades, these were ignored as useless or "junk" DNA. McClintock suggested that these mobile elements could have some kind of regulatory role governing switching *on* and *off* of genes.

Almost at the same time when McClintock did work on jumping genes, Roy Britten and Eric Davidson (during 1969) speculated that TEs are also vital in generating cell types and biological structures based on the location of TEs in the genome. They further hypothesized that this might explain the necessity of cells, tissues and organs in a biological system. If every single gene was expressed at all the time, the plant would be an undifferentiated matter. Speculations of both McClintock and Britten and Davidson were not accepted by the scientific community. Now, scientists realize TEs make up of almost 40% of the genome and carry out regulatory role.

24.1.9 DNA Microarrays (DNA Chip or Biochip)

DNA microarrays is a process by which minuscule amounts of hundreds or thousands of DNA sequences are arranged on a single microscope slide created by robotic machines. Upon activation of a gene, mRNA is produced. mRNA is the template for creating proteins. mRNA thus produced is complementary to the DNA sequence from where it is produced. Hence, the mRNA can bind to the DNA strand from which it was produced. To determine which genes are turned on and which are turned off in a given cell, a researcher must first collect the messenger RNA molecules present in that cell. The enzyme reverse transcriptase enzyme (RT) can generate complementary cDNA to mRNA. cDNA thus formed will be with fluorescent nucleotides. Next, the labelled cDNAs are added onto a DNA microarray slide. Such labelled cDNAs will hybridize to their synthetic complementary DNAs attached on the microarray slide, emitting fluorescence. Fluorescence is measured by a special scanner for each spot on the microarray slide. When a gene is very active, many mRNA molecules are produced. So when more labelled cDNAs, when hybridize with the DNA on the microarray slide, it produces bright fluorescence. In this way, when a gene is less active, it gives dimmer fluorescent spots. If the gene is inactive, no fluorescence will be produced (Fig. 24.3). There are two main applications of DNA microarrays: the determination of gene expression level and the analysis of the genomic DNA. They are briefly discussed here.

Gene Expression During cell life cycle, some genes are actively transcribed and some are not. When a protein is needed in high amounts, the gene in question is activated and efficiently transcribed to produce large amounts of mRNA. Some genes that are responsible to produce proteins involved in the basic cellular processes are always active. Some genes are more tissue-specific. To know the specific function of a gene, when and where the gene is getting activated is to be known. While technique, like Southern blotting, can only deal with very few genes, DNA microarrays can determine the expression level for the whole genome simultaneously.

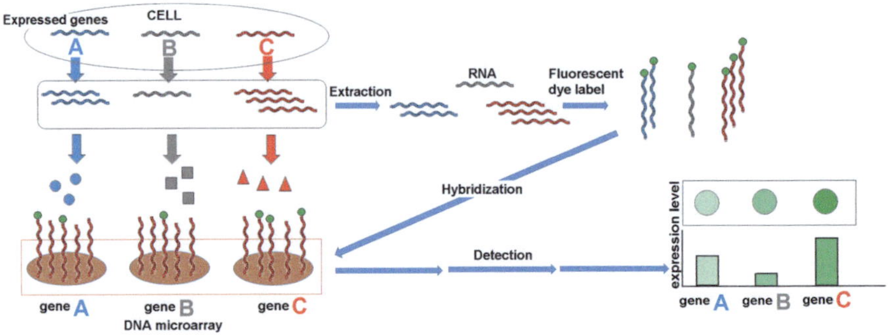

Fig. 24.3 DNA microarrays

Analysis of Genomic DNA Analysis of genomic DNA is done with SNPs or through analysis of deleted/amplified regions. SNPs differ with different members of a species. For instance, two different DNA fragments from two individuals may have sequences . . .GGT**C**ACC. . . and . . .GGT**A**ACC. . . There is an SNP with two alleles C/G. Specific microarrays are designed for SNP genotyping, i.e. the determination of the alleles of SNP in one individual. DNA microarrays can be used to analyse copy number variation. In principle, DNA is composed of equal amounts of paternal and maternal DNAs. Hence, each gene is present in two copies. But, due to aberration in DNA replication, there is a chance that a fragment of a chromosome is lost, leaving only one copy of a gene. Sometimes, a DNA fragment may be copied more than once that leads to amplification of a chromosomal region. DNA microarrays are very useful for the analysis of genomic DNA.

24.2 Genomics-Assisted Breeding

Conventional breeding is based on phenotypic selection that resulted in high-yielding commercial varieties. But this is labour intensive, time-consuming, less efficient and dependent on environment. With emergence of genomics, the focus shifted from phenotype-based to genotype-based selection. Breeding efficiency could be improved through marker-assisted selection (MAS). MAS strategies developed are:

(a) Marker-assisted backcrossing or introgression of major genes or quantitative trait loci (QTL)
(b) Enrichment of favourable alleles in early generations
(c) Selection for quantitative traits using markers at multiple loci

A whole genome could be analysed now with high-density SNP markers through whole genome sequencing and maker development. The complex traits can be

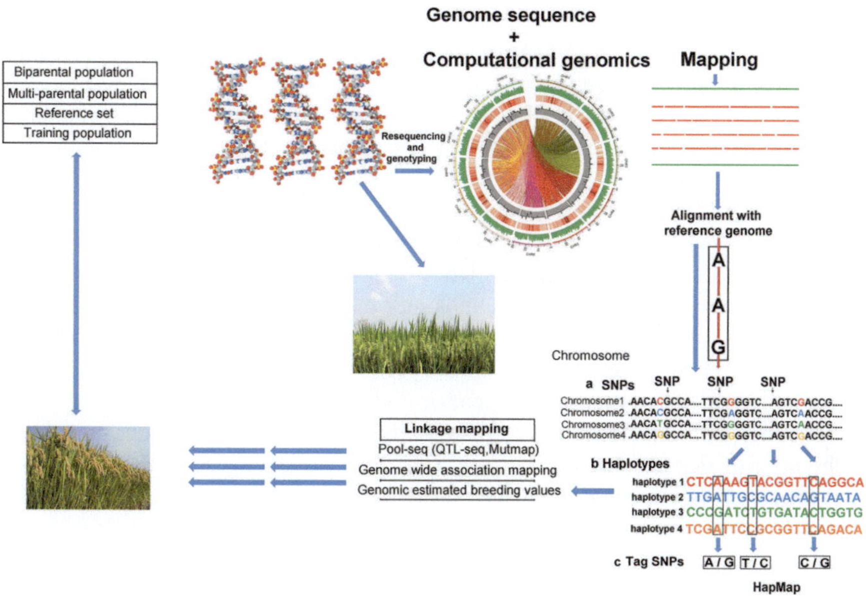

Fig. 24.4 Scheme for genomics-assisted breeding
The figure illustrates a roadmap for the utilization of various genetic and genomic resources for deploying genomics-assisted breeding (with rice as an example). In order to accelerate the existing breeding efforts, the strategy has been given in the figure which will be followed in the coming years (figure representative)

analysed through whole genome and transcriptome sequencing that gives a bridge between phenotype and the genotype. Genomics-assisted breeding (GAB) has become a powerful strategy for plant breeding. GAB enables the integration of genomic tools with high-throughput phenotyping that facilitates prediction of phenotype from genotype (Fig. 24.4). GAB is with high accuracy, direct improvement, short breeding cycle and high selection efficiency. The ultimate goal of GAB is to find the best combinations of alleles (or haplotypes), optimal gene networks and specific genomic regions to facilitate crop improvement.

24.2.1 Genome Sequencing and Sequence-Based Markers

DNA fingerprinting methodologies like RFLPs, RAPDs and SSRs are often labour intensive and time-consuming and impractical to be implemented on a large scale. Most of these markers are not localized in the target gene region and fail to exhibit any impact. Of late, SNPs became popular because of their abundance and ability to be detected with high-throughput methods.

The sequences of crop genomes are useful for exploring genome organization and gaining insight into genetic variation via the re-sequencing of different accessions. A total of 278 maize lines, including public US and elite Chinese lines, were re-sequenced and resulted in the identification of >27 million SNPs. With the initiation of the "3000 Rice Genomes Project", a large panel of rice accessions has been re-sequenced with an average of 14× sequencing depth, resulting in >18.9 million SNPs. In wheat, a combined strategy using methylation-sensitive digestion of genomic DNA and next-generation sequencing was carried out for high-throughput SNP discovery, resulting in ~23,500 SNPs. Whole genome re-sequencing was conducted in barley and soybean. Sequence-based markers associated with rare elite alleles will facilitate positional cloning and crop breeding.

The whole genome re-sequencing data generates high-throughput unlimited SNP genotyping technologies, such as DNA chips, to detect genome-wide DNA polymorphisms. Two chip-based technologies have been widely used, namely, the GeneChipTM microarray technology from Affymetrix (Santa Clara, CA, USA; www.affimetrix.com) and the BeadArrayTM technology from Illumina (San Diego, CA, USA; www.illumina.com). Other newly developed commercial genotyping platforms including EurekaTM from Affymetrix® and Infinium from Illumina also depend on high-density SNP markers. In maize, large-scale SNP genotyping array has been established using more than 800,000 SNPs. Such SNPs were evenly distributed across the maize genome.

24.2.2 High-Throughput Phenotyping

Plant phenotyping remains a big challenge in this era of high-throughput plant genome analysis. Conventional phenotyping does not provide accurate prediction of complex quantitative traits. Thus, high-throughput phenotyping platforms (HTPPs) became essential for plant phenomics. HTPP facilitates non-destructive phenotyping and high-efficiency data recording and processing. Rapid progress was made towards HTPPs due to technological advances in computing and robotics, light detection and ranging (LiDAR), unmanned aerial vehicle remote sensing, etc. An International Plant Phenomics Network was set up for high-throughput phenotyping via robotic, non-invasive imaging across the life cycle of small, short-lived model plants and crops. Plant height, leaf length, width and angle were measured on a phenotyping platform in the greenhouse, which was developed by the integration of LiDAR, high-resolution camera and hyperspectral imager. Dynamic growth traits from the seedling to tasselling stage were quantified using a HTPP from a maize RIL population in the greenhouse.

Field phenotyping with the development of novel sensors, image analysis, robotics, etc. has benefited plant breeding (Table 24.2). Still, large-scale accurate phenotyping is still infant. It is also inefficient for estimating association of genotype and phenotype under highly variable environments. Physiological breeding based on HTPPs together with genomic selection is beneficial in many ways. But for traits like disease resistance, where artificial inoculation is required to induce disease infestation, low-cost and accessible data managements are urgently needed. Renovated technique will certainly assist further application of HTTP in genome-assisted breeding to benefit crop breeders.

Table 24.2 High-throughput phenotyping platforms

Technology	Trait	Condition
Imaging	Plant growth and chlorophyll fluorescence	C
Camera	Leaf growth	C
Spectroradiometer	Drought tolerance	C
Imaging	Leaf area	F
Visual	Root architectural traits	F
Camera	Presence of rice bugs	F
Hydraulic push press	Root depth and distribution	F
Sensor	Canopy height	F

C controlled conditions, *F* field conditions

24.2.3 Marker-Trait Association for Genomics-Assisted Breeding

Almost all agronomically and economically important traits are controlled by multiple QTL. QTL detection is of great relevance to marker-assisted breeding. Linkage mapping delineates genetic basis of quantitative trait loci. So far, a huge number of QTLs have been identified using this method. Bioinformatics together with genetic information gave way to meta-QTL analysis.

Genome-wide mapping through utilizing high-density SNP markers led to emergence of the new genome-wide association study (GWAS – association of genomic regions to traits). GWAS helps to dissect complex traits. By combining high-throughput phenotypic and genotypic data, GWAS provides insights into the genetic architecture of complex traits in maize. Through GWAS, a total of 26 loci were detected to be associated with oil concentration in maize kernels. This data can be used for marker-based breeding for oil quantity and quality. In rice, QTLs associated with chilling tolerance were identified through GWAS, set as useful markers for chilling tolerance improvement.

Genomic selection (GS – a form of marker-assisted selection in which genetic markers covering the whole genome are used so that all QTLs are in linkage disequilibrium with at least one marker) predicts genomic-estimated breeding values (GEBVs). GS is another promising breeding strategy for rapid improvement of complex traits. Even for traits with low heritability, correlations were found between genomic-estimated and true-breeding values. GS was proved to be advantageous for complex traits, like grain yield. The other advantages with GS are shortening the selection cycle and generation of reliable phenotypes. GS has been applied to several traits in maize, barley, bread wheat and rice. Data obtained from six maize segregating populations predicted higher levels of grain moisture and grain yield (0.90 and 0.58, respectively), and accurate predictions were made across several locations. Similar predictions were made in wheat for *Fusarium* head blight resistance. Though costly, GS is superior to marker-assisted recurrent selection for improving complex traits.

Table 24.3 Isolated genes associated with important traits in staple cereals

Cereal species	Trait
Maize	Zein storage protein
	Resistance to the domestication flowering time
	Photoperiod sensitivity
	Resistance to head smut
	Drought tolerance
	Male sterility
	Resistance to southern leaf blight, grey leaf spot and northern leaf blight
Rice	Resistance to *Xanthomonas oryzae* pv. *oryzae*
	Grain size
	Bacterial streak disease
	Blast resistance
	Grain chalkiness
	Resistance to rice stripe
	Chilling tolerance
	Thermotolerance
Wheat	Leaf rust disease resistance
	Grain protein and iron content
	Stripe rust resistance
	Grain width, thousand-kernel weight, polyploidization and evolution
	Wheat rust, powdery mildew
	Leaf width, flowering time and chlorophyll

24.2.4 From Genotype to Phenotype

Phenotype corresponds to genotype in a linear manner. To date, a large number of QTLs have been identified by linkage mapping and GWAS, and several genes with major effects have been functionally validated by both gain-of-function and loss-of-function approaches. It is possible to predict phenotypes from genotypes through rapid genome sequencing methods coupled with whole genome transcription profiling. There are several QTLs associated with yield-related traits and resistances to abiotic and biotic stresses (Table 24.3).

24.2.5 Post-transcriptional Gene Silencing (PTGS)

Gene silencing can occur either transcriptionally or post-transcriptionally. Post-transcriptional gene silencing (PTGS) is an RNA-based immune mechanism that gives protection against virus and foreign gene invasion. PTGS pathway is embedded in cellular regulatory networks. In plants, PTGS was first detected in transgenic plants where expression of both transgenes and their endogenous counterparts was disrupted. The expression of most endogenous genes does not trigger PTGS.

Cellular double-stranded RNAs (dsRNAs) are the main functionaries in PTGS. These dsRNAs are recognized and processed into 20–22-nucleotide (nt) RNA duplexes by Dicer family proteins. One strand of the small RNAs, such as small interfering RNA (siRNA) duplexes processed by DCL2 (Dicer-like 2) and DCL4 and microRNA (miRNA) duplexes processed by DCL, can be loaded into the Argonaute (AGO)-containing RNA-induced silencing complex (RISC), resulting in mRNA cleavage or translational inhibition (Fig. 24.5). Additional round of siRNA production is needed to amplify primary PTGS effect. The target transcripts are multiplied through the involvement of RNA-dependent RNA polymerases

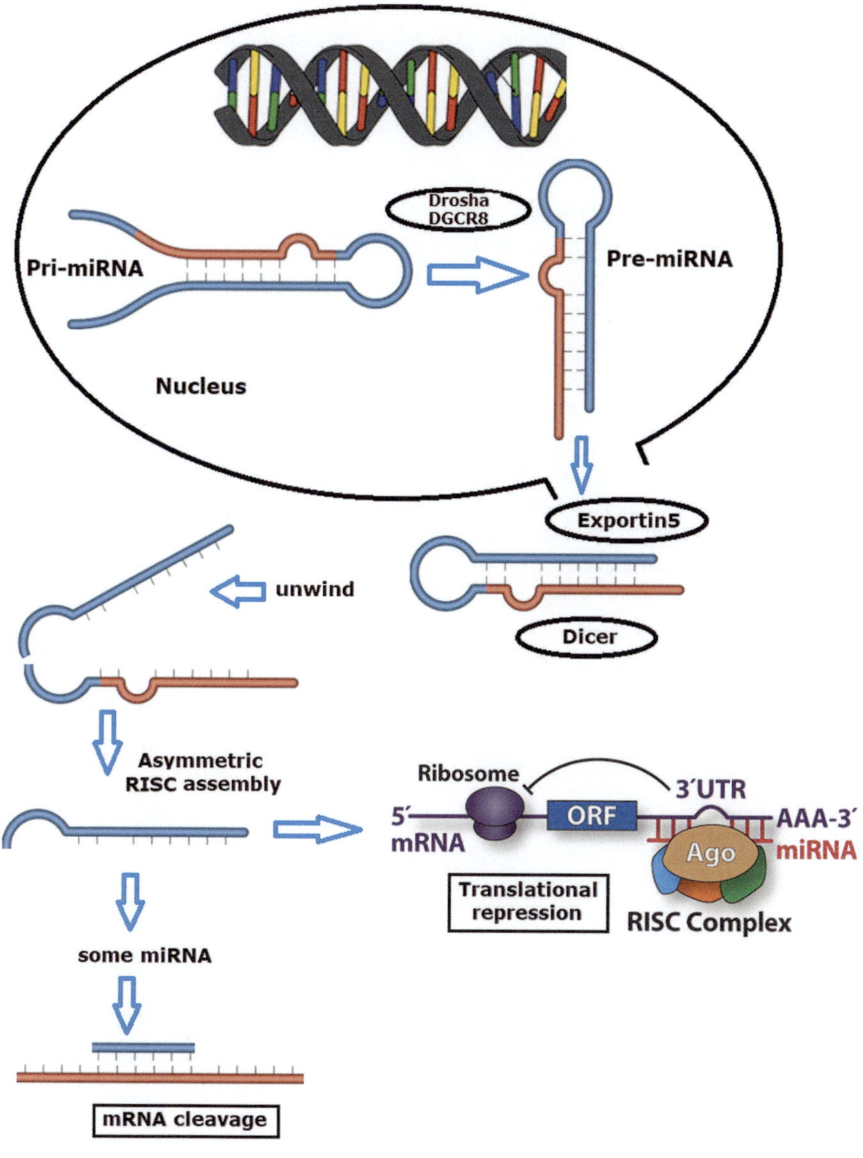

Fig. 24.5 Production of miRNA, translational repression and PTGS

(RdRPs). This process is referred to as secondary siRNA biogenesis. It is noteworthy that a subset of the secondary siRNAs, known as epigenetically activated siRNAs (easiRNAs), is actively involved in the defence of plants.

Genome-assisted breeding (GAB) has great potential but with bottlenecks. Foremost is the establishment of high-throughput phenotyping platforms in the field. Higher costs and limited phenotyping capabilities are the other disadvantages. Data management and bioinformatics usage are other major challenges. Epigenetic phenomena such as DNA methylation, genomic imprinting, maternal effects, RNA editing, etc. are to be addressed more vehemently. Epigenetics research has advanced further, but mechanisms governing epigenetic phenomena are to be understood well. In the coming years, it is believed that extensive implementation of MAS and GS either alone or in combination will help to improve plant breeding at genomic level (see Box 24.1). The emergence of systems biology is one such step forward.

Box 24.1 Genomic Features for Future Breeding

Genomics has explosively altered the scope of plant breeding with information on ordered genes and their epigenetic states with high precision and accuracy. Genetic maps in the beginning were made up of sparse markers, like anonymous markers based on simple sequence repeats (SSR) or restriction fragment length polymorphisms (RFLP). For example, if a phenotype of interest was affected by genetic variation within the SSR1-SSR2 interval, the complete region would be selected with little information on its gene content and variation. Whole genome sequencing of a closely related species enabled projection of gene content. Through conserved gene order across species (synteny), breeders could find out the presence of specific genes. While whole genome sequencing facilitated putative gene function and precise genomic positions, RNA-seq or microarrays allowed expression levels to be monitored in different tissues under varied environments. On the other hand, re-sequencing of varieties can identify high density of SNP markers across genomic intervals that enables genome-wide association studies (GWAS), genomic selection (GS) and more defined marker-assisted selection (MAS) strategies.

ENCODE (Encyclopedia of DNA Elements)-level analyses can provide new data to predict phenotype from genotype. The goal of ENCODE is to build a comprehensive parts list of functional elements in the genome, including elements that act at the protein and RNA levels, and regulatory elements that control cells and circumstances in which a gene is active. Another information layer is relating to functional aspects like flowering time in response to day length and over-wintering (Fig. 24.6). Such networks are identified in *Arabidopsis* and rice. Evolutionary mechanisms like gene duplication and domestication can be mapped to networks. Such "systems

(continued)

Box 24.1 (continued)
breeding" techniques use diverse genomic information to predict phenotype from genotype, thus helping to address food security.

Development of chromatin immunoprecipitation (ChIP) facilitates identification of discrete regions of the genome bound for specific proteins as also identification of transcription factor binding events (putative cis-regulatory elements) in entire genomes. Comparison of protein-DNA binding maps identifies regulatory differences and change in gene expression across species. ChIP experiments help to establish the effect of divergence of binding events on species-specific gene expression (Fig. 24.7).

24.3 The New Systems Biology

Systems biology is the computational and mathematical modelling of complex biological systems. It is a holistic approach that leads to an understanding that networks that form the whole of living organisms are more than the sum of their parts. A collaboration of biology, computer science, engineering, bioinformatics, physics and others gives prediction on how these systems can change over time and environments.

Over the last three decades, efficiency attained in DNA sequencing through next-generation sequencing technologies and the changeover of such technologies becoming more cost-effective made studies on systems biology more efficient. Parallel to this, gene transfer and genome editing gave broad support to such studies. Genomics-assisted breeding (GAB) tracks a trait of interest and has the provision to integrate such genomic region into a given phenotype. Mapping genetic marker-associated QTLs would assist breeders to select genotypes inheriting alleles in favourable combination. Traits with low heritability can be selected in this way.

Gene transfer involves introduction of DNA sequences into a target genome. Inserted sequence can be from same species (*cisgenesis*) or from different species (*transgenesis*). While considering multiple sequences inserted at different loci, backcrossing in germplasm of interest will be limiting (see Chaps. 22 and 23). Gene editing allows direct and targeted editing of gene sequences leading to total or partial expression of the gene. CRISPR-Cas9 technology (see Chap. 22) caused a paradigm shift during the last 5 years in the domain of genetic modification of plants. CRISPR-Cas9 needs to reach its full potential; however, its precision and cost-effectiveness keep it more promising to bypass conventional breeding constraints.

A collection of molecular regulators (genes, RNA and proteins) makes the gene regulatory network (GRN). This network directly or indirectly interacts with each other to collectively influence a biological process. The most common way to represent GRN is through graphs. A graph is mathematically defined as a set of nodes, and edges linking those nodes, where nodes represent molecular regulators.

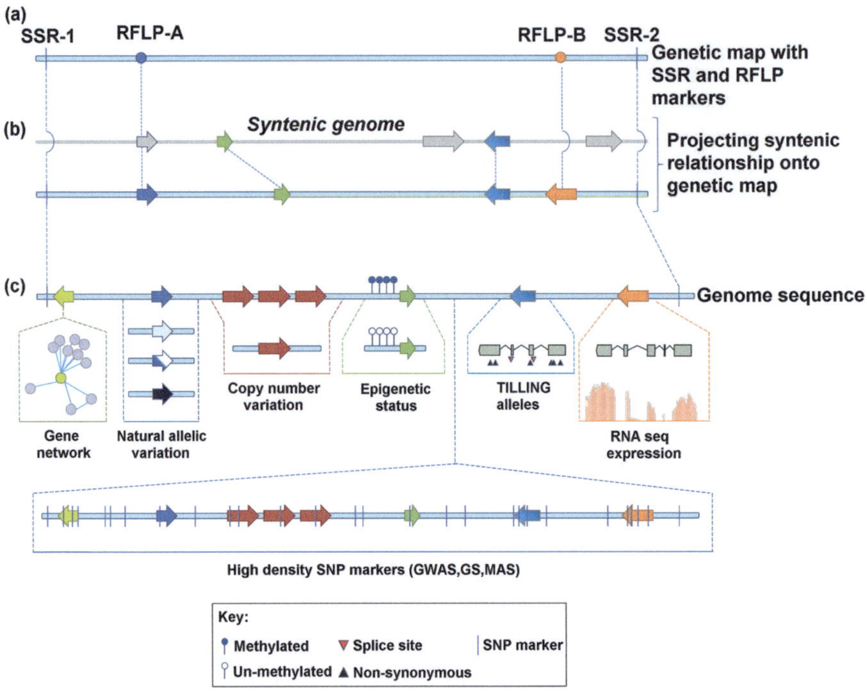

Fig. 24.6 The impact of whole genome sequencing on breeding. (**a**) Initial genetic maps consisted of few and sparse markers, many of which were anonymous markers (simple sequence repeats (SSR)) or markers based on restriction fragment length polymorphisms (RFLP). For example, if a phenotype of interest was affected by genetic variation within the SSR1-SSR2 interval, the complete region would be selected with little information about its gene content or allelic variation. (**b**) Whole genome sequencing of a closely related species enabled projection of gene content onto the target genetic map. This allowed breeders to postulate the presence of specific genes on the basis of conserved gene order across species (synteny), although this varies between species and regions. (**c**) Complete genome sequence in the target species provides breeders with an unprecedented wealth of information that allows them to access and identify variation that is useful for crop improvement. In addition to providing immediate access to gene content, putative gene function and precise genomic positions, the whole genome sequence facilitates the identification of both natural and induced (by TILLING) variation in germplasm collections and copy number variation between varieties. Promoter sequences allow epigenetic states to be surveyed, and expression levels can be monitored in different tissues or environments and in specific genetic backgrounds using RNA-seq or microarrays. Integration of these layers of information can create gene networks, from which epistasis and target pathways can be identified. Furthermore, re-sequencing of varieties identifies a high density of SNP markers across genomic intervals, which enable genome-wide association studies (GWAS), genomic selection (GS) and more defined marker-assisted selection (MAS) strategies

Most of the time, a given gene and its subsequent RNAs and proteins are considered together, and the “gene” terminology is used as a shortcut, and edges indicate direct or indirect regulatory interactions between these elements (Fig. 24.8).

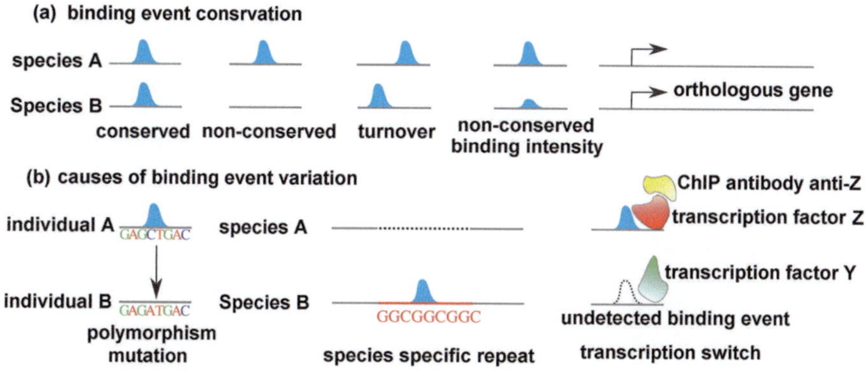

Fig. 24.7 (**a**) Cartoons of ChIP peak signals representing binding events near a target gene. (**b**) Variation in cis can potentially alter a DNA motif recognized by a transcription factor and render it unrecognizable and lead to a loss of a binding event. Between species, the appearance of a repeat element or other lineage-specific sequences can create new binding events. Changes of the transcription factor that regulates a given gene can occur during evolution. As ChIP targets specific transcription factors, such changes might be undetected, leading to a false loss of binding event

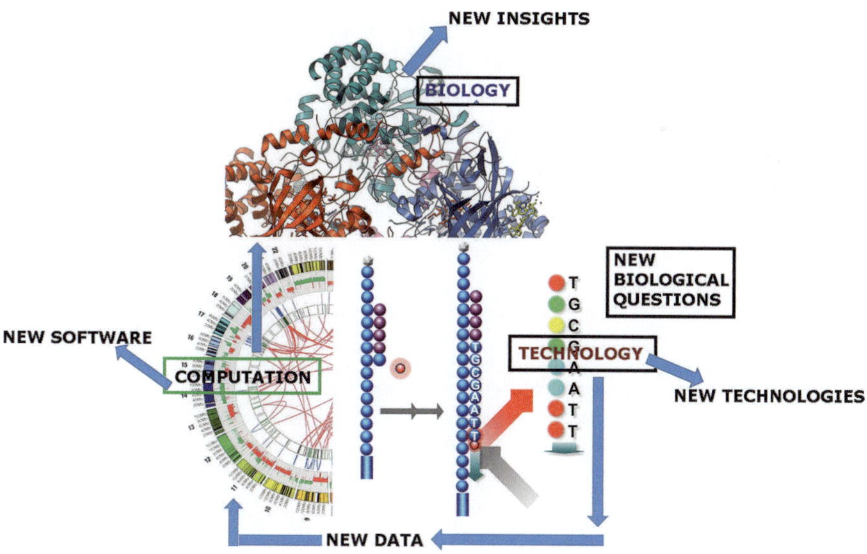

Fig. 24.8 Scheme showing emergence of systems biology (figure representative)

The organization of edges within a graph defines its topology. Edges can be either directional or non-directional; in the first case, the interaction of a given Node A on another Node B is differentiated from the interaction of Node B on A, whereas in the second case, the two are equal. The subsequent graphs are considered as directed or undirected, respectively. In addition, edges can be weighted, that is, associated with positive or negative values, to quantitatively model the positive or negative regulatory interaction between genes.

The high-throughput "-omics" have been defined as the method used to characterize and quantify at once the thousands of biological molecules playing a role in the structure, function and dynamics of an organism. Many high-throughput -omics methods are available, ranging from genomics, epigenomics, transcriptomics, proteomics, interactomics, to metabolomics. The data sets generated by the different -omics methods are often conceptualized as describing different "layers" of a biological system.

As a cell's behaviour is based on the integration of environmental and endogenous signals by its internal GRN, tracking the state of this GRN through time is prime through mechanistic modelling approaches. Considering that RNA extraction using generic methods is now feasible for a wide range of plant species, transcriptomics is the most pragmatic choice as an input for GRN-state tracking and top-down modelling approaches. RNA sequencing (RNA-seq) and DNA microarrays allow exhaustive and quantitative exploration of RNA populations. Spatial resolution down to the cell level can be accessed through laser capture microdissection, while time series, which are crucial to capture relevant information about biological processes involving a notion of temporality (e.g. gene expression changes within minutes during hormonal signalling, while flower organogenesis processes take hours to take place), mostly rely on the experimental design. With the advances in sequencing technologies, the data point cost is ever-decreasing. This leads to an easier access to a wealth of information, as well as to the accumulation of transcriptomic data sets that can be used as cross-resources. Hope systems biology will certainly revolutionize genomics in the years to come.

Further Reading

Bolger ME et al (2014) Plant genome sequencing – applications for crop improvement. Curr Opin Biotechnol 26:31–37

Chakradhar T (2017) Genomic-based-breeding tools for tropical maize improvement. Genetica 145:525–539. https://doi.org/10.1007/s10709-017-9981-y

Kang YJ et al (2015) Translational genomics for plant breeding with the genome sequence explosion. Plant Biotechnol J:1–13. https://doi.org/10.1111/pbi.12449

Ronald PC (2014) Lab to farm: applying research on plant genetics and genomics to crop improvement. PLoS Biol 12:e1001878. https://doi.org/10.1371/journal.pbio.1001878

Songstad DD et al (2017) Genome editing of plants. Crit Rev Plant Sci 36:1–23. https://doi.org/10.1080/07352689.2017.1281663

Zhang X, Zhu Y, Wu H, Guo H (2016) Post-transcriptional gene silencing in plants: a double-edged sword. Sci China Life Sci 59:271–276. https://doi.org/10.1007/s11427-015-4972-7

Maintenance Breeding and Variety Release 25

Keywords
Breeder's trials · designing field trials · crop registration · cultivar/variety maintenance · DUS testing · types of expression of characteristics · DUS descriptors for major crops · generation system of seed multiplication

Improved cultivars are usually more uniform than the local cultivars grown and maintained by the farmers. Such cultivars are to be multiplied so that it can be distributed to the farmers. As a repeated process, through multiplication, seed should be available at the start of each growing season. Every multiplication cycle commences from the stock seed of the variety, the "breeder seed" (BS). This BS is expected to maintain genetic purity (true-to-type). During maintenance and multiplication, there may be contamination and even complete loss of the improved traits. Prevention of contamination gets top most priority during maintenance.

25.1 Breeder's Trials

The primary purpose of breeder's trials is evaluation of the performance of the final set of genotypes so that the breeder can take a decision as to which genotype to be released as a cultivar. This evaluation can be done under two stages. The first stage is the preliminary yield trial (PYT). This consists of large number of entries (10–20 genotypes) and starts at an earlier generation (e.g. F_6, depending on the objectives and method of breeding). These entries may be planted in fewer rows per plot (e.g. two rows without borders) and fewer replications (2–3) than would be used in the final trial, the advanced yield trial (AYT). Superior genotypes are identified for more detailed evaluations in this AYT (second stage). AYT is conducted for several years over different environments, using more replications and plots with more rows and with borders rows. It is also subjected to more detailed statistical analysis.

P. M. Priyadarshan, *PLANT BREEDING: Classical to Modern*,
https://doi.org/10.1007/978-981-13-7095-3_25

Breeder's trials vary in scope, and many are limited to within the state or mandate region. Private/commercial breeders use to conduct regional, national and even international trials through established networks. Public breeders may have wide networks for trials (e.g. Potato Breeding Network of International Potato Centre – CIP). In terms of management, BS follows two ways – research managed and farmer managed.

25.1.1 Designing Field Trials

PYTs will have more entries than AYTs. Locations must be representative of the target region where the variety is to be released. They are not randomly selected. Sites are limited to where collaborators (e.g. institutes, research stations, universities) or farmers are willing to participate in the project. The total number of sites is variable (about 5–10), but it depends on the extent of variability in the target region (see Chaps. 7 and 20 for accounts on statistical layouts and GE interactions, respectively).

25.1.2 Crop Registration

After the formal release of the variety, it may be registered. In the USA, this voluntary activity is coordinated by the Crop Science Society of America (CSSA). In India, it is by the National Bureau of Plant Genetic Resources. In Canada, it is at Canadian Food Inspection Agency. According to the CSSA, crop registration is designed to inform the scientific community of the attributes and availability of the new genetic material and to provide readily accessible cultivar names or designations for a given crop. Further, crop registration helps to prevent duplication of cultivar names. Complete guidelines for crop registration may be obtained from the CSSA.

What Can Be Registered? Normally, over 50 crops and groups of crops may be registered. Sub-committees used to be established to review the registration manuscripts for various crops. Hybrids may not be registered. Eligible materials may be cultivars, parental lines, elite germplasm, genetic stocks and mapping populations. The cultivar to be registered must have demonstrated its utility and provide a new variant characteristic (e.g. disease or insect resistance).

Variety Protection In addition to registration, a breeder may seek legal protection of the cultivar in one or several ways as discussed in detail in Chap. 15. A common protection, the Plant Variety Protection, or the Plant Breeders' Rights, is a sui generis (of its kind) legal protection.

25.2 Cultivar/Variety Maintenance

The mode of reproduction is the determining factor for the genetic makeup of varieties. Henceforth, the crops can be classified into four categories:

(a) Typical cross-pollinating crops
(b) Self-pollinating crops with a substantial amount of outcrossing
(c) Typical self-pollinating crops with very little outcrossing
(d) The vegetatively reproduced crops

Open-pollinated species like maize are genetically narrowed populations, with high frequencies of the desired genes. They are hard to maintain. Improved cultivars of crops of category b, like quinoa (*Chenopodium quinoa*) and faba bean (*Vicia faba*), are difficult to maintain. Improved cultivars of crops of category c, like wheat, barley, *Hordeum vulgare* and common bean, consist of very similar desirable genotypes, and maintaining is fairly simple. Improved cultivars of crops of the last category, such as potato, are a clone, and its genetic purity is easily maintained. However, to upkeep them and free from pathogens, especially viruses, is very difficult.

25.2.1 Maintenance of a Cultivar

Each multiplication cycle has to start from its basic stock seed, the breeder's seed. Storing sufficient amount of seed under low temperatures keeps the seeds viable. The amount stored must be sufficient to start many multiplication cycles. This demands for a huge storage space for crops with low multiplication rates. Under many circumstances, this is not a feasible option. If storage is not possible, maintenance selection is the appropriate way to maintain a cultivar.

Maintenance Selection The maintenance selection starts with a small plot containing a number of spaced plants, derived from the BS. The plants must be well spaced to allow for individual plant assessment and for the harvest of sufficient seed per plant, especially important for crops with a low multiplication rate such as potato, common bean, faba bean, barley and wheat. A fair number of healthy plants of the cultivar type are selected and marked for progeny testing. Plants with a seed-borne disease are removed. The seeds of the marked plants are harvested per plant and sown in small plots the next season, the first-cycle progenies (Fig. 25.1). Only progeny plants that have the required uniformity are selected, and the seed is bulked per progeny. Even if only one or two plants deviate phenotypically, including being infected by a seed-borne pathogen, the whole progeny should be discarded. In cross-pollinating crops, the purity cannot be maintained for long. For this, the seeds are stored under optimal conditions. Under maintenance selection, the cultivar can change genetically as negative or positive. Either positive or

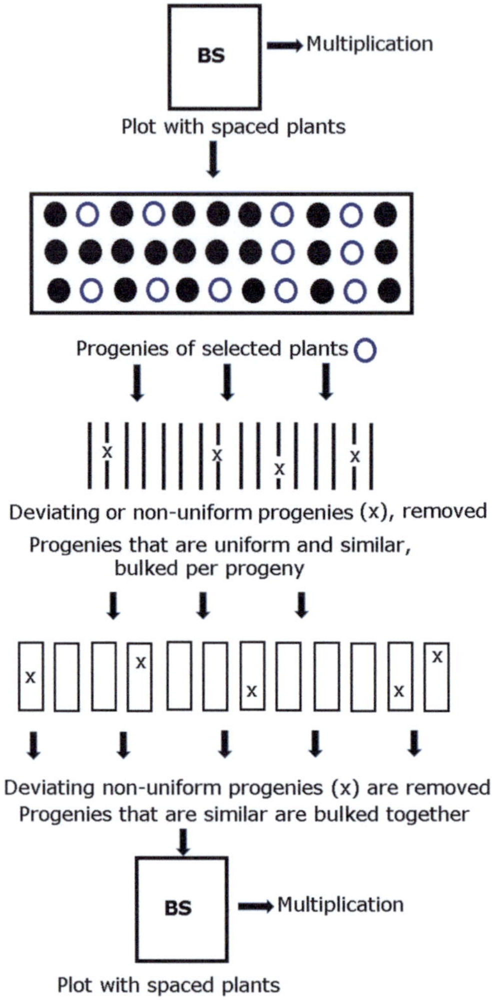

Fig. 25.1 Maintenance selection, general scheme, starting from the bag of breeder seed (BS)

negative way will be preferred depending on the balance between the contaminating forces and the selection pressure against such forces.

An improved cultivar is a gene pool where the genes are reshuffled into a new set of genotypes under each generation. The maintenance selection of strong genotypes can neutralize these negative effects. After each cycle of maintenance selection, the BS will be improved than the previous one. Repeated maintenance selection will ensure improvement over time provided progeny size is kept fairly large (Fig. 25.2).

The case of cross-pollinating crops is different based on the fact whether the progenies are assessed before or after flowering. If assessed after flowering,

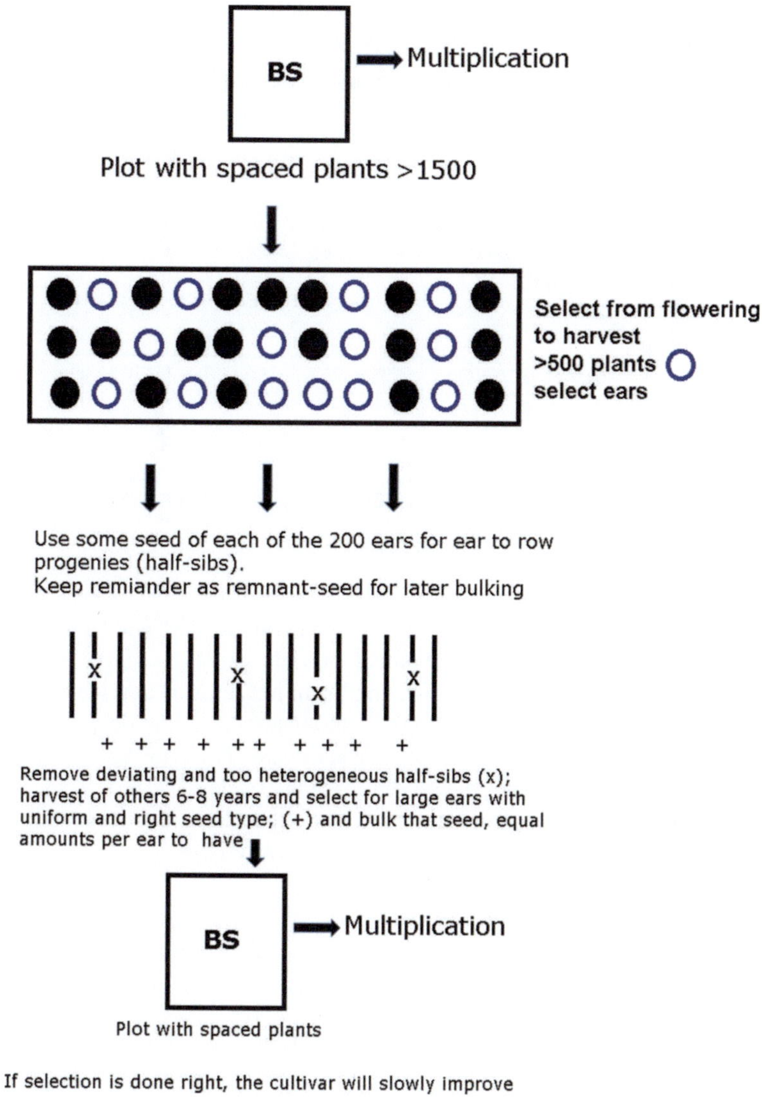

Fig. 25.2 Maintenance selection of a maize cultivar

pollination by undesirable plants cannot be prevented. The traits to be assessed before flowering are usually those relating to the vegetative growth. Selection for increased yields of such traits tends to be negatively associated with traits related to the generative growth complex, i.e. seed yield. A fairly strong natural selection occurs due to this negative association. In spinach (*Spinacia oleracea*), leaf yield is positively associated with late bolting and negatively with seed yield, which results in a strong natural selection towards earlier bolting during the maintenance and seed production of late spinach cultivars (Fig. 25.3).

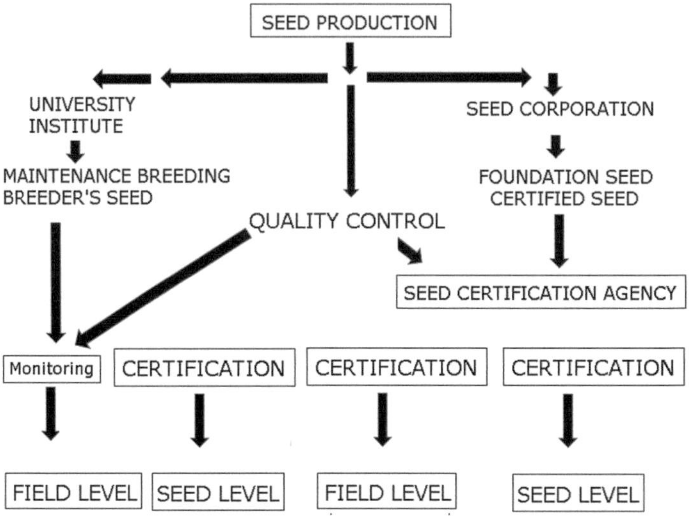

Fig. 25.3 Scheme for seed production

If assessment is done before flowering, the selection intensity will have to be very strong so that within progenies selection for the right genotype can be undertaken. When the assessment is done after flowering (as in maize), it is advisable to use the remnant seed approach. Maize owes high multiplication rate, and only seeds from a small part per ear are sown in the first progeny cycle. The remnant seed from the selected plants is used to plant the second progeny cycle. The plot in the second cycle can be larger to accommodate sufficient seeds. In order to ensure strong selection, the number of ears to start with shall be fairly large.

25.3 DUS Testing

DUS (distinctness, uniformity, stability) testing determines whether a newly bred variety differs from existing varieties within the same species (the distinctness), whether the characteristics used to establish distinctness are expressed uniformly (uniformity) and that these characteristics do not change over subsequent generations (stability). DUS tests are for granting of Plant Breeders' Rights, a form of intellectual property rights (IPR) designed to safeguard the investment incurred in breeding varieties. DUS is being overseen by the Protection of Plant Varieties and Farmer's Rights Authority, which is available in every country. This body is constituted as per UPOV (International Union for the Protection of New Varieties of Plants, Geneva) Convention guidelines.

25.3.1 Test Guidelines and Requirements

The UPOV Convention Article 7(1) of the 1961/1972 and 1978 Acts and Article 12 of the 1991 Act requires that a variety be examined for compliance with the distinctness, uniformity and stability criteria. The 1991 Act of the UPOV Convention clarifies that "In the course of the examination, the authority may grow the variety or carry out other necessary tests, cause the growing of the variety or the carrying out of other necessary tests, or take into account the results of growing tests or other trials which have already been carried out". UPOV has established specific Test Guidelines for a particular species, or other group(s) of varieties, in conjunction with the basic principles contained in the General Introduction, should form the basis of the DUS test.

To attain a variety capable of protection, the same must be clearly defined. This is a prerequisite for examination of DUS criteria for protection. All Acts of the UPOV Convention have established that a variety is defined by its traits and that those traits are the basis for examination of a variety through DUS norms.

The following are the requirements for DUS testing:

(a) Representative plant material: The material to be submitted for the DUS testing is to be representative. In the case of specially propagated varieties (like hybrid and synthetic), the material to be tested must be from the final stage in the cycle of propagation.
(b) General health of submitted material: The plant material must be healthy, vigorous and devoid of pests and disease infestation. In case of seed, it must have higher germination capacity.
(c) Factors affecting expression of the characteristics: This may be affected by pests and disease, chemical treatment (e.g. growth retardants or pesticides), effects of tissue culture, different rootstocks and scions taken from different growth phases of a tree, etc.

In most countries, variety testing is administered by an official authority (e.g. Protection of Plant Varieties and Farmer's Rights Authority in India), although the breeders participate in the growing tests to varying degrees.

25.3.2 Types of Expression of Characteristics

The different ways of expression of characteristics is to be understood properly to use characteristics for DUS testing. The different types of expression are:

(a) Qualitative characteristics like those that are expressed in discontinuous states, e.g. sex of plant like dioecious female, dioecious male, monoecious unisexual and monoecious hermaphrodite. These states are self-explanatory and independently meaningful. As a rule, the characteristics are not influenced by environment.

(b) Quantitative characteristics where the expression of variation is from one extreme to the other. The expression can be recorded on a one-dimensional, continuous or discrete, linear scale. The range of expression is divided into a number of states for the purpose of description (e.g. length of stem: very short, short, medium, long, very long). The division is expected to have even distribution across the scale. The states of expression should, however, be meaningful for DUS assessment.
(c) Pseudo-qualitative characteristics like whose expression is at least partly continuous, but varies in more than one dimension (e.g. shape: ovate, elliptic, circular, obovate) and cannot be adequately described by just defining two ends of a linear range. In a similar way to qualitative (discontinuous) characteristics – hence the term "pseudo-qualitative" – each individual state of expression needs to be identified to adequately describe the range of the characteristic.

25.3.3 DUS Descriptors for Major Crops

Bioversity International (a CGIAR concern) is the nodal agency for the documentation of plant genetic resources. Biodiversity International collaborates with other organizations like the International Union for the Protection of New Varieties of Plants (UPOV); Organisation Internationale de la Vigne et vin (OIV), France; the World Vegetable Centre (AVRDC), Taiwan; CGIAR Centres; Instituto Nacional de Investigación Agropecuaria (INIA), Uruguay; French Agricultural Research Centre for International Development (CIRAD) and Institut national de la recherché agronomique (INRA), France; and a number of universities and research organizations for coordinating information on plant genetic resources. Descriptor lists have been an important element of Biodiversity's germplasm documentation activities almost since the establishment of IBPGR in the 1970s (the name International Bureau of Plant Genetic Resources has been changed later to Biodiversity International) and the production of the first descriptor list in 1977.

Minimum Descriptors: The original objective of descriptor was to provide a minimum number of characteristics to describe a crop. But these descriptors lacked the appropriate internationally accepted definitions and descriptor states needed for consistency. This lack of compatibility seriously hampered data exchange between collections.

Comprehensive Lists of Descriptors: The idea of minimum lists was revisited in 1990, and a new approach was developed. Comprehensive lists of descriptors were produced including all descriptors for characterization and evaluation (e.g. Descriptors for Sweet Potato developed in collaboration with AVRDC and CIP in 1991). The comprehensive descriptor lists also included a number of standard detailed sections (e.g. site environment and management) that were common across different crop descriptor lists and that provided users with options to choose from. This improved compatibility between documentation systems and the ease of information exchange.

Highly Discriminating Descriptors for International Harmonization: It was recognized that each curator utilized only those descriptors that were useful for the maintenance and management of their collection. Consequently, the descriptor lists were further revised in 1994 in order to provide users with more comprehensive lists but at the same time containing a minimum set of highly discriminating descriptors, which were flagged in the text with asterisks (*) (e.g. in Descriptors for Barley (*Hordeum vulgare* L)) (please see https://www.bioversityinternational.org/fileadmin/_migrated/uploads/tx_news/Descriptors_for_barley__Hordeum_vulgare_L.__333.pdf).

25.4 Generation System of Seed Multiplication

There are four generally recognized classes of seeds.

Nucleus seed: This is the 100% pure seed at genetic and physical levels from basic nucleus seed stock. This seed is not certified by any agency.

Breeder seed: This is the progeny of the nucleus seed multiplied in large area under the supervision of plant breeder and monitored by a committee. It is with 100% physical and genetic purity. A golden yellow colour certificate is issued for this category of seed by the producing agency.

Foundation seed: Progeny of breeder seed is handled by recognized seed producing agencies in public and private sectors under the supervision of seed certification agency in such a way that its quality is maintained according to the prescribed seed standard. A white colour certificate is issued for the foundation seed by seed certification agencies.

Certified seed: Progeny of foundation seed is produced by registered seed growers under the supervision of seed quality as per Indian Seed Certification Standards. A blue colour certificate is issued by seed certification agency for this category of seed. Size of tag is 15 cm length and 7.5 cm breadth.

Truthfully labelled seed (TL): When a seed is sold based on the result of the laboratory established by the producer, then the seed is considered as TL seed, e.g. seed produced and sold by many private agencies. The price of TL seed is always lower than the certified seed offered by government sector. Seed rejected due to genetic impurity or presence of objectionable disease, pest or weed is not labelled as truthful.

Registered seed: In USA mainly for autogamous crops, the generation between foundation and certified seed is considered as registered seed, which is not a commercial class. Registered seeds are labelled by purple colour tag.

Seed certification: It is a process designed to ensure the availability of high-quality seeds to the general public with physical identity and genetic purity. It is legally sanctioned system for quality control of seed multiplication and production.

The Association of Official Seed Certifying Agencies (AOSCA), formerly known as the International Crop Improvement Association, is a trade organization based in the USA. Founded in 1919, its function is to develop and promote certified varieties of seed for agricultural use. AOSCA assists clients in the production, identification, distribution and promotion of certified classes of seed and other crop propagation materials. Its membership currently includes seed certifying agencies across the USA and member countries including Canada, Australia, New Zealand, South Africa, Argentina, Chile and Brazil. Likewise, every country is having its own seed certifying agencies.

Further Reading

Biodiversity International (2007) Developing crop descriptor lists. Bioversity technical bulletin no. 13

Cooke RJ, Reeves JC (2003) Plant genetic resources and molecular markers: variety registration in a new era. Plant Genet Resour: Charact Util 1:8187. https://doi.org/10.1079/PGR200312

Garrett KA et al (2017) Resistance genes in global crop breeding networks. Phytopathology 107:1268–1278. https://doi.org/10.1094/PHYTO-03-17-0082-FI

Guidelines for the conduct of tests for Distinctiveness, Uniformity and Stability. Protection of Plant varieties and Farmer's Rights Authority, Government of India

Wani SH et al (2013) Intellectual property rights system in plant breeding. Jour Pl Sci Res 29 (1):112–122